Ecological Studies

Analysis and Synthesis

Edited by

W.D. Billings, Durham (USA) F. Golley, Athens (USA)
O.L. Lange, Würzburg (FRG) J.S. Olson, Oak Ridge (USA)
H. Remmert, Marburg (FRG)

Volume 39

Resource Use by Chaparral and Matorral

A Comparison of Vegetation Function in Two Mediterranean Type Ecosystems

Edited by P.C. Miller

With 118 Figures

Springer-Verlag
New York   Heidelberg   Berlin

P.C. Miller
Professor of Biology
Department of Biology
College of Sciences
San Diego State University
San Diego, California 92182

Library of Congress Cataloging in Publication Data
Resource use by chaparral and matorral.

(Ecological studies; v. 39)
Bibliography: p.
Includes index.
1. Chaparral ecology. 2. Matorral ecology. 3. Primary productivity (Biology) 4.
Botany—California—Ecology. 5. Botany—Chile—Ecology. 6. Primary productivity
(Biology)—California. 7. Primary productivity (Biology)—Chile. I. Miller, Philip C. II.
Series.
QK938.S57R47   581.5′264′097   81-8786   AACR2

9 8 7 6 5 4 3 2 1
ISBN 978-1-4612-5898-8   ISBN 978-1-4612-5896-4 (eBook)
DOI 10.1007/978-1-4612-5896-4

Foreword

The comparative study of mediterranean type ecosystems has gained considerable momentum during the past two decades. Modern studies on these systems date from the work of Ray Specht, who studied the dynamics of the heath vegetation of southern Australia. The results of these studies first appeared in 1957 (Specht and Rayson, 1957) and were summarized in 1973 (Specht, 1973). Specht followed this detailed work, which pointed to the central role of nutrients in limiting the productivity of the Australian heath, with a general comparison of the structural features of woody plant communities in mediterranean type ecosystems of Australia, southern France, and southern California (Specht, 1969a,b). The comparative studies emphasized remarkable convergent features of these ecosystems, particularly in relation to structural features affecting primary production. Naveh (1967) also did comparative studies focusing on grassland types that occur in the mediterranean climatic zones of California and Israel. About this same time, independent studies by Mooney and Dunn (1970 a,b) and Mooney et al. (1970) made preliminary structural and functional comparisons of the vegetations of mediterranean type ecosystems in California and Chile in an attempt to derive an evolutionary model explaining the basis of their convergent natures. Much of the knowledge of these ecosystems up to 1973 was summarized in volume 7 of *Ecological Studies, Mediterranean Type Ecosystems: Origin and Structure* (di Castri and Mooney, 1973). This volume builds on its series predecessor in many ways.

The initiation of the International Biological Program (IBP) in the late 1960's brought a whole new dimension to the study of mediterranean type ecosystems. The principal thrust of the IBP was the determination of the productive capacities of the earth's ecosystems. Within this framework, a program was initiated in southern France on carbon, water, and nutrient cycling in a series of mediterranean type ecosystems concentrating on *Quercus ilex* evergreen forests and *Quercus coccifera* evergreen scrub (garrigue) (Lossaint, 1973; Eckardt et al., 1975).

Under the auspices of IBP, scientists from the United States and Chile collaborated in a program to compare mediterranean type ecosystems in southern California and central Chile. Rather than comparing the productive capacities of the communities of these areas, a detailed analysis was made of the structural characteristics of the analogous ecosystems. Study sites were matched as closely as possible in terms of climate, topographic position, and geological substratum. Selected groups of organisms were studied in detail to test for structural and behavioral similarities at the matched sites. This research was designed to answer the question, "Do unrelated organisms assume comparable structural attributes when subjected to similar physical selective environments?" The research indicated that the primary producers showed comparable structural attributes, but that higher trophic levels and higher levels of community organization showed progressively less similarity (Mooney, 1977a; Thrower and Bradbury, 1977; Cody and Mooney, 1978).

After the IBP, a new study was initiated by Miller and co-workers, the results of which are the subject of this book. Using the same Californian and Chilean study sites, which were compared earlier for structural attributes, they made an in-depth analysis of the resources which are utilized by the primary producers. They further assessed the functional relationships between the principal plant species and these resources. Finally, through the use of innovative simulation models, they were able to link quantitatively the patterns of availability of resources, i.e., of water, light, and nutrients, with such processes as transpiration, photosynthesis, and nutrient uptake. Built into these models were the relationships between plant morphological features and plant function. Thus, quantitative simulations could be made not only of the interrelations of existing form and function, but also form and resource availability could be varied to determine the nature of their interaction.

The quantitative approach to the study of relationships of plants with their environment and with each other is a powerful tool and is applicable to most ecological problems, including those related to the study of the basis of convergent evolution. This book illustrates this approach. The detailed information that is necessary for this kind of ecology can be available only through the collaborative efforts of many scientists working together toward a common goal. For many ecological problems, an integrated research effort offers the only possibility for making progress. Clearly, this quantitative-integrated approach is the greatest legacy from the International Biological Program.

Harold Mooney
Stanford, California

REFERENCES

di Castri, F., and Mooney, H.A. (1973) Mediterranean type ecosystems: Origin and structure. New York: Springer-Verlag, 405 pp.

Cody, M.L., and Mooney, H.A. (1978) Convergence versus nonconvergence in mediterranean-climate ecosystems. Annu. Rev. Ecol. Syst. 9, 265-321.

Eckardt, F., Heim, G., Methy, M., and Sauveyon, R. (1975) Interception de l'énergie rayonnante, exchanges gazeaux et croissance dans une foret mediterranêne a feuillage persistant (Quercetum ilicis). Photosynthetica 9, 145-156.

Lossaint, P. (1973) Soil-vegetation relationships in mediterranean ecosystems of south-
    ern France, pp. 199-210. In F. di Castri and H.A. Mooney (eds.). Mediterranean
    type ecosystems: Origin and structure. New York: Springer-Verlag

Mooney, H.A. (ed.) (1977a) Convergent Evolution in Chile and California Mediter-
    ranean Climate Ecosystems. Stroudsburg, Pa.: Dowden, Hutchinson, and Ross.

Mooney, H.A., Dunn, E.G., Shropshire, F., and Song, L. (1970) Vegetation compari-
    sons between the mediterranean climate areas of California and Chile. Flora *159*,
    480-496.

Mooney, H.A., and Dunn, E.L. (1970a) Convergent evolution of mediterranean climate
    evergreen sclerophyll shrubs. Evolution *24*, 292-303.

Mooney, H.A., and Dunn, E.L. (1970b) Photosynthetic systems of mediterranean cli-
    mate shrubs and trees of California and Chile. Am. Nat. *104*, 447-453.

Naveh, Z. (1967) Mediterranean ecosystems and vegetation types in California and
    Israel. Ecology *48*, 445-459.

Specht, R.L. (1969a) A comparison of the sclerophyllous vegetation characteristic of
    mediterranean type climates in France, California, and southern Australia. I. Struc-
    ture, morphology, and succession. Aust. J. Bot. *17*, 277-292.

Specht, R.L. (1969b) A comparison of the sclerophyllous vegetation characteristic of
    mediterranean type climates in France, California, and southern Australia. II. Dry
    matter, energy, and nutrient accumulation. Aust. J. Bot. *17*, 293-308.

Specht, R.L. (1973) Structure and functional response of ecosystems in the mediter-
    ranean climate of Australia, pp. 113-120. In F. di Castri and H.A. Mooney (eds.).
    Mediterranean type ecosystems: Origin and structure. New York: Springer-Verlag

Specht, R.L., and Rayson, P. (1957) Dark Island Heath (Ninety-Mile Plain, South
    Australia). III. The root systems. Aust. J. Bot. *5*, 103-114.

Thrower, N.J.W., and Bradbury, D.E. (eds.) (1977) Chile-California mediterranean
    scrub atlas: A comparative analysis. Stroudsburg, Pa. Dowden, Hutchinson and
    Ross, 237 pp.

Preface

Mediterranean type ecosystems have held the attention of plant geographers and plant ecologists for many years because of the similarity of vegetation form in the five widely disjunct regions of the world, which have similar climates of a unique type. Early plant geographers (Griesebach, 1872, Schimper, 1898) pointed out and described these broad similarities. Walter (1973) elaborated on some of the temperature and water requirements of mediterranean shrubs. Mooney and co-workers in the late 1960's began a set of measurements comparing physiological attributes of shrub species from California and Chile. The comparison of the mediterranean scrub regions in these countries continued during the International Biological Program, which emphasized the structural similarities of the vegetation related to primary production, topography, soils, and fauna (Mooney, 1977a, Thrower and Bradbury, 1977). A preliminary synthesis of structural and functional characteristics was developed in the First International Conference on Mediterranean Type Ecosystems held in Valdivia, Chile (Mooney, 1973). A second international conference was held at Stanford, California in August 1977 (Mooney and Conrad, 1977). The structural similarities were extended to other mediterranean regions by Cody and Mooney (1978). Synthesis of various structural and descriptive parameters are currently beginning to appear in the literature on heathlands and related shrublands (Specht, 1979) and on mediterranean type shrublands (Specht, *in press*). These recent syntheses emphasize that the structural similarities between the five disjunct mediterranean scrub regions have developed with different intensities of selection by climatic and soil nutrient factors in each regions. The selection process during the past 10,000 years has been one of eliminating species in the flora, rather than evolving new species as the mediterranean type climate developed.

Ecological theory often includes assumptions of optimization or maximization of resource use. According to theory the development of the vegetation in these disjunct mediterranean type ecosystems should lead to forms and functions which optimize resource use. Because mediterranean scrub regions are semiarid and nutrient poor but allow the development of vegetative canopies which cast shade on lower leaves and the soil, the resources that should be optimized include water, nitrogen and light energy. Thus, the similarity in vegetation form should relate to the use of water, nitrogen, and light energy. However, the concept of optimization has its opponents. The theoretical problem is whether the current resource use is optimal in a local or global sense. Resource use may be high relative to the current species characteristics, but higher resource uses may be possible if radically different species characteristics are available.

Following the International Biological Program, we wished to test the theory of optimization or maximization of resource use by capitalizing on the accumulated comparative information on the mediterranean scrub regions of California and Chile. At the onset of this research we felt that we could not evaluate whether the use of water, nitrogen, and light energy was a local or global optimum. Therefore, we proposed the more testable hypothesis, that if the resources available were similar (which could be expected in regions with similar climates), and if the vegetations were using resources optimally in each region, the resource use should be similar in mediterranean scrub regions of the two countries. The resource use was expressed as the ratio of water transpired to precipitation, nitrogen taken up to nitrogen mineralized, and solar irradiance converted carbohydrate to incoming solar irradiance. The resource uses were expressed as dimensionless efficiencies. We hypothesized that the annual totals and seasonal progression of resource use efficiencies should be similar in mediterranean scrub regions in California and Chile.

This volume summarizes 4 years of research on climate, vegetation distribution, phenology and growth, energy exchange, water relations, carbon balance and allocation, and plant and ecosystem nutrient relations. Our work should be relevant to persons interested in information on mediterranean type ecosystems as well as in the applicability of ecological theory to natural systems.

Many persons have aided in this research and deserve credit. Danilo Vucetich S. has permitted us to use his land at Fundo Santa Laura in Chile for the past 9 years. Wilbur and Joan Fitch and Duncan and Dorothy Miller have cooperated with us in using their land at Echo Valley in California for the same period. Field work on vegetation patterns was aided by Adriana Hoffmann, Vera Komárkova, and Mercedes Lawrence. Field and laboratory work on climate, microclimate, and plant water relations has been aided by David Albright, Peter Alpert, Sandra Araya, Albert Damm, James Ehleringer, Hiram Estay, Linda Fehringer, Juan Giliberto, Nina Hemphill, John Hom, Barry Hynum, Richard Kendall, Jeffrey Miller, Juan Domingo Molina, Russell Moore, Carole Murray, Edward Ng, Steven Oberbauer, Fernando Riveros de la Puente, Eda Roberts, Andres Seguy, William Smith, Lee Stuart, and Ali Valamanish. Field and laboratory work on photosynthesis, respiration, and carbon allocation has been aided by Lisa Harvey, Celia Hemphill, John Hom, James Houpis, Mercedes Lawrence, Eric Limbach, William Lowell, Louise Meyer, Ray Sievers, and Mark Sytsma. Field and laboratory work on growth dynamics has been aided by Maria-Ester Aljaro, Guacolda Avila, Barbara Ellis, Julia Etchegaray, Kathleen Fishbeck, Luis González, Alicia Hoffman, Adriana Hoffmann, Ann Jackson, William Jow, Gregory McMaster, and Robert Mangan. Field and laboratory work on nutrients has been aided by Sylvia Barkley, Jay McKendrick, Philip Rundel, and Theodore St. John. Drafting was done by the staff at the Institute of Arctic and Alpine Research, University of Colorado, and by Patsy Miller. Beth Sigren and Gay Wilkins suffered through retyping the innumerable drafts of this volume, and Martha Poole coordinated and completed the final editing. Daniel Botkin, Dale Cole, George Cox, George Innis, and Harold Mooney reviewed the manuscript in part or in whole, and made helpful suggestions. We deeply appreciate everyone's help.

The research was funded largely by the National Science Foundation grants DEB75-19491 and DEB77-13944. We also thank the San Diego State University Foundation and members of the College of Sciences and the Departments of Biology and Botany and Universidad Católica de Chile for their assistance in this research and its publication.

Philip C. Miller

Contents

Contributors

GRAY, JOHN T.

Department of Biological Sciences, University of California, Santa Barbara, California 93106

HAJEK, ERNST

Laboratorio de Ecología, Instituto de Ciencias Biológicas, Universidad Católica de Chile, Santiago, Chile

JACOBSON, MARTHA B.

Systems Ecology Research Group, San Diego State University, San Diego, California 92182

KRAUSE, DAVID

Systems Ecology Research Group, San Diego State University, San Diego, California 92182

KUMMEROW, JOCHEN

Systems Ecology Research Group and Department of Botany, San Diego State University, San Diego, California 92182

LAWRENCE, WILLIAM

Systems Ecology Research Group, San Diego State University, San Diego, California 92182

MARTÍNEZ, JOSE'

Laboratorio de Botánica, Instituto de Ciencias Biológicas, Universidad Católica de Chile, Santiago, Chile

MILLER, PHILIP C.

Systems Ecology Research Group and Department of Biology, San Diego State University, San Diego, California 92182

MONTENEGRO, GLORIA   Laboratorio de Botánica, Instituto de Ciencias
                            Biológicas, Universidad Católica de Chile, Santi-
                            ago, Chile

MUSTAFA, JAMIL   Ecole Superieure de grande Cultures du Kef,
                            Kef, Tunisia

OECHEL, WALTER C.   · Systems Ecology Research Group, San Diego
                            State University, San Diego, California 92182

POOLE, DENNIS K.   Systems Ecology Research Group, San Diego
                            State University, San Diego, California 92182

RICHARDS, SUSAN P.   Systems Ecology Research Group, San Diego
                            State University, San Diego, California 92182

ROBERTS, STEPHEN W.   Systems Ecology Research Group, San Diego
                            State University, San Diego, California 92182

SCHLESINGER, WILLIAM H.   Department of Biological Sciences, University
                            of California, Santa Barbara, California 93106

SHAVER, GAIUS R.   Systems Ecology Research Group, San Diego
                            State University, San Diego, California 92182

STEWARD, DEBORAH   Institute of Arctic and Alpine Research, Uni-
                            versity of Colorado, Boulder, Colorado 80302

STONER, WAYNE A.   Systems Ecology Research Group, San Diego
                            State University, San Diego, California 92182

WEBBER, PATRICK J.   Institute of Arctic and Alpine Research, Univer-
                            sity of Colorado, Boulder, Colorado 80302

1. Conceptual Basis and Organization of Research

PHILIP C. MILLER

In the first chapter we present the objective of the research, which was to test a hypothesis about community theory in mediterranean shrub ecosystems using techniques from primary production research. This chapter also describes the organization of the research.

1.1. General Background

It often has been implicitly assumed in community theory that competition and niche specialization among the species in a community results in optimal or maximal resource use by the community over ecological time, given an initial set of species abundances and a defined species pool (Clements and Shelford, 1939; Odum and Pinkerton, 1955; Elton, 1958; Cohen, 1966, 1967, 1970; Lewontin, 1969; Miller, 1969a; Simpson, 1969; MacArthur, 1970, 1972; Schoener, 1971; Katz and Bartnick, 1974; Rosenzweig, 1974; Orians, 1975). For these concepts to be applied to plants, the resources considered should include light, water, and inorganic nutrients. To the extent that the form and function of vegetation relate to resource use, evolution should produce a vegetation with specific physiognomy and physiology under given environmental conditions, which is only limited by the evolutionary potential of the organisms concerned. This implies that vegetations in similar environments converge toward a single pattern of resource utilization.

The similarity of vegetation form in widely disjunct regions of the world has long attracted the attention of plant geographers and ecologists (von Humboldt, 1806; Warming, 1909; Schimper, 1898; Köppen and Geiger, 1930; Raunkiaer, 1934; Walter, 1973). As climatic data became available with the settlement of mediterranean type regions, indices based on means and monthly values for temperature and precipitation were developed to show climatic similarity (Köppen and Geiger, 1930; Aschmann, 1973a; di Castri, 1973; Nazar et al., 1966; Walter, 1973). Mediterranean type vegetation and climates occur in five disjunct regions of the world, all centered around

32.5° latitude: around the Mediterranean Sea, on the southwest coast of Africa, in southwestern and southern Australia, in southern California, and in central Chile (Figure 1.1). Although the vegetation in all five mediterranean regions is broadly similar, variations occur within each region following trends that are also broadly similar. This similarity of form can be taken as an example of convergence for optimal resource use.

The assumption of convergence toward optimal resource use should not be accepted uncritically. Innis (1974) questioned whether the optimization concept leads to empirical studies. Odum (1960) showed that primary production in an old-field succession did not increase through several years, although species were being replaced. Whittaker (1965) suggested that productivity-diversity curves of vegetation types in the Great Smoky Mountains resulted from particular patterns of resource partitioning by species; he did not find any simple relationships between productivity and patterns of resource utilization. MacArthur (1957) developed several models of community structure based on a random pattern of utilization of available resources, but the predictions from these models were poor fits to many available data. As pointed out by Smith et al. (1975), a problem in community theory dealing with competition for resources is that the theory is largely based on Lotka-Voltera type equations, in which the resources are implicit in the coefficients. Smith and his coworkers suggested that models should keep the resources explicit and separate from species parameters. Major (1973) and Walter (1973) questioned whether the vegetation in the mediterranean regions has converged because of climatic factors alone.

The relations between vegetation form and patterns of resource utilization are not clearly understood. No naturally occurring ecosystem has been studied so that the spatial and temporal patterns of form and physiological activity of the different species can be related to the availability and utilization of resources.

A complete test of the assumption of optimal or similar resource use in natural ecosystems is not practical. However, the heritage of research on primary production processes can be used to integrate a broad data base and compare the resources of different species and vegetations under given environmental conditions. While community ecologists were developing theories based on abstract principles, the processes underlying primary production were being measured and formalized by agricultural scientists (Figure 1.2). This formalization began in 1953 when Monsi and Saeki developed some of the effects of canopy architecture on primary production and published a mathematical theory of canopy photosynthesis validated in several vegetation types. Since then the theories of canopy processes, including photosynthesis, transpiration, energy exchange, radiation penetration, and turbulent exchange of heat and water, have been developed to a high degree of sophistication. Mathematical, physical, and physiological techniques for analyzing and synthesizing the interactions of climate, microclimate, stand structure, water, and primary production now exist; these can be used to analyze quantitatively some of the interrelations between the climate and the

---

Figure 1.1. World map of the areas with mediterranean climates (from Aschmann, 1973b).

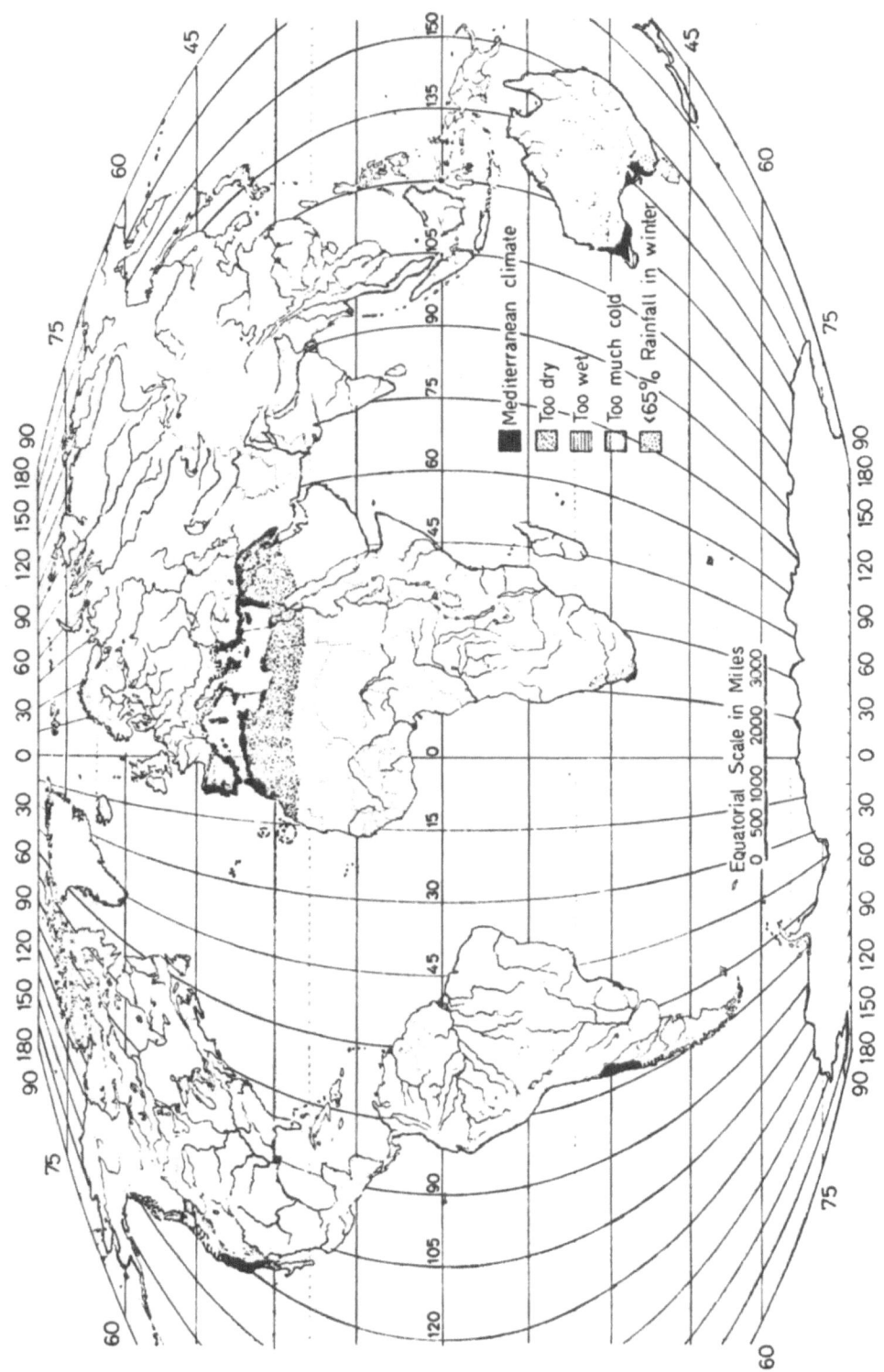

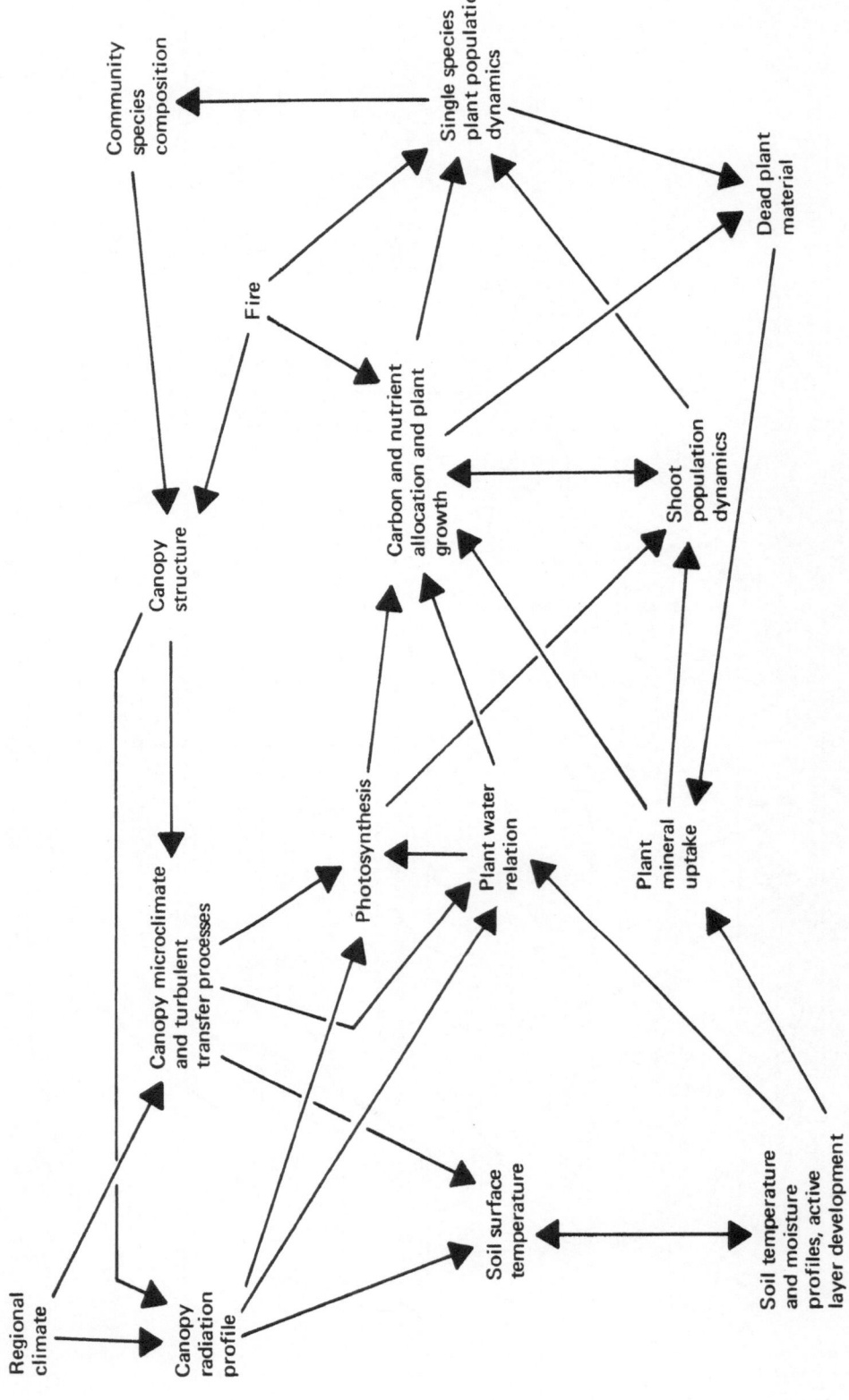

structure and function of the vegetation (Monsi and Saeki, 1953; Davidson and Philip, 1958; Anderson, 1964, 1966; Monteith, 1965; de Wit, 1965; Duncan et al., 1967; Waggoner et al., 1969; Lemon et al., 1971; Lommen et al., 1971; Murphy and Knoerr, 1972; Stewart and Lemon, 1972; Miller et al., 1976a; Tenhunen et al., 1976; Cunningham and Reynolds, 1978; and others). Models of soil physical processes are equally sophisticated (de Wit and van Keulen, 1972; van Keulen, 1975; van Keulen et al., 1975; Zartman et al., 1976). Models of soil chemical processes are approaching the same level of refinement (Nye and Tinker, 1969; Beek and Frissel, 1973; Frissel and Reinger, 1974; Jungk and Barber, 1974; Mengel and Barber, 1974; Phillips et al., 1976; Hunt, 1977; Cole et al. 1977; Reuss and Innis, 1977). Models of plant growth are more recent and less sophisticated but are developing predictive power (Brouwer and de Wit, 1969; de Wit et al., 1970; Fick et al., 1975; Holt et al., 1975; McKinion et al., 1975; Morgan, 1976; Seligman et al., 1975; Lawrence et al., 1978; Sauer, 1978; Stoner et al., 1978a, Loomis et al., 1979). These models, when adapted to naturally occurring species, can be used to develop a quantitative understanding of the interaction between vegetation structure and function in the use of light, water, and inorganic nutrients. The models then can be used to synthesize information on various physiological processes, integrate the processes in a meaningful way, and generalize the behavior of the species to new environments based on the physiological characteristics of the species. Through simulations, the utilization of resources by vegetation in a diversity of environments can be assessed, and the quantitative significance of structural and functional characteristics of species can be ascertained.

Primary production models are useful in testing current ecological concepts. Mooney and Dunn (1970a) and Mooney (1972) have described a theoretical framework within which plant allocation patterns can be quantified in terms of carbon costs and benefits and can be considered in a community context. Orians and Solbrig (1977) developed the cost-benefit approach with application to desert plant adaptations. Miller and Mooney (1976) used a production simulation model to show that the seasonal course of the vertical distribution of leaf biomass corresponded with a pattern that should optimize production and water use. Miller (1979) predicted the shift from shrubs with deciduous leaves to shrubs with evergreen leaves along an elevation-climate gradient using similar methods. Miller and Stoner (1979) used simulation models to ascertain the interactions between canopy characteristics and environment. Thus, the state of knowledge of primary production processes is at a point where this knowledge can be used to test some of the assumptions of community theory in natural vegetation.

1.2. Objective of Research

The general objective of this research was to examine the hypothesis that vegetation in regions of similar environments, despite different phylogenetic histories, displays similar patterns of resource utilization and similar patterns of nutrient and carbon allo-

Figure 1.2. Flow diagram of interactions between processes affecting the primary production.

cation to vegetative and reproductive structures. The hypothesis was derived from the ideas of community function held by community theorists and plant geographers and was meant to be an analogue of hypotheses proposed by MacArthur (1970). However, the conclusions regarding resource use were expected to differ from commonly expressed community theory because (1) plant, not animal, populations were being considered; (2) several resources, i.e., water, light, and nitrogen, were being considered simultaneously; (3) the underlying methodology of this research differed from the traditional methods of community ecologists; and (4) the natural environment was more variable than is usually assumed in community theory.

Resources available to the vegetation are either available directly from the physical environment, i.e., solar energy and water, or are partly controlled by the physical and chemical environment, i.e., available nitrogen and other inorganic nutrients. The efficiency of resource use by the vegetation depends on the ability of the species composing the vegetation to capture and incorporate these resources, which changes through evolution or succession; the efficiency also depends on the rate of supply of these resources by the environment, which changes with physical, chemical, and biological conditions.

Resources available to the vegetation should include at least water, light energy, carbon, nitrogen, and other inorganic nutrients. The hypothesis related to water was that seasonal patterns of the partitioning of incoming precipitation to interception losses, stemflow, throughfall, runoff, evaporation from the soil, and transpiration are similar in the mediterranean type ecosystems of southern California and central Chile and that the physiological controls on water use by species are similar in these vegetations (Chapter 6). The hypothesis for light energy was that the seasonal pattern of partitioning of incoming solar irradiance into intercepted irradiance, reflected irradiance, irradiance absorbed by leaves and soil, net infrared irradiance, convection, and transpiration is similar in the mediterranean type ecosystems of southern California and central Chile; that the mechanisms leading to similar vegetation form are similar (Chapter 5); and that the efficiencies of conversion of light energy to stored energy are similar (Chapter 7). The hypothesis for carbon was that the seasonal patterns of allocation to productive, supportive, absorptive, and reproductive structures and the biochemical costs of producing these structures are similar in vegetation with similar form (Chapters 4 and 8). The hypotheses for nitrogen were that the seasonal pattern of nutrient allocation to different plant structures is similar and that the pattern of nutrient division in the plant, litter, and soil compartments is similar in vegetation types of similar form (Chapters 9 and 10). Inasmuch as resource utilization and allocation by the vegetation result from the combined patterns of resource utilization by the individual species, the hypothesis was that the individual species in similar vegetation may differ in these patterns and that the similarities in patterns of resource utilization in the vegetation as a whole are due to variations in the number of species present and the relative abundances of individual species (Chapter 12).

Resource utilization was defined as the quantity of resource taken up by the species or vegetation and used, i.e., the quantity of water taken up by the species or vegetation and transpired, the quantity of light energy absorbed and converted into carbohydrate, and the quantity of inorganic nutrient taken up and incorporated in tissues. Resource use efficiency was defined as the dimensionless ratio of the quantity taken up to the quantity that became available to the vegetation. For water, the resource use efficiency was the ratio of transpiration to precipitation, which was reduced by subsoil drainage, soil surface evaporation, or interception losses. For light, the resource use efficiency was the ratio of the energy converted to carbohydrate to the incoming solar irradiance, which was reduced by interception by stems, high penetration of light through the canopy to the soil surface, low leaf absorptance, high resistances to carbon dioxide diffusion, and low quantum efficiency. For nitrogen, the resource use efficiency was the ratio of nitrogen taken up by the vegetation to the amount of nitrogen made available by mineralization, biological organism fixation, and dry- and wet-fall. This ratio was reduced by low absorbing root densities, low rates of uptake per unit of root, or low rates of mycorrhizal infection. The strategy of the species composing the vegetation is to minimize extraneous losses and to capture and use these resources more fully by altering their structure, abundance, and seasonal patterns of growth of leaves, stems, and roots.

1.3. Methodology

The general hypothesis could not be tested by direct field or response garden experiments because the deep-rooted chaparral and matorral shrubs are difficult to transplant intact and are slow to recover from transplantation. The time required for the plant and soil processes to adjust to the new conditions would have been too long to be practicable. The year-to-year variation in climate and soil processes is an additional constraint on assessing resource use in a short time period of observations. Therefore, the hypothesis was tested in simulation experiments, as well as in direct field measurements.

The underlying methodology of this project was to employ primary production process models to synthesize data, which were related to the utilization of light energy, water, and nitrogen on the vegetation-soil system, and to use these models to generalize from the data to diurnal, seasonal, and annual patterns of resource utilization (Figure 1.3). A model of water relations, photosynthesis, and heat exchange in the canopy (Canopy Process Simulator, CAPS) and models of soil physical processes, which simulate diurnal patterns, were used to integrate processes of carbon dioxide exchange, plant-soil water relations, and plant and soil heat exchange (Miller, 1972a, b; Miller et al., 1976a; Miller and Stoner, 1979). A model of plant growth and soil nutrient processes, which also included the carbon and water balance of the vegetation and soil, simulated seasonal patterns of resource use and plant responses in 1-day time steps. This model (The Mediterranean Ecosystem Simulator, MEDECS) was the principal mechanism for calculating the seasonal progression of resource use and the efficiencies of resource use by the vegetation. MEDECS included models of plant growth, decomposition of soil organic matter, and inorganic nutrient release in the soil and provided a means of relating growth of leaves, stems, and roots to photosynthesis, respiration, carbohydrate storage, nitrogen storage, nitrogen release in the soil, nitrogen uptake by the plant, soil water availability, and temperature (Miller et al., 1978).

The goal of the research was to apply existing models to ecological theory, rather than to develop or expand additional process models. However, after the research

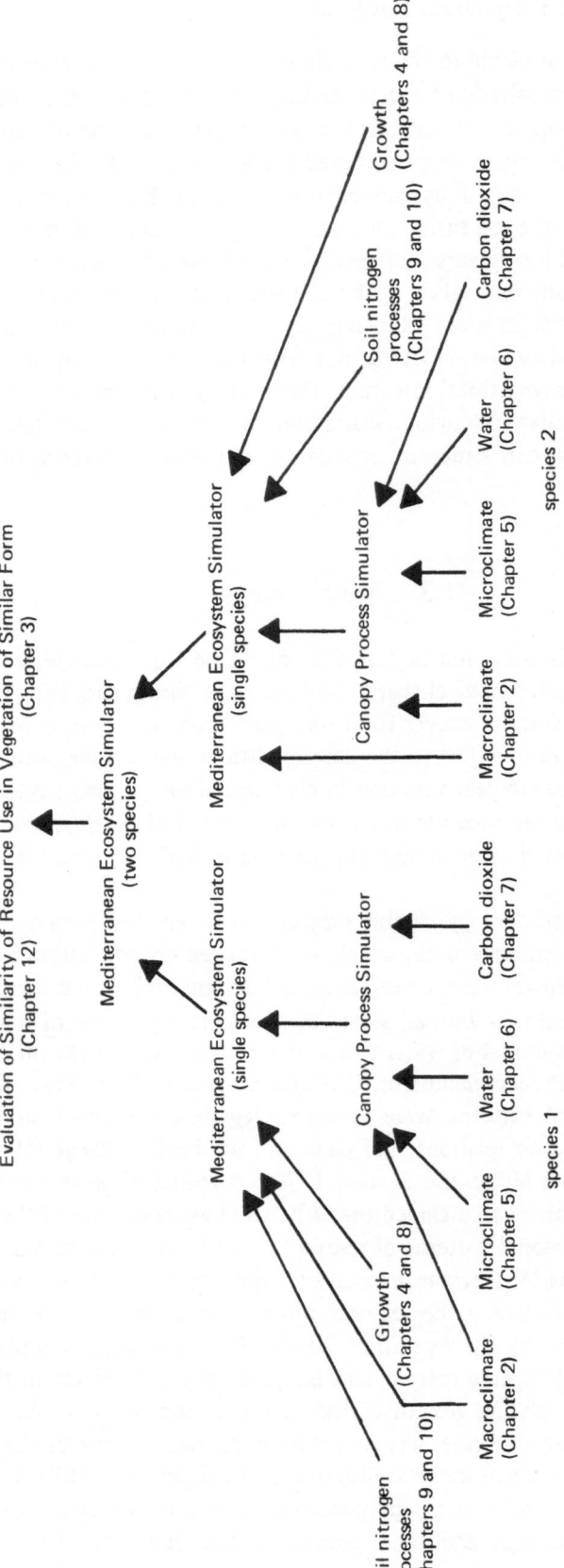

Figure 1.3. Diagram of the hierarchical procedure for synthesizing data from the several projects on climate, microclimate, soils, and plant physiology to evaluate the hypothesis of the similarity of resource utilization in two ecosystems of similar environments but different phylogenetic histories.

began, existing models were found to be inappropriate for considering the simultaneous utilization of light, water, and nitrogen by shrubs. Thus, MEDECS was developed
from concepts of van Keulen (1975) and Harpaz (1975) and from plant process concepts of Miller et al. (1976b) and Stoner et al. (1978b). MEDECS was validated with
field data and by a comparison of output with output from the daily model.

The hypotheses were examined in the vegetation of the mediterranean type ecosystems of southern California and central Chile because: (1) the differences in the
phylogenetic histories of California and Chile were well documented (di Castri and
Mooney, 1973; Mooney et al., 1974; Mooney, 1977a; Thrower and Bradbury, 1977);
(2) the similarity in the physical environments in the two regions was well documented (di Castri, 1973; Thrower and Bradbury, 1977; Miller et al., 1978); and (3)
there existed a body of comparable information on the two regions and a pair of
research sites matched in elevation, distance from the coast, geology, and topography,
which were not duplicated in any other two mediterranean type ecosystems (Mooney,
1977a; Thrower and Bradbury, 1977).

Prior to 1970, when the United States Origin and Structure of Ecosystems Program
began as part of the International Biological Program (IBP-OSE), studies on the photosynthesis, carbon balance, and patterns of allocation of Californian and Chilean shrubs
were being carried out by Mooney and co-workers (Harrison, 1971; Mooney and Hays,
1973; Dement and Mooney, 1974; Mooney and Bartholomew, 1974; Mooney and
Chu, 1974; Morrow and Mooney, 1974; Mooney et al., 1975). The mediterranean
scrub subproject of the IBP-OSE had the objective of documenting the structural similarity in the mediterranean regions of California and Chile (Mooney, 1977a; Thrower
and Bradbury, 1977). Studies were completed on climate, microclimate, soil moisture
and temperature, geology, geomorphology, human use, regional patterns of plant
growth forms, qualitative and quantitative phenology, vegetation and canopy characteristics, insects, lizards, and small mammals. The selection of the structural properties
of the vegetation to be measured was based on their importance in primary production
processes. At the conclusion of the project, information on the air and soil temperatures, air humidities, and solar and infrared irradiances were available from the research
sites in both countries. The water relations of some Californian shrubs had been characterized. Three years of quantitative and qualitative phenology data had been collected on major species in both countries. Simulations of canopy processes including
energy exchange, transpiration, and photosynthesis were made for *Heteromeles
arbutifolia*, using the measured climates in both regions and physiological data available prior to the IBP (Miller and Mooney, 1976).

Researchers tested the hypotheses in the research reported here through a 4-year
period, from September 1975 to January 1980, by refining and modifying existing
models of primary production processes for the mediterranean type ecosystems of
southern California and central Chile in order to calculate the seasonal patterns of
resource utilization by the predominant species. In southern California, research sites
were maintained to study montane, inland, and coastal vegetation and climate (Table
1.1). In central Chile, only an inland site was maintained. All sites were established
during the IBP-OSE. In California, the montane site was located at the San Diego
State University Mount Laguna Observatory (116°26' W, 32°50' N), 67 km east of San
Diego. The inland site was located at Echo Valley (116°40' W, 32°55' N), 45 km east

Table 1.1. Research Sites in Southern California and Central Chile

Sites	Distance From Coast (km)	Elevation (m)	Annual Precipitation (mm)	Annual Mean Temperature (°C)	Dominant Vegetation Type[a]	Dominant Plant Growth Forms
				California		
Mount Laguna	62	2030	449[b]	11.5b	Montane coniferous forest	Evergreen and deciduous trees
Echo Valley	45	1000	476[c]	13.4[c]	Chaparral	Schlerophyllous evergreen shrubs
Torrey Pines State Park	1	110	200[d]	15.8[e]	Coastal sage scrub/chaparral/coastal pine	Malacophyllous semideciduous shrubs and sclerophyllous evergreen shrubs
Camp Pendleton Marine Corps Base	2	90	201[f]	16.1[f]	Coastal sage scrub	Malacophyllous semideciduous shrubs
				Chile		
Fundo Santa Laura	75	1000	593[g]	12.4[c]	Matorral	Schlerophyllous evergreen shrubs and malacophyllous semideciduous shrubs

[a] From Munz (1974).
[b] For the periods Mar 1971-Dec 1974, Sep 1975-Sep 1978.
[c] For the period May 1971-June 1979.
[d] For the period May 1971-Dec 1974.
[e] For the periods Apr 1971-Dec 1974, Sep 1975-Jan 1977.
[f] For the period July 1971-Jan 1977.
[g] For the period July 1971-Jan 1977.
Note: The years 1977-1978 and 1978-1979 had above-average annual precipitation in southern California. Thus, Mount Laguna, Torrey Pines State Park, and Camp Pendleton Marine Corps Base are low relative to Echo Valley.

of San Diego and 10 km north of Descanso on an east-facing slope in the foothills of the coastal mountains (Figure 1.4a). Coastal sites were located at the Camp Pendleton Marine Corps Base (117°30' W, 33°20' N), 80 km north of San Diego, and at the Torrey Pines State Park (117°15' W, 32°55' N), 20 km north of San Diego. In Chile, the inland site was located at Fundo Santa Laura (71°00' W, 33°04' S), 30 km northwest of Santiago on the east flank of the Coast Ranges (Figure 1.4b). Most of the experiments and field measurements were made at the Echo Valley and Fundo Santa Laura sites. These sites have similar environments and vegetation, but the vegetation is of different phylogenetic origin. The Camp Pendleton, Torrey Pines State Park, Echo Valley, and Mount Laguna sites have different environments but related phylogenetic histories and are within 100 km of each other.

The experimental design was "T" shaped. Predominant species at Echo Valley and Fundo Santa Laura were characterized in situ so their patterns of resource utilization and allocation could be simulated. The predominant species at the coastal and montane sites were characterized from the literature and from measurements at the inland research sites in order to simulate the resource utilization of the vegetation at these sites. The comparison of resource use between Echo Valley and Fundo Santa Laura formed the stem bar of the "T" and the comparison between the coastal, inland, and montane sites in southern California formed the top of the "T." If the general hypothesis is true, water utilization, primary productivity, and nitrogen utilization should be more similar between Echo Valley and Fundo Santa Laura than between the coastal, inland, and montane sites in southern California. This experimental procedure was an approximation to the more exact test in which the vegetation of the coast and mountains would be characterized similarly to the vegetation of the inland research sites. However, the more exact test was not carried out because of the cost of characterizing a third or fourth vegetation type in situ. Simulations were also run to ascertain the potential effects of differences in the environment, vegetation form, and physiology in southern California and central Chile on resource use in these two mediterranean type ecosystems. Comparisons between sites also were made with data in the literature on water utilization, primary productivity, and nutrient cycling. Except for a few simulation experiments for the arctic and alpine tundra (Miller et al., 1976a; Tieszen, 1979), simulation models have not been applied in this way to ecological research.

The simulation models used data on the regional macroclimate obtained from the research sites and standard weather stations as driving variables. As preliminary validations, the models were used to calculate the relative water, carbon, and nitrogen balance of species at Echo Valley and Fundo Santa Laura. The calculated values were compared with measured values. A model of phenological controls was developed from data collected in 1976-1978 at Echo Valley and Fundo Santa Laura (Chapter 4) and was used to predict the phenological patterns recorded during the IBP-OSE (Mooney, 1977a; Thrower and Bradbury, 1977) for the species at the coastal and montane sites. Satisfactory results from the preliminary validations of MEDECS indicated that the models were valid, so MEDECS was used in the simulation experiments for comparisons of resource use among Echo Valley, Fundo Santa Laura, Camp Pendleton, and Mount Laguna. The test of similarity was that the seasonal patterns of resource use

(a)

(b)

Figure 1.4. Photographs of the Echo Valley Research Site (a) and Fundo Santa Laura Research Site (b) looking northwards. The area of intensive research at Echo Valley is in the center of the photograph, in the ravine which rises to the west (left). The area of intensive research at Fundo Santa Laura is in the lower left corner of the photograph (from Thrower and Bradbury, 1977).

and the annual total resource use at Echo Valley and Fundo Santa Laura were more similar to each other than to the seasonal patterns and annual totals at Camp Pendleton and Mount Laguna.

1.4. Conclusions and Discussion Regarding Hypotheses

The conclusions of this research, based on measurements and simulations, were that the uptake and uptake efficiency of water, solar energy, and nitrogen differed by species, slope, and country, such that there was a greater difference between the uptake and uptake efficiency of chaparral on the pole- and equator-facing slopes at Echo Valley than between the chaparral on the pole-facing slope at Echo Valley and the matorral at Fundo Santa Laura (Table 1.2).

The utilization of water differed by species, by slope exposure, and by country. Water utilization increased by species in the order *Rhus ovata, Adenostoma fasciculatum, Ceanothus greggii,* and *Arctostaphylos glauca* at Echo Valley; and in the order *Satureja gilliesii, Colliguaya odorifera,* and *Lithraea caustica* at Fundo Santa Laura. At Echo Valley, chamise chaparral, predominant on equator-facing slopes, used precipitation less efficiently than did mixed chaparral, which predominates on pole-facing slopes. The vegetation on the pole-facing slope at Echo Valley used all the annual precipitation; the vegetation on the equator-facing slope at Echo Valley and at Fundo Santa Laura did not.

The utilization of light energy differed by species, by slope exposure, and by country. Solar energy utilization by species increased in the order *R. ovata, A. fasiculatum, C. greggii,* and *A. glauca* at Echo Valley; and *L. caustica, S. gilliesii,* and *C. odorifera* at Fundo Santa Laura. At Echo Valley, production by chamise chaparral on equator-facing slopes was one-third that of the mixed chaparral on the pole-facing slopes. Incoming solar radiation on the equator-facing slope is about 1.3 times that on the pole-facing slope. Thus, light utilization efficiency in chamise chaparral is about 50% that of mixed chaparral. Production by shrubs at Echo Valley is slightly greater than the production at Fundo Santa Laura. Production by herbs is negligible at Echo Valley and relatively high at Fundo Santa Laura. Leaf longevity of shrubs is less at Echo Valley than at Fundo Santa Laura; however, leaf longevity of the total vegetation is similar because of the shorter longevity of the herb leaves at Fundo Santa Laura. At Echo Valley, the vegetation is at a steady state and production efficiency will not be increased without changes in species composition. At Fundo Santa Laura, solar energy efficiency may be increased by the growth of the vegetation.

The utilization of nitrogen and phosphorus differed by species, by slope exposure, and by country. Nitrogen and phosphorus use by species increased in the order *A. fasciculatum, R. ovata,* and *A. glauca* at Echo Valley and *C. odorifera* and *L. caustica* at Fundo Santa Laura. At Echo Valley, nitrogen levels in soil and plant material were lower in chamise chaparral on equator-facing slopes than in mixed chaparral on pole-facing slopes. Vegetation at Echo Valley appears to be nitrogen limited; vegetation at Fundo Santa Laura appears to be less limited by either nitrogen or phosphorus. Efficiencies for nitrogen and phosphorus utilization are higher than those for water and

Table 1.2. Summary of Resources and Resource-Use Efficiency for Mature Vegetation in the Mediterranean Shrub Regions of California and Chile[a,b]

	Echo Valley		Fundo Santa Laura	
Variable[c]	Pole-facing Slope	Equator-facing Slope	Pole-facing Slope	Equator-facing Slope
Resource Uptake				
P_{net} (g dry wt m^{-2} yr^{-1})	1646 ± 118	1107 ± 55	1521 ± 41	1564 ± 19
Tran (mm yr^{-1})	350 ± 24	257 ± 7	369 ± 12	351 ± 15
N_{up} (g N m^{-2} yr^{-1})	3.6 ± 0.5	1.9 ± 0.3	4.4 ± 0.1	4.7 ± 0.3
Resource Uptake Efficiency				
P_{net}/Solar (J J^{-1})	0.0052 ± 0.0004[d]	0.0028 ± 0.0001	0.0048 ± 0.0001[d]	0.004 ± 0.0001
Tran/Ppt (mm mm^{-1})	0.79 ± 0.05	0.60 ± 0.02[d]	0.61 ± 0.02[d]	0.60 ± 0.02[d]
N_{up}/N_{rel} (g N g N^{-1})	0.98 ± 0.08[d]	0.72 ± 0.08[e]	0.91 ± 0.01[d,e]	1.02 ± 0.05[d]
Growth Efficiency				
Growth/P_{net} (g dry wt g dry wt^{-1})	0.22 ± 0.01[d]	0.33 ± 0.01	0.22 ± 0.01[d]	0.22 ± 0.01[d]
Growth/Tran (g dry wt kg H$_2$O^{-1})	1.05 ± 0.04[d]	1.43 ± 0.07	0.91 ± 0.06[d]	0.97 ± 0.09[d]
Growth/N_{up} (g dry wt g N^{-1})	108 ± 7[d]	201 ± 24	77 ± 7[d,e]	72 ± 7[e]

[a] At Echo Valley, mixed chaparral occurs in the pole-facing slope and chamise chaparral occurs on the equator-facing slope.
[b] Means and standard errors are given using sample sizes of 5.
[c] P_{net} = net photosynthesis; Tran = transpiration; N_{up} = nitrogen taken up by the vegetation; Solar = solar irradiance; Ppt = precipitation; N_{rel} = nitrogen released in mineralization; and Growth = aboveground growth.
[d,e] Values in a row with the same letter are not statistically different.

light utilization. The conclusions regarding nutrient utilization are preliminary because of the paucity of complete data.

Three patterns of resource use efficiency occurred. Limitation by water, solar irradiance, and nitrogen are indicated by relatively high resource-use differences for all three resources on the pole-facing slopes at Echo Valley and Fundo Santa Laura. Limitation by only one resource, nitrogen, is indicated on the equator-facing slope at Fundo Santa Laura. Only modest limitation by these resources is indicated by the relatively low resource-use efficiencies for all three resources in the chamise chaparral on the equator-facing slope at Echo Valley. Chamise chaparral is the most abundant chaparral species as the chaparral recovers after fire (Hanes, 1971). Although chamise chaparral has low resource-use efficiencies, it has a high growth efficiency so that its growth is comparable to that of mixed chaparral.

At Echo Valley and Fundo Santa Laura, spring is the period of greatest activity in both soil and plant processes, including decomposition, mineralization, nutrient uptake, photosynthesis, transpiration, growth of shoots and roots, and reproduction.

Characteristics of mediterranean vegetation, including broad leaves, sclerophylly, and evergreenness, are selected for by climate and nutrient conditions. When the seasonal progression of temperature and precipitation in mediterranean regions creates conditions favorable for photosynthesis before conditions are favorable for growth and where the length of the growing season is shorter than the time required to regain the costs of constructing a leaf, evergreenness predominates. Where the length of a period unfavorable for photosynthesis is such that the cost of maintaining a leaf is greater than the cost of replacing the leaf, deciduous-leaved plants predominate. With nutrient deficiencies, evergreenness predominates. However, the nutrient deficiency can be induced by deficiencies in the parent material, by low rates of mineralization because of poor quality organic substrate, or by high leaching rates. Thus, a site on which evergreenness has been selected for by climate may appear nutrient limited, especially for nitrogen. A site on which evergreenness is selected for by deficiencies in the parent material will appear nutrient limited, but primarily for phosphorus or cations, and secondarily for nitrogen. Sites with evergreen vegetation in mediterranean regions of Australia and South Africa are parent-material limited, but similar sites in mediterranean regions of California, Chile, and the Mediterranean basin are primarily climatically limited and secondarily nutrient limited.

The dissimilar resource-use efficiencies reported in this research may be characteristic of periodically disturbed communities like the chaparral, where requirements of assuring successful reestablishment after the disturbance reduces the ability of the plants to capture resources and where alternate life history strategies, i.e., resprouting shrubs versus seed reproducing shrubs, are possible. Environmental factors and species physiology critical to survival or reinvasion do not necessarily lead to increased resource use. In mature chaparral stands, resprouting species have lower resource-use efficiencies than do seed-reproducing species.

2. Resource Availability and Environmental Characteristics of Mediterranean Type Ecosystems

PHILIP C. MILLER and ERNST HAJEK

This chapter describes regional and seasonal patterns of the mediterranean type climate in southern California and central Chile as they influence the utilization of water, solar irradiance, and inorganic nutrients by the vegetation.

2.1. Introduction

2.1.1 General Setting

The seasonal progressions of temperature, precipitation, and solar irradiance determine the availability of resources for the vegetation. The fundamental resources studied in this research, which the plant species must capture and utilize, are water, solar irradiance, and soil nutrients, especially nitrogen. The availability and use of these resources are constrained by the temperatures of the air and the soil.

The mediterranean type climate is characterized by cool, wet winters with low solar irradiance and hot, dry summers with high solar irradiance (Figure 2.1). Aschmann (1973b) defined the mediterranean type climate as one with a total annual precipitation of 300 to 900 mm, at least 65% of which falls during the 6 winter months (November through April in the Northern Hemisphere and May through October in the Southern Hemisphere), with daily mean air temperatures below 0°C no more than 3% of the year, and with at least 1 month with an average temperature below 15°C. Aschmann's definition of mediterranean type climate is somewhat narrower than that of Köppen and Geiger (1930) or Meigs (1953). The difference between definitions affects the areal distribution of the climatic types, but the areas included in this study were well within the boundaries of any definition of mediterranean climate.

The seasonal pattern of the mediterranean type climate is the result of changes in the position of the high pressure zones that occur at about 30° latitude north and

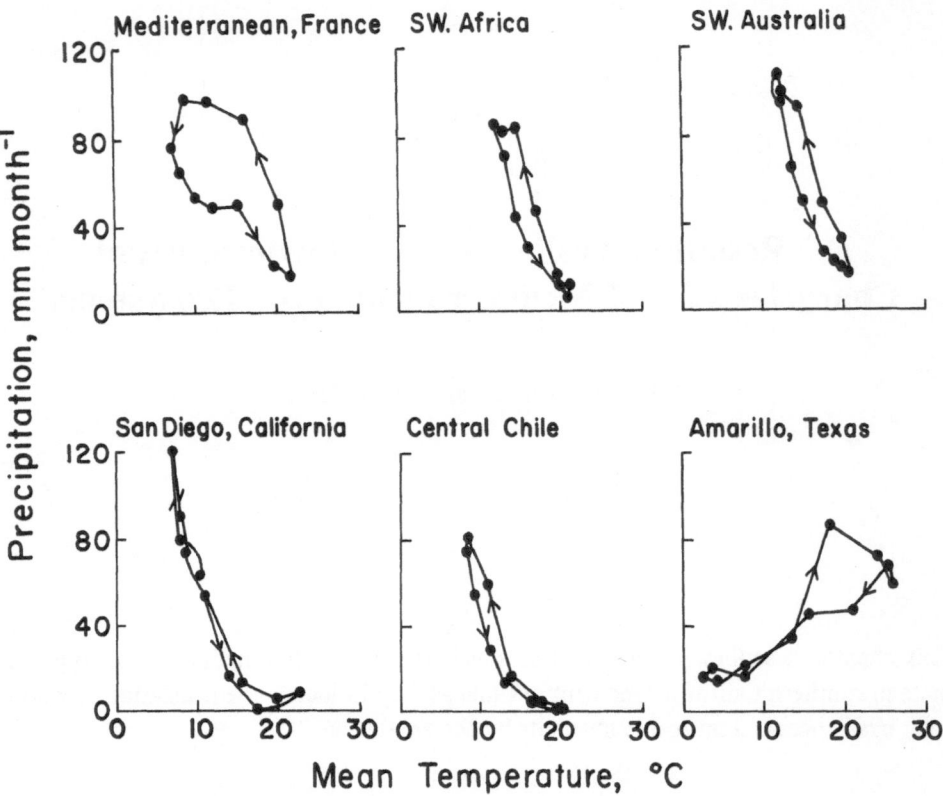

Figure 2.1. Seasonal progression of monthly mean air temperatures (°C) and monthly precipitation (mm) in five regions of the world with mediterranean type climates and at Amarillo, Texas, an area with similar annual precipitation but with summer rainfall and grassland vegetation (after Miller et al., 1977).

south of the equator. As the sun moves north, the high pressure zone moves north; as the sun moves south, the high pressure zone moves south. The high pressure zone controls the equatorward migration of the polar frontal zone. The passage of solar fronts in winter when the high pressure zone is near the equator causes the winter rains. In summer, the passage of moisture-laden tropical air masses is inhibited by the high pressure zone, cool offshore water, and land masses lying to the east. Thus, throughout the world, regions with predominantly winter rain and summer drought occur on the polar side of the high pressure zone, usually on the southwest side of the continents.

Although the climates of southern California and central Chile are broadly similar, minor differences occur because of the difference in the general circulation patterns. The equatorward shift of the high pressure zone in the Northern Hemisphere is less than in the Southern Hemisphere. Thus, summer rains occur rarely in central Chile but are occasional in southern California, because the location of the Northern Hemisphere high pressure zone is less effective in blocking the summer tropical storms from moving northward off the west coast of Baja California on through the Gulf of California into southern California. High pressure cells develop over the deserts in the southwestern United States, especially in late summer and fall, and push dry desert air, which warms adiabatically as it descends, into southern California in what are called Santa Anas.

During a Santa Ana, the weather is hot, clear, dry, and often windy, and fire danger is extremely high. Most of the wildfires in southern California occur during a Santa Ana. The movement of air into Chile from the deserts east of the Andes is blocked by the Andes. Thus, Chile does not experience analogous hot, dry conditions. Because of the greater expanse of water in the Southern Hemisphere relative to that in the Northern Hemisphere, climates in the Southern Hemisphere are cloudier, colder, and more maritime at a given latitude than at a similar northern latitude (Sellers, 1965).

The mediterranean type climate in southern California and central Chile is perhaps a recent development. Axelrod (1973) argued that the winter rain, summer drought pattern developed in both these areas during the general cooling trend of the last 10,000 years and was a change from a previous climate with year-round precipitation. Heusser (1978) concluded that the summer drought pattern in southern California formed within the last 2500 years. Because the mediterranean type climate is determined by the global circulation pattern, Raven (1973) concluded that these climates could never have been contiguous between the Northern and Southern Hemispheres. Thus, the current similarities in the vegetation of these regions must have developed independently.

2.1.2 Sources of Data

Climatic comparisons were made between southern California and central Chile using data from standard weather stations with at least 20-year records and data collected from the project research sites. Isopleth maps were prepared for monthly absolute maximum and minimum, mean maximum and minimum, and mean temperatures, and monthly precipitation for the region between 32°32' and 32°25' N latitude and between 117°25' and 116°30' W longitude in southern California and the region between 32°30' and 33°15' S latitude and between 71°45' and 70°45' W longitude in central Chile (Hynum, 1974). Temperature data were from 25 stations in California and 25 stations in central Chile. Precipitation data were from 127 stations in California and 163 stations in Chile. The Californian temperature data were taken from the United States National Oceanic and Atmospheric Administration (1950-1970). Californian precipitation data were taken from the San Diego County Department of Sanitation and Flood Control (1966-1973). Most of the Chilean temperature and precipitation data were taken from summaries compiled by the Oficina Meteorológica de Chile (1930-1970), with additional data from the Ministeria de Agricultura. In both countries, the stations are concentrated along the coast and are sparse at higher elevations in the interior.

Temperature and precipitation, together with solar irradiance, soil moisture, and air humidity, were measured at the project research sites: Echo Valley, Camp Pendleton Marine Corps Base, Torrey Pines State Park, and Mount Laguna Observatory in California, and at Fundo Santa Laura in Chile. Solar irradiance was measured with a Belfort pyrheliograph. Air temperature and humidity were measured with a hygrothermograph in a standard weather screen, precipitation with a rain gauge 0.05 m in diameter supplemented at Echo Valley with a rain gauge 0.20 m in diameter, and wind speed and direction with a cup anemometer and wind vane. Soil moisture was measured with a neutron probe at three to four access tubes on north, south, and ridgetop exposures.

Soil moisture was measured from a depth of 0.3 m to bedrock, which was at a depth of 0.9 to 1.2 m. In California, the neutron probe was calibrated with soil from Echo Valley (Ng, 1974; Ng and Miller, 1980). Moisture-tension curves and soil thermal properties were measured on the same soils. Most of the instruments in California were installed by March 1971 and were run fairly continuously thereafter until July 1978. Instruments in Chile were installed in November 1971 and run until July 1978.

Soils at the research sites were analyzed for several physical and chemical properties by Thrower and Bradbury (1977). In addition, samples of soil in the depths of 0 to 15 cm were collected at each ordination plot (Chapter 3). Analyses on these soils were performed at the Sedimentology Laboratory, Institute of Arctic and Alpine Research (Boulder, Colorado) and the Alaskan Agricultural Experiment Station (Palmer, Alaska). Nutrient levels were based on exchangeable cations and total nitrogen and phosphorus and were expressed as parts per million of dry soil.

2.2. General Description of Regions

2.2.1 Topography, Geology, and Soil Nutrients

The regimes of mediterranean type climates in southern California and central Chile are located in areas of similar tectonic history and share several characteristics (Mooney, 1977a, Thrower and Bradbury, 1977). In both regions, the coastal plain extends inland about 20 km. Foothills give rise to mountains with elevations reaching about 2000 m. Both Echo Valley and Fundo Santa Laura are located at an elevation of 1000 m on generally east-facing slopes dissected by drainage to the east. However, Echo Valley is situated in a relatively shallow basin within the massive, gradually ascending tabular upland of the west slope of the Peninsular Ranges of southern California, while Fundo Santa Laura is on a regionally east-facing slope within a series of ridges that forms the broken crest of the Coast Ranges in Chile. Farther to the east in Chile lie the Central Valley and the Andes. The Mount Laguna research site at 2030 m elevation is located close to the eastern escarpment of the mountains where the land surface descends rapidly to the inland desert. The Torrey Pines State Park research site is located at the same elevation as and about 1 km inland from bluffs of 110 m elevation at the shore of the Pacific Ocean. The Camp Pendleton research site is located about 2 km from the shore at 90 m elevation on low hills that face south.

The predominant parent rock of both Echo Valley and Fundo Santa Laura consists of quartz diorite (Mooney, 1977a; Thrower and Bradbury, 1977). In addition, there are outcrops of gabbro at Echo Valley and andesite at Fundo Santa Laura. At both sites, the soils are poorly developed, shallow, commonly stony, gravelly textured throughout the profile, and neutral to slightly acid. Soils developed on quartz diorite at Echo Valley are sandy loams, while those at Fundo Santa Laura are coarse textured, granular, loamy sands. The clay contents are similar. The soils are permeable and well drained. Soils at Echo Valley are classified in the eroded Cienaba series (U.S. Department of Agriculture, 1973) in which the surface soil may be an old subsoil. The andesite parent rock at Fundo Santa Laura contains abundant kaolinite and weathers to produce soils with over three times as much clay as soils from the quartz diorite. Associated subsoils have nearly four times as much free iron oxide as do the quartz diorite soils. Soils from the gabbro rocks at Echo Valley have intermediate concentrations of free iron oxides.

At Echo Valley and Fundo Santa Laura, all soils were characterized by a rather high calcium-magnesium complex and were at least three-fourths base saturated (Thrower and Bradbury, 1977). Exchangeable bases, especially calcium and magnesium, were more plentiful in the andesitic soils. The cation exchange capacity of soils developed from quartz diorite was about half that of soils developed from gabbro at Echo Valley and from andesite at Fundo Santa Laura. The higher cation exchange capacity of the andesite and gabbro soils indicated a higher nutrient-storage capacity. The andesite soils also had a greater water-holding capacity owing to small mean grain size. However, on the quartz diorites, the percentage of organic matter at the two sites was similar. Total nitrogen in the soil was similar on the three parent materials at the two sites, but nitrate-nitrogen was higher in soils at Echo Valley than at Fundo Santa Laura.

The most conspicuous difference between the soils of southern California and central Chile was the difference in nutrient levels. The Californian soils contained significantly lower quantities of potassium, calcium, magnesium, nitrate, and phosphorus than did the Chilean soils. The amount of ammonium in the soils was not significantly lower in the soils from central Chile (Chapter 3).

2.2.2 Fire and Disturbance

Both chaparral and matorral are subject to disturbance. Chaparral is periodically disturbed by fire. Fire in chaparral characteristically removes most of the small-diameter, aboveground, live and dead plant material and deposits inorganic nutrients in an available and leachable state on the soil surface, thus potentially increasing the rates of nutrient cycling and the potential for nutrient loss. A major effect of fire is to mineralize nutrients in litter and live plant material. These nutrients are for the most part deposited on the soil surface in soluble forms, which are available for uptake by surviving plant species and seedlings. However, 75% of the aboveground plant nitrogen or 10% of the nitrogen above 10 cm soil depth may be lost during the fire (DeBano et al., 1977; DeBano and Conrad, 1978). Potassium and phosphorus are also subject to leaching losses after the fire. Soil pH is increased (Ahlgren and Ahlgren, 1960), favoring bacteria and actinomycetes relative to fungi and perhaps creating a more favorable environment for nitrogen fixation and nitrification (Buckman and Brady, 1969). Fire also affects shrub architecture by favoring stump-sprouting or crown-sprouting forms (Keeley, 1977; Keeley and Zedler, 1978). Grazing, on the other hand, favors arborescent forms.

While the mediterranean areas of California have undoubtedly always been subjected to fire, the present fire frequency may be lower than before the advent of European man. Lightning is thought to have ignited chaparral stands adjacent to both low-elevation grasslands and higher elevation forests. Vogl (1977) speculated that grassland fires exposed low-elevation chaparral to fire almost every year. However, in the center of the chaparral distribution, fires probably burned through southern California chaparral at average intervals of 25 to 50 years (DeBano et al., 1967; Dodge, 1975) when sufficient fuel had accumulated to carry the fire. These fires in grasslands and chaparral probably burned over large areas uninterrupted by freeways, buildings, and cleared land and were of lower intensity than today. Mooney (1977b) estimated that presently an average of 2.8% of the chaparral in San Diego County burns each year, resulting in a fire cycle of about 35 years. In southern California, humans are responsible for 99% of all acreage burned (Keeley, 1977). It is probable that the fire cycle has been recently altered by human activities and that the vegetation has not fully adapted to the extant fire cycle. A decrease in fire frequency results in changes in the temperature of the burn, which may affect seed and sprouting shrub survival, soil litter layers, and soil microorganisms.

A fire in a young chaparral stand may not consume the young, moist tissue that remains as standing dead in the new stand. This may increase the probability of the stand burning again, increasing the fire frequency and favoring resprouting perennials over nonsprouting species, i.e., grasses, forbs, and aggressively introduced species, over woody species (Vogl, 1977).

A decreased fire frequency can lead to decadent stands, which recover more slowly following fire, and to hot fires with hot spots in areas of fuel buildup, which effectively kill both seeds and resprouting chaparral shrubs. Today, fires occur primarily during the drought period, when environmental conditions slow postburn recovery. Historically, fires may have been extinguished by rains, which increased postburn recovery (Vogl, 1977). Keeley (1977) suggested that a long fire cycle, more than 100 years, may select for chaparral shrubs that are obligate seeders because of the large seed pool in the soil and because of the reduced number of surviving resprouting shrubs with which to compete.

Echo Valley and Fundo Santa Laura have been relatively undisturbed through the 15 years previous to the study. The Echo Valley site burned in 1952 in a widespread fire. Fundo Santa Laura was cut for wood and charcoal until about 1962 when the owner set the area aside as a natural reserve.

2.3. Wind, Temperature, and Atmospheric Humidity

2.3.1 Wind

Throughout the year, the wind speeds measured at Echo Valley were about one-half to one-third those measured at Fundo Santa Laura and Camp Pendleton (Miller et al., 1977; Thrower and Bradbury, 1977). Wind speeds were lower at Camp Pendleton than at Fundo Santa Laura. The generally higher wind speeds measured at Fundo Santa Laura were probably related to the massive Andean mountains and the cold offshore water temperatures. At all three research sites, wind speeds were slightly higher in summer than in winter. At Echo Valley, the predominant wind direction was from the northwest; however, a daily pattern was apparent. At night, the wind directions were often from the east-northeast changing to the south for a short period during the day and then to the northwest or north as air moved in from the ocean. Speeds increased in the afternoon as wind direction shifted to the northwest. Characteristically, the daily course included light winds at night and higher winds in the afternoon. At Fundo Santa Laura, the winds consistently came from the west throughout the day and night, and the diurnal pattern was for high winds beginning about noon and tapering off late in the afternoon. The differences in the pattern of directional wind shifts were probably related to topographic differences between the two research sites.

2.3.2 Air Temperature

Annual mean air temperatures were about 2°C higher in southern California than in central Chile at similar elevations (Figure 2.2). In each country, the annual mean temperatures at coastal and inland locations were similar when the stations were at similar elevations. Mean maximum and absolute maximum temperatures were higher in southern California, which reflects the influence of the Santa Ana. Continental conditions predominate in southern California; marine air predominates in central Chile. The

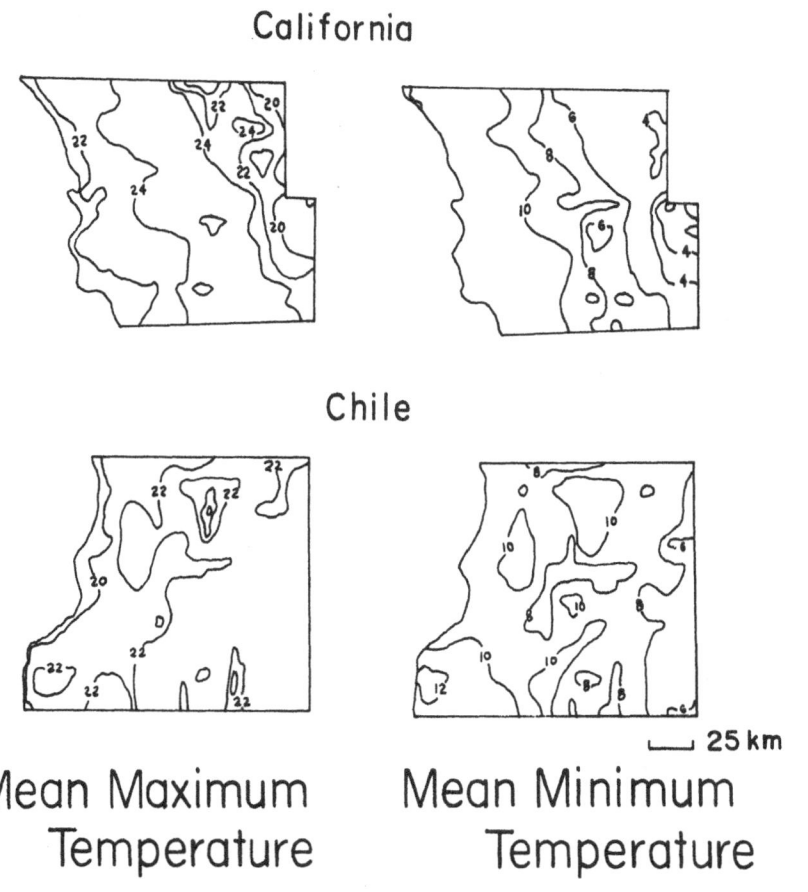

Figure 2.2. Isotherms of mean maximum temperature and mean minimum temperature (°C) in southern California and central Chile (from Miller et al., 1977).

mean maximum temperatures along the coast were about 3°C higher in southern California than in central Chile; they were about 6°C higher inland in southern California. The absolute maximum temperatures along the coast were 12°C higher and inland were 7°C higher in southern California than in central Chile.

Mean minimum temperatures along the coast were similar in southern California and central Chile, but inland stations were 1°C warmer in southern California. Absolute minimum temperatures at the coast were about 2°C lower in southern California, and inland stations were 5°C lower than in central Chile. Temperatures below 0°C occurred throughout southern California and in inland central Chile but occurred only occasionally along the central Chilean coast. The minimum temperatures in central Chile showed the influence of the cold air drainage down the major river valleys from the higher elevations in the Andes. Along the river valleys, such as the Rio Aconcagua and Rio Maipo, minimum temperatures were lower than the minimum temperatures measured at stations above the valley bottom. Temperatures around Valparaiso tended to be higher than other temperatures along the central Chilean coast, possibly because of the protection from the down-slope wind by the Coast Ranges.

The seasonal patterns of temperature differed slightly in the two regions. Throughout central Chile, minimum temperature occurred in July and maximum temperatures in January. In southern California, minimum temperatures occurred in January. Maximum temperatures occurred in July in the mountains and in August at the lower coastal stations. The maximum temperatures were dominated by the radiation regime in the mountains and by the adiabatic warming of the air during Santa Anas in late summer at the lower elevations. Monthly mean temperatures in the mountains were below 0°C from October to May but were always above freezing from June to September. Throughout central Chile, minimum temperatures occurred in July, and maximum temperatures occurred in January. In both countries, diurnal and seasonal variation in air temperature increased from the coast to the interior, but the increase was more pronounced in southern California.

Annual mean maximum air temperatures increased with elevation up to about 400 m in southern California based on data from stations with 20-year records (Miller et al., 1977). Above 400 m, the temperatures decreased at a rate of 5.4°C per 1 km change in elevation (Figure 2.3). Mean minimum air temperatures decreased with elevation more rapidly below 400 m than above. The minimum temperatures showed greater deviation from the elevational trends than did the maximum temperatures because of the effects of local topography on cold air drainage at night. Along the coast, mean temperatures during the daylight period were above 15°C throughout the year. Temperatures above 17.5°C occurred only during the summer drought. With increasing elevation, mean daytime temperatures during the winter decreased such that, at about 1100 m, temperatures were below 10°C. At higher elevations, temperatures became warmer late in the spring. Mean daily temperatures followed a similar pattern but were 2° to 5°C lower. Thus, moderate temperatures occurred in winter along the coast in spring at midelevations, and in late spring and summer at higher elevations.

The annual mean temperatures were 13.4°C at Echo Valley and 12.1°C at Fundo Santa Laura (Table 2.1). During the year, daily mean temperatures were between 8° and 20°C at Echo Valley and between 7.5° and 17°C at Fundo Santa Laura. Air temperatures were more variable diurnally and seasonally at Echo Valley than at Fundo Santa Laura. The mean annual temperatures were 15.9°C at Camp Pendleton and

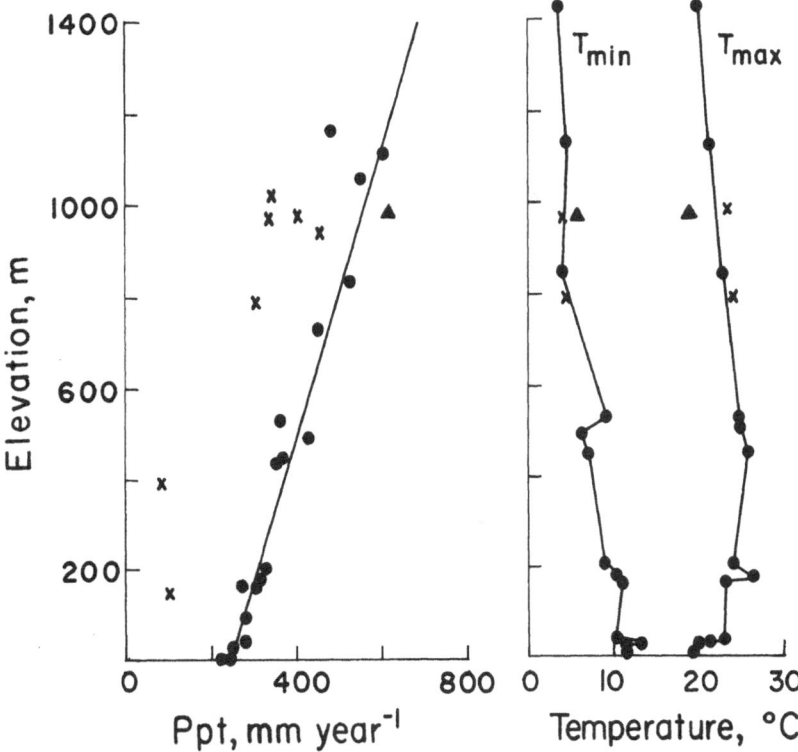

Figure 2.3. Elevational distributions of annual precipitation and mean annual maximum and minimum temperatures for southern California and central Chile: coastal-facing slope (●) and inland (X) in southern California and Fundo Santa Laura (▲) (after Miller et al., 1977).

11.3°C at Mount Laguna. On the basis of the annual mean air temperature, Fundo Santa Laura was equivalent to a site in southern California at about 1500 m. When Duncan's test (Bancroft, 1968) was used for evaluating the significance of the annual mean temperatures given the year-to-year variability, Mount Laguna and Fundo Santa Laura were not significantly different (0.95 level). Fundo Santa Laura and Echo Valley were not significantly different, but Mount Laguna differed from Echo Valley, and Camp Pendleton differed from all the other research sites.

During the respective summers, mean minimum temperatures at both sites were near 10°C, but diurnal variations in air temperature of 20°C were common at Echo Valley, while variations were rarely greater than 10°C at Fundo Santa Laura (Table 2.1). In winter, minimum air temperatures were lower at Echo Valley, although maximum temperatures were higher. During the period 1971-1978, more extreme temperature conditions, i.e., a greater number of days with a mean maximum temperature above 30°C or a mean minimum temperature below 0°C, occurred at Echo Valley than at Fundo Santa Laura. However, the air temperatures were moderate throughout the year at both research sites with subfreezing and extremely high temperatures rarely occurring. Temperatures throughout the winter were favorable for photosynthesis but were less favorable for growth.

Table 2.1. Summary of macroclimate characteristics at the Mount Laguna, Echo Valley, and Camp Pendleton research sites in California and the Fundo Santa Laura research site in Chile

Characteristic	Oct	Nov	Dec	Jan	Feb	Mar	Apr	May	June	July	Aug	Sep	Mean	Annual Total
Mount Laguna (2030 m) Apr 1971-July 1978														
Solar max. (mJ m^{-2} day^{-1})	19.0	13.2	11.6	12.2	15.7	19.2	23.0	24.6	25.0	25.1	22.6	20.5	19.2	7150 ± 184
Solar mean (mJ m^{-2} day^{-1})	13.2	10.1	7.8	7.8	10.5	13.7	18.9	20.9	21.8	19.6	19.1	17.4	15.1	5602 ± 54
Precipitation (mm mo^{-1})	17.2	42.8	55.7	58.5	67.3	71.9	27.3	15.7	3.6	13.9	51.2	23.4	449	444 ± 89
Potential evapotranspiration (mm day^{-1})	1.7	0.8	0.5	0.5	1.0	1.2	2.0	2.5	3.8	4.0	2.8	3.7		1182 ± 36
Relative humidity, midday (%)	53	58	62	63	60	52	46	45	44	45	42	48		
Air temperature max. (°C)	16.8	12.1	8.4	6.6	8.4	9.7	15.1	18.0	25.4	28.6	25.6	23.5	16.5	
Air temperature min. (°C)	7.0	3.8	1.5	0.7	0.9	1.1	3.2	6.2	12.1	16.0	13.7	11.6	6.5	
Annual mean temperature (°C)													11.3	
Echo Valley (1000 m) Apr 1971-July 1978														
Solar max. (mJ m^{-2} day^{-1})	19.6	15.9	14.5	15.2	18.2	21.8	24.9	26.1	27.7	26.4	24.4	23.3	21.5	7883 ± 159
Solar mean (mJ m^{-2} day^{-1})	14.8	11.7	9.7	10.9	12.6	15.4	19.0	20.9	23.2	22.3	21.4	18.0	16.7	6184 ± 113
Precipitation (mm mo^{-1})	43.1	48.5	65.2	57.0	114.2	59.8	38.1	16.2	3.6	9.0	8.3	11.1	476.	476 ± 60
Potential evapotranspiration (mm day^{-1})	1.8	1.2	0.8	0.9	1.2	1.4	1.9	2.3	3.0	3.7	3.0	2.5		1129 ± 29
Relative humidity, midday (%)	30	32	39	36	35	40	32	31	28	27	26	30		
Air temperature max. (°C)	23.8	19.4	16.4	15.9	16.6	16.3	18.5	21.6	27.8	31.8	31.2	27.9	22.3	
Air temperature min. (°C)	5.3	2.4	1.8	1.2	1.0	2.6	1.6	5.0	6.9	10.1	9.3	8.7	4.7	
Annual mean temperature (°C)													13.4	
Camp Pendleton (110 m) July 1971-Dec 1974[a]														
Solar max. (mJ m^{-2} day^{-1})	17.0	13.3	11.2	14.2	15.0	19.5	24.2	25.0	24.2	25.8	23.0	20.4	19.4	7079 ± 176
Solar mean (mJ m^{-2} day^{-1})	12.6	10.5	8.1	9.7	11.8	13.3	19.1	16.9	18.5	19.3	17.9	15.3	14.4	5234 ± 226
Precipitation (mm mo^{-1})	14.0	37.0	38.5	22.8	22.6	49.2	3.9	3.3	6.4	0.2	0.0	3.0	202	202 ± 60

Characteristic	Apr	May	June	July	Aug	Sep	Oct	Nov	Dec	Jan	Feb	Mar	Mean	Annual Total
Potential evapotranspiration (mm day^{-1})	1.8	1.0	0.9	1.0	1.4	1.8	2.2	2.5	2.6	3.2	2.8	2.5		1117 ± 52
Relative humidity, midday (%)	60	65	67	67	68	66	59	61	66	63	65	64		
Air temperature max. (°C)	21.0	17.6	16.1	14.9	16.8	16.6	19.3	20.1	21.3	23.3	24.4	23.3	19.6	
Air temperature min. (°C)	13.5	11.1	10.1	9.0	9.9	10.2	10.8	12.5	14.5	15.9	16.7	15.9	12.5	
Annual mean temperature (°C)													15.9	
Fundo Santa Laura (1000 m) Nov 1971-July 1978														
Solar max. (mJ m^{-2} day^{-1})	18.7	12.7	9.5	11.6	17.1	21.1	26.2	27.8	29.4	29.0	26.8	22.9	21.0	7640 ± 125
Solar mean (mJ m^{-2} day^{-1})	12.9	8.1	6.3	7.9	11.2	14.5	20.2	24.3	26.4	26.0	23.7	19.0	16.7	6125 ± 197
Precipitation (mm mo^{-1})	13.8	80.6	142.8	181.4	83.7	30.0	36.7	23.1	0.0	0.0	0.0	0.4	593	590 ± 101
Potential evapotranspiration (mm day^{-1})	2.1	1.4	0.8	0.9	1.4	2.0	2.9	3.4	3.7	3.8	3.5	3.0		1303 ± 24
Relative humidity, midday (%)	46	54	60	61	53	49	48	44	44	45	44	43		
Air temperature max. (°C)	20.1	16.3	12.2	11.3	12.7	14.7	16.6	19.9	21.8	23.6	23.6	22.2	18.3	
Air temperature min. (°C)	7.4	6.3	4.3	3.3	3.3	3.8	4.8	6.8	8.7	10.9	10.6	8.9	6.6	
Annual mean temperature (°C)													12.1	

[a]Site burned in Dec 1975.

2.3.3 Soil Temperature

The mean annual soil temperatures at 0.02 m depth on the ridgetops were 18.8°C at Echo Valley and 19.8°C at Fundo Santa Laura. These near-surface soil temperatures were often over 55°C from April to August during a dry year but only reached 38°C during a wet year at Echo Valley. At Fundo Santa Laura, they reached 44°C. Minimum soil temperatures at 0.02 m were usually higher at Echo Valley than at Fundo Santa Laura but were above 0°C at both sites in all months except July 1973 at Fundo Santa Laura. These temperatures were generally above 10°C during the day throughout the year. At 0.32 m depth, the mean annual soil temperature was 16.8°C at Echo Valley and 15.9°C at Fundo Santa Laura, varying between 8° and 27°C at Echo Valley and 8° and 24°C at Fundo Santa Laura. The topographic position and vegetation cover influenced the soil surface temperatures at both research sites (Chapter 5).

2.3.4 Heat Sum

Heat sums accumulated on degree days above 10°C after 1 October were higher at Echo Valley than at Fundo Santa Laura and were higher at the soil surface than in the air. An air temperature heat sum of 500 degree days was reached in late June at Mount Laguna, by early June at Echo Valley, between December and April at Camp Pendleton, and in late June at Fundo Santa Laura. The year-to-year variability at Camp Pendleton was large because the winter temperatures were normally close to 10°C. Compared to the date of a heat sum of 500 degree days, heat sums of 1000 degree days occurred 1.5 months later at Mount Laguna, slightly over 1 month later at Echo Valley, 2.5 to 3.5 months later at Camp Pendleton, and 2.5 months later at Fundo Santa Laura. Soil surface heat sums reached 1000 degree days 2 months earlier than the air temperature heat sums at the bottom of the canopy at Echo Valley and 3 months earlier than air temperature heat sums at the bottom of the canopy at Fundo Santa Laura. The soil surface heat sums reached 1000 degree days 4.5 months earlier than air temperature heat sums at the top of the canopy at Fundo Santa Laura. With respect to heat sums, Fundo Santa Laura was similar to Mount Laguna and Echo Valley. Camp Pendleton was dissimilar to the other three stations. The heat sums at the soil surface at Echo Valley were similar to the heat sums in the air at Camp Pendleton, although the heat sums were accumulated rapidly in the spring at Echo Valley and slowly through the winter at Camp Pendleton.

2.3.5 Atmospheric Humidity

In southern California, measured relative humidities were usually 100% at night at all sites except near the coast where they often were lower than 100% at night because of the warm night temperatures. Midday relative humidities and vapor pressures were similar during winter and during summer at Mount Laguna and Fundo Santa Laura and were lowest at Echo Valley. Vapor pressures increased 1 to 2 mb through the course of the day and varied throughout the year with the air temperatures.

2.4. Seasonal Patterns of Water Availability

2.4.1 Precipitation

In the mediterranean regions of southern California and central Chile, precipitation was 350 to 650 mm yr^{-1} where evergreen shrubs predominate (Figure 2.4). Precipitation was 250 to 350 mm yr^{-1} along the coast where drought deciduous shrubs occur and 650 to 800 mm yr^{-1} at the top of the coastal mountains where forests occur (Miller et al., 1977). This general pattern of annual precipitation and vegetation forms seems to hold for the mediterranean regions of Israel and South Africa as well (Cooke, 1965; Shachori and Michaeli, 1965). In South Africa and Australia, shrublands instead of trees often occur with relatively high precipitation because of low nutrient availability or low soil water-holding capacity (Specht, 1979).

Precipitation increased with elevation at a rate of about 313 mm yr^{-1} per 1 km change in elevation at stations with 20-year records located along a transect from the coast inland in San Diego County (Fig. 2.3). Along the coast, precipitation was about 240 mm yr^{-1}. Stations on the lee side of the mountains received 75 to 100 mm yr^{-1} less precipitation than coastal-facing stations at similar elevations, while stations on the lee side of ridges on the coastal slope received 50 to 75 mm yr^{-1} less. The elevational increase of precipitation in the Coast Ranges in central Chile may be similar to that in southern California, but most of the high-elevation stations in Chile are in the foothills of the Andes and in the rain shadow of the coastal mountains. The few stations in the high elevations of the Coast Ranges in central Chile support the general trends. During the 8-year study period, precipitation averaged 476 mm yr^{-1} at Echo Valley and 595 mm yr^{-1} at Fundo Santa Laura (Table 2.1). Large annual variations occurred. On

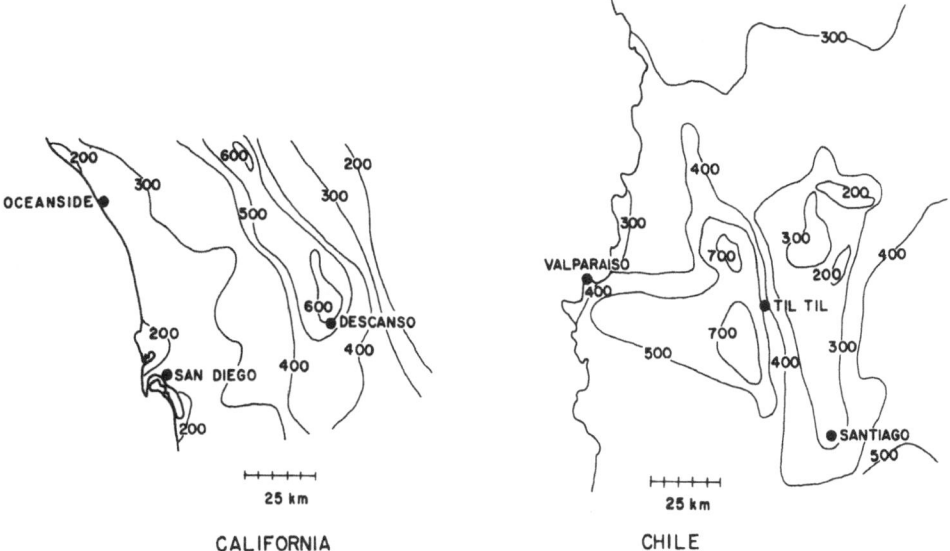

Figure 2.4. Isopleths of annual precipitation (mm) for southern California and central Chile (from Miller et al., 1977).

the basis of annual precipitation, Fundo Santa Laura is equivalent to a site at 1400 m elevation in southern California. Precipitation averaged 202 mm yr^{-1} at Camp Pendleton and 449 mm yr^{-1} at Mount Laguna. At Mount Laguna, the influence of the desert was apparent in the low annual precipitation value, which appears inconsistent with the pine-oak forest that occurs there. The records at both Camp Pendleton and Mount Laguna did not include the water years 1977-1978 and 1978-1979 during which above-average precipitation occurred. A comparison of the annual precipitation for the years with simultaneous measurements shows that Mount Laguna received about 10% more precipitation than Echo Valley, while Camp Pendleton received about half the precipitation of Echo Valley. On the basis of Duncan's test (Bancroft, 1968), the annual precipitation at Camp Pendleton, Echo Valley, and Mount Laguna was similar and differed from that at Fundo Santa Laura. Precipitation at Echo Valley, Mount Laguna, and Fundo Santa Laura was similar and differed from that at Camp Pendleton. The significance of the difference was reduced by the large year-to-year variability and short period of record.

Precipitation increases toward the poles in both southern California and central Chile. However, at equivalent latitudes, southern California receives less precipitation (Miller et al., 1977). Precipitation along the coast in southern California at 32.5° N was similar to that of La Serena in Chile at 30.2° S. The Californian inland precipitation regime at 32.5° N was most similar to that of Ovalle at 30.3° S, the only Chilean inland station for which precipitation data were available. The year-to-year variation in precipitation was greater in central Chile, perhaps because of the steeper latitudinal gradient (Miller et al., 1977).

In regions with mediterranean type climates, the water year can be divided into three periods: (1) winter when precipitation occurs and soil moisture levels are high; (2) spring when little precipitation occurs, but soil moisture levels are relatively high; and (3) summer when little precipitation occurs and little soil moisture is present (Wallén and de Brichambaut, 1962; Cochemé and Franquini, 1967; Cochemé, 1968; Wallén, 1968). In southern California, winter rains usually begin in November with most of the precipitation falling in December, January, and February. In central Chile, the winter rains usually begin in April, with the greatest amount falling in June. In southern California, about 17% of the annual precipitation is received after 15 March, while in Chile, about 9% of the annual precipitation is received after 15 September. Thus, more precipitation usually occurs in the spring in southern California than in central Chile. This spring season precipitation is readily available for photosynthesis and growth and may affect the length of the season of active growth. The difference in the yearly distribution of precipitation is also true at other stations in mediterranean regions of the Northern and Southern Hemispheres. In addition to the spring precipitation in southern California, precipitation frequently occurs during the summer months at higher elevations. In central Chile, most of the precipitation occurred during the winter months at all elevations.

2.4.2 Soil Moisture

Mediterranean regions experience dry soils for 1 to 4 months during the summer (Mooney and Parsons, 1973; Ng, 1974; Miller and Poole, 1979; Ng and Miller, 1980). Miller (1979) described a pattern of soil drought and air temperature along a transect

from 0 to 2000 m in southern California based on calculations of precipitation and potential evapotranspiration (Figure 2.5). Plant water potentials were measured along the
transect as a check against the calculations (Miller and Poole, 1979). The summer
drought began at the coastal and desert edges of chaparral distribution and progressed
toward the center. The pattern of drying was affected by the vegetation cover. Miller
and Poole (1979) suggested that at the edges of the distribution, water use was about
200 mm yr^{-1} per unit leaf area index (LAI), which occurred with precipitation of
about 400 mm yr^{-1}. With higher precipitation amounts, the leaf area index increases
until the resulting transpiration rate is decreased to about 200 mm yr^{-1} LAI^{-1}. This
implies that chaparral vegetation in southern California, without disturbance, develops
until the balance between the water use per leaf area and the duration of the summer
drought is similar throughout the distribution of the chaparral. Such an explanation

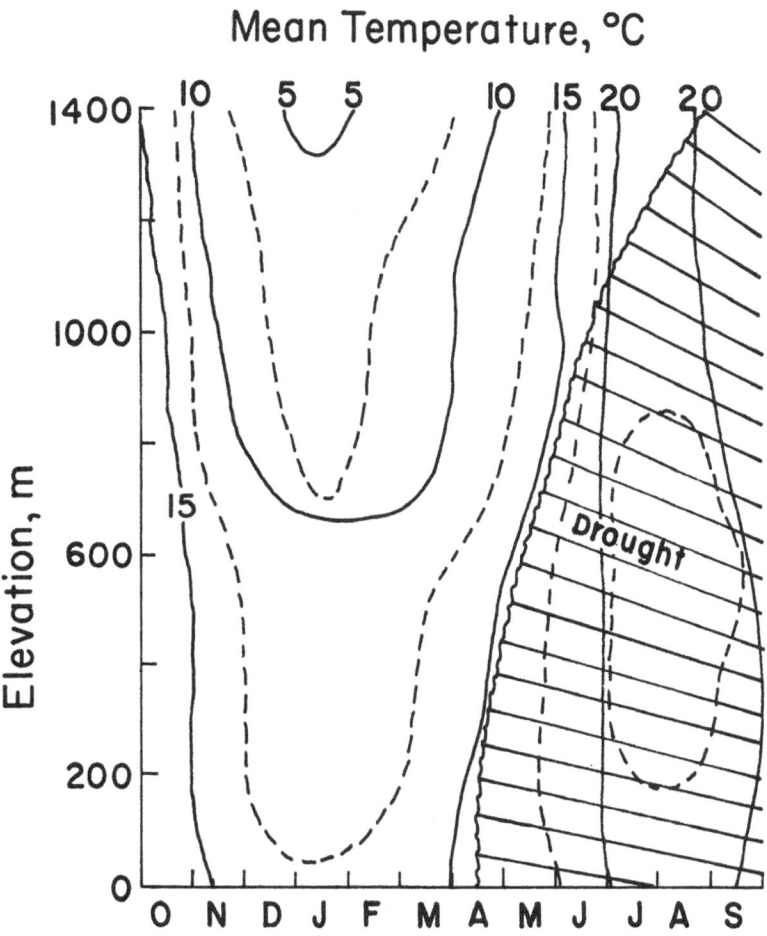

Figure 2.5. Seasonal progression of mean daily temperature and drought at different
elevations on the coastal-facing slope of the mountains in southern California (from
Miller, 1979).

was consistent with the patterns of water and soil drought measured under the vegetation of pole- and equator-facing slopes at Echo Valley (Ng and Miller, 1980). The desert-facing slopes at 2000 m were the driest sites studied, although they received some summer rain.

Profiles of soil moisture from surface to bedrock, which occurs at about 1 m, were measured for 3 years at Camp Pendleton, 4 years at Mount Laguna, 8 years at Echo Valley, and 6 years at Fundo Santa Laura. The seasonal progression of soil moisture showed similar patterns throughout the year at all the research sites (Figure 2.6). When the rains began in the fall, soil moisture soon increased throughout the soil profile. Moisture increased in lower soil levels with precipitation greater than 25 mm in a storm. Moisture in the soil depths of 0.3 to 1.0 m remained relatively high throughout the winter and spring at all sites. The patterns of soil moisture were reflected in patterns of plant water potential. Predawn plant water potentials rose rapidly after the first rain in the fall and remained high throughout the winter (Chapter 6). As the soil dried in the spring, the surface layers dried first. Similar patterns of soil moisture were measured at the San Dimas Forest and Range Experiment Station in southern California (Mooney et al., 1973) and in Israel (Shachori et al., 1967). Soil moisture decreased to minimum values by the end of June at Camp Pendleton and on the pole-facing slope at Echo Valley, by early July on the equator-facing slope at Echo Valley, and by mid-July at Mount Laguna. At Echo Valley, the minimum soil moisture values were reached at nearly the same time of the year in every year of measurement, regardless of annual precipitation (Figure 2.7). Summer precipitation at Mount Laguna did not penetrate to depths below 0.3 m. At Echo Valley, soil water potentials below 0.3 m were above -10 bars throughout the year on the ridgetop and equator-facing slopes but were below -10 bars on the pole-facing slope, which had a greater vegetation cover (Miller and Poole, 1979; Ng and Miller, 1980). Similar to the earlier drying of the soil in the spring at Camp Pendleton than at Echo Valley, predawn plant water potentials decreased earlier at Camp Pendleton than at Echo Valley (Miller and Poole, 1979; Chapter 6). Soil moisture was low during the summer at both Echo Valley and Fundo Santa Laura. At low soil moisture levels, soil water potentials changed rapidly with small changes in soil water (Ng, 1974; Chapter 6).

Soil moisture differed with topographic position and vegetation cover. Ng (1974) and Ng and Miller (1980) reviewed soil moisture measurements in southern California. Contrary to the general trend of higher soil moisture being associated with greater vegetation cover on pole-facing slopes, measurements of soil moisture at Echo Valley through 5 years indicated less soil moisture throughout the summer and earlier soil drying in the depths of 0.3 to 1.0 m on the pole-facing slope (Ng and Miller, 1980). These patterns were confirmed after additional neutron probe access tubes were installed on the pole- and equator-facing slopes (Miller and Poole, 1979; Chapter 6). The onset and intensity of summer drought also was supported by the seasonal pattern of plant water potentials and leaf conductances at Echo Valley (Poole and Miller, 1975; Miller and Poole, 1979) and by trends in xylem tensions in oak woodlands in the mediterranean regions of central California (Griffin, 1973; Syvertsen, 1974).

During the 1976-1977 water year, the sixth year of measurements at Echo Valley, the soil moisture (0.3- to 1.0-m depth) on the two slopes was similar and low; precipitation was uniformly distributed throughout the winter. No one month received a large amount, and water never penetrated below 0.3 m depth. By fall 1977, the chapar-

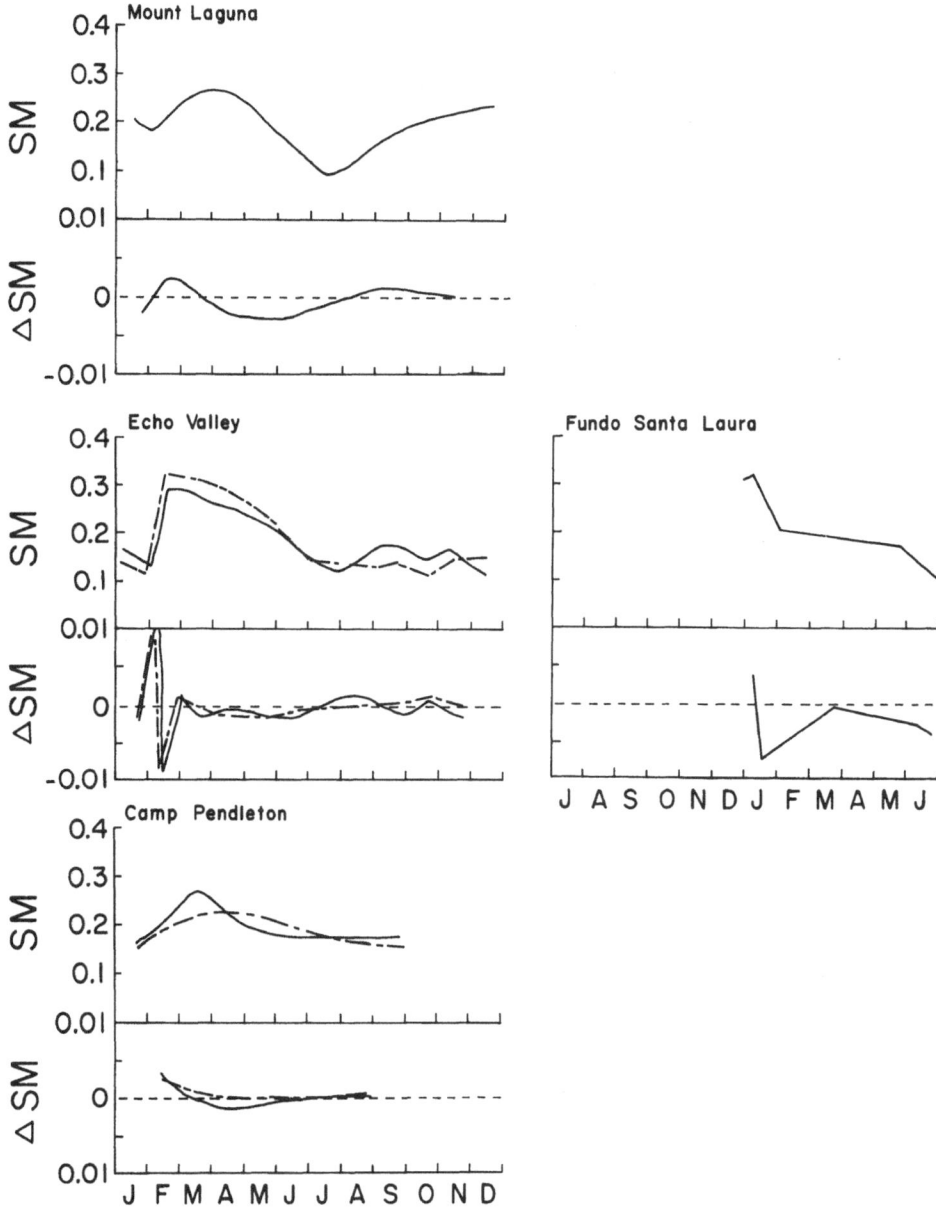

Figure 2.6. The seasonal progression of soil moisture (SM, g water cm⁻³) and the change between measurements in soil moisture (ΔSM, g water cm^{-3} day^{-1}) at Mount Laguna, Echo Valley, and Camp Pendleton on the equator-facing slope (dashed lines) and the pole-facing slope (solid lines) and at Fundo Santa Laura.

ral plants were showing the effects of a prolonged summer drought with increased leaf drop and leaf curling, not because of low annual total precipitation, which was near normal, but because of the lack of a large amount of rain in any one month to wet the full soil profile. Regardless of the precipitation in the preceding winter, the soils become dry between mid-June and the end of June at Echo Valley.

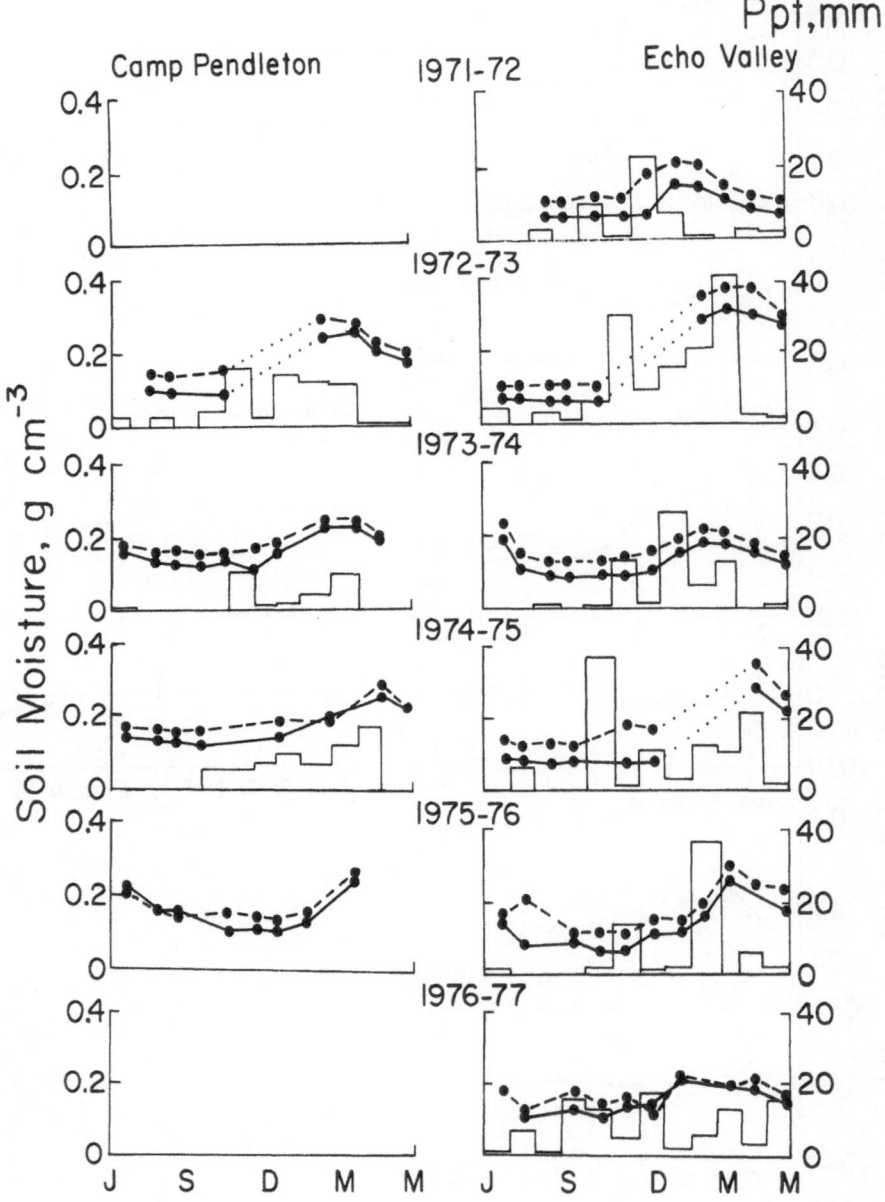

Figure 2.7. Average soil moisture contents in the 0.3 to 0.1 m depth on the pole-facing (solid lines) and equator-facing (dashed lines) slopes and monthly precipitation (bars) at Camp Pendleton and Echo Valley (from Miller and Poole, 1979).

At Fundo Santa Laura, higher soil moisture on the equator-facing slope than on the pole-facing slope also was recorded (Giliberto and Estay, 1978; Figure 2.8). Measurements indicated that the equator-facing slope in Chile received more convectional cooling and fog in winter than the pole-facing slope and was wetter in winter and drier in summer. Vegetation cover was similar on both slopes.

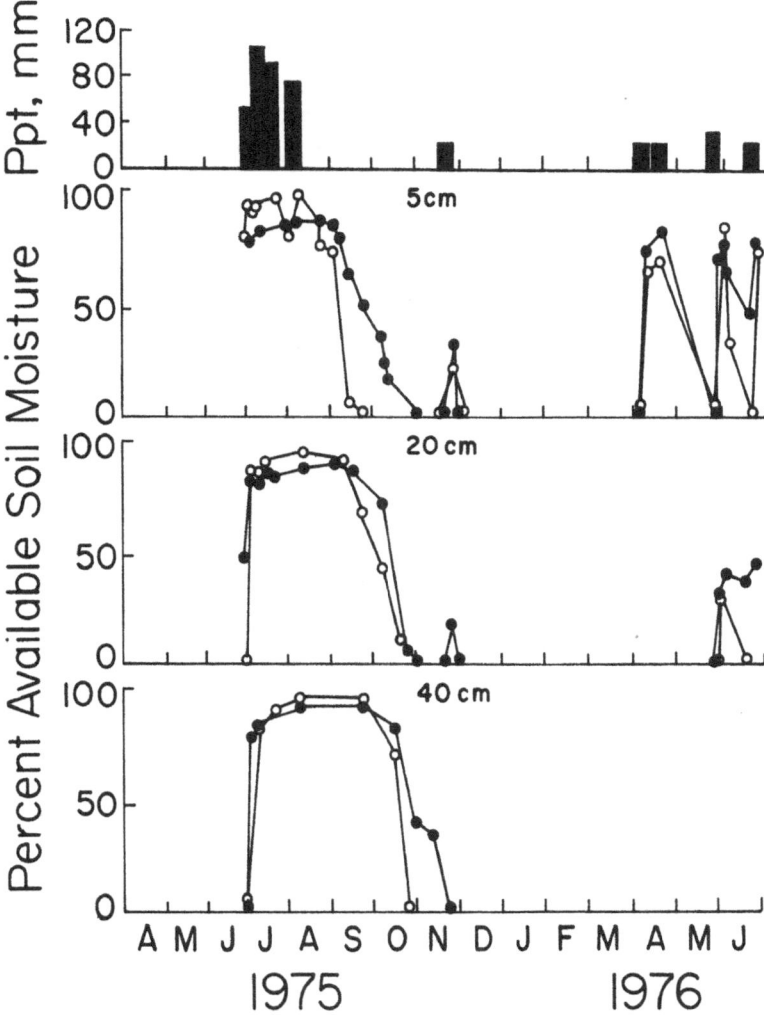

Figure 2.8. Precipitation (Ppt) and soil moisture for the pole-facing (●) and equator-facing (o) slopes at Fundo Santa Laura at different depths, expressed as a percentage of available moisture in the soil (from Giliberto and Estay, 1978).

2.5. Seasonal Patterns of Solar Irradiance

The annual receipt of solar irradiance as measured on the pyrheliographs at Echo Valley and Fundo Santa Laura was similar: 6184 ± 113 MJ m^{-2} yr^{-1} at Echo Valley and 6129 ± 197 MJ m^{-2} yr^{-1} at Fundo Santa Laura (Table 2.1). The difference is well within measurement error and annual variation. However, the seasonal distribution of solar irradiance differed. Above the atmosphere, solar irradiance is less in summer and greater in winter in the Northern Hemisphere than the Southern Hemisphere because the earth is farther from the sun on 21 June than on 21 December (Sellers, 1965). Theoretically, the difference is about 2.1 MJ m^{-2} day^{-1} in winter and about 2.5 MJ m^{-2} day^{-1} in summer at 32.5° latitude (Frank and Lee, 1966). Maximum measured irradiances were

5 MJ m^{-2} day^{-1} greater at Echo Valley than at Fundo Santa Laura during winter and 1.2 MJ m^{-2} day^{-1} less at Echo Valley than at Fundo Santa Laura during the summer. Mean intensities were 3.4 MJ m^{-2} day^{-1} greater during winter and 3.3 MJ m^{-2} day^{-1} less at Echo Valley than at Fundo Santa Laura. Thus, the measured intensities were consistent with the theoretical trends. At both Echo Valley and Fundo Santa Laura, receipt of less than 4 MJ m^{-2} day^{-1} was common in winter but was more common at Fundo Santa Laura because of the reduced irradiance above the atmosphere and the greater cloudiness at Fundo Santa Laura during the winter.

Solar irradiance varies from coast to mountains in southern California. Terjung et al. (1969) calculated the solar and net irradiance over the southern California region with standard equations. Solar irradiance measured at Camp Pendleton, Echo Valley, and Mount Laguna indicated lower intensities along the coast and at 2000 m than at intermediate elevations (Table 2.1). At Camp Pendleton, the annual total was 0.81 that at Echo Valley, while at Mount Laguna the annual total was 0.91 that at Echo Valley. Solar irradiance at Camp Pendleton was depressed in May and June because of the influx of fog and clouds from the Pacific Ocean. Considering only annual total solar irradiance, Fundo Santa Laura at 1000 m is probably equivalent to elevations between about 600 and 1600 m in southern California. Duncan's test (Bancroft, 1968) for similarity indicated that the solar irradiances at Camp Pendleton and Mount Laguna were not significantly different and that the solar irradiances at Mount Laguna, Fundo Santa Laura, and Echo Valley were not significantly different. In winter, Echo Valley received the highest solar irradiance and Fundo Santa Laura the lowest; in summer, Fundo Santa Laura received the highest solar irradiance and Camp Pendleton the lowest.

Cloud cover, indicated in these measurements by the ratio of the mean monthly measured solar irradiance to the maximum measured solar irradiance within the same time period, was slightly lower at Fundo Santa Laura than at Echo Valley and Mount Laguna and was highest at Camp Pendleton. Cloud cover in winter was highest at Fundo Santa Laura of all four stations but in summer was lowest at Fundo Santa Laura. Camp Pendleton had the least variable cloud cover throughout the year.

2.6. Environmental Influences on the Length of the Growing Season and Summer Drought

2.6.1 Potential Evapotranspiration

Potential evapotranspiration was calculated with the Penman equation (Penman, 1948; Miller, 1979), which was evaluated for arid regions by Stanhill (1961) and Omar (1968) and was used to indicate agro-climatic zones in mediterranean desert regions (Wallén and de Brichambaut, 1962; Cochemé and Franquini, 1967; Cochemé, 1968; Wallén, 1968). McAlpine (1970) suggested the utility of this approach in regions where detailed analyses were not possible because of data limitations. Potential evapotranspiration and precipitation were balanced monthly. The potential evapotranspiration depends on net irradiance and the drying power of the air. Net irradiance depends on the incoming and reflected solar and infrared irradiance. The albedo of the vegetation and ground was assumed to be 0.13 (Chapter 5). The measurements of solar irradiance at the research sites gave values along the elevational gradient. Solar irradiance was assumed to increase from the low elevations near the coast to the top of the inversion at about 400 m because of the reduction of the coastal fog. Above the inversion, solar irradiance was assumed constant up to 1400 m, above which it was assumed to decrease with increasing elevation because of clouds and fog. Net infrared irradiance was calculated with the Brunt equation (Brunt, 1932), which expresses a relation between infrared irradiance from the sky and vegetation, air temperature, vapor pressure, and cloud cover. Net infrared irradiance became more negative from the coast inland because of decreasing cloud cover and air temperature. The contribution of net total irradiance became more negative from the coast inland because of decreasing cloud cover and air temperature. The contribution of net total irradiance to potential evapotranspiration decreased above and below 400 m elevation. The drying power of the air depends on the vapor pressure deficit and wind. The mean daily vapor pressure of the air was calculated as the saturation vapor pressure at the minimum air temperature. The effect of errors in estimating vapor pressure was small because net radiation dominates the potential evapotranspiration. Wind was assumed to be equal at Camp Pendleton and Mount Laguna, both of which are exposed, and was interpolated for elevations between these sites and Echo Valley.

In the calculations for southern California, potential evapotranspiration was greatest at about 400 m elevation and decreased toward both the coast and the mountains. Potential evapotranspiration was reduced by the high cloud cover near the coast and the influence of the inversion layer. Uncertainties regarding wind speeds at high elevations led to uncertainties regarding the potential evapotranspiration at these locations.

The seasonal progression of potential evapotranspiration at the research sites (Table 2.1) indicated that the seasonal variation was greatest at Mount Laguna (0.5 to 4.0 mm day^{-1}) and least at Camp Pendleton (0.9 to 3.2 mm day^{-1}). Echo Valley and Fundo Santa Laura were similar (0.8 to 3.8 mm day^{-1}). With respect to annual potential evapotranspiration, Fundo Santa Laura (1303 mm yr^{-1}) was most similar to Mount Laguna (1183 mm yr^{-1}). At Fundo Santa Laura, the greater wind speeds more than compensate for the lower air temperature in calculating potential evapotranspiration. Soil water calculated from the potential evapotranspiration and the seasonal progression of precipitation decreased earlier than measured soil moisture at the four sites, but the sequence of drying at the sites was similar in the calculations and measurements. The length of the period without both precipitation and soil moisture, i.e., the period of drought, was also calculated. In southern California, the drought period was 5.5 months at the coast and decreased to less than 1 month in the mountains. At Mount Laguna, the length of drought was negligible because soil moisture decreased to minimum levels at about the time summer rains began. The calculated length of drought was about 3.5 months at Echo Valley, similar to the drought measured by soil moisture and plant water potentials (Poole and Miller, 1975; Ng and Miller, 1980), but at Camp Pendleton, the calculated length of drought was 1 month longer than that described by Poole and Miller (1975).

2.6.2 Soil Depth

The seasonal distribution of rainfall in the mediterranean climate affects transpiration, photosynthesis, and growth. The period of high transpiration, high photosynthesis, and growth is controlled by temperatures through the winter and early spring and by the loss of soil moisture in early summer. Thus, the length of the growing period depends on the time required for evapotranspiration to remove all available water from the soil once temperatures become favorable. The length of time depends on the amount of water held in the soil at the beginning of the growing season, the precipitation during the period, and the rate of evapotranspiration, which is related to the potential evapotranspiration. Thus, the length of the growing season (λ) can be expressed as

$$\lambda = [(\theta - \theta_{min})z(1 - f) + P] / [(PET)(\phi)] \tag{2.1}$$

where θ is soil water content at the beginning of the period in grams per cubic centimeter excluding rocks in the soil, z is soil depth, f is the fraction of rocks in the profile, P is precipitation in this period, PET is potential evapotranspiration, and ϕ is the ratio of actual water removal to potential evapotranspiration. Assuming that 17% of the annual precipitation occurs in the growing season in southern California and about 9% occurs in central Chile, the growing season should lengthen about 9 days per 100-mm increase in annual precipitation in southern California and about 4.5 days per 100-mm increase in central Chile. A particular year may deviate markedly from this pattern because most of the annual precipitation usually falls in two storms, which can occur at any time during the winter, including during the growing season. The length of the growing season should decrease about 12 days per 10% increase in soil rockiness and lengthen about 10 days per 0.1-m increase in soil depth. In general, the soil drought should occur at nearly the same time of year regardless of annual precipitation, a generalization supported by the Echo Valley data.

The soil depth, by affecting the amount of water in the soil at the beginning of the growing season, affects the length of the growing season and the length of the dry season, which in turn can affect the distribution of the evergreen and drought semideciduous growth forms. Because precipitation increases with elevation, interrelations between elevation, precipitation, and soil depth can be approximated (Figure 2.9). The interrelations indicate that with increasing precipitation or elevation, evergreen shrubs can be expected on shallower soils and that drought semideciduous shrubs may be less widespread in southern California than in central Chile.

2.7. Hypothetical Plant Responses

The geographic pattern of evergreen and deciduous plant growth forms is related to patterns of water availability and temperature via the influence of temperature and water availability on the carbon balance of these plant growth forms. Miller and Mooney (1976) suggested that the segregation of plant growth forms was due to the relative lengths of drought periods in coastal and inland locations and to the carbon costs of maintaining or recreating leaves. In their analysis, the evergreen form predominated

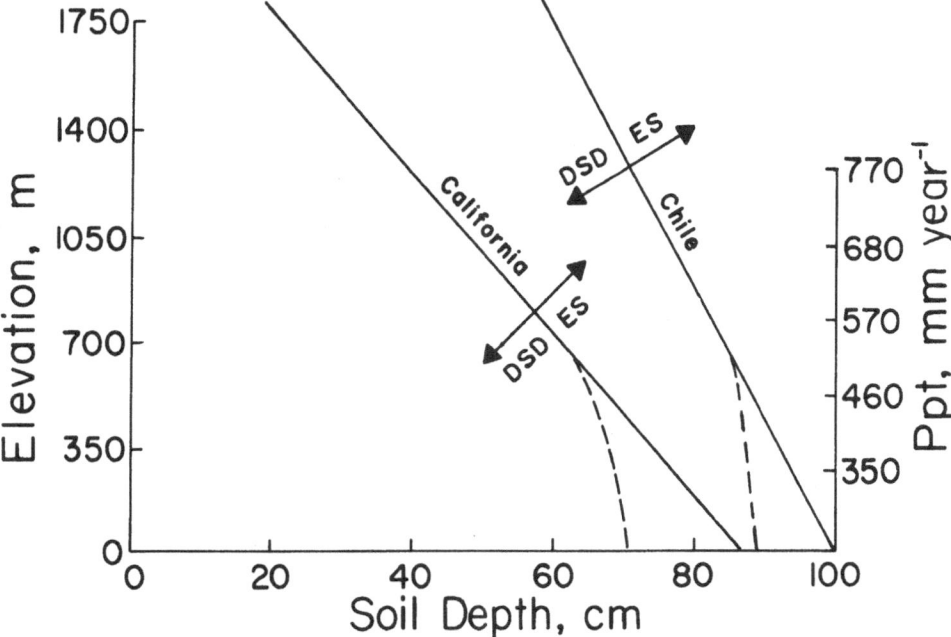

Figure 2.9. Hypothetical distribution of drought semideciduous shrubs and evergreen shrubs along gradients of elevation, precipitation, and soil depth in California and Chile. The solid lines represent combinations of soil depth and either elevation or precipitation at which the length of the drought calculated with Eq. 2.1 equals 100 days. The long dashed lines indicate the effect of increased rates of soil evaporation at the lower precipitation amounts (Chapter 12). Drought semideciduous shrubs should occur with longer drought length, which would occur with shallower soils, lower elevations, or lower precipitation amounts.

where the carbon costs of maintaining leaves through the unfavorable period was less than the carbon costs of recreating leaves after the drought if they were dropped. The deciduous form predominated where the carbon cost of maintaining leaves through the unfavorable period was greater than the cost of recreating the leaves if they were dropped.

This analysis was extended (Miller, 1979) by developing the interplay between temperature, precipitation, potential evapotranspiration, soil moisture, photosynthesis, and growth. During the drought, photosynthesis should be reduced because of stomatal closure (Harrison, 1971; Poole and Miller, 1975), and growth should cease. Using simplified relations for temperature and water effects on photosynthesis and growth, Miller (1979) calculated the annual dry weight gain for evergreen and drought deciduous shrubs from the elevational distribution of temperature, precipitation, and summer drought in southern California. Dry weight production increased with elevation when evergreen shrubs were used in the calculations. Production by drought deciduous shrubs was inhibited at midelevations by low temperatures for leaf growth during the winter when soil moisture was available. The warmer temperatures at the coast permitted growth through the winter and spring. The length of the growing season, constrained in winter by temperature and in early summer by the onset of drought, decreased from the coast to midelevations. At high elevations, above 1400 m, growth was possible throughout the summer, although clearly inhibited in winter (Figure 2.10). At all elevations, growth of low-growing plant forms is greater during the winter because the soil surface temperatures are higher than the air temperatures. Deciduous shrubs were inhibited at midelevations because of the short growing season.

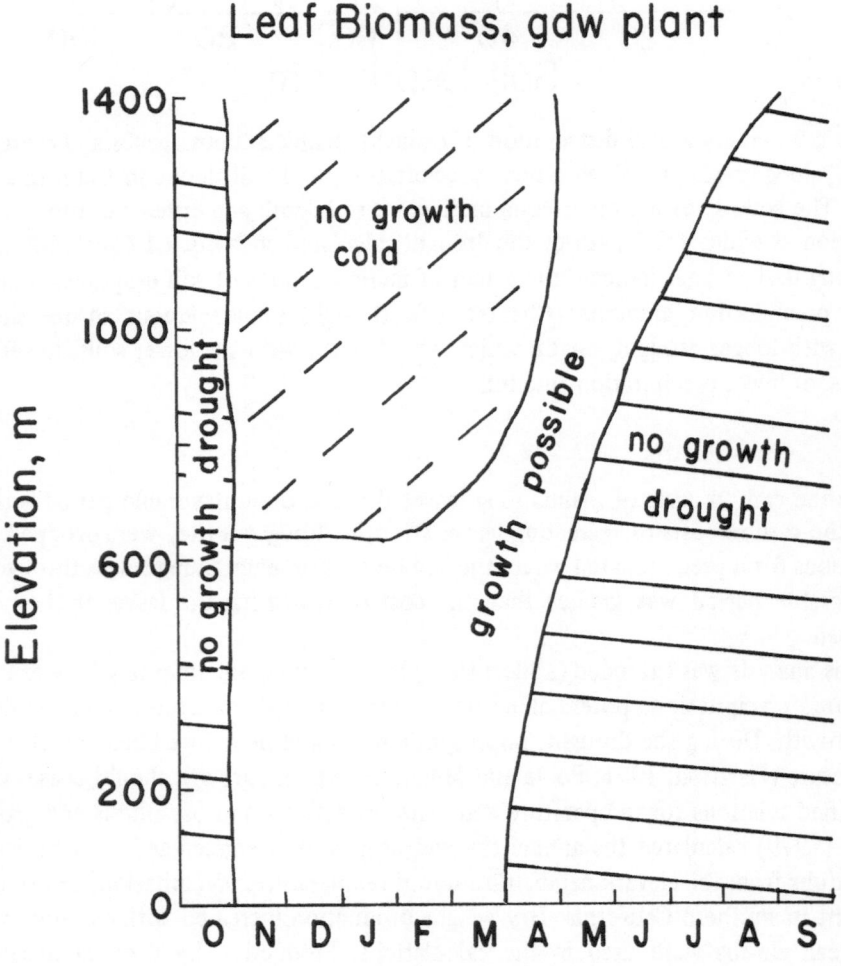

Figure 2.10. Hypothetical distribution of bioclimate along an elevational gradient on the coastal-facing side of the mountains in southern California.

2.8. Conclusions

Based on various climatic characteristics, including temperature, precipitation, solar irradiance measurements, and potential evapotranspiration, Fundo Santa Laura at 1000 m is equivalent to sites at elevations of 1400 to 1500 m in southern California, between Echo Valley and Mount Laguna (Table 2.1). Fundo Santa Laura is cooler and receives more precipitation than Echo Valley but receives similar solar irradiance. Potential evapotranspiration at Fundo Santa Laura is higher than Echo Valley, Mount Laguna, or Camp Pendleton. The influence of soil depth and seasonal progression of precipitation on water availability indicates that drought semideciduous shrubs should be more common at midelevations in central Chile than in southern California.

3. The Plant Communities and Their Environments

DEBORAH STEWARD and PATRICK J. WEBBER

This chapter describes the major shrub communities and their environments in the mediterranean type ecosystems of southern California and central Chile. It presents information on the response of major species within each community to environmental gradients and compares the communities of each country in terms of growth forms.

3.1. Introduction

3.1.1 Background

Mediterranean type ecosystems, which occur in climates with cool, rainy winters and hot, dry summers, are found in several parts of the world. Schimper (1898), who was the first to describe the mediterranean type vegetation in different parts of the world, found a nearly constant relationship between the dominance of sclerophyllous vegetation and climate. He wrote that "mild temperate districts with winter rain and prolonged summer drought are the home of evergreen xerophilous woody plants, which, owing to the stiffness of their thick leathery leaves, may be termed sclerophyllous woody plants." Schimper further observed that, in the absence of disturbance, this sclerophyllous vegetation formed dense and continuous scrublands. The mediterranean type vegetation of California is referred to as chaparral, which is a regional term specific to the southwestern section of the United States. Physiognomically similar vegetation types in other mediterranean type ecosystems are called matorral (Chile), heath (southwest Australia), fynbos (South Africa), and maquis (true Mediterranean) (Lossaint, 1973). The purpose of this chapter is to describe and compare the vegetation in mediterranean-type ecosystems of southern California and central Chile.

3.1.2 Description of Regions Studied

The area sampled in southern California is in San Diego County ($32°5'$ N, $116°20'$ W), reaching north to the Ortega Highway (Highway 74) in Riverside County and ranging from sea level to 2000 m (Figure 3.1). The vegetation in southern California is commonly divided into eight types: coastal sage scrub, chaparral, oak woodland, montane coniferous forest, valley grassland, pinyon-juniper woodland, desert scrub transition, and desert scrub. The sampling of shrub communities was emphasized, based on vegetation types from Cooper (1976), and included coastal sage scrub, desert scrub transition, and chaparral, and to a lesser degree, montane coniferous forest. Ravine and riparian communities were not sampled.

The chaparral of southern California has evolved with fire. Many chaparral species are efficient and rapid root sprouters. In such sprouting species, new growth often occurs within a few weeks after fire (Hanes, 1971). As Sampson (1944) pointed out, this rapid regeneration of brush after fire is a strong deterrent to production of herbaceous vegetation. Mature California chaparral is noted for its lack of herbaceous cover, which has been attributed to allelopathic inhibitors released by shrub members of the community (Muller, 1966). The lack of herbaceous cover could also be attributed to the adverse water conditions characteristic of the chaparral ecosystem. Grazing and human disturbance, although important in California, have had less impact in chaparral than in other mediterranean type ecosystems. Present-day grazing is restricted to moderate cattle grazing, which has little effect on the shrubby vegetation of the chaparral. The most significant aspect of chaparral degradation through human disturbance is the prospect of sprawling urbanization of chaparral habitats (Aschmann, 1973a).

The area sampled in central Chile is in Santiago Province ($32°5'$ to $34°5'$ S, $70°$ to $72°$ W). It extends from Pichidangui to Cartagena along the coast and extends eastward into the mountains toward La Disputada and San Jose de Maipo and covers an elevational range from sea level to 2000 m. It includes the coastal plain, the Cordillera de la Costa, and the lower reaches of the Andean cordillera but excludes ravines, quebradas, streamsides, and the large Central Valley. The Central Valley was not sampled due to heavy human impact on the vegetation, and the "thorn-tree savannah" was not sampled because its status as a natural vegetation unit could not be verified. The eastern vegetation boundary of the study was the xeromorphic cushion plant vegetation of the high Andes and the western boundary was the sand dune vegetation of the Pacific coastal strand.

The vegetation of the mediterranean zone of central Chile is generally divided into five major types: coastal shrub, matorral, thorn-tree savannah, Andean matorral, and xerophytic shrubs and cushion plants. Of these types, the coastal shrub, matorral, and Andean matorral intergrade markedly and have many species in common. This intergradation is probably due in part to the thermic homogeneity of central Chile, providing an environment whose chief variant is moisture, which can differ greatly from site to site (di Castri, 1973).

3.1.3 Field Sampling Methods

In order to characterize the shrub communities in southern California and central Chile, 130 stands along a transect from the coast to an elevation of 2000 m were sampled in each country during the early part of the growing seasons of 1976, March to May for California and September to November for Chile. Sample stands were selected on the basis of maturity and lack of recent disturbance. Uniform stands with no obvious structural boundaries and no variation in stratification visible within the stand were chosen for analysis (Westhoff and Maarel, 1973). The sample plots were between 100 and 200 m^2.

The Braun-Blanquet method of vegetation sampling was chosen to describe the stands because it is the fastest and most efficient method of vegetation description for large areas (Moore et al., 1970). Ordinations of stands on the basis of their similarities have been presented by Matuszkiewicz and Traczyk (1958) and by Dagnelie (1960). Cover values for each species were used to calculate Bray and Curtis (1957) indices of similarity and these were used to construct, by the average linkage of pairs method (Sokal and Sneath, 1963), a dendrogram of the sampled stands.

Slope stability, site moisture, sign of fire, animal disturbance, and insect damage were ranked subjectively on ten-point scales. The endpoints of the scale were determined by the most extreme conditions possible for the character in question. For example, the endpoints of the moisture scale were continual presence of water in the site and, at the other end of the scale, extreme desert site moisture conditions. The midpoint of the scale was represented by a well-drained, mesic site. Direct gradient analysis was used based on the ordination of samples in relation to the complex environmental gradients of site moisture and elevation (Whittaker, 1970, 1973). The gradients consist of many individual factors that change together, such as precipitation, temperature, wind velocity, humidity, soil characteristics, and insolation.

3.1.4 Comparison of the Environments in the Regions Studied

Elevation and slope inclination were measured at each sample site (Table 3.1). There was no significant difference in the mean elevation of sample stands in California and Chile. Mean slope inclinations showed no significant difference between the two countries, although the samples in central Chile were taken on somewhat steeper slopes. In California, most samples were taken on flat sites, while in Chile most samples were from slopes with an inclination between 15° and 25°. This difference can be attributed to land use practice. In central Chile, flat areas tend to be farmed, inhabited, highly overgrazed, or a combination of these.

Slope stability depended on slope inclination, substrate, and vegetation cover. Based on subjective measurements, slopes in southern California were more stable than in central Chile. In addition to a higher mean slope inclination, the mean shrub vegetation cover was lower in Chile (mean = 86%) than in California (mean = 106%), which may have increased erosion and runoff in central Chile. However, the percent cover of rock and bare soil was higher in southern California than in central Chile, but this did not appear to be the deciding factor in slope stability.

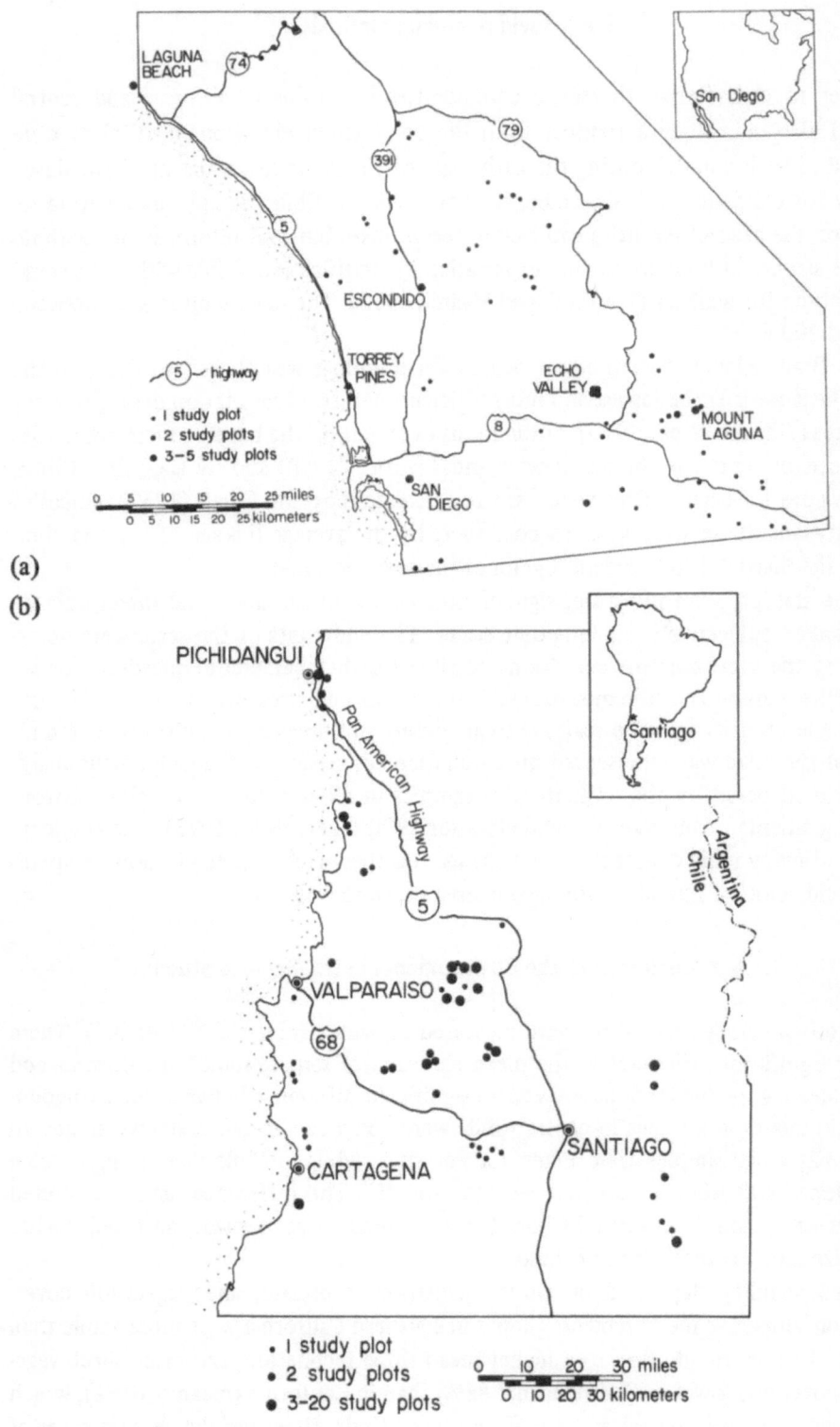

Figure 3.1. (a) Location of study sites in San Diego County and vicinity. Echo Valley, Mount Laguna, and Torrey Pines were areas in which sampling was concentrated. (b) Location of study sites in central Chile.

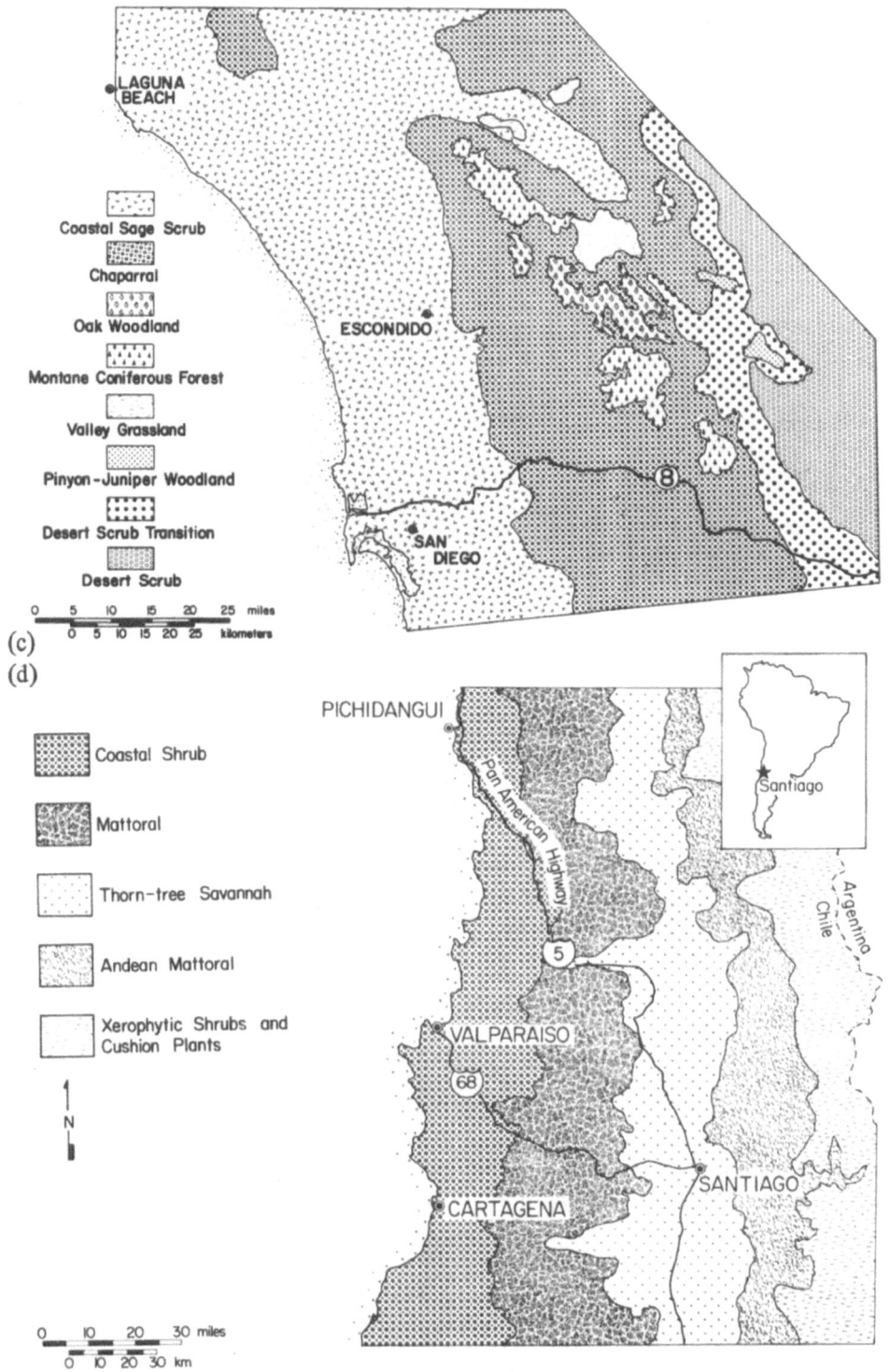

Figure 3.1. (c) Vegetation map of San Diego County and vicinity. Adapted from Cooper (1976); vegetation types follow Munz (1974). (d) Vegetation map of central Chile. Constructed from topographic maps, literature review, and a generalized vegetation map presented in Schilling (1975).

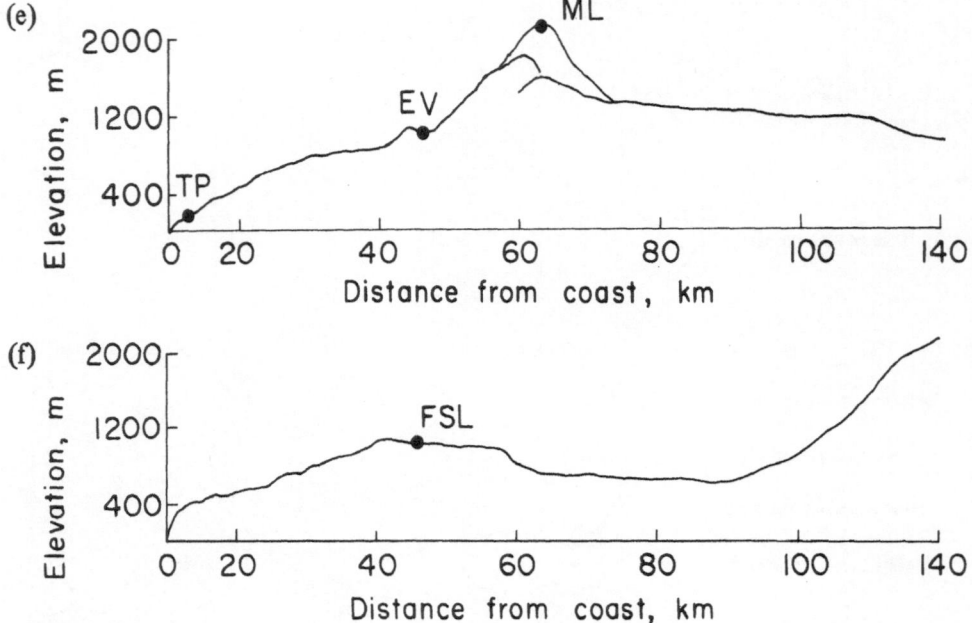

Figure 3.1. (e) Transect from coast inland in San Diego County. (f) Transect from coast inland of central Chile.

Table 3.1. Mean ± standard error of the primary environmental factors recorded at sites sampled in southern California and central Chile [a]

Parameters	California $\overline{X} \pm SE$	(N)	Chile $\overline{X} \pm SE$	(N)	t [b]
Elevation (m)	799.9± 43.3	(130)	841.0± 46.7	(130)	0.65
Slope inclination (°)	16.4± 2.6	(130)	18.5± 0.9	(130)	0.78
Slope stability	6.9± 0.1	(130)	5.4± 0.2	(129)	−7.60 [c]
Site moisture	5.4± 0.1	(130)	5.0± 0.1	(130)	−2.09 [d]
Fire	3.9± 0.2	(130)	2.4± 0.2	(130)	−6.97 [c]
Animal disturbance	4.0± 0.2	(130)	5.4± 0.2	(130)	4.87 [c]
Insect damage	2.5± 0.1	(130)	3.6± 0.1	(129)	6.96 [c]

[a] Slope stability, site moisture, fire, animal disturbance, and insect damage values were based on a 10-point scale where 1 represents a minimum value and 10 represents a maximum value.
[b] t is Student's t-test.
[c] Significant at 0.01 level.
[d] Significant at 0.05 level.

The criteria used for determining site moisture were condition of the vegetation, condition of the soil, presence or absence of plants restricted to wet sites, and condition of the litter layer. Weather conditions during the sampling period may have introduced a sampling bias. If the weather was cold and rainy, there was a possibility of overestimating site moisture, and in hot, dry weather, there was a possibility of the converse occurring. Nonetheless, there appeared to be a trend toward drier sites in central Chile. This trend was borne out by statistical analysis; the mean site moisture value for central Chile was 5.0 and for southern California was 5.4 (t-value using pooled variance estimate = -2.09, $P \leqslant 0.038$). The dryness of the sites sampled in central Chile may explain the lower cover of shrubs, but the openness of the shrub vegetation also may have biased the moisture rating to the drier side. Site moisture was used as a major gradient against which growth-form performance was compared.

Because of important differences that were noted between the areas sampled in southern California and central Chile, care should be exercised in assuming vegetation similarities between the two countries or drawing sweeping conclusions about their high degree of convergence. One of the major differences between southern California and central Chile is the frequency, extent, and the intensity of fire in the shrub ecosystems. Signs of fire, including fire scars, burned stumps, charcoal pieces on the ground, and occasionally a fire layer of burned material a few centimeters under the surface of the ground, were much more prominent in the sites sampled in southern California than in central Chile.

A second major difference between the chaparral and matorral was the amount of animal disturbance found in the sample sites. Data collected in southern California and central Chile corroborated the opinion of other researchers that central Chile was more heavily grazed than southern California (see e.g., Aschmann, 1973a; Aschmann and Bahre, 1977). The amount of insect damage to the plants, determined using a scale based on number of insect-damaged leaves per branch, also indicated a heavier herbivore pressure on the plants sampled in central Chile.

Although litter coverage was the same in samples from both countries, there was a marked difference in the coverage of rocks and bare soil, both of which were significantly higher in southern California (Table 3.2). The greater amount of bare soil in southern California was due primarily to a lack of herbaceous cover; the greater degree of rockiness was not so easily explained. The percent soil organic matter was significantly higher at the sites sampled in central Chile than in southern California. The higher soil organic matter in central Chile might reflect the accelerated recycling of material from the more prevalent deciduous shrubs and subshrubs and the presence of herbaceous material, which was an important understory component in central Chile. The difference in organic matter was also reflected in the difference in the water-holding capacity of the soils. Chilean soils were more absorptive and had a significantly higher percentage of hygroscopic water. The pH values of the soil samples from both countries showed no significant difference.

Table 3.2. Mean ± standard error and Student's *t-test* for selected soil factors at sites sampled in southern California and central Chile.[a]

Soil factors	California X̄ ± SE (N)		Chile X̄ ± SE (N)		t
Litter (%)	50.8± 4.5	(130)	51.5± 4.6	(126)	0.22
Rock (%)	11.1± 1.0	(129)	6.8± 0.6	(128)	-3.18[b]
Bare soil (%)	24.4± 2.1	(129)	13.1± 1.2	(129)	-5.55[b]
Pebble (%)	7.4± 0.6	(127)	8.0± 0.7	(125)	0.62
Granule (%)	7.1± 0.6	(128)	5.8± 0.5	(125)	-2.19[c]
Organic matter (%)	6.4± 0.6	(128)	8.5± 0.8	(126)	3.37[b]
Water absorbed (%)	50.1± 4.4	(128)	53.4± 4.8	(126)	1.68
Air dry water content (%)	1.2± 0.1	(128)	2.0± 0.2	(126)	6.11[b]
pH	6.1± 0.5	(128)	6.2± 0.6	(126)	1.33
Potassium (ppm)	160.9± 14.6	(121)	219.2± 19.5	(126)	3.48[b]
Calcium (ppm)	1059.5± 93.9	(129)	1823.4±162.4	(126)	5.14[b]
Magnesium (ppm)	192.9± 17.0	(129)	286.0± 25.5	(126)	3.86[b]
Nitrate (ppm)	16.1± 1.4	(129)	25.8± 2.3	(126)	3.26[b]
Ammonia (ppm)	9.5± 0.8	(129)	10.9± 1.0	(126)	1.43
Phosphorus (ppm)	23.9± 2.1	(127)	62.5± 5.6	(126)	7.86[b]

[a] Percentage of litter, rock, and bare soil refer to cover of ground surface. Water absorbed is percentage by weight of water the soil can absorb, and air dry water content is the percentage of weight (after air drying) of the soil sample that is water.
[b] Significant at 0.01 level.
[c] Significant at 0.05 level.

3.2. Species Abundance and Phyletic Similarity

A ranking of the shrub species encountered (Table 3.3) showed the predominance of two species, *Adenostoma fasciculatum* and *Quercus dumosa*, in southern California and one species, *Lithraea caustica*, in central Chile. Other common species were *Eriogonum fasciculatum* in southern California and *Colliguaya odorifera* in central Chile.

The mean number of shrub species per stand was higher in southern California, 9.6, than in central Chile, 7.1. However, 120 woody plant species were encountered in the stands sampled in central Chile and only 95 species in the stands sampled in southern California. The percent ground cover of shrubs was significantly higher ($P \leqslant 0.001$) in the chaparral, with a mean value of 106, than in the matorral, with a mean value of 86. The distribution of the percent shrub cover with elevation indicated that cover values were somewhat higher along the coast in southern California, while in central Chile the values were higher at higher elevations.

Herb cover differed significantly between the two regions sampled (Figure 3.2). The mean herb cover in southern California was 8.6%. The stands sampled in central Chile had a mean herb cover of 36.7%. The sparse herb cover in southern California was composed mainly of native grasses, a few ephemeral dicots, *Marah macrocarpus* a perennial vine, and *Paeonia californica* a perennial caespitose herb. The Chilean herb understory was dominated by mediterranean ruderal species and contained numerous geophytes, ferns, and vines.

The species list for the areas sampled in southern California and central Chile showed little phyletic affinity. Few species were found to occur in both countries except for ruderals, and genera common to both countries were rare. Except for the Anacardiaceae, families that dominated the vegetation communities of southern California did not do so in central Chile. Mooney et al. (1970) and Parsons and Moldenke (1975) tabulated the floristic affinities of the chaparral and matorral and reached the same conclusion (Figure 3.3).

3.3. Vegetation Communities in Southern California and Central Chile

3.3.1 Southern California

Adenostoma fasciculatum dominated the shrub communities of southern California. *Adenostoma fasciculatum* was most prevalent at midelevations but was also an important component of the coastal sage scrub and was present from the montane coniferous forest eastward into the desert scrub transition. Ten species, *A. fasciculatum, Q. dumosa, Eriogonum fasciculatum, Salvia mellifera, Artemisia californica, Ceanothus greggii, Arctostaphylos glauca, Arctostaphylos glandulosa, Rhus ovata,* and *Ceanothus leucodermis,* constituted 70% of the total relative shrub cover in the southern California shrub communities (Table 3.3).

Cover values for the shrub vegetation as a whole ranged from 40% to 150%. The lower cover values were found most often on equator-facing slopes at low elevations, while denser cover was common on pole-facing slopes at higher elevations. Excessive drainage in the soils, as well as extreme wind exposure, often reduced the cover and height of the shrub vegetation. Alternatively, protected sites, which retained more moisture and received less insolation, tended to have denser and taller shrubs. The average cover of shrub vegetation in the area studied in southern California was slightly greater than 100%, which is in agreement with Schimper's (1898) original characterization of typical mediterranean type vegetation. Shrubs at low elevations, particularly along the coast, often grew in tightly packed, mixed-species clumps.

By using Bray and Curtis indices of similarity of 55%, nine vegetation types were recognized. Of these nine, the most typical vegetation type was that dominated by *A. fasciculatum*, to such an extent that other species were relatively unimportant (Figure 3.4, type A). The *A. fasciculatum* vegetation type was followed in importance by stands of *A. fasciculatum, C. greggii,* and *Q. dumosa* (type E), which were common at midelevations. *Quercus dumosa* dominated the vegetation (type I) in cool, moist sites at higher elevations. Stands with *S. mellifera* and *A. californica* occurred at approximately the same mean coverage (types F and G) and were typical representatives of the coastal sage scrub vegetation. The remaining types (B, C, D, and H) occurred less frequently. By using a percentage of similarity of approximately 40 as the lowest common similarity, three broad vegetation types were recognized, which correspond to coastal sage scrub, chamise chaparral, and mixed chaparral. These three types have generally been recognized as major chaparral vegetation types (Horton, 1941).

Table 3.3. Indices of abundance for 29 most common woody plant species of the 95 encountered in southern California and for 25 of the 120 species encountered in central Chile. Species nomenclature for California follows Munz (1974) and for Chile follows Muñoz Pizarro (1966).

Species	Family	Relative cover	Relative frequency	Importance index
Southern California				
Adenostoma fasciculatum H. & A.	Rosaceae	26.57	7.40	33.97
Quercus dumosa Nutt.	Fagaceae	11.89	5.28	17.17
Eriogonum fasciculatum Benth.	Polygonaceae	5.00	6.18	11.18
Salvia mellifera Greene	Lamiaceae	5.10	3.17	8.27
Ceanothus greggii Gray var. *perplexans* (Trel.) Jeps. [*C. p.* Trel.]	Rhamnaceae	4.44	3.40	7.84
Artemisia californica Less	Asteraceae	4.30	1.90	6.20
Arctostaphylos glauca Lindl.	Ericaceae	3.43	2.60	6.03
Ceanothus leucodermis Greene	Rhamnaceae	3.23	2.03	5.26
Arctostaphylos glandulosa Eastw.	Ericaceae	2.73	2.50	5.23
Rhamnus crocea Nutt. in T. & G.	Rhamnaceae	1.16	3.82	4.98
Rhus ovata Wats.	Anacardiaceae	1.59	3.09	4.76
Lonicera subspicata H. & A.	Caprifoliaceae	0.75	3.50	4.25
Eriophyllum confertiflorum (DC.) Gray	Asteraceae	0.55	3.58	4.13
Galium angustifolium Nutt.	Rubiaceae	0.69	3.25	3.94
Ceanothus oliganthus Nutt. in T. & G.	Rhamnaceae	2.22	1.63	3.85
Rhus laurina Nutt. in T. & G. [*Malosma l.* Nutt. ex Abrams.]	Anacardiaceae	1.25	2.44	3.69
Yucca whipplei Torr.	Agavaceae	0.65	3.01	3.66
Heteromeles arbutifolia M. Roem.	Rosaceae	1.58	1.95	3.53
Cercocarpus betuloides Nutt. ex T. & G.	Rosaceae	1.51	1.71	3.22
Haplopappus squarrosus H. & A.	Asteraceae	0.68	2.52	3.20
Ceanothus crassifolius Torr.	Rhamnaceae	1.60	1.30	2.90

Table 3.3. (Continued)

Species	Family	Relative cover	Relative frequency	Importance index
Salvia apiana Jeps.	Lamiaceae	0.53	2.03	2.56
Lotus scoparius (Nutt. in T. & G.) Ottley	Fabaceae	0.38	2.20	2.58
Xyloccus bicolor Nutt.	Eriaceae	1.24	1.22	2.46
Ceanothus verrucosus Nutt. in T. & G.	Rhamnaceae	1.74	0.65	2.39
Prunus ilicifolia (Nutt.) Walp.	Rosaceae	0.47	1.54	2.10
Rhus integrifolia (Nutt.) Benth. & Hook	Amaranthaceae	0.78	1.14	1.92
Ambrosia chenopodiifolia (Benth.) Payne.	Asteraceae	1.16	0.41	1.57
Quercus kelloggii Newb.	Fagaceae	0.90	0.73	1.63
	Central Chile			
Lithraea caustica (Mol.) H. et Arn.	Anacardiaceae	18.15	8.20	26.35
Colliguaya odorifera Mol.	Euphorbiaceae	7.00	6.04	13.04
Cryptocarya alba (Mol.) Looser	Lauraceae	7.07	3.24	10.31
Trevoa trinervis Miers	Rhamnaceae	5.47	3.56	9.03
Satureja gilliesii (Grah.) Brig.	Lamiaceae	4.01	4.64	8.65
Quillaja saponaria Mol.	Rosaceae	4.62	3.78	8.40
Baccharis rosmarinifolia	Asteraceae (Astereae)[a]	2.16	3.56	5.72
Flourensia thurifera (Mol.) DC.	Asteraceae (Heliantheae)[a]	3.90	1.72	5.62
Kageneckia oblonga R. et Pav.	Rosaceae	2.72	2.37	5.09
Gochnatia fascicularis D. Don	Asteraceae (Mutisieae)[a]	1.73	2.37	4.10
Retanilla ephedra (Vent.) Brongn.	Rhamnaceae	2.16	1.51	4.12
Podanthus mitiqui Lindl. in Loud.	Asteraceae (Heliantheae)[a]	1.73	2.37	4.10
Peumus boldus Mol.	Monimiaceae	2.56	1.29	3.86
Ephedra andina Poepp.	Ephedraceae	0.67	2.91	3.59
Eupatorium salvia Colla	Asteraceae (Eupatorieae)[a]	1.60	1.94	3.54

Table 3.3. (Continued)

Species	Family	Relative cover	Relative frequency	Importance index
Baccharis concava Pers.	Asteraceae (Astereae)[a]	1.49	1.94	3.43
Chusquea cumingii Nees	Poaceae	2.40	0.75	3.15
Colletia spinosa Lam. emend. Suess.	Rhamnaceae	1.10	2.05	3.15
Escallonia pulverulenta (R. et Pav.) Pers.	Escalloniaceae	1.42	1.40	2.82
Azara petiolaris (Don) Johnston	Flacourtiaceae	1.80	0.86	2.66
Mutisia subulata	Asteraceae (Mutisieae)[a]	0.67	1.72	2.40
Mutisia latifolia	Asteraceae (Mutisieae)[a]	0.77	1.62	2.39
Adesmia aborea Bert.	Papaveraceae	0.76	1.62	2.39
Lepechinia sabviae (Lind.) Epl.	Lamiaceae	1.20	1.08	2.28
Puya violacea (Brongn.) Mez.	Bromeliaceae	1.43	0.75	2.18

[a] Considering the large number of Asteraceae in Chile it was thought to be more informative to name additionally the subfamilies in parentheses.

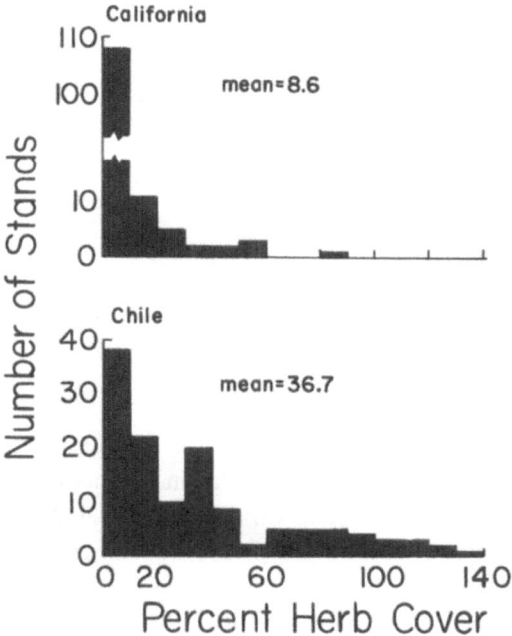

Figure 3.2. A histogram of the percent herb cover in the stands sampled in southern California and central Chile.

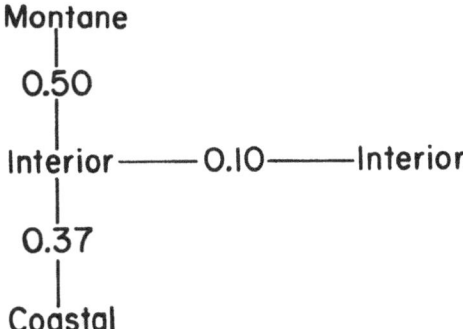

Figure 3.3. Generic affinities of the woody vegetation in different vegetation types in southern California and central Chile. Each similarity coefficient equals the number of genera common to the two sites divided by the sum of genera found at both sites (after Parsons and Moldenke, 1975).

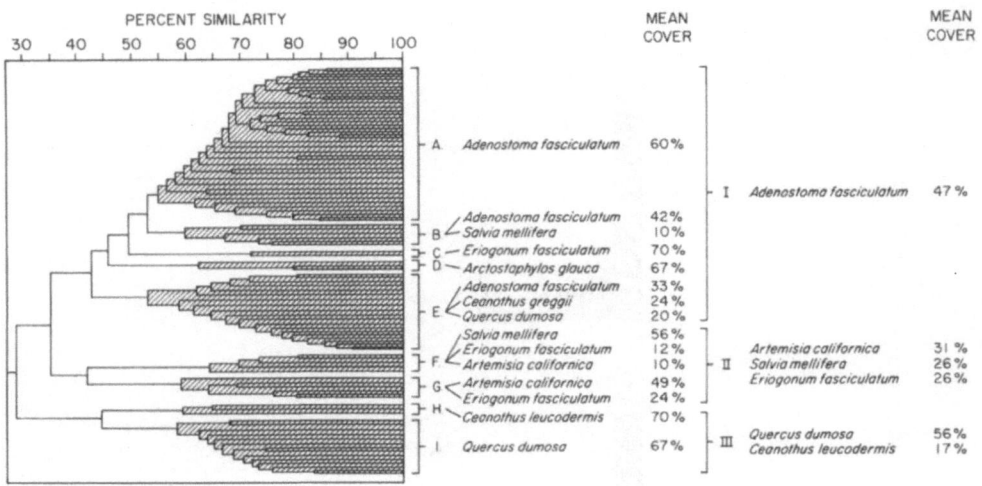

Figure 3.4. Dendrogram of stand relationships in southern California based on the average linkage of pairs methods. Stands with alphabetical designations were selected at a level of 55% similarity, while vegetation types with roman numeral designations were linked at 40% similarity. The species names given reflect the dominant member(s) of the stand; the numbers following represent their mean percentage of absolute cover within the stand.

3.3.2 Central Chile

Lithraea caustica, a typical broad evergreen sclerophyllous shrub, dominated the matorral of central Chile, occurring in over 60% of the samples and with an importance index of 26 (Table 3.3). Of the 120 shrub species encountered, 25 species accounted for three-quarters of the total importance indices of the vegetation and nine species, together with *L. caustica*, accounted for over 50% of the relative cover. The shrubs in order of their importance were *L. caustica*, *C. odorifera*, *Cryptocarya alba*, *Trevoa trinervis*, *Satureja gilliesii*, *Quillaja saponaria*, *Baccharis rosmarinifolia*, *Flourensia thurifera*, *Kageneckia oblonga*, *Gochnatia fascicularis*, *Retanilla ephedra*, and *Podanthus mitiqui*. Because of their importance, most of these species were examined individually for their roles in the landscape.

Shrub cover for the vegetation as a whole ranged from 40% to 116% with an average of 87%. The mean height of the shrub vegetation was 1.4 m, but shrub height depended on the environmental characteristics of the site. Equator-facing slopes tended to have a lower and more open shrub vegetation, while pole-facing slopes often supported taller and denser vegetation.

Herbaceous cover was high in the matorral. Particularly striking was the diversity of geophytes, including *Alstroemeria* spp., *Hippeastrum* spp., *Sisyrinchium junceum*, *Solenomelus pedunculatus*, *Leucocoryne ixioides*, and *Pasithea coerulea*. One suggestion in recent papers was that allelopathy was not as strong in central Chile as in southern California (Muller, 1966; Montenegro et al., 1978). Ferns, notably the *Adiantum* spp., were conspicuous even in fairly dry environments. Ruderal species, such as

Bromus spp. and *Erodium cicutarium*, were also well established. Their prominence in the landscape was probably due to the high grazing pressure in central Chile where undisturbed sample sites were difficult to locate. Woody vines such as *Mutisia* spp. and *Tropaeolum* spp. were frequent in the matorral of the Coast Ranges.

In the dendrograms of the stands, 82 of the 130 original stands were selected to represent the vegetation. The rest were eliminated because they were singleton or rare stands that had extremely low affinity with the majority of the other samples. The vegetation types were composed of stands with at least 45% similarity. The most common vegetation type in central Chile was dominated, to the virtual exclusion of other shrubs, by *L. caustica* (Figure 3.5, type A). *Lithraea caustica* was found throughout the elevational range of the samples but most commonly on sites at midelevations with moderate amounts of moisture. Other vegetation types (B and C) included significant amounts of *L. caustica*; however, both type B, which was dominated by a mixture of *L. caustica* and *Q. saponaria*, and type C, which was dominated by *C. alba*, were found in moister habitats than type A. *Colliguaya odorifera* (in types D and E) dominated the vegetation on dry habitats and occurred in significant amounts (in type F) in the vegetation found on disturbed sites. *Flourensia thurifera* (in type G) dominated almost exclusively on pole-facing slopes at low elevations.

3.4. Distribution of Species Along Gradients of Aspect, Elevation, and Moisture

3.4.1 Chaparral Species

Plant distribution depends on the available moisture, temperature, light and nutrients, all of which are modified by slope aspect and elevation. The distribution of ten chaparral species, which used percent cover values, was plotted by polar coordinates according to elevation, aspect, and inclination. Isolines were drawn around percent cover values. In southern California, *A. fasciculatum* and *E. fasciculatum* occurred on all aspects of every elevation sampled (Figure 3.6). However, *E. fasciculatum* was more common at low elevations and aspects other than north-facing; *A. fasciculatum* had greatest cover values at midelevations on southeast- to west-facing slopes. *Quercus dumosa* was restricted to north-facing slopes below 600 m. At about 600 m elevation, *Q. dumosa* was found on all slopes, but its highest concentration was on north-facing slopes. Of the two low elevation dominants, *S. mellifera* appeared most frequently on southeast-facing hillsides, while *A. californica* was found most often on southwest slopes. *Rhus ovata* and *A. glauca* were midelevation species and did not occur on west- and southwest-facing slopes. *Ceanothus greggii* was found on all aspects at midelevations but more so on east-facing slopes. *Arctostaphylos glandulosa* was generally considered to be a species more common to higher elevations but extended to lower elevations on north-facing slopes. *Ceanothus leucodermis* was an upper-elevation species that was found above 900 m but usually did not dominate a sample stand.

The same ten California species were plotted by using percent cover values within complex environmental gradients of elevation and moisture. *Adenostoma fasciculatum*

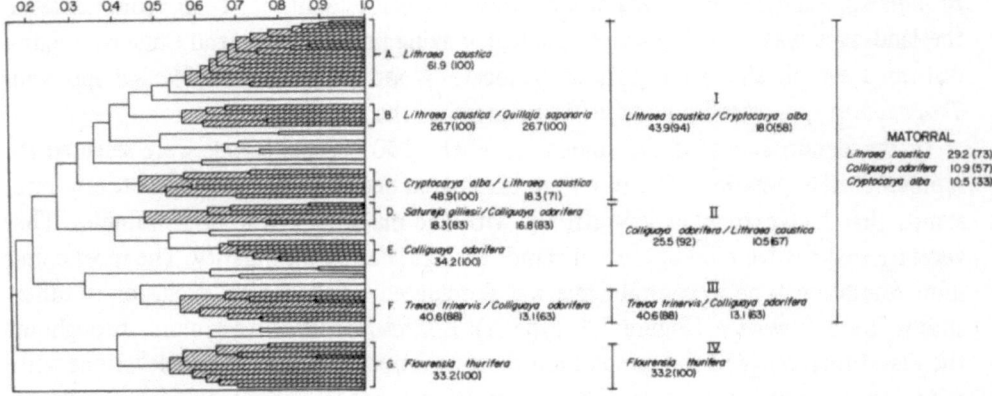

Figure 3.5. Dendrogram of stand relationships in central Chile. Linkage at 45% was considered significant to illustrate vegetation groupings of primary magnitude. Singleton stands (stands having very little similarity with the rest of the sample) were deleted for purposes of clarity.

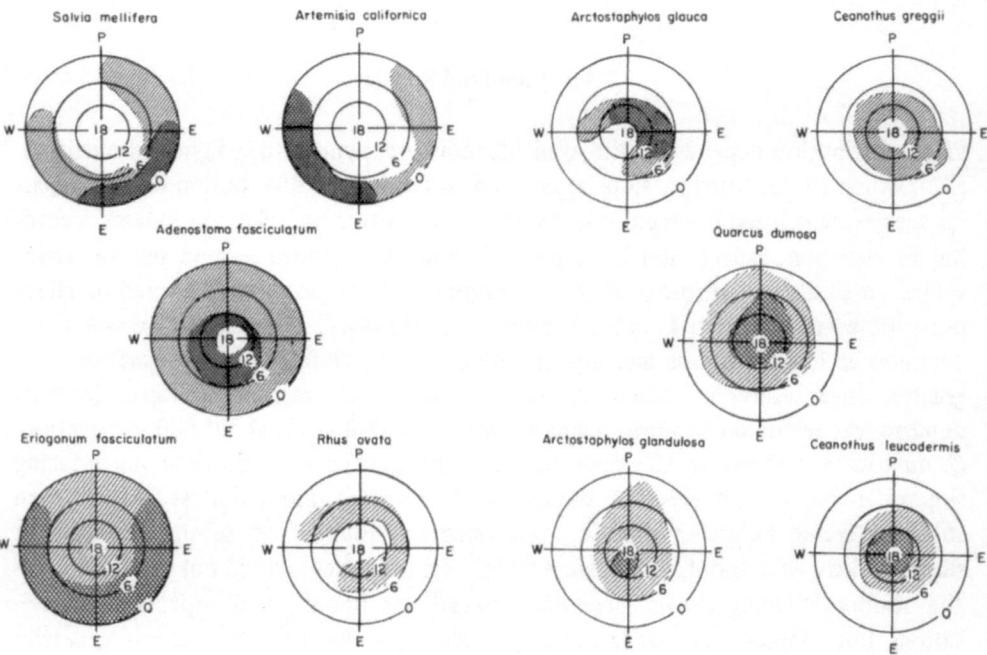

Figure 3.6. Polar diagrams of percent cover plotted as a function of elevation and aspect for 10 important Californian shrubs. Hatching and crosshatching represent relative, not absolute, amounts of lower and higher cover, respectively.

again showed its ability to grow over a wide range of environmental gradients, although it clearly predominated in slightly dry sites at midelevations (Figure 3.7). *Eriogonum fasciculatum* was present at a low percent cover throughout the gradient of elevation and moisture, which supports the idea that the species is a weedy invader of disturbed sites outside the mature coastal sage scrub. *Quercus dumosa* was disturbed on dry-to-moist sites at low-to-high elevations but grew best with increased moisture at elevations from 600 to 1500 m. Both *A. fasciculatum* and *Q. dumosa* had extensive ranges and high cover values despite the dissimilarity of leaf type between the two species. *Adenostoma fasciculatum* is a narrow sclerophyllous shrub, while *Q. dumosa* is a broad sclerophyllous shrub. *Salvia mellifera* and *A. californica* dominated the low-elevation landscape with *Artemisia* occurring in greater amounts at sites with slightly more moisture. *Rhus ovata* and *A. glauca* had similar distributions in midelevation-high moisture sites, but *R. ovata* was more common at slightly lower elevations than was *A. glauca. Arctostaphylos glandulosa* and *C. leucodermis* were usually found in moist sites at higher elevations.

The shrubs could be grouped into two major categories. Those found on drier sites were *A. fasciculatum, E. fasciculatum, S. mellifera,* and *A. californica,* and those pre-

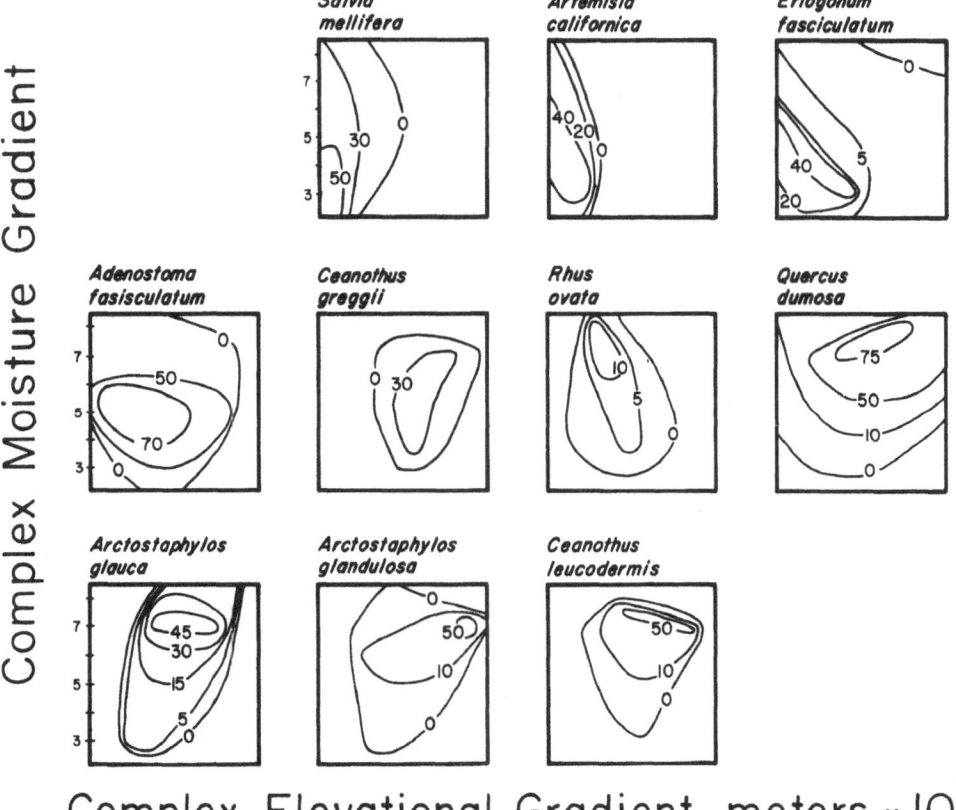

Figure 3.7. Distribution of common species in southern California by percent cover values within a complex environmental gradient of elevation and moisture.

ferring more moist sites were *Q. dumosa, R. ovata, A. glauca, C. greggii, A. glandulosa,* and *C. leucodermis* (Figure 3.7). Some of the species that shared common elevational distribution, such as *Adenostoma* and *Quercus,* had markedly different moisture preferences. Of the principal species of midelevation chaparral, *A. fasciculatum* and *C. greggii,* the former was a wide-ranging species found in greatest abundance on dry sites, while the latter was less widely distributed and most abundant on mesic sites. However, at midelevations, these two species grew in close association.

3.4.2 Matorral Species

By means of percent cover values, the distribution of ten matorral species was also plotted by polar coordinates according to elevation, aspect, and inclination. In central Chile, *L. caustica, T. trinervis,* and *C. odorifera* were widely distributed on all elevations, aspects, and inclinations sampled (Figure 3.8). The distribution of *B. rosmarinifolia* was somewhat misleading because, although widely distributed, *Baccharis* was not as important as the other nine species sampled. *Podanthus mitiqui* and *F. thurifera* occurred chiefly on equator-facing slopes. *Quillaja saponaria, C. alba,* and *K. oblonga* occurred at lower elevations on more pole-facing slopes and at higher elevations on equator-oriented slopes. *Satureja gilliesii* is restricted in distribution to midelevations.

When percent cover values of ten Chilean species were plotted within complex gradients of elevation and moisture, the Chilean species had broader distribution patterns than the California species (Figure 3.9). The two major species of matorral, *L.*

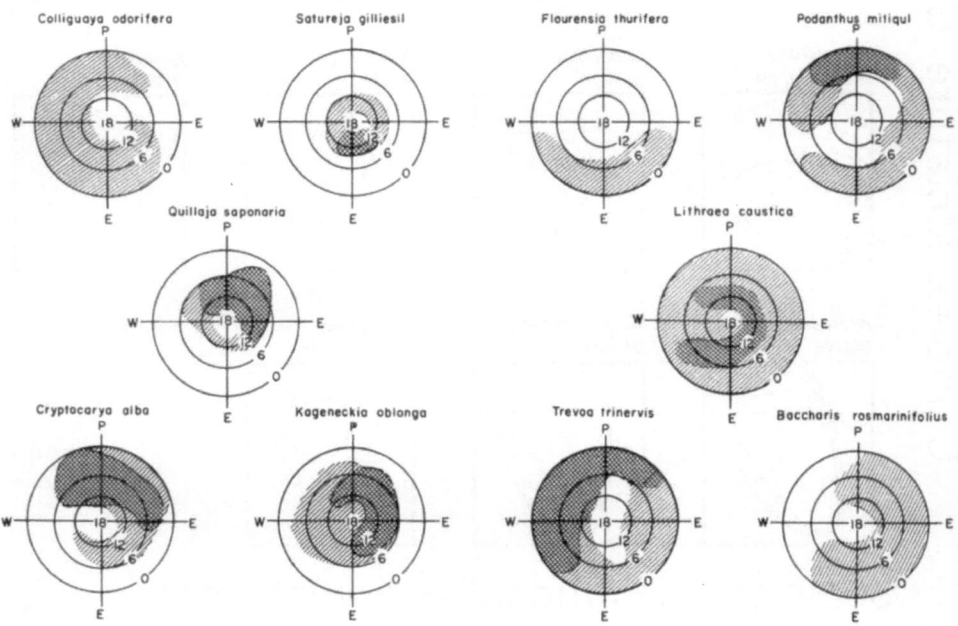

Figure 3.8. Polar diagrams of percent cover plotted as a function of elevation and aspect for 10 important Chilean shrubs. Hatching and crosshatching represent relative, not absolute, amounts of lower and higher cover, respectively.

caustica and *C. odorifera*, were segregated by moisture despite a common elevational range. *Lithraea caustica* occurred on sites with moderate moisture availability, while *C. odorifera* was more common on dry sites. Species often found together, such as *C. alba*, *Q. saponaria*, and *K. oblonga*, showed discrimination between their optimal distributions through moisture requirements. At low elevations, *P. mitiqui* occurred on moister sites and *F. thurifera* on drier sites. *Trevoa trinervis* occurred on sites with moderate moisture over a wide elevational range. *Satureja gilliesii* and *B. rosmarinifolia* were found on dry sites at higher elevations. None of the most abundant Chilean species were restricted to high elevations.

3.5. Distribution of Growth Forms Along Gradients of Aspect, Elevation, and Moisture

Five simplified growth forms for the chaparral, i.e., broad sclerophyllous shrub, narrow sclerophyllous shrub, broad malacophyllous shrub, narrow malacophyllous shrub, and succulent, and seven growth forms for the matorral, i.e., broad sclerophyllous shrub, narrow sclerophyllous shrub, broad malacophyllous shrub, narrow malacophyllous shrub, succulent, photosynthetic stem, and photosynthetic stem and spine, were used to illustrate the response of growth forms to environmental gradients of aspect, elevation, and moisture in each country.

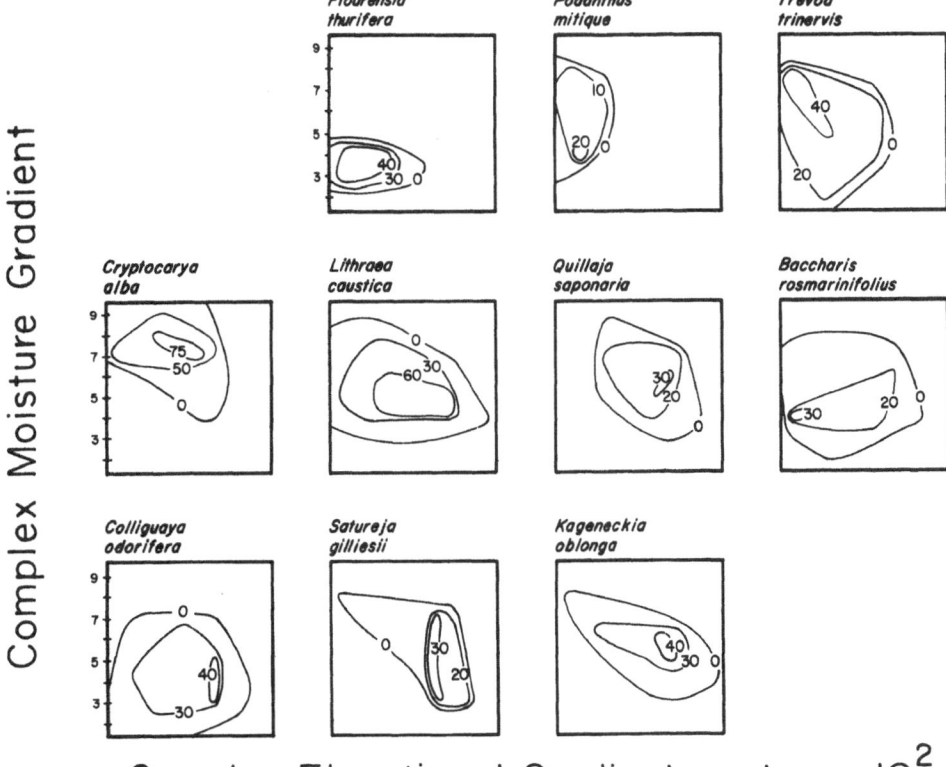

Figure 3.9. Distribution of common species in central Chile by percent cover values within complex environmental gradients of elevation and moisture.

3.5.1 Chaparral

Polar diagrams of percentage of cover values were plotted as a function of elevation and aspect for five growth forms commonly found in the chaparral of southern California. Broad sclerophyllous shrubs appeared to dominate sites at mid- to high elevations, particularly those sites oriented to the North Pole (Figure 3.10). At low elevations, a high cover of broad sclerophyllous shrubs occurred only on pole-facing slopes. However, broad sclerophyllous shrubs were present to some degree throughout the range of aspects and elevations. The most widespread member of this group was *Q. dumosa*. These findings were consistent with earlier research (Mooney et al., 1974; Parsons and Moldenke, 1975; Parsons, 1976b).

Narrow sclerophyllous shrubs also occurred over a wide range of environmental and topographic conditions. The narrow sclerophyllous shrubs reached their peak of distribution at midelevations on west-oriented slopes and were not important at the upper extreme of the elevational gradient.

Both broad and narrow malacophyllous shrubs predominated at low elevations in the coastal sage scrub community. Winter deciduous shrubs were represented in the diagram of broad malacophyllous shrubs by a patch of high occurrences at high elevations. Winter deciduous shrubs included such species as *Rhus trilobata* and, to a certain extent, *Cercocarpus betuloides*. The succulents, which included *Yucca* spp.,

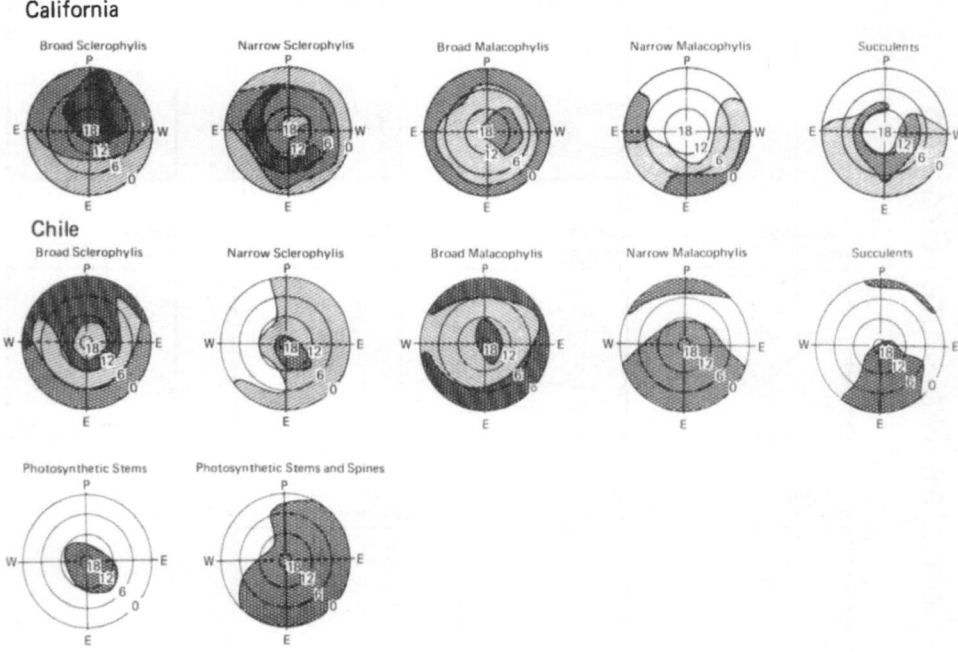

Figure 3.10. Polar diagrams of percent cover plotted as a function of elevation and aspect for growth forms in southern California and central Chile. Hatching and cross-hatching represent relative, not absolute, amounts of lower and higher cover, respectively.

Opuntia spp., and *Dudleya* spp., were not important dominants but served as indicators for the xeric coastal environments and the midelevation desert scrub transition at the eastern edge of the study area.

The distribution of major growth forms (Figure 3.11) along the elevational gradient followed distinct patterns. Broad sclerophyllous shrubs, the typical growth form of the mediterranean scrub, dominated the landscape at all elevations, but particularly above 1000 m, where *Q. dumosa* and the various *Arctostaphylos* species were common. The narrow sclerophyllous shrub form, which is composed primarily of *A. fasciculatum*, was present at almost all elevations. It was most prevalent between 300 and 1200 m but was not an important component of the vegetation above 1350 m. Broad malacophyllous shrubs, represented by the various *Salvia* species, were most frequent at low elevations in the coastal communities. However, this growth form was found at all elevations, and the bimodal appearance of the curve probably resulted from the presence of winter deciduous shrubs at high elevations. The narrow malacophyllous shrubs, of which *A. californica* was the most prominent, were restricted to low elevations. Succulent distribution in San Diego County was found to be bimodal, with a high on the coast where *Dudleya* spp. and *Opuntia littoralis* were found and a smaller peak at 1600 m in the desert scrub transition zone.

Major growth forms responded to moisture availability differently from one another (Figure 3.12). The coverage of broad sclerophyllous shrubs increased with increasing moisture within the limits set by the arbitrary moisture scale. This response was similar to the response of broad sclerophyllous shrubs to the complex elevational gradient,

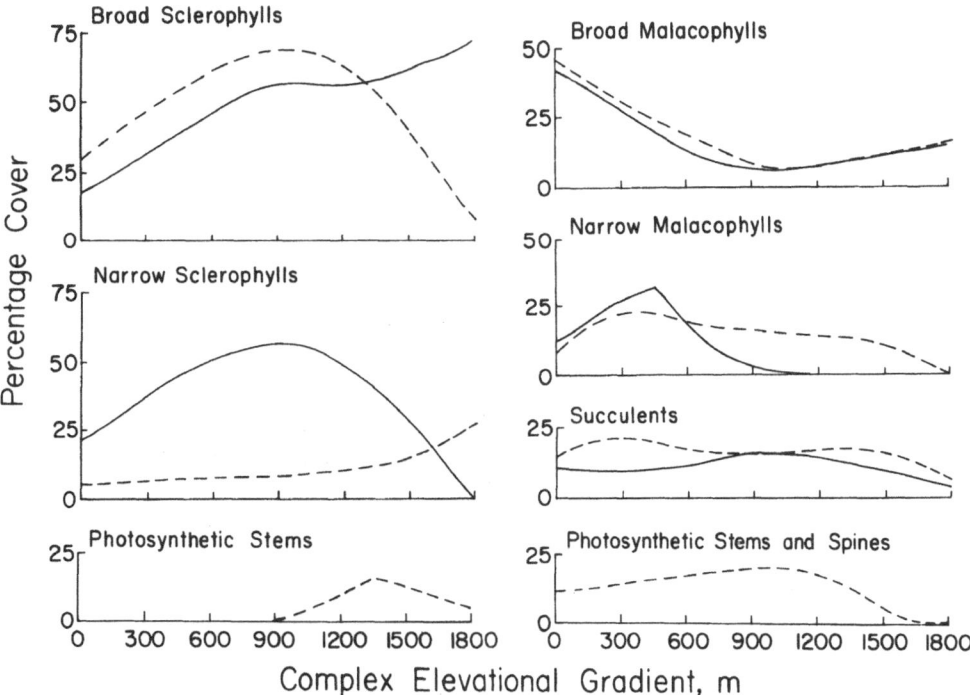

Figure 3.11. The distribution by percentage of cover along a complex elevational gradient of major growth forms in southern California (–) and in central Chile (- -).

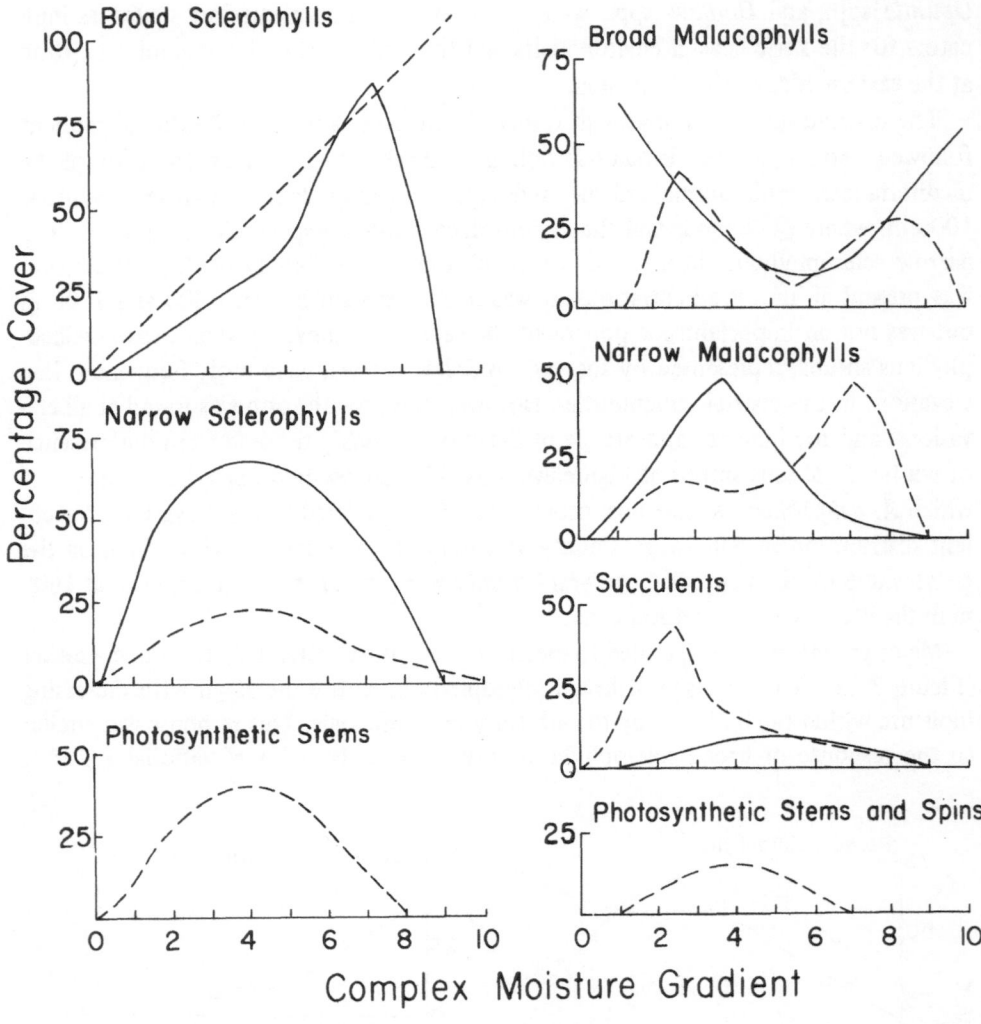

Figure 3.12. The distribution by percentage of cover along a complex moisture gradient of major growth forms in southern California (–) and in central Chile (- -).

indicating that increasing elevation was accompanied by a concomitant increase in moisture availability. As would also be expected by their generalized distribution, the narrow sclerophyllous shrubs were most prevalent on slightly dry-to-moderate sites. The distribution of broad malacophyllous shrubs over the complex moisture gradient was bimodal, reflecting the occurrence of drought deciduous shrubs on dry sites and winter deciduous shrubs (e.g., *C. leucodermis*) on more mesic sites, which were common at higher elevations. The narrow malacophyllous shrubs responded predictably to the moisture gradient and were most common on rather xeric sites. Because of the prevalence of *Yucca* spp. over the range of environments sampled, the succulents as a whole were not as strongly oriented to xeric sites as tradition would dictate. However, *Yucca* frequently occurred in xeric or disturbed microsites in generally mesic communities.

3.5.2 Matorral

Polar diagrams of percent cover values were plotted as a function of elevation and aspect for the seven growth forms studied in central Chile. In central Chile, large areas in the 600- to 800-m elevation were not sampled because of disturbance and agricultural development, making it difficult to construct continuous curves of growth-form performance (Figure 3.10). Broad sclerophyllous shrubs were widely distributed and were most frequently found on the pole-facing slopes. Narrow sclerophyllous shrubs reached their best development at high elevations on east-facing slopes. Broad malacophyllous shrubs were present throughout the sample area, predominantly at low elevations. Narrow malacophyllous shrubs and succulents showed a similar distribution and were more frequent on equator-facing slopes. Photosynthetic stem shrubs were most common at high elevations. The photosynthetic stem and spine shrubs were typically found in moderate abundance on disturbed sites throughout central Chile. They, as well as all other growth forms except broad sclerophyllous and broad malacophyllous shrubs, were not generally found on west-facing slopes at low elevations. This condition may be attributed to the influence of a coastal fog zone, which encourages the growth of more mesic-loving species.

In central Chile, the response of growth forms to the elevational gradient was not as well defined as in southern California. The importance of sclerophyllous shrubs, particularly broad sclerophyllous shrubs, increases across the elevational gradient. Mooney et al. (1974) observed increasing sclerophylly across an elevational gradient in Baja California. The increase of sclerophylly across an elevational gradient is indicative of the moisture regime of the Chilean Central Valley. The increase in cover values of narrow sclerophyllous shrubs at high elevations was chiefly due to the presence of two narrow-leaved species of *Colliguaya, C. integerrima* and *C. salicifolia*, which freely interbreed. The importance of both broad and narrow malacophyllous shrubs was greatest at the lower elevations (Figure 3.11). The slight increase in importance of broad malacophylls at high elevations reflected the presence of *Nothofagus obliqua* in the Coast Ranges. As the importance of malacophyllous shrubs declines, the importance of sclerophyllous shrubs increases. Succulents were most frequent around 300 m because of the low-elevation xeric conditions in central Chile. The slightly increased importance of succulents at high elevations reflects the dry conditions in the foothills of the Chilean Andes. The distribution of shrubs with photosynthetic stems and no spines, primarily *Ephedra andina*, was restricted to high elevations, which suggested that something other than moisture controlled their distribution. The wide elvational tolerance of shrubs with photosynthetic stems and spines was based primarily on the distribution of *T. trinervis* and *Colletia spinosa*, both of which are species common on disturbed sites.

Growth-form distribution was closely tied to moisture variability in central Chile (Figure 3.12). Broad sclerophyllous shrubs predominated across the moisture gradient but were found particularly at sites with high moisture ratings. The dominance of broad sclerophyllous shrubs reflected their importance in all the habitats that support shrub growth in central Chile. Narrow sclerophyllous shrubs were most important at sites with intermediate moisture conditions. Both broad and narrow malacophyllous shrubs showed bimodal distributions along the site moisture gradient, the occurrence of which probably reflects two moisture strategies. Drought deciduous plants are typical of dry, low-elevation sites and are probably malacophyllous in order to conserve energy, carbon, and inorganic nutrients during leaf production. Malacophylls may return nutrients to the soil more rapidly after leaf drop than do sclerophylls. The presence of malacophylls under high moisture conditions reflects the high transpiration capacity of leaves that are not heavily cuticularized. Succulents predominated at sites with low moisture ratings. Shrubs with photosynthetic stems and shrubs with photosynthetic stems and spines were usually found on sites with intermediate moisture conditions.

Narrow sclerophyllous shrubs, broad malacophyllous shrubs, and photosynthetic stemmed shrubs were found at high elevations in the Coast Ranges where marine influences reduce climatic fluctuation and provide increased moisture, particularly in the form of fog. At comparable elevations in the Andean cordillera, the climate is harsher and more subject to variations in temperature and moisture availability. The vegetation at 1800 m in the Andes has adapted to xeric conditions in contrast to the equable and mesic conditions found at similar elevations in the Coast Ranges.

3.6. Structural Similarity of Vegetation in Southern California and Central Chile

The distribution based on the percentage of absolute cover of five major growth forms, broad sclerophyllous shrubs, narrow sclerophyllous shrubs, broad malacophyllous shrubs, narrow malacophyllous shrubs, and succulents, found in both southern California and central Chile were compared along complex gradients of elevation and moisture (Figure 3.13). The distribution of broad sclerophyllous shrubs, which domi-

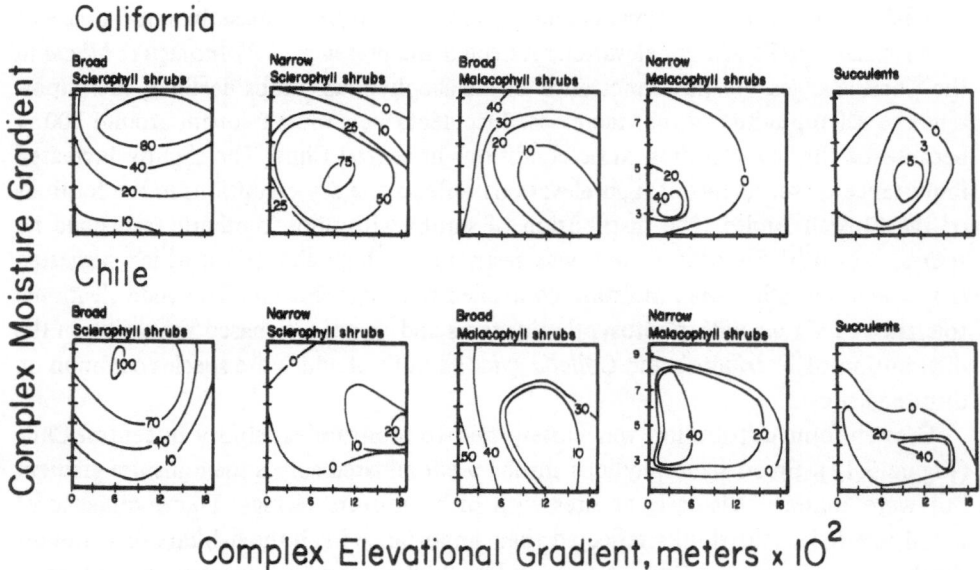

Figure 3.13. Two-factor ordinations of the major shrub growth-form types found in southern California and central Chile. Isolines connect equal percentages of absolute cover.

nated the landscape in the mediterranean type shrub ecosystems of southern California and central Chile, was similar. Broad sclerophyllous shrubs showed a peak in distribution at midelevations with moderately high moisture ratings and a pattern of distribution that radiated concentrically outward. The narrow sclerophyllous shrubs differed markedly in their respective distribution, which can be attributed to the widespread and dominant occurrence of *A. fasciculatum* in the California chaparral. The distribution of broad malacophyllous shrubs was similar in southern California and central Chile, even to the extent of having a bimodal distribution that was a function of elevation and moisture. Broad malacophyllous shrubs were most likely to be found at low elevations where both mesic and drought deciduous malacophyllous shrubs occurred and at sites of both high elevation and high moisture. Broad malacophyllous shrubs did not appear to be highly successful in low-moisture sites at mid- to high elevations. The distribution of narrow malacophyllous shrubs was similar in both countries, with a center at low elevations. However, in Chile, narrow malacophyllous shrubs occurred over a wider range of moisture ratings at low elevation and also extended to higher elevations. Succulents shared an affinity for low-moisture sites in both countries but differed markedly in their importance. Absolute cover values of >40% were measured in central Chile, while in southern California, cover values were about 5%. Included in this growth-form designation were the *Yucca* in southern California and the *Puya* spp. in central Chile. The widespread occurrence and importance of the *Puya* spp. was mainly responsible for high values in central Chile.

In southern California, the greatest number of species were in the broad sclerophyllous shrub growth form. About half the species had long leaves; the other half had short leaves. The predominate growth form, in terms of absolute percent cover, was the narrow sclerophyllous shrub because of the wide distribution of *A. fasciculatum*. Communities dominated by broad sclerophyllous shrubs were common in the chaparral. The broad and narrow malacophyllous shrubs formed less frequent but distinct communities.

The shrub communities in central Chile were dominated by the broad sclerophyllous shrubs in terms of both number of species in the growth form and absolute percent cover of the growth form. Deciduous malacophyllous shrubs and spiny shrubs formed communities of lesser importance that were distinctly different from communities in southern California.

3.7. Conclusions

The species studied in detail in this research (Chapters 4-9) were the most common species in the chaparral of southern California and the matorral of central Chile and thus defined those plant communities. Low floristic affinities indicated the different phyletic histories of species in the two regions. Shrub cover was higher in southern California than in central Chile, although shrub diversity was higher in central Chile. Herb cover was lower in the chaparral of southern California than in the matorral of central Chile.

Although the shrub communities of southern California and central Chile appeared to be quite similar, there were major differences, particularly in the dominant growth form. Narrow sclerophyllous shrubs were abundant in the shrub communities of southern California, but the diversity of species in this growth-form category was low. In contrast, Chilean shrub communities were dominated by broad sclerophyllous shrubs of many species. Communities dominated by broad sclerophyllous shrubs were common but were not the most frequent communities in southern California. Photosynthetic stem and spine shrubs played an important role in the Chilean shrub communities, whereas these types were rare in southern California. Results of the sampling indicated that species and growth forms within each country segregate along gradients of aspect, elevation, and moisture. However, the between-country comparison indicated that similar growth forms were distributed similarly along moisture gradients. The distribution of growth forms along elevational gradients was unclear because of the difference in topography between the continuous elevational gradient from the coast to the Peninsular Ranges in southern California and the more abrupt changes from the coastal mountains to the Central Valley to the Andean foothills in central Chile.

4. Biomass, Phenology, and Growth

Jochen Kummerow, Gloria Montenegro, and David Krause

In this chapter, shrub structure, biomass, phenology, and growth at Echo Valley and Fundo Santa Laura are analyzed and compared. The growth dynamics of shoot and root systems are studied and the similarity of these growth processes in both countries assessed.

4.1. Introduction

The structural similarity between the Californian and the Chilean mediterranean shrub vegetation has been demonstrated and the difficulties of assessing degrees of similarity discussed (Mooney, 1977a; Thrower and Bradbury, 1977). However, a static, species-to-species comparison of morphological structures in two ecosystems with only remotely related phylogenetic histories has its shortcomings. Comparable morphological traits may have evolved under the pressure of similar environmental conditions; yet these traits may not have the same ecological importance in both ecosystems. The following example illustrates this point. The midelevation chaparral in southern California is dominated by the narrow sclerophyllous shrub *Adenostoma fasciculatum*. Narrow sclerophyllous shrubs also have evolved in the central Chilean matorral. *Satureja gilliesii* is the physiognomically ideal Southern Hemispheric counterpart of *A. fasciculatum* (Sierra Ráfols, 1977). However, an anatomical analysis identifies *Satureja gilliesii* as a summer semideciduous malacophyllous species. In addition, it is of only limited importance in the matorral in terms of ground coverage and frequency. *Baccharis rosmarinifolia* is truly a narrow sclerophyllous shrub but is of even less floristic significance. The species-to-species comparison shows structural similarities, but structural similarities may be irrelevant in terms of vegetation dynamics, resource use, and allocation patterns in the two ecosystems.

This chapter assesses the relative degree of similarity in growth processes of chaparral shrubs of southern California and of matorral shrubs of central Chile. The hypothesis was that similar patterns of carbon allocation to shoot and root systems evolved in the two vegetations, which have only a very distant phylogenetic relationship, yet are exposed to similar environmental conditions. The similar patterns are not based on the growth dynamics of pairs of analogous shrub species from the chaparral and matorral, but on comparable resource-use patterns of the shrub community as a whole. Shrub structures and root-to-shoot biomass ratios in the vegetation at Echo Valley and Fundo Santa Laura are compared, and the seasonal courses of phenological events, including detailed information on leaf, stem, and root growth, are analyzed in order to test the hypothesis.

4.2. Biomass

4.2.1 Shrub Structure

The degree of structural similarity between the shrub types of the mediterranean climatic regions of southern California and central Chile was assessed by Giliberto et al. (1977) and Mooney (1977a). The research sites at Echo Valley and Fundo Santa Laura were closely matched regarding geological, topographic, and climatological features (Chapter 2), and detailed biological information was available for each area (Thrower and Bradbury, 1977). Six dominant shrub species from Echo Valley, six from Fundo Santa Laura, and three different shrub species from Camp Pendleton were analyzed.

Adenostoma fasciculatum, *Arctostaphylos glauca*, *Ceanothus greggii*, *Quercus dumosa*, *Heteromeles arbutifolia*, and *Quercus agrifolia* were studied at Echo Valley; *Trevoa trinervis*, *Satureja gilliesii*, *Colliguaya odorifera*, *Quillaja saponaria*, *Lithraea caustica*, and *Cryptocarya alba* at Fundo Santa Laura; and *Encelia californica*, *Salvia mellifera*, and *Artemisia californica* at Camp Pendleton. Five individual shrubs per species, representative in form and size, were harvested after diameter and height were recorded. Leaf and stem dry weights were measured and the canopy weight per unit of canopy volume calculated (Table 4.1). Only shrub-sized individuals of *Q. agrifolia* and *C. alba* were studied because harvesting tree-sized individuals proved to be impractical.

The shrub species studied at Echo Valley and Fundo Santa Laura had similar shrub architecture (Figure 4.1). The Chilean shrubs had a slightly larger canopy volume than did the Californian shrubs. However, the average canopy weight per volume, which is an indicator for branch density, was higher at Echo Valley (Table 4.1). Shrubs from Echo Valley and Fundo Santa Laura were more similar to each other in their general architecture than were shrubs from Echo Valley and Camp Pendleton. Evaluation of these data should take into account that the calculation of average canopy weights per volume was somewhat misleading because one of the smaller chaparral shrubs, *A. fasciculatum*, dominated the vegetation over wide areas; whereas, the largest species, *Q. agrifolia*, was limited in its distribution to the more mesic ravine bottoms and thus covered a much smaller percentage of the total area. The structure of *S. mellifera* differed from that of the other two coastal sage shrubs. *Salvia mellifera* may not be an ideal representative of the coastal sage scrub because it occurred over a broad range of habitats and was found inland in the chaparral as well.

Specific leaf weights of the Californian and Chilean species showed a marked similarity. Arrangement of the species in a sequence from arid to moister habitats (Parsons,

Table 4.1. Shrub structure characteristics of chaparral species at Echo Valley, coastal sage scrub at Camp Pendleton, and matorral at Fundo Santa Laura

Echo Valley

	Adenostoma fasciculatum[d]	California Arctostaphylos glauca	Ceanothus greggii	Quercus dumosa	Heteromeles arbutifolia	Quercus agrifolia
Canopy weight per volume (g m^3)[a]	2262	2847	2590	2018	2930	3312
Leaf specific weight (mg cm^2)[b]	25.3	27.0	37.0	14.0	23.0	13.0
Leaf area index (m^2 m^2)	2.1	3.6	1.0	2.5	2.8	3.6
Leaf angle (degrees)[c]	58	50	53	43	46	46

Camp Pendleton

	Encelia californica[d]	Salvia mellifera	Artemisia californica
Canopy weight per volume (g m^3)[a]	464	1141	1196
Leaf specific weight (mg cm^2)[b]	7.0	9.2	8.0
Leaf area index (m^2 m^2)	0.7	2.4	0.8
Leaf angle (degrees)[c]	49	48	46

Fundo Santa Laura

	Trevoa trinervis[d]	Chile Satureja gilliesii	Colliguaya odorifera	Quillaja saponaria	Lithrae caustica	Cryptocarya alba
Canopy weight per volume (g m^3)[a]	1057	1248	1843	1466	1621	1754
Leaf specific weight (mg cm^2)[b]	8.0	8.0	27.0	21.0	26.0	20.0
Leaf area index (m^2 m^2)	2.3	1.4	1.0	1.8	1.1	4.4
Leaf angle (degrees)[c]	26	51	47	34	27	25

[a] Each value is the mean of five harvested shrubs.
[b] Leaves were collected in equal numbers from all canopy levels from 5 harvested shrubs; total leaf samples = 70.
[c] Angles for all canopy levels (N = 300-600/species) were measured in reference to a horizontal line at the point of leaf attachment to the stem.
[d] Shrubs are arranged along an aridity gradient (Parsons, 1973) from the xeric (left) to the more mesic habitats (right).

CAMP PENDLETON

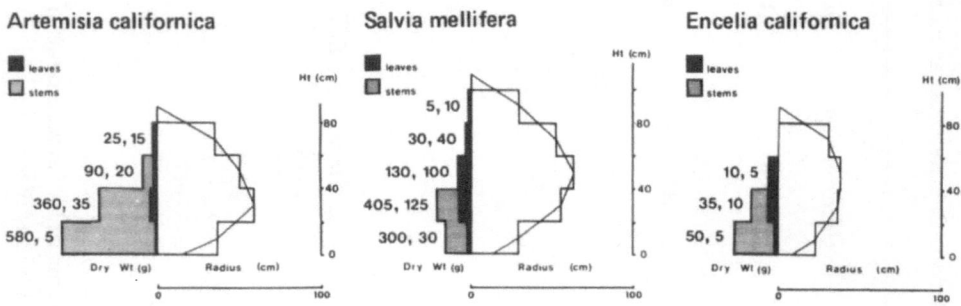

ECHO VALLEY

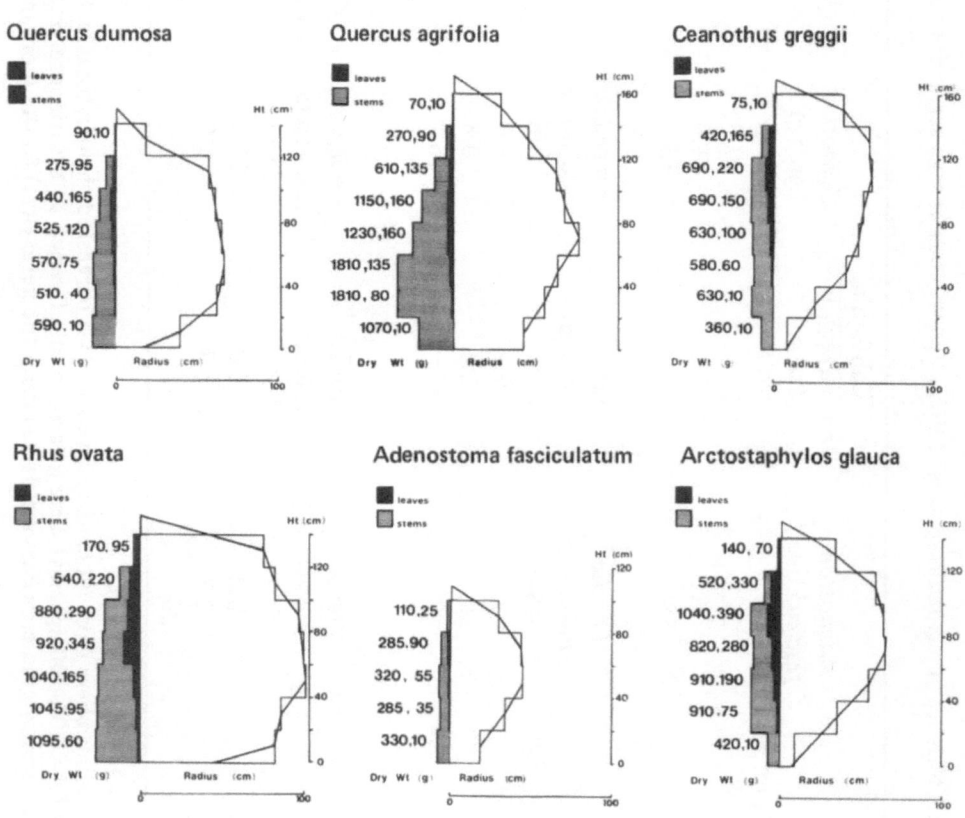

Figure 4.1. Structure diagrams of shrubs from Camp Pendleton, Echo Valley, and Fundo Santa Laura. Numbers on the left of the diagrams are average dry weights (g) of stems (digits left of comma) represented by hatched areas and of leaves (digits right of comma) represented by solid dark areas. The right side of the diagrams indicates the shrub dimensions (radius in cm). All values are given for 20-cm-deep canopy levels. Each diagram is the mean of five harvested shrubs, representative in size and form (after Giliberto et al., 1977).

FUNDO SANTA LAURA

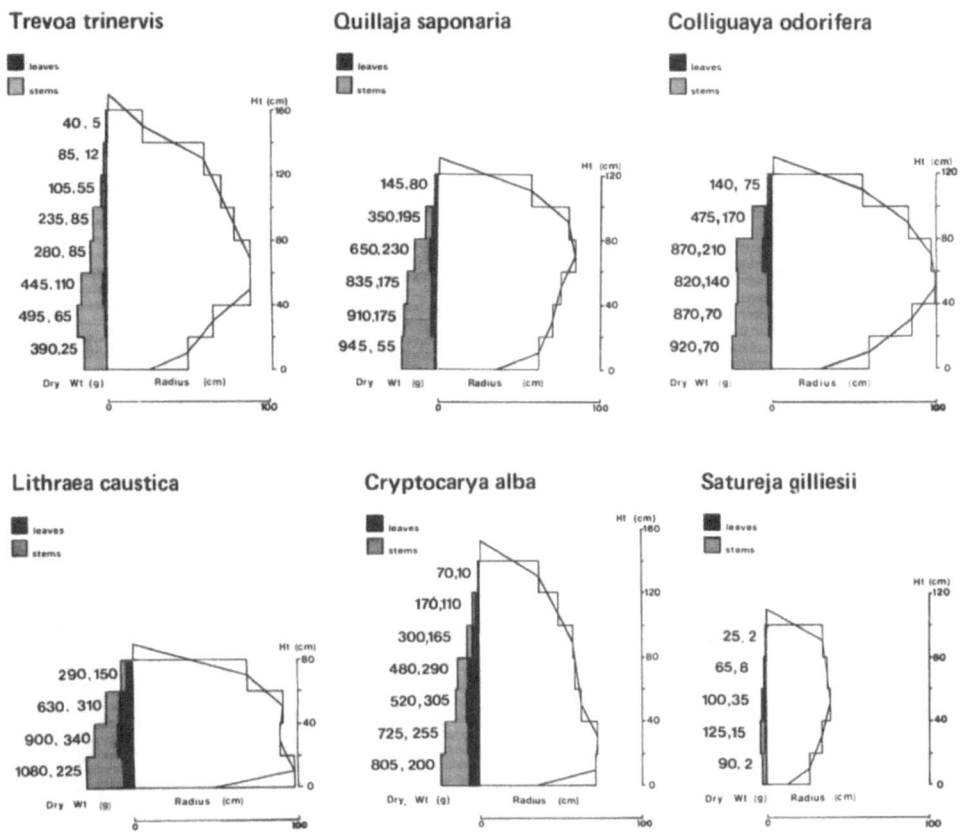

Figure 4.1. (Continued)

1973) revealed a pattern of decreasing leaf specific weight toward the mesic sites in California (Table 4.1). *Trevoa trinervis* and *S. gilliesii* did not follow this pattern, in spite of growing on the arid end of the gradient. These species are summer deciduous and in this respect are similar to the two shrubs of the Californian coastal sage scrub. *Quercus agrifolia* and *C. alba* occupied very similar habitats, namely, the bottoms of smaller valleys and ravines, in California and Chile, respectively. The surprisingly low specific leaf weight of *Q. dumosa* may be related to its deep-reaching root system. Xylem pressure potential measurements showed that *Q. dumosa* had higher pressure potentials than did *C. greggii* var. *perplexans* or *A. fasciculatum* (Krause and Kummerow, 1977b). Interestingly, the two more or less summer deciduous species from Fundo Santa Laura, *T. trinervis* and *S. gilliesii*, most closely matched the specific leaf weights of the three shrubs from the coastal sage scrub.

Leaf area indices were similar at Echo Valley and Fundo Santa Laura. The relatively high leaf area index for *T. trinervis* was related to the deciduousness of this species and was similar to that of *S. mellifera*. *Satureja gilliesii* and *C. odorifera* resembled the California coastal sage scrub species in size and degree of deciduousness. Thus, their low leaf area index was not unexpected and reflected the typically dry sites where these species occurred. The high leaf area indices measured in *C. alba* reflected the fact that the individuals characterized were stump sprouts, which typically had very dense foliage. The leaf area index for mature *C. alba* was estimated as 3 to 4.

Leaf angles were calculated as the mean of 20 leaves from each 20-cm canopy stratum measured in reference to a horizontal line. Plant species with leaves having the lowest average leaf angle occurred at the moistest sites, except for the summer deciduous *T. trinervis*, which with a mean leaf angle of 26.3° occurred on the dry end of the habitat gradient. Mean leaf angles did not vary significantly between coastal sage scrub species at Camp Pendleton and chaparral species at Echo Valley.

Significant structural similarities exist between the evergreen sclerophyllous shrubs of Echo Valley and Fundo Santa Laura. Shrubs from the Californian coastal sage scrub area were more structurally similar to the summer deciduous shrubs in the matorral at Fundo Santa Laura than to the shrubs in the chaparral at Echo Valley. Inasmuch as structural similarities can be judged as circumstantial evidence for similar carbon allocation patterns, the hypothesis was supported that similar patterns have evolved in the chaparral of southern California and the matorral of central Chile.

4.2.2 Standing Crop

4.2.2.1 Aboveground Shrub Biomass. The standing crop of aboveground phytomass can vary widely depending on topographic position, stand age, and season. Areas in southern California that were classified by Dodge (1975) as light brush had standing crops of 2.2 to 3.4 kg m^{-2}, medium brush areas had 3.4 to 6.7 kg m^{-2}, and for areas of heavy brush, values as high as 6.5 to 9.0 kg m^{-2} were recorded. Dodge's data were similar to data from Echo Valley, Camp Pendleton, and Fundo Santa Laura. The aboveground standing crops in evergreen communities at Echo Valley and Fundo Santa Laura were 2.8 and 1.2 kg m^{-2}, respectively, while the coastal sage scrub at Camp Pendleton had a standing crop of only 0.7 kg m^{-2} (Mooney et al., 1977). The values were averages for the sites and included both pole- and equator-facing slopes. The standing crop of an *A. fasciculatum-C. greggii* stand at Echo Valley was 1.9 kg m^{-2}, including 1.0 kg m^{-2} of *A. fasciculatum*, 0.8 kg m^{-2} of *C. greggii*, and 0.1 kg m^{-2} of *Q. dumosa*. The data showed that, even in short distances, considerable differences could be found.

4.2.2.2 Belowground Shrub Biomass. Compared with the relative abundance of information on aboveground standing crops, information on belowground biomass is limited. However, data were available from a 70-m^2 root excavation plot at Echo Valley (Kummerow et al., 1977, 1978a). A total of 0.64 kg m^{-2} of roots from four different species was harvested. This value did not include the major part of the fine root fraction, those roots with a diameter less than 1 mm, which were lost during the hydraulic excavation. The fine root fraction may add an additional 0.5 kg m^{-2} to the reported root biomass of 0.64 kg m^{-2}. Therefore, the total root biomass in the 70-m^2 plot could be 1.14 kg m^{-2}. The excavated plot was in a poorly developed area of chaparral. Shrub coverage was estimated at 60%; soil depth was barely 60 cm; and only a small amount of organic matter was present in the uppermost 30 cm. On a different site in the same general area, Miller and Ng (1977) found similar root biomass values of 0.5 to 0.6 kg m^{-2}, excluding the fine root fraction.

A similar set of data was available from Fundo Santa Laura where excavation of an 18-m^2 plot produced 6.9 kg m^{-2} of roots (Hoffmann and Kummerow, 1978). However, this value was not representative because the excavated site had a high level of soil fertility and favorable moisture conditions. Miller and Ng (1977), working on a site with shallow soil at Fundo Santa Laura, found only 0.29 kg m^{-2} of roots, not including the fine root biomass. Recent work carried out in a more representative area at the same research site included values for the fine root fraction and indicated substantially higher values reaching a mean of 2 kg m^{-2} in the ground shaded by canopy (Kummerow et al., *unpubl. data*). No root biomass data exist from Camp Pendleton or any other chaparral or matorral area that would allow generalizations over a broader geographical area. Thus, the existing data have to be used cautiously and interpretations should take these limitations into account.

4.2.2.3 Biomass of Herbaceous Plants. The productivity of the herbaceous understory in all mediterranean areas is highly variable from year to year and appears to be dependent on annual precipitation. Biomass data were collected by means of enclosures established at Echo Valley and Fundo Santa Laura. The enclosures eliminated grazing cattle and deer but did not keep out small mammals such as rabbits and mice. The results showed large variability in the biomass production of herbaceous plants (Table 4.2). The data confirmed the visual impression that the herbaceous cover at Fundo Santa Laura was substantially denser than at Echo Valley. There was general correlation between total rainfall and biomass production but not a linear relationship because rainstorm distribution, as well as rainfall totals, may influence production. In 1972, Fundo Santa Laura received 868 mm of rain. Most of the rain fell in heavy rainstorms early in the winter while the spring, September-October, remained relatively dry. In 1977, total precipitation was similar, but storms continued well into November when higher mean temperatures stimulate herbaceous plant growth. In 1977, the biomass production of herbaceous plants was twice the amount produced in 1972. The higher shrub productivity measured at Echo Valley was, at least partially, balanced by the higher productivity of herbaceous plants at Fundo Santa Laura.

Table 4.2. Biomass production of herbaceous plants at Echo Valley and Fundo Santa Laura

	1972	1973	1974	1976	1977
		Echo Valley			
Biomass dry wt (g m^{-2})	nd[a]	10[b]	28[b]	nd	nd
Precipitation (mm)	nd	330	416	nd	nd
		Fundo Santa Laura			
Biomass dry wt (g m^{-2})	100[b]	55[b]	nd	97[c]	202[c]
Precipitation (mm)	868	412	nd	349	851

[a] nd = no data.
[b] Biomass values after Sierra Ráfols (1977).
[c] Biomass values after Montenegro et al. (1978).

4.2.2.4 Root-to-Shoot Biomass Ratios. Root-to-shoot biomass ratios of species at Echo Valley and Fundo Santa Laura tended to substantiate Barbour's Hypothesis (1973) that the root-to-shoot ratio remains below unity in warm desert areas (Table 4.3). In cold deserts, values ranged from 5 to 9 (Rodin and Bazilevich, 1967; Caldwell and Fernandez, 1975). Values for *L. caustica* and *C. alba*, however, were considerably higher than unity, 2.9 and 2.3, respectively. The excavated plot at Fundo Santa Laura may have once been a *Lithraea-Cryptocarya* grove of large trees like the *Q. agrifolia* groves in the moister valleys and ravines of southern California. The *Lithraea-Cryptocarya* grove may have been cut for charcoal, as several charcoal ovens are close by. Charcoal production stopped about 20 years ago. After *L. caustica* and *C. alba* were cut, vigorous stump sprouts formed the typical multistemmed shrubs of the area. This may have led to the high observed root-to-shoot ratios (Hoffmann and Kummerow, 1978).

Documentation of substantial seasonal changes in fine root biomass (Section 4.3.5) makes reported values of root-to-shoot ratios somewhat questionable. Ratios should not be compared without considering the season during which the data were collected, because fine root biomass values may only be valid for the season of the fine root extraction. In a recent study at Fundo Santa Laura, Jaksic and Montenegro (1979) measured root-to-shoot ratios of 0.58 for annual herbs at the time of maximum standing crop.

Table 4.3. Root-to-shoot biomass ratios (R:S) of mature plants at Echo Valley and Fundo Santa Laura

Species	Country	N	R:S	Source
Adenostoma fasciculatum	Calif.	4	0.6	Miller and Ng, 1977
Arctostaphylos glauca	Calif.	4	0.9	Miller and Ng, 1977
Ceanothus greggii	Calif.	2	0.3	Miller and Ng, 1977
Heteromeles arbutifolia	Calif.	1	0.7	Miller and Ng, 1977
Colliguaya odorifera	Chile	2	0.7	Miller and Ng, 1977
Satureja gilliesii	Chile	4	0.7	Miller and Ng, 1977
Adenostoma fasciculatum	Calif.	3	0.6	Kummerow et al., 1977
Arctostaphylos pungens	Calif.	2	0.4	Kummerow et al., 1977
Haplopappus pinifolius	Calif.	1	0.6	Kummerow et al., 1977
Eriogonum fasciculatum	Calif.	10	0.8	Kummerow, *unpubl. data*
Colliguaya odorifera	Chile	4	0.6	Hoffmann and Kummerow, 1978
Cryptocarya alba	Chile	2	2.7	Hoffmann and Kummerow, 1978
Lithraea caustica	Chile	2	2.9	Hoffmann and Kummerow, 1978
Mutisia sp.[a]	Chile	9	0.3	Hoffmann and Kummerow, 1978
Trichocereus chiloensis	Chile	1	0.6	Hoffmann and Kummerow, 1978

[a] Climber on *Lithraea* and *Cryptocarya*

4.3. Seasonal Biomass Dynamics

4.3.1 Phenology and Prediction of Phenological Events

Detailed phenological observations were made from 1971-1976 at Echo Valley and Fundo Santa Laura (Hoffmann et al., 1977; Mooney et al., 1977). These observations formed the data base for an examination of the predictability of phenological events in four major chaparral species, *A. fasciculatum, A. glauca, C. greggii,* and *Rhus ovata.* Previous work using regression analyses had established strong correlations between phenology and specific environmental parameters for several perennial species (Kasku-rewicz and Fogg, 1967; Benacchio and Blair, 1972; French and Sauer, 1974; Taylor, 1974; Wielgolaski, 1974).

Of the seasonal fluctuations of the essential resources, such as light, water, and nutrients, in mediterranean type ecosystems, water was the most unpredictable. Light and, to a somewhat lesser degree, mean temperature fluctuated seasonally in a fairly predictable pattern. However, water and, to a certain degree, nutrient availability ap-peared to be quite unpredictable. Seasonal development of the shrubs at Echo Valley appeared to be largely dependent on the limiting effect of water once a warming trend released the shrubs from a forced winter dormancy. Unusual summer rains in 1977 in California were intensive enough to disrupt the drought-induced dormancy of many shrubs. The resulting growth initiated by these plants was arrested in November, not by lack of water but by low temperature. Thus, variations in the time pattern and in-tensity of phenological events, such as spring growth initiation and flowering, were controlled by water availability and temperature.

Between February 1976 and August 1977, biweekly soil moisture measurements were taken at Echo Valley in the 10- to 40-cm depth, the zone of highest root density, because conventional rainfall data, expressed as seasonal or monthly means, did not re-flect storm frequency and rainfall intensity. In addition, air temperature and relative humidity were monitored daily and total precipitation was measured weekly.

A correlation between phenophases and factors of the physical environment was not expected to be synchronous for all the species. Species react differently to en-vironmental stimuli, creating the broad mosaic of resource utilization patterns that characterize a multispecies plant community. The mean date of spring growth initi-ation varied by species. *Ceanothus greggii* initiated stem elongation by 20 March, fol-lowed by *A. glauca* (17 April) and *A. fasciculatum* (21 April); *R. ovata* did not begin growth until early June (3 June). These dates are based on observations of 10 shrubs per species from 1973 to 1976. The standard deviations for these dates were 37, 33, 13, and 12 days, respectively. The later a species initiated growth in the spring the nar-rower became the range of time for the initiation of stem elongation. In *R. ovata,* growth initiation seemed to be unrelated to the available soil moisture. However, *R. ovata,* a deep-rooting species with access to water sources unavailable to the other shrubs, maintained a xylem pressure potential of -20 bars during the driest period of early fall when the three other species approached xylem pressure potentials of -50 bars (Poole and Miller, 1975).

78   J. Kummerow, G. Montenegro, and D. Krause

Data on *A. fasciculatum* and *A. glauca* illustrate species differences in the dependence of flowering on soil moisture (Figure 4.2). In both species, flowering depended on the availability of soil moisture during April and May, but in one species, flowering depended on the soil moisture just preceding flowering, while in the other species, flowering depended on the soil moisture during April and May a year before. *Adenostoma fasciculatum*, which flowers in early summer, and *A. glauca*, which generally

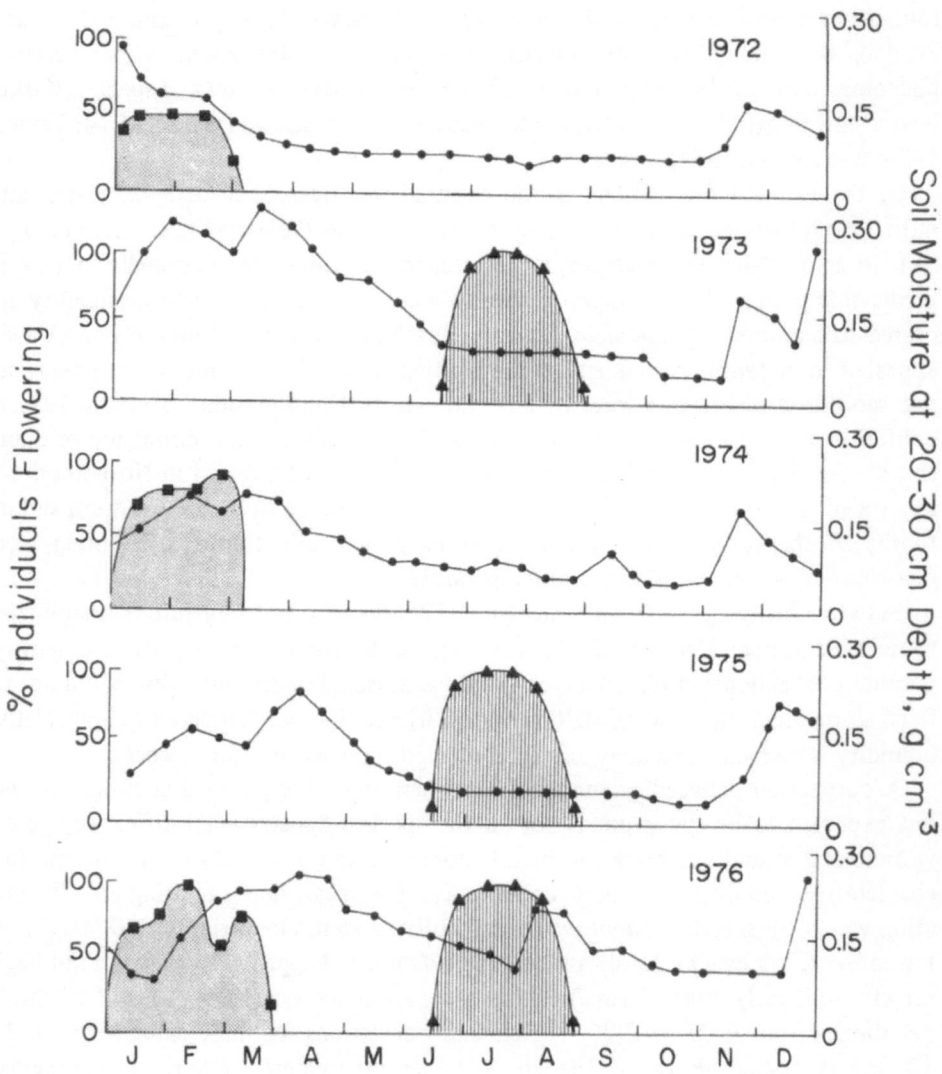

Figure 4.2. Flowering of *A. fasciculatum* (▲ and lines under curve), and *A. glauca* (■ and dots under curve) and soil moisture at 20 to 30 cm depth. Soil moisture measured by the neutron attenuation method (Visvalingam and Tandy, 1972). Mean (●) of measurements from three access tubes for the neutron probe. Flowering expressed as percentage of 10 individuals of both species. The same individuals were monitored over the 5-year study period (Jow, *unpubl. data*).

flowers from January to March, failed to flower after soil moisture levels during April and May fell below 0.12 g cm^{-3} in 1972 and 1974. In *C. greggii*, which flowers in April and May, flower bud differentiation occurred during the spring about 12 months before flowering (Fishbeck and Kummerow, 1976). These flower buds remained dormant for approximately 9 months. In *A. glauca*, flower bud differentiation follows the pattern of *C. greggii* with differentiation occurring about 10 months before flowering. In *A. fasciculatum*, flowers differentiated, developed, and passed through anthesis within a 3-month period between spring and summer, relying on the current accumulation of carbohydrate, mineral nutrients, and the availability of water. Dissections of flower buds of Chilean species have shown similar differences in flowering strategies (Figure 4.3). In *C. odorifera* and *Saphora macrocarpa*, flower buds differentiate well

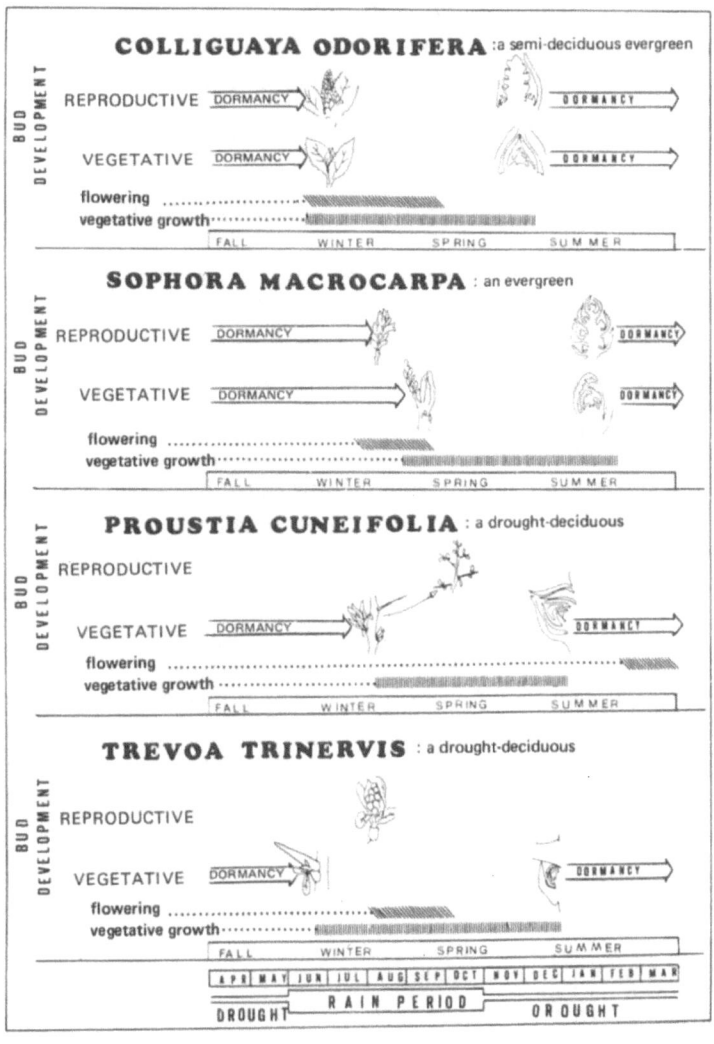

Figure 4.3. Seasonal development and dormancy periods of buds in four shrub species at Fundo Santa Laura (Hoffmann et al., *unpubl. data*).

before flowering, while in *T. trinervis* and *Proustia cuneifolia*, flower buds differentiate just before flowering. Thus, any correlation between precipitation and flowering has to account for these differences between species. In both the chaparral and the matorral, similar patterns of flower differentiation and flowering have evolved, an indication of a similar temporal range of carbon allocation to reproductive efforts.

Regression analyses on the percentage of the population that has initiated elongation or flowering as a function of warming hours (sum hours $\geqslant 15°C$ from the day in the preceding fall after which the soil moisture $\geqslant 0.1$ g cm^{-3}) account in all but one case (*C. greggii* flowering) for more than 90% of the total variation in the occurrence of these phenological events (Table 4.4). Increasing day length, sum precipitation, and sum hours $\geqslant 0°C$ are other variables that show significant correlation with the onset of these phenological events.

In conclusion, warming hours (sum hours $\geqslant 15°C$) and a soil moisture threshold value in late fall are the most useful tools to predict initiation of the two key phenological events, i.e., onset of stem elongation and flower initiation. Preliminary evidence for a similar dependence of growth and flowering of Chilean matorral shrubs has been found.

4.3.2 Shoot Elongation and Leaf Expansion

Common shrubs at Echo Valley and Fundo Santa Laura with distant taxonomic relationships showed closer structural similarities to each other than did chaparral and coastal sage scrub growing only 50 km apart (Parsons and Moldenke, 1975). However, comparisons of the growth of chaparral and matorral shrubs are static when based on harvests of individual shrubs and the dry weight of their component parts, e.g., leaves, stems, and roots, in midsummer at the end of the growing season. The procedure does not allow comparison of growth dynamics during the growing season.

Growth patterns of shrubs were studied in detail between February 1976 and August 1977 at Echo Valley and between October 1975 and April 1977 at Fundo Santa Laura (Montenegro et al., 1979; Table 4.5). In both research areas, identical methods were used. Four shrub species, *A. fasciculatum, C. greggii, A. glauca,* and *R. ovata*, were monitored at Echo Valley and six species, *S. gilliesii, T. trinervis, C. odorifera, Q. saponaria, C. alba,* and *L. caustica*, were monitored at Fundo Santa Laura. Ten individual shrubs of each species were selected. Care was taken to include shrubs on both pole- and equator-facing slopes. One apparently healthy branch was tagged on each individual shrub at the height of the last leaf formed during the previous growing period. After spring growth began, the lengths of the shoots distal to the tags and the length of each leaf on the new shoots were measured at weekly intervals. Sylleptic axillary shoots were included in the measurements when they reached a minimum length of 0.5 cm.

The resulting data sets from Echo Valley and Fundo Santa Laura allowed comparison of growth patterns within each site and between the two sites. Individual growth curves of shoots showed the expected sigmoid form (Montenegro et al., 1979). However, a comparison of the growth dynamics of individual shoots in the two research areas revealed important differences. The total leaf areas per shoot and mean shoot length produced in one growing season were much higher at Fundo Santa Laura

Table 4.4. Flower initiation and onset of stem elongation for a given percentage of the population as a function of warming hours ($\Sigma h \geq 15°C$) in four chaparral species

Species	Minimum days to initiation	Range of initiation period (days)	Regression equations[a]	N	R^{2b}
		Flower initiation			
Adenostoma fasciculatum	193	50	$Y_1 = -144.6 + (0.13) X_1$	5	0.940
Arctostaphylos glauca	0	122	$Y_1 = -25.6 + (0.19) X_1$	6	0.896
Ceanothus greggii	77	87	$Y_1 = -66.9 + (0.20) X_1$	4	0.748
Rhus ovata	134	54	$Y_1 = -69.4 + (0.11) X_1$	6	0.985
		Onset of stem elongation			
Adenostoma fasciculatum	127	61	$Y_1 = -111.5 + (0.20) X_1$	7	0.931
Arctostaphylos glauca	76	118	$Y_1 = -41.1 + (0.12) X_1$	8	0.935
Ceanothus greggii	96	77	$Y_1 = -50.0 + (0.17) X_1$	6	0.943
Rhus ovata	163	65	$Y_1 = -81.8 + (0.10) X_1$	6	0.961

Note: Regression for the rate of accumulation of warming hours: $\log X_1 = 2.16923 + (0.0045535) X_2$; $N = 16$; $R^2 = 0.985$.
Y_1 = percentage of population in which flower initiation or stem elongation has been initiated.
X_1 = Σ h during which temperature $\geq 15°C$ from the time soil moisture ≥ 0.100 g cm^{-3} in the preceding fall.
X_2 = days from date in preceding fall when soil moisture ≥ 0.100 g cm^{-3}.

[a] In order to predict when a given percentage of the population has initiated flowering or stem elongation (in terms of accumulated warming hours) use the following form of the equations:
X_2 (days to flowering or elongation) = $\{\log[(\%$ population flowering or elongating $- Y_1 - $ intercept)b_1] $- 2.16923\}/b_2$, where Y_1 $-$ intercept is the y intercept of the regression for a given species; b_1 is the slope of the regression for a given species; and b_2 is the slope of the regression for the accumulation of warming hours.

[b] All regressions are significant at $\alpha = 0.005$ except the flower initiation regression for *Ceanothus greggii* ($0.20 < \alpha > 0.10$).

Table 4.5. Leaf area per shoot, growth rates, shoot lengths, and the length of growth periods of four chaparral shrub species at Echo Valley and six matorral shrub species at Fundo Santa Laura ($\overline{X} \pm SE$)

Species	Total leaf area per shoot and per cm^2	Rate of leaf area growth (cm^2 day^{-1})^a	Total shoot length in season (cm)	Rate of shoot elongation (cm day^{-1})^a	Date of growth initiation	Duration of growth days
Southern California						
Adenostoma fasciculatum	2.5 ± 0.45(7)	0.03 ± 0.01(7)	5.2 ± 1.70(7)	0.09 AB ± 0.03(7)	09 April	69 ± 26.08(7)
Arctostaphylos glauca	33.0 ± 6.50(17)	0.82 A ± 0.13(17)	6.9 ± 1.24(17)	0.13 B ± 0.02(17)	30 April	48 ± 3.64(17)
Ceanothus greggii	7.9 ± 1.52(20)	0.12 B ± 0.02(20)	4.2 ± 0.83(20)	0.06 A ± 0.01(20)	12 April	68 ± 3.58(20)
Rhus ovata	43.5 ± 24.05(6)	1.19 AB ± 0.43(6)	2.9 ± 1.18(6)	0.08 AB ± 0.06(6)	06 June	31 ± 5.31(6)
Central Chile						
Satureja gilliesii	14.9 ± 2.92(4)	0.08 ± 0.02(4)	25.9 ± 3.25(4)	0.14 ± 0.12(4)		178 ± 3.85(4)
Trevoa trinervis	18 ± 5.33(2)	0.095 ± 0.04(2)	21.6 ± 7.40(2)	0.12 ± 0.04(2)		169 ± 16.97(2)
Colliguaya odorifera	33 ± 6.74(10)	0.22 ± 0.04(10)	8.4 ± 1.49(10)	0.05 ± 0.01(10)		151 ± 8.10(10)
Quillaja saponaria	37 ± 8.19(10)	0.43 ± 0.21(10)	5.3 ± 1.14(10)	0.05 ± 0.01(10)		86 ± 7.24(10)
Cryptocarya alba	117 ± 16.10(9)	1.18 ± 0.14(9)	12.2 ± 1.77(9)	0.12 ± 0.16(9)		99 ± 3.9(9)
Lithraea caustica	93 ± 37.49(7)	0.93 ± 0.33(7)	13.8 ± 5.75(7)	0.12^b ± 0.05(7)		100 ± 6.24(7)

Note: Number of observations is in parentheses.

^{a} Growth rates with the same letters (A, B) are not statistically different (ANOVA). The rates are calculated on the basis of the total growth period. Mean values are from 10 branches, each branch from a different shrub.

^{b} Includes insect predated shoots.

than at Echo Valley. These differences were related to the length of the growing period in the two research sites. The Chilean evergreen sclerophyllous shrubs showed active growth over a period of 3 months, while the summer deciduous shrubs grew over a period of 5 to 6 months. In California, *A. fasciculatum* had a growth period of only 69 days, which was the longest growth period for any of the Californian species. *Rhus ovata* completed its vegetative growth in only 31 days.

At Fundo Santa Laura, *S. gilliesii* and *T. trinervis*, which are summer deciduous, and *C. odorifera*, which is semideciduous, initiated growth during the winter as soon as precipitation provided adequate soil moisture. In fact, for *C. odorifera* and *S. gilliesii*, the degree of deciduousness depended on the available soil moisture (Hoffmann and Hoffmann, 1976; Montenegro et al., 1979). If watered biweekly, *T. trinervis* did not shed its foliage, thus showing a behavior similar to that of southern California's *Ceanothus leucodermis*, which can be considered facultatively deciduous.

The long growing period of *T. trinervis* and *S. gilliesii* was not only related to the available soil moisture but also to shoot morphology. In *S. gilliesii*, relatively large winter leaves were formed after the first rains and later much smaller summer leaves grew on axillary short shoots (Montenegro et al., 1979). A similar leaf dimorphism was shown for *Ononis natrix* under comparable climatic conditions in Israel (Orshan, 1938). The complex shoot structure in *T. trinervis* was analyzed by Hoffmann (1972) who demonstrated that leaf expansion on typical short shoots was soil moisture dependent but was restrained by low temperature.

The summer deciduous shrub growth form, which has large fractions of its total foliage displayed on short shoots, was only poorly represented in the Californian chaparral but was found more frequently in the drier coastal sage scrub zone. The short shoot characteristic of *A. fasciculatum* differed from the Chilean species in that the first leaf generated on the elongating long shoots was short-lived and negligible in regard to leaf area. These first leaves were replaced after 1 to 3 months by foliated short shoots that emerged from their axils. The more definite short shoot leaves may reach an age of 3 years (Jow et al., 1980).

Earlier phenological studies documented the fact that the seasonality of the vegetation was considerably more pronounced at Echo Valley than at Fundo Santa Laura (Hoffmann et al., 1977), which was reflected by the relatively long growing periods at Fundo Santa Laura. However, a comparison of leaf area growth rates and shoot elongation rates showed that these were quite similar at the two sites. The Chilean species produced higher leaf areas per shoot and longer individual shoots but required more time for the process, which raised the question of why the net shrub productivity at Fundo Santa Laura was substantially lower than at Echo Valley (Mooney et al., 1977). The answer can only be speculative. Perhaps the difference in productivity between the two sites was due to a smaller number of larger individual shoots per square meter of canopy projection at Fundo Santa Laura or a larger number of smaller individual shoots at Echo Valley. Or the difference may be because the Chilean matorral had a substantially higher herbaceous plant productivity than did the Californian chaparral.

Growth rates for individual shoots and leaf areas were comparable at Echo Valley and Fundo Santa Laura. However, the longer growing period at Fundo Santa Laura resulted in higher values for leaf areas per shoot and longer mean shoot lengths. Data from shoot growth dynamics studies were not suitable to calculate biomass production per square meter, but could be used to estimate the seasonal progress of leaf and shoot growth.

4.3.3 Secondary Root and Shoot Growth

Abundant information is available concerning the dynamics of aboveground shoot and leaf production in mediterranean ecosystems (Mooney et al., 1977). Less is known about belowground production (Kummerow et al., 1977, 1978a), and there is very little quantitative data regarding the secondary stem and root growth. However, understanding carbon allocation patterns in the chaparral, or any other ecosystem dominated by trees or shrubs, requires quantitative information on the volume of secondary growth in stems and roots.

Wilson (1975) modeled the distribution of secondary thickening in tree roots. In his model, the amount of photosynthate allocated to secondary thickening was regulated by the rates of transport, storage, respiration, and growth. The carbon balance presented by Caldwell et al. (1977) for *Ceratoides lanata* and *Atriplex confertifolia*, two cold desert shrubs, did not include secondary growth. Carbon flux into secondary growth could only be estimated as one-third of the new growth in leaves and twigs. In *H. arbutifolia*, an evergreen species of Californian chaparral, 80% of terminal branch production was represented by leaves, 10% by young shoots, and 10% by reproductive parts (Mooney and Chu, 1974). However, no estimates on the carbon commitment to older stems and roots were given.

Seasonal fluctuations of vascular cambium activity, which is a qualitative measure of secondary growth, were studied in Echo Valley and Fundo Santa Laura (Avila et al., 1975). Riveros de la Puente (1973) showed the positive correlation between yearly precipitation and the width of the annual growth layers in stems and roots of *Q. saponaria* and *C. alba*. He quantitatively documented that a dry winter reduced the year ring width in the stems as well as in the larger roots.

In spite of the diffuse porousness of the chaparral wood (Fishbeck and Kummerow, 1976) and frequent false year rings, 3- to 20-year-old stems were reliably aged (Kummerow and Giliberto, *unpubl. data*). A series of regression coefficients were produced to estimate stem age based on stem diameter (Table 4.6). The Fundo Santa Laura species had higher secondary stem growth rates than did the Californian species. A branch of *Q. dumosa*, 1.5 to 2.0 cm in diameter, was 14.5 years old, while the branch of the same size class of *Q. saponaria* was only 10.4 years old.

A detailed analysis was made of the amount of secondary wood produced in the stems and roots of a single shrub of *C. odorifera* during the years 1972-1975 (Avila et al., 1978). The shrub was 10 years old, about 1 m tall, 0.6 m wide, and had the typical symmetric form of this species when growing in full sun without immediate competitors. After the entire shrub was harvested including the root system, branches and roots were subdivided into diameter classes, and the total length of each diameter class was recorded.

The estimate of the secondary wood production in stems and roots was based on the principle that the annual growth layers, the wood produced in a single growing season, represented superimposed cylinders with a measurable wall thickness. The measurements indicated that the yearly growth increments were larger close to the stem and root base and smaller in the distal parts of stems and roots, which followed the general pattern shown by Wilson (1975). The volume of annual xylem production was determined by summing the annual growth layers for all stems and roots for each

Table 4.6. Predictions from regression analyses of stem age versus stem diameter

	Echo Valley			Fundo Santa Laura		
Species	Stem diameter (yr cm^{-1})	SE		Species	Stem diameter (yr cm^{-1})	SE
Adenostoma fasciculatum	11.9	0.88		Satureja gilliesii	6.6	0.49
Quercus dumosa	8.9	0.33		Colliguaya odorifera	6.1	0.42
Ceanothus greggii	6.8	0.29		Lithraea caustica	5.9	0.23
Heteromeles arbutifolia	6.1	0.24		Quillaja saponaria	5.5	0.24
Arctostaphylos glauca	4.7	0.15		Kageneckia oblonga	5.3	0.23
Ceanothus leucodermis	4.6	0.22		Trevoa trinervis	4.1	0.37
Quercus agrifolia	4.1	0.18		Cryptocarya alba	3.2	0.22
Rhus ovata	3.5	0.14				

Note: One branch was cut at the base and subdivided into size classes from each of nine randomly selected shrubs of each species in both ecosystems. Age estimates are based on year ring counts. Shrubs are arranged according to decreasing numbers of year rings per centimeter of stem diameter. Thus, the fastest growing shrubs are at the bottoms of the two columns.

of 4 years. The weight density of *C. odorifera* wood, 0.98 g cm^{-3}, was used to convert the actually measured cubic centimeters into grams (Table 4.7). In 1973, the shrub had a total leaf and twig production of 240 g, secondary stem xylem production of 20 g, and secondary root xylem production of 25 g. Leaf and twig production measured in 1973 should be increased by about 20% in order to account for the secondary growth in thickness. The amount of carbon used annually for bark growth may be as high as 5% to 10% of the new shoot production (Avila et al., 1978); more studies are needed to substantiate these figures. *Colliguaya odorifera* is a relatively slow growing shrub, and the mean annual ring width calculated from the branch size classes came close to the measured values in the harvested shrub. This supported the assumption that the estimated 20% of new shoot growth could be used as a mean value of secondary growth in a general carbon allocation scheme for chaparral shrubs. Measurements on a small specimen of *C. alba* growing in the same general area gave similar results.

The quantitative analysis of secondary growth of one or two matorral shrubs is certainly insufficient for far-reaching conclusions. However, the linear relationship between stem age and diameter in chaparral shrubs suggested that a similar relationship may exist between root age and diameter.

The seasonal dynamics of vascular cambium activity was described in detail for four pairs of morphologically similar shrubs from Echo Valley and Fundo Santa Laura

Table 4.7. Width of growth layers and total secondary growth production in a 10-year-old shrub of *Colliguaya odorifera*

| Year | Mean width of year rings (mm) | | | | | | | | Annual xylem production (g) | |
| | Near shoot (root) base | | | | 10-20 cm from shoot (root) tip | | | | | |
	Shoot	SE	Root	SE	Shoot	SE	Root	SE	Stems	Roots
1975	0.51	(0.11)	0.53	(0.02)	0.49	(0.04)	0.43	(0.10)	14.9	19.2
1974	0.35	(0.04)	0.51	(0.15)	0.41	(0.07)	0.43	(0.07)	20.3	24.9
1973	0.48	(0.07)	0.52	(0.11)	0.34	(0.05)	0.38	(0.04)	29.3	30.6
1972	0.53	(0.09)	0.56	(0.08)	0.31	(0.09)	0.36	(0.07)	60.0	40.0

Note: Year ring width values are the mean of 40 measurements on 8 different roots and shoots. SE = standard error (after Avila et al., 1978).

(Avila et al., 1975). Most of the shrubs maintained cambium activity in spring and summer over a period of 4 to 6 months and passed through autumn and winter with a resting vascular cambium. The exceptions were *R. ovata* and *L. caustica*, which showed continuous cambium activity throughout the year with a maximum of cell production in spring, March and November, respectively.

4.3.4 Fine Root Biomass and Length Growth

Several recent root studies have offered a relatively comprehensive picture regarding root biomass, root depth, and horizontal extension of root systems in summer dry areas (Hellmers et al., 1955b; Hanes, 1965; Miller and Ng, 1977; Kummerow et al., 1977; Hoffmann and Kummerow, 1978). The accumulated information permits one to conclude that root systems in mediterranean climate regions are plastic in their response to environmental conditions. For example, the chaparral species *A. fasciculatum* has been found with roots reaching 4 m in depth but also has been shown to compete successfully with other shrub species on soils of only 0.6 m depth (Hellmers et al., 1955b; Kummerow et al., 1977). Shachori et al. (1967) found evidence for roots 9 m deep in the Israeli maquis, while Miller and Ng (1977) reported shrubs in the Chilean matorral with roots as shallow as 0.5 m. In general, the horizontal spread of the roots was much wider than the crown diameter. For a small area in the southern Californian chaparral, it was shown that even the open, i.e., unshaded, area between individual shrubs was densely occupied by shrub roots (Kummerow et al., 1977).

The above studies also confirmed that root-to-shoot biomass ratios are generally below unity (Barbour, 1973). In cold deserts, the situation seems to be different; root-to-shoot ratios in the order of 5 to 9 have been reported (Rodin and Bazilevich, 1967; Caldwell and Fernandez, 1975). However, most of these values were obtained from root excavations in which considerable loss of fine roots was unavoidable. Fine root studies are generally pursued by means of soil core extractions, a tedious and time-consuming procedure. The simpler method of carbon-14/carbon-12 dilution used by Caldwell and Camp (1974) to assess seasonal belowground biomass turnover rates overcomes many technical difficulties, but this method cannot provide information regarding root diameter classes, root length, and distribution.

4.3.5 Fine Root Production in the Chaparral

The core sampling technique was used to assess fine root production and fine root density at Echo Valley (Kummerow et al., 1978a). These studies revealed that fine root densities cannot be estimated from hydraulic soil excavations. Hydraulic soil excavations produced an average value of 60 g m^{-2} of fine roots, while the core extractions produced an average of more than 1.4 kg m^{-2} live and dead fine roots at the time of the highest root densities in November and 260 g m^{-2} in February, the period with the lowest root densities. The fine root estimates from this work were probably too low by the factor of four because the hydraulic excavations were made in January-February.

The mean results based on the study of four major shrub species showed a clear seasonal pattern of fine root biomass from low values in late winter to a maximum in summer (Figure 4.4). Beginning in July, a trend of diminishing root dry weights became noticeable. However, in September-October, this trend was unexpectedly reversed and the fine root biomass increased again, perhaps due to an unusually heavy rainfall of 33 mm in the third week of August 1977 (Kummerow et al., 1978a).

4.3.6 Dead Fine Roots

The difficulty of distinguishing living from dead fine roots has been mentioned repeatedly (Caldwell and Fernandez, 1975; Reynolds, 1975; Edwards and Harris, 1977; Persson, 1978). Because the tetrazolium technique (Knievel, 1973) did not produce satisfactory results, the fine root samples from Echo Valley were sorted into living and dead fractions by hand with the aid of dissection microscopes. About equal amounts

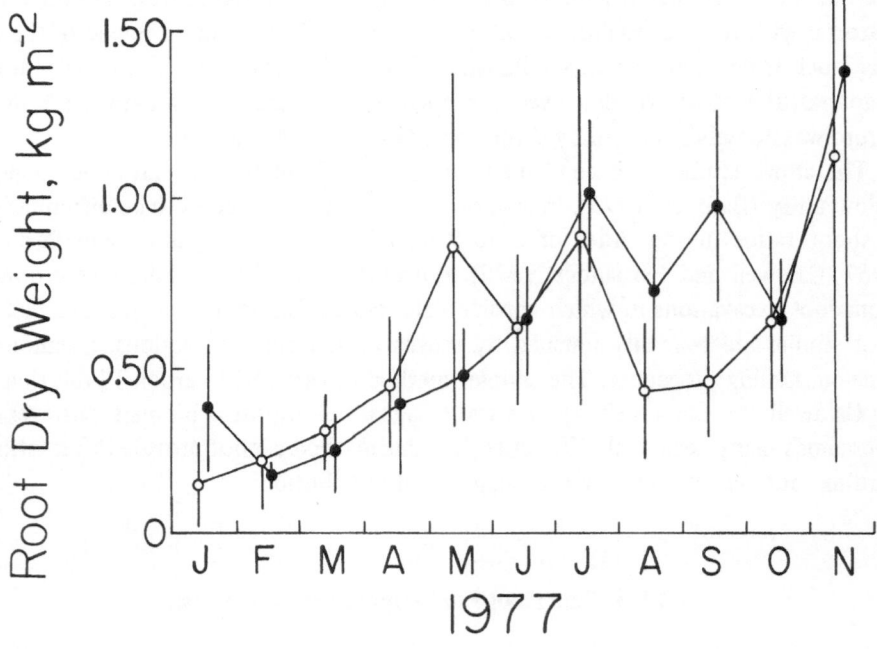

Figure 4.4. The monthly course in 1977 of live (o) and dead (●) fine root dry weights (kg m^{-2}) in the chaparral of Echo Valley, California. Each data point is the mean of roots from soil cores collected during that particular month. Sample size varied but averaged 38 cores per month. The 125 cm^3 cores were extracted from the layer at a depth of 10 to 20 cm, the zone of highest fine root densities. Soil depth and vertical fine root distribution were considered for the extrapolation to square meter. Samples contained roots from *A. fasciculatum, A. glauca, C. greggii,* and *R. ovata.* Vertical bars are 95% confidence intervals (after Kummerow et al., 1978a).

of living and dead roots were found in samples collected during winter and spring. In summer and fall, more dead than living roots were collected. By late summer and fall, a considerable amount of dead fine roots had accumulated in the soil (Figure 4.4). The accumulation of dead roots may be because summer decline of soil moisture down to 0.05 g cm^{-3} slowed down or perhaps suspended the decomposition processes.

4.3.7 Fine Root Length and Surface Area

At Echo Valley, the total extension of living fine roots per cubic decimeter of soil volume, calculated as the mean value from four species, reached 6.1 m during the summer. Assuming an average depth of the soil profile of 0.6 m and a fine root distribution of 45% in the 0- to-20-cm layer, 45% in the 21- to-40-cm layer, and 10% in the 41- to-60-cm layer (Kummerow et al., 1977), the measured root length from the 10- to-20-cm layer was multiplied by a factor of 4.4 in order to obtain the value for the complete soil core of 60 cm depth. The mean root length of 6.1 m dm^{-3} of soil volume thus translated into 2.68 km m^{-2} of shrub roots (Figure 4.5). Miller and Ng (1977) calculated that such a root density corresponded to an average distance between roots of 1.4 cm. However, observations of root distribution in the soil cores from Echo Valley indicated that fine roots were clumped together with much shorter distances from root to root and larger distances between fine root clusters. Mycorrhizal infections were frequently observed with all species; the largest amount of hyphae was found around the roots of *A. glauca*.

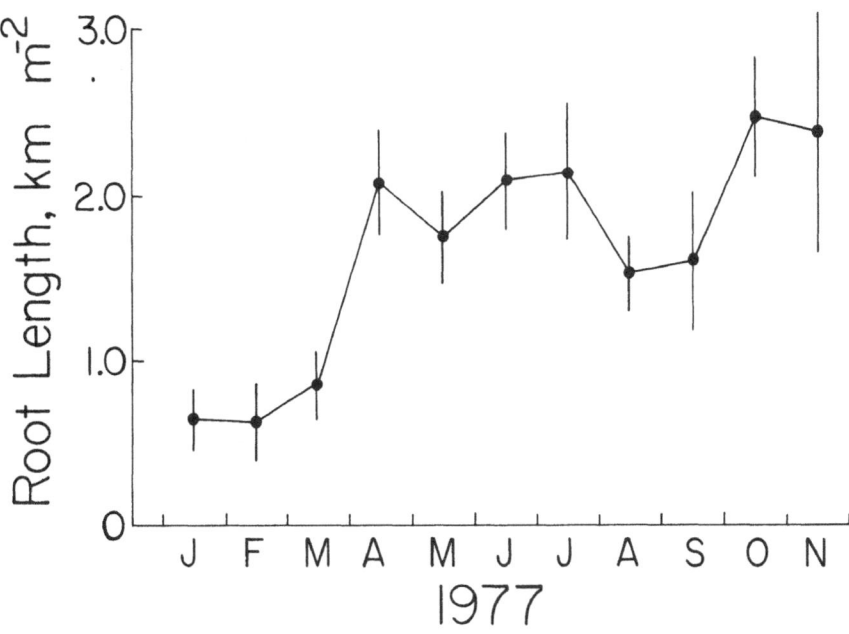

Figure 4.5. The monthly course of total fine root length (km m^{-2}). Values based on length measurements of fine roots after extraction from the soil cores.

The mean value of fine root diameters for the four species analyzed was 0.35 mm, with a range from 0.09 to 0.89 mm ($N = 660$). No significant differences were observed between species and seasons. By using data on the mean root diameter and total root length per square meter, the fine root surface per unit of ground area was estimated. A root area of 0.2 m^2 m^{-2} was calculated for late winter and an area of 2.9 m^2 m^{-2} for summer at the time of the maximum fine root length. These data supported speculations regarding the existence of a deciduous fine root system (Reynolds, 1975). Unfortunately, data on fine root dynamics in a summer deciduous vegetation were not available.

Similar studies on the growth dynamics of fine roots have been carried out in a 15-to-20-year-old pine forest in central Sweden (Persson, 1978) and in a mixed deciduous forest at Oak Ridge, Tennessee, dominated by a 50-year-old stand of *Liriodendron tulipifera* (Edwards and Harris, 1977). Results of these studies were compared with values for the chaparral. Data on grass root systems were not comparable because fibrous root systems represent a very different picture.

The stand of *Pinus sylvestris* with an understory heath vegetation composed mainly of *Calluna vulgaris* and *Vaccinium vitis-idaea* had a combined annual production of fine roots of 356 g m^{-2} (Persson, 1978). The corresponding value from the mixed deciduous forest was 1170 g m^{-2} (Edwards and Harris, 1977). The estimated annual production of fine roots in the chaparral reached 700 g m^{-2}, based on summation of differences of seasonal maximum and minimum biomass and accounting for lower root densities in the unshaded areas between shrubs. In all three ecosystems, roughly an equal amount of dead fine roots was produced, and annual fluctuations in the fine root biomass were in the same order of magnitude. Belowground litter production was higher than previously estimated. Leaf litter production was 402 g m^{-2} in the deciduous forest, 150 g m^{-2} in the pine forest, and 213 g m^{-2} in the chaparral at Echo Valley. In all three cases, the belowground litter production was substantially higher than the litterfall from the aboveground vegetation.

4.3.8 Fine Root Surface Area-to-Leaf Area Ratio

Root-to-shoot biomass ratios are considered a measure of resource allocation patterns in different plant species. However, the conventional root-to-shoot ratios do not discriminate between the living and dead material in shoot and root systems. Field observations at Echo Valley indicated that dead roots, of various size classes, tended to decompose quite readily during the moister seasons of the year. Leaves and flowers had a relatively fast turnover rate in the form of litter; yet the major standing dead stems remained intact until a fire recycled the inorganic material accumulated in the woody stems. The greater number of dead stems was reflected in the lower root-to-shoot ratios of mature chaparral stands.

Under these circumstances, it is questionable whether root-to-shoot ratios have more than a very limited informative value. Anderson et al. (1972), following the same line of reasoning, used a root-to-leaf biomass ratio in analyzing the root excavation data from Garcia-Moya and McKell (1970). Anderson and co-workers showed that the ratio of the partitioning of dry weight between the primary sites of water gain (roots) and water loss (leaves) was generally greater than one. However, this ratio is still unsatisfactory because the root dry weight included the secondary wood production of all major roots, and major roots do not contribute to water absorption. Fine roots with a diameter of less than 1 mm are the primary sites of water absorption. The mean fine root diameter of all species collected at Echo Valley was 0.35 mm (N = 660). Fine roots may account for as much as 50% of the total root biomass in the chaparral (Kummerow et al., 1977, 1978a). Seasonal trends in fine root densities may or may not be matched by seasonal changes in leaf areas of the respective species. On the other hand, leaf dry weight may not be the best measure of area available for water loss in a plant because leaf densities are highly variable between chaparral species (Fishbeck and Kummerow, 1976). It seems more reasonable to assess the total fine root surface and estimate the ratio of fine root surface area to leaf area in different chaparral species. Seasonal changes in fine root densities and seasonal trends of leaf litter production make it probable that such a ratio would fluctuate and depend to a large degree on environmental conditions.

Information on fine root surface areas and corresponding leaf areas was obtained from four cultivated chaparral shrubs grown through a growing season in redwood boxes measuring 75 X 60 X 20 cm. The shrubs were exposed to different watering regimes and harvested at the end of the growing season when total leaf area, leaf dry weight, stem dry weight, total root dry weight, fine root dry weight and length, and fine root diameter were measured (Table 4.8).

Although the number of shrubs measured was too small to permit far-reaching conclusions, a trend was evident. With increasing aridity of the rooting medium, the fine root surface area-to-leaf area ratios increased, which may indicate that a relatively higher proportion of carbohydrate was allocated to the root system. This explanation was consistent with observations by Davidson (1969), who found increases in the total nonstructural carbohydrate content in the roots of *Lolium perenne* with decreases in soil moisture

A major unresolved problem is the complex question regarding the physiological importance of root hairs. In *A. fasciculatum*, root hairs were found in variable quantities and distribution. In *Yucca whipplei*, root hairs were never observed, and in *A. pungens* and *A. glauca*, although root hairs were not observed, analysis of soil cores from the field revealed dense mats of mycorrhizae.

4.3.9 Root Growth Rates

Information on growth rates of roots in mediterranean climatic regions is scarce. The measurement of root growth rates in situ is difficult, and most of the quantitative information has been collected from root observation boxes, chambers, tunnels, and other constructions with slanted glass walls against which root growth could be observed in a nondestructive way. Oppenheimer (1936) found daily growth rates from 6 to 9 mm for container-grown oaks and pines, *Quercus aegilops*, *Quercus coccifera*, *Pinus halepensis*, and *Pinus pinea*. These techniques and results were reviewed by Lyr and Hoffmann (1967). Recently, an innovative technique using a duodenoscope was introduced (Sanders and Brown, 1978).

Table 4.8. Fine root surface area (FRSA), leaf areas (LA), FRSA/LA ratios, and root-to-shoot ratios (R:S) of container-grown chaparral shrubs

Watering regime	Fine root biomass (g)	Fine root diam (mm)	Extension of 1 g fine roots (m)	Total fine root surface/plant (m²)	Total leaf area/plant (m²)	FRSA/LA	R:S ratio
Arctostaphylos pungens							
H	691	0.17	8.6	3.17	0.54	5.86	0.61
H	317	0.17	10.5	1.78	0.38	4.67	0.88
M	491	0.17	11.2	2.94	0.40	7.42	
L	543	0.20	9.5	3.24	0.22	14.80	1.73
Arctostaphylos glauca							
H	785	0.24	7.1	4.20	0.80	5.23	0.66
M	599	0.17	8.7	2.78	0.29	9.47	1.63
Yucca whipplei							
H	290	0.45	3.5	1.39	0.33	4.17	0.30
H	242	0.45	3.5	1.20	0.36	3.33	0.23
L	469	0.50	4.8	3.54	0.48	7.32	0.29
Adenostoma fasciculatum							
H	92	0.24	6.7	0.46	0.08	5.81	0.71
M	242	0.21	10.1	1.61	0.13	12.70	0.78

Note: Water regime: (H) weekly watering to saturation, (M) watering to saturation after appearance of water stress symptoms, and (L) plants exposed to severe water stress. Four of six plants died during the experiment. Shrubs were 1.5 years old at time of planting. Duration of experiment was 8 months.

Observations on chaparral and matorral species grown in 1-m-tall, redwood observation boxes showed considerably lower values than those reported by Oppenheimer (1936). Root growth rates of 2- to 3-year-old shrubs from eight species ranged from 4.9 mm day^{-1} for *Q. saponaria* to 1.4 mm day^{-1} for *L. caustica*. An exceptionally high rate of 10.2 mm day^{-1} was recorded for *A. fasciculatum*. However, in this case the roots belonged to 6-month-old seedlings. The values were means from 30 to 50 measurements per species made in 48-h intervals during a three-month period. The regression lines showed that over a limited time (100 to 120 days) root growth rates were basically linear even when small fluctuations became visible (Figure 4.6). Deviations from the linear root growth rates were probably the result of temperature fluctuations in the root box environment.

Root box values cannot substitute for field data. However, they showed that root growth rates of chaparral and matorral species fell into the same order of magnitude. The shrubs in the observation boxes were grown in a soft rooting medium of fine sand and peat moss under an adequate water and fertilizer regime. They showed excellent aboveground growth. Values for root growth may be high compared to natural conditions. The recorded growth rates included that fraction of the fine roots that may have, in addition to absorbing functions, the task of penetrating into new soil areas and thus have somewhat larger, but always less than 1 mm, diameters. Fine penetrating roots may stay alive longer than the short-lived rootlets that may not survive a growing season or may even die after a couple of days or weeks (Head, 1973).

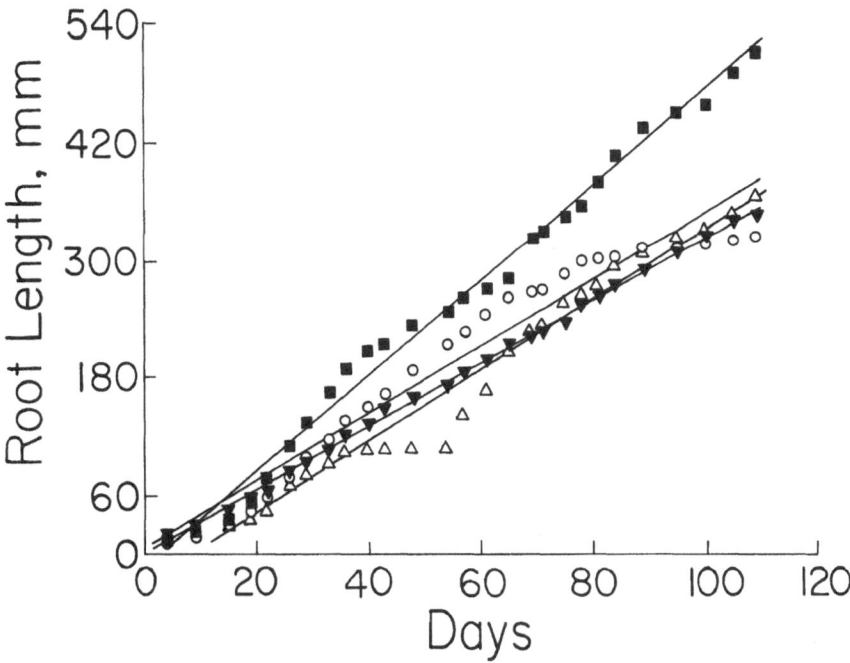

Figure 4.6. Lines of best fit for root growth with time. The data points for *Q. saponaria* (■) and *Y. whipplei* (●) are close to the regression lines. The data points for *C. odorifera* (○) and *R. laurina* (△) fit less closely probably in response to changes in the root box temperatures. Californian and Chilean species show similar root growth rates.

4.4. Annual Biomass Budget of Chaparral Shrubs

The accumulated information on growth and litterfall at Echo Valley and Fundo Santa Laura offered the possibility of estimating mean annual biomass production and loss for an average shrub based on the mean of eight different species in both countries. Estimates of this nature may be distorted by large errors due to annual fluctuations of biomass production (Mooney et al., 1977). Conventionally presented biomass production values would be improved by the addition of fine root biomass values.

Estimates of fine roots in chaparral and matorral species were based on one year, 1977-1978, of intensive soil core extractions at Echo Valley and Fundo Santa Laura. It appeared that aboveground productivity per square meter at Echo Valley was higher than at Fundo Santa Laura. However, root biomass values were substantially higher at Fundo Santa Laura. Unfortunately, the values on fine root densities in Chile were

Table 4.9. Annual biomass production and litterfall at Echo Valley and Fundo Santa Laura

	Production[a] ($g\ m^{-2}\ yr^{-1}$)	Percentage of total	Litter[b] ($g\ m^{-2}\ yr^{-1}$)
	Echo Valley		
Leaves	473	32.7	273
Young stems	96	6.6	15
Secondary growth, total	114	7.9	
Stems	68		
Roots	46		
Fine roots (diam < 1 mm)[c]	700	48.3	700
Flowers, fruits	65	4.5	65
Total	1448	100	1053
	Fundo Santa Laura		
Leaves	283	14.4	213
Young stems	86	4.4	23
Secondary growth, total	74	3.8	
Stems	44		
Roots	30		
Fine roots (diam < 1 mm)[d]	1500	75.7	1500
Flowers, fruits	38	1.9	38
Total	1981	100	1774

Note: Values are means of eight dominant shrub species. Data for leaves, young stems, flowers, and fruits after Mooney et al. (1977). Secondary growth estimated after Avila et al. (1978) and fine roots after Kummerow et al. (1978a).
[a] Means for 1971-1973.
[b] 1973 California and 1972-1973 Chile.
[c] Root extractions in 1977.
[d] Root extractions in 1978-1979.

collected in 1978-1979 after a winter with an unusually high rainfall of 864 mm, while the aboveground biomass production values were means from 1971-1973, years with precipitation of about 350 mm. The Californian root values were collected in 1977 with 389 mm of annual rainfall. A detailed analysis of root and stem secondary growth in *C. odorifera* (Avila et al., 1978) and unpublished measurements of secondary root and stem growth in eight Californian and eight Chilean shrubs (Table 4.9) showed that secondary root and stem growth was 7.9% of the total biomass produced per square meter at Echo Valley and only 3.8% of the total biomass at Fundo Santa Laura. These values are appropriate for the area measured and will vary with varying shrub cover.

4.5. Conclusions

The analysis of shrub structures demonstrated the close similarity between Californian and Chilean shrubs. Shrubs from Echo Valley and Fundo Santa Laura were more similar to each other than were shrubs in the chaparral and coastal sage scrub in southern California, thus confirming results obtained earlier by Parsons and Moldenke (1975) using other parameters in the same vegetation. Summer deciduous shrubs in the Chilean matorral were similar in many respects to the species of the Californian coastal sage scrub. Aboveground standing crop values at Echo Valley and Fundo Santa Laura were similar, but belowground biomass values were different, perhaps a result of an inadequate sample size. The difference in root biomass per square meter in the two areas should be interpreted cautiously. The fine root densities under matorral shrub canopies at Fundo Santa Laura may be higher than root densities at Echo Valley and a higher fraction of the available carbon may be allocated to the root system in the matorral. The question of whether this speculated higher carbon allocation to roots in Chilean matorral species could account for the smaller aboveground biomass measured at Fundo Santa Laura can only be determined by additional research.

The comparison of biomass production at Echo Valley and Fundo Santa Laura became even more complex when differences in herbaceous plant production were taken into consideration. The lower net biomass production of the Chilean shrubs was, at least partially, balanced by the five-to-eight times greater herbaceous production at Fundo Santa Laura than at Echo Valley.

The phenological studies demonstrated that warming hours and precipitation were the dominant environmental factors controlling the timing of phenological events in the chaparral. Comparison of the dynamics of leaf and stem growth showed that the young shoots on Californian and Chilean evergreen sclerophyllous shrubs developed with comparable growth rates. Similar growth rates could be caused by similar patterns of carbon allocation. Seasonal production of fine roots at Echo Valley was barely half that at Fundo Santa Laura. The seasonal course of fine root densities per square meter indicated a low density in winter and a maximum density in early summer with decline during the following fall. A comparison of the growth rates of fine roots under semi-controlled conditions revealed a similarity of rates in a group of Californian and Chilean evergreen sclerophylls. However, fine root surface estimates of chaparral shrubs at Echo Valley produced a ratio of root area index to leaf area index of about unity (Kummerow et al., 1978a), meaning that about 1 m^2 of leaf area was supported by 1 m^2 of root surface area. This estimate disagrees with results from the container-grown plants where a 3-to-15-times higher root surface versus leaf area was found.

Overall, the following conclusions were drawn. Growth rates of leaves, stems, and roots at the two research areas were similar. Absolute biomass production above-ground was higher at Echo Valley than at Fundo Santa Laura. The substantially higher production of herbs in Chile balanced, at least partially, the Chilean shrub biomass deficit. The fine root biomass values for Fundo Santa Laura appeared higher than those at Echo Valley, but the root studies in Chile had not progressed to a point that allowed definite conclusions. The hypothesis that similar patterns of carbon allocation to root and shoot systems have evolved under similar environmental conditions received strong support, yet could not be proven conclusively.

5. Microclimate and Energy Exchange

Philip C. Miller, Ernst Hajek, Dennis K. Poole, and Stephen W. Roberts

This chapter describes the seasonal course of energy exchange between the vegetation and the environment, with an emphasis on the partitioning of solar irradiance. The chapter also discusses the effect of plant form on energy exchange and microclimate in relation to the climatic selection of similar plant forms in mediterranean type regions.

5.1. Introduction

5.1.1 Background and Objectives

The similarities of vegetation form in regions with mediterranean type climates indicate the importance of climate in shaping the vegetation. In order to correlate climate and vegetation, climatic similarity has been assessed by indices based on monthly values for temperature and precipitation (Köppen and Geiger, 1930; Nazar et al., 1966; Aschmann, 1973b; di Castri, 1973; Walter, 1973). However, other factors also affect the vegetation in these regions; light limits production throughout much of the year (Mooney and Dunn, 1970b), and nutrients are generally low (Chapter 10). The vegetative similarities, including the predominance of shrubs with hard, flat, small evergreen leaves and relatively low leaf area indices, should be important responses of the vegetation to the mediterranean type climate. Vegetative characteristics vary within mediterranean climatic regions in southern California and central Chile (Chapters 2 and 3) in response to topographic and microclimatic variation. The vegetation is generally more closed on pole-facing slopes and more open on equator-facing slopes. At lower elevations and low annual precipitation, semideciduous malacophyllous shrubs occur with increased frequency. At midelevations, evergreen sclerophyllous shrubs predominate. At high elevations and high precipitation, evergreen forests usually occur

(Hanes, 1965; Carter, 1973a, b, c; Parsons, 1973, 1976b; Armesto and Martinez, 1978; Ng and Miller, 1980). The broad correlations and complex interactions indicate that a detailed examination should clarify some of the mechanisms of interactions between the forms of the mediterranean scrub vegetation and the climatic properties.

Climate influences the vegetation through the processes of energy exchange (Budyko, 1956; Gates, 1962, 1965), which determine plant temperature and, therefore, rates of water loss and uptake, respiration, photosynthesis, growth, and phenological events. Plants can influence these processes by their structure and physiology. Broad patterns of correspondence between climate and vegetative structure may exist because of interactions between the climate, vegetation structure, microclimate, energy exchange processes, and physiological processes. Of the energy exchange processes, i.e., solar and infrared irradiance, evaporation, convection, and conduction, solar irradiance is of primary importance because it is the source of energy that is ultimately exchanged by the other processes and because it is the source of energy that the plant must capture and convert into storable and usable forms. The plant must intercept and absorb solar irradiance for photosynthesis, but the plant must compensate for the potentially high temperatures and high rates of water loss that can result from high solar energy absorption.

The climate-vegetation interactions depend on several environmental variables and several plant variables. The environmental variables include solar irradiance directly from the sun and indirectly from the sky, infrared irradiance from the sky and from the ground, wind speed, air temperature and humidity, and soil surface temperature. The plant variables, which can be changed by natural selection, include vertical distribution of foliage, leaf width, stem diameter, leaf and stem inclination, leaf and stem indices, leaf and stem absorptances, and leaf conductances to water loss.

The incoming solar irradiance is partitioned to that reflected back, that absorbed in the vegetation canopy, and that absorbed in the soil. The solar irradiance absorbed in the canopy is exchanged by net infrared irradiance, evaporation, and convection. Net infrared irradiance and convection can contribute energy to the plant, but even when the absorbed solar irradiance is moderate, a loss of energy through these processes occurs that balances the energy received by the plant canopy.

The interception and absorption of solar irradiance are fundamental processes. Interception depends on the canopy architecture, which includes the inclination of leaves and stems, the leaf area index, and the stem area index. Interception also depends on the fraction of incoming solar irradiance that is diffuse sky irradiance. Absorption depends on leaf properties, the ratio of leaf area index to stem area index, and the clustering of leaves. Absorption of solar irradiance increases plant temperatures and transpiration rates. Increased transpiration rates can shorten periods of photosynthesis within a day because of midday and afternoon stomatal closure or shorten the period of seasonal activity by hastening the drying of the soils in the summer. The reduced photosynthesis may lead to stresses on the plant carbon balance. In winter when air temperatures are low, increased leaf temperatures can lead to increased respiration and reduced net photosynthesis. Absorbed solar irradiance is lost by infrared irradiance, convection, and transpiration. Convection depends partly on leaf width, and transpiration depends on leaf conductance.

The objective of this chapter is to evaluate the hypotheses that at Echo Valley and Fundo Santa Laura: (1) the incoming solar irradiance is similarly partitioned into components that are reflected from the canopy, absorbed in the canopy, and absorbed by the soil; (2) the solar irradiance absorbed in the canopy is similarly partitioned into net infrared irradiance, evaporation, and convection; and (3) the structural characteristics of the vegetation canopy in mediterranean regions show the greatest similarity in those characteristics that heat transfer theory indicates will most affect plant temperature, transpiration, and carbon balance.

5.1.2 Methods

At Echo Valley, the pole-facing slope was covered with a closed canopy of mixed chaparral comprised largely of *Ceanothus greggii, Quercus dumosa*, and *Arctostaphylos glauca*. The equator-facing and ridgetop exposures were covered with more widely spaced *Adenostoma fasciculatum*. At Fundo Santa Laura, the pole- and equator-facing exposures, where the instruments were located, were covered with *Lithraea caustica*. The ridgetop was sparsely covered with *Colliguaya odorifera, Satureja gilliesii*, and *L. caustica*. In both countries, the inclination and azimuth of the sites where the instruments were located were carefully matched. The inclination from horizontal and azimuth were 19° and −130° for the pole-facing exposures, 12° and −20° for the ridgetops, and 27° and +10° for the equator-facing exposures. The azimuth is given as degrees from the south in the Northern Hemisphere and from the north in the Southern Hemisphere, with east equal to −90° (Brooks, 1959; Gates, 1962). Incoming solar total and diffuse irradiance were measured with two Moll-Gorzynski solarimeters, one with a shading band (Horowitz, 1969). The solarimeters were located centrally on a ridgetop. Infrared irradiance from the sky was measured with a Fritschen type net radiometer, one hemisphere of which was covered with an insulated metal dome. The temperature of the dome was measured, and the irradiance onto one side of the net radiometer thermopile was calculated with the blackbody radiation law. Infrared irradiance from the sky was calculated by subtracting the total solar irradiance from the total hemispherical irradiance. Soil temperatures were measured at depths of 0.02, 0.05, 0.10, 0.20, and 0.30 m with thermocouples wired to record the average temperature of five points at each depth. Air temperatures and humidities were measured with aspirated psychrometers located at the top and bottom of the canopies. The soil thermocouples and psychrometers were installed on the pole- and equator-facing slopes and on the ridgetop. The microclimatic instruments were installed in April 1971 at Echo Valley and in December 1972 at Fundo Santa Laura and maintained for 1.5 years. All of these microclimate sensors were scanned hourly at 5 min before the hour (solar time), on the hour, and 5 min after the hour. Weekly summaries were made of the mean hourly values, only those weeks being calculated for which at least 4 days of good data were available. In 1978, solar and net radiometers were installed on each slope at Echo Valley, above and beneath the canopy, to measure directly the incoming and intercepted irradiances. These radiometers, along with measurements of air and soil temperatures, were monitored from April 1978 through April 1979 (Miller and Poole 1981). Throughout these periods, solar irradiance, as measured on the Moll-Gorzynski solarimeters, was consistently about 80 cal cm^{-2} day^{-1} more than that measured on the pyrheliographs used in the macroclimate stations (Chapter 2). The Moll-Gorzynski solarimeter values were assumed to be more accurate because they were regularly calibrated against a 50-junction Eppley$^{®}$ pyranometer.

5.2. Topographic Influence on Microclimates

5.2.1 Air Temperature and Humidity

Air temperatures measured on the slopes at Echo Valley and Fundo Santa Laura supported the similarities in the macroclimatic data from these two sites. In the respective winters, air temperatures on the slopes were similar at Echo Valley and Fundo Santa Laura. In the respective summers, mean minimum temperatures were 1° to 4°C higher and mean maximum temperatures were 7° to 10°C higher at Echo Valley (Figure 5.1). Temperatures below 0°C occurred on all slopes but were common only on the equator-facing slope at Echo Valley.

Air temperatures above the canopy were within 1°C of each other on the different slopes at each research site. Maximum air temperatures, which occurred during the day when the air was well mixed, were similar on the slopes and at the macroclimate weather stations, but minimum air temperatures were higher on the slopes because cold air did not accumulate on the slopes. At Echo Valley, air temperatures in the canopy were higher on the pole-facing slope than on the ridgetop and equator-facing slope. The differences were greater during the spring and summer than during the winter and were probably related to the higher leaf area on the pole-facing slope,

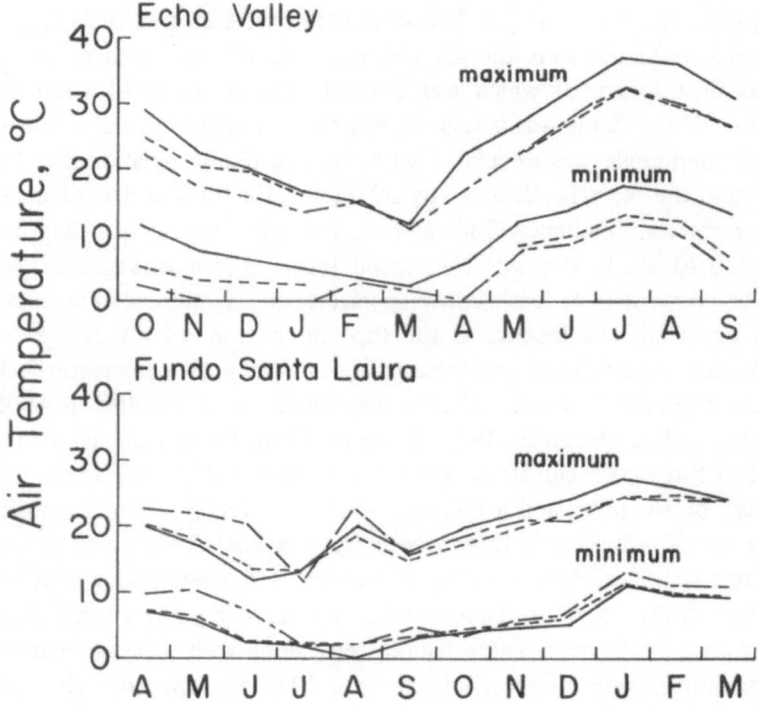

Figure 5.1. Seasonal progression of mean maximum and mean minimum air temperature on the different topographic positions at Echo Valley and Fundo Santa Laura. Pole-facing slope (—), ridgetop (- - -), and equator-facing slope (— · —).

which increased the interception of radiation and decreased the free exchange of air, the availability of water, and transpirational cooling. At Fundo Santa Laura, temperatures on the equator-facing slope were usually above temperatures on the ridgetop and pole-facing slope.

Air temperatures at the bottom of the canopy were within 1°C of temperatures at the top of the canopy but were slightly warmer in open canopies, such as those on the ridgetop and the equator-facing sites at Echo Valley, and were slightly cooler with closed canopies, such as those on the pole-facing slopes. The absolute humidity at Fundo Santa Laura was lower than at Echo Valley because of the lower air temperatures at Fundo Santa Laura and was similar on both pole- and equator-facing slopes at each research site.

5.2.2 Soil Temperature

Annual mean soil temperatures were higher than annual mean air temperatures at both Echo Valley and Fundo Santa Laura (Figure 5.2). Soil temperatures at 0.02 m depth were above 40°C during 5 months of the year beginning in May on the equator-facing slope at Echo Valley. During these months, the diurnal variation of soil temperatures at 0.02 m was about 40°C. Soil temperatures over 50°C were common in July and August at Echo Valley on the ridgetop and equator-facing slope, but soil temperatures never reached 30°C and were usually 20° to 25°C in July-September on the pole-facing slope. Diurnally, temperatures varied less on the pole-facing slope than on the equator-facing slope. Soil surface temperatures were never below 1°C on either slope. In April 1973 at Echo Valley, temperatures were about 1°C on the equator-facing slope and about 7°C on the pole-facing slope. The soil temperatures on the pole-facing slope at Echo Valley and the pole- and equator-facing slopes at Fundo Santa Laura were more similar and more uniform throughout the year than temperatures on the equator-facing slope at Echo Valley. During the Chilean summer, soil surface temperatures at Fundo Santa Laura were about 22°C in the middle of the day and 18° to 19°C at night, and temperatures over 45°C were not recorded. The lowest surface temperatures measured were near 6°C on the pole-facing slope. In the winter at Fundo Santa Laura, the pole-facing slope was cooler than the equator-facing slope, while in the summer the temperatures were almost identical. Minimum soil temperatures on the equator- and pole-facing slopes were similar in analogous months at both research sites, although at Echo Valley minimum soil temperatures were slightly warmer in summer and slightly cooler in winter on the equator-facing slope. The soil temperatures at 0.02 m depth on the equator- and pole-facing slopes were more variable through the day and through the season at Echo Valley than at Fundo Santa Laura. However, the surface temperatures were more similar at Echo Valley and Fundo Santa Laura than were the air temperatures, which were 5° to 10°C higher during the summer at Echo Valley.

The difference in temperature varied with season and vegetative cover (Figure 5.3). In winter with low vegetative cover, such as on both ridgetops and the equator-facing slope at Echo Valley, mean daily soil temperatures were 1° to 4°C above mean daily air temperature; while with high vegetative cover, such as on both pole-facing slopes

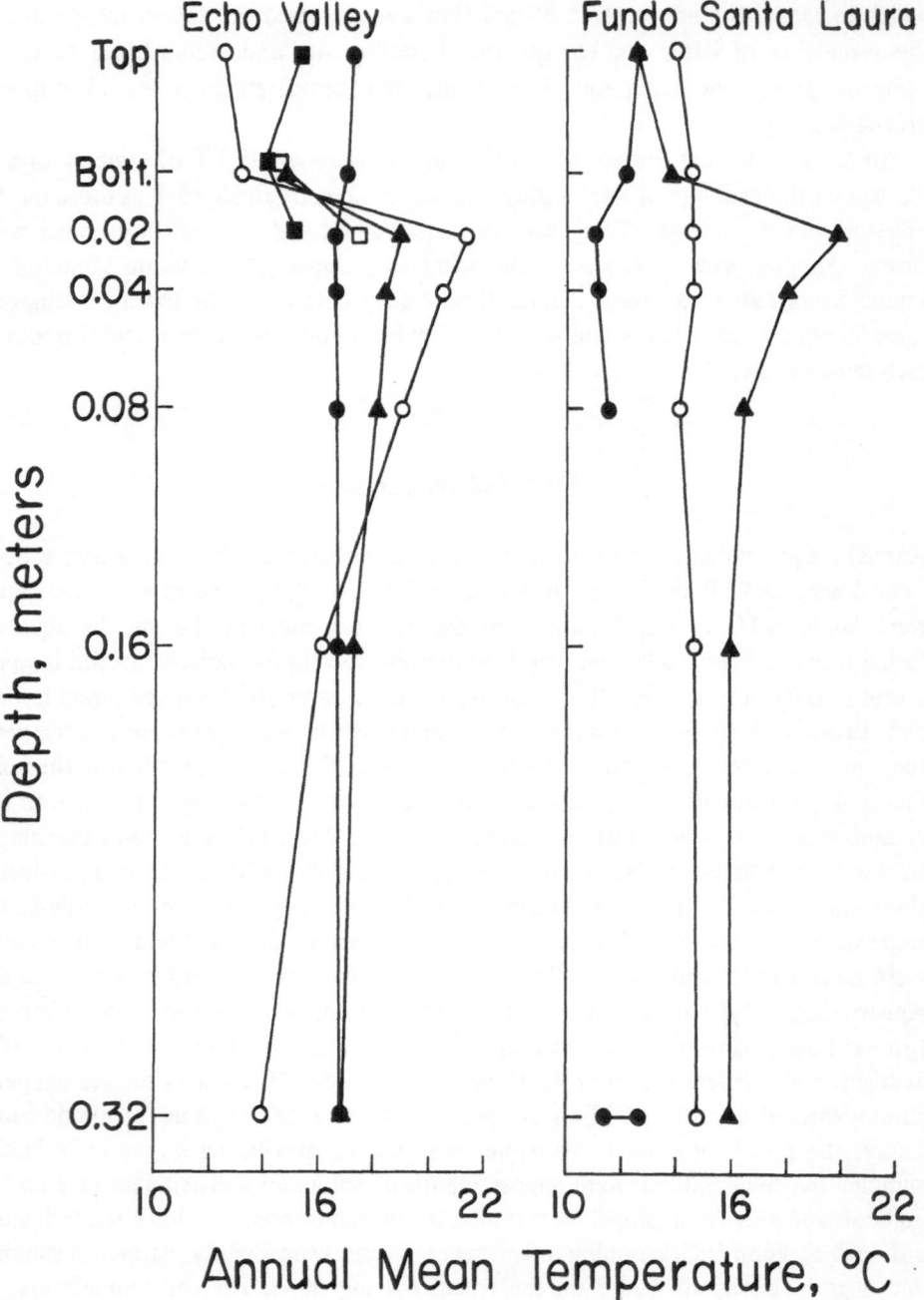

Figure 5.2. Vertical profiles of annual mean air temperature in 1972 at the top and bottom of the canopy and soil temperatures at five soil depths at Echo Valley and Fundo Santa Laura on the pole-facing slope (●), equator-facing slope (○), and ridgetop (▲). Data for 1979 (Poole, *unpubl. data*) are shown as pole-facing slope (■) and equator-facing slope (□).

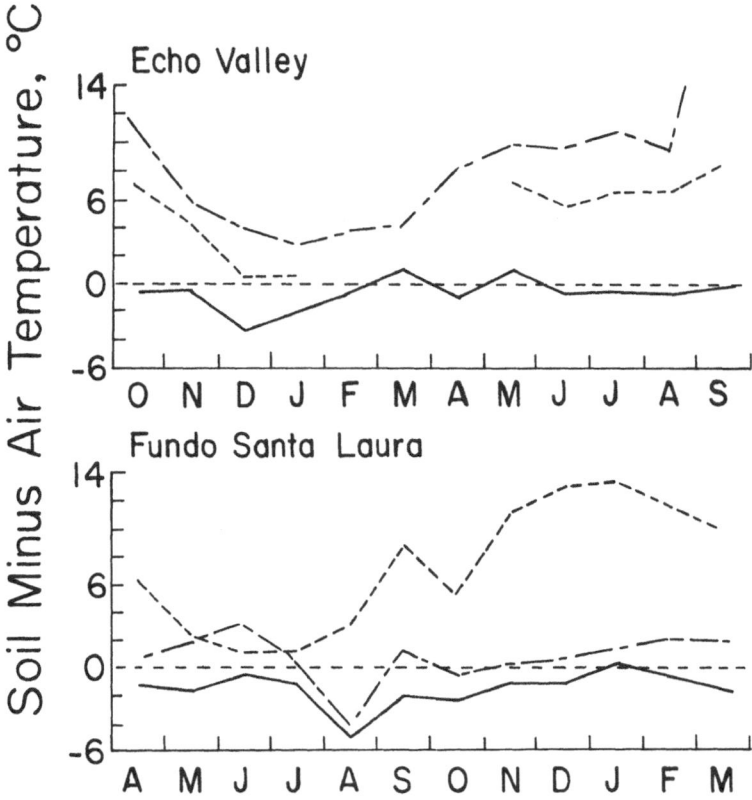

Figure 5.3. Seasonal progression of the differences between soil and air temperature on three topographic positions at Echo Valley and Fundo Santa Laura in 1973. Horizontal scale has been reorganized to begin in the fall. Pole-facing slope (–), ridgetop (- - -), and equator-facing slope (– - –).

and on the equator-facing slope at Fundo Santa Laura, the soil temperatures were 0° to 5°C below the air temperature. In summer with low vegetative cover, mean soil temperatures were 7° to 13°C above mean air temperatures, while with high vegetative cover, soil temperatures were from –2° to 1°C of mean air temperature. During Santa Ana weather conditions, with the characteristically hot, dry winds, air temperatures at Echo Valley were well above soil temperatures. In the closed canopy on the pole-facing slope, solar irradiance, which otherwise would have reached and warmed the soil, was intercepted in the canopy and dissipated by infrared radiation and convection. In winter, solar irradiance was relatively low on both slopes, and soil and air temperatures became more similar. Evaporation from the soil surface also reduced soil temperatures. After rains in the fall, soil temperatures became more similar to air temperature, and soil temperatures on the different slopes converged. As the soil surface dried, the temperatures diverged. Steeply inclined slopes and those that, because of obstructions, did not receive direct solar irradiance throughout the daylight period had relatively low soil temperatures. Both pole-facing slopes received less solar irradiance than the equator-facing slopes because of horizontal effects (next section), but the shading from the horizon was less at Echo Valley than at Fundo Santa Laura.

5.3. The Partitioning of Solar Irradiance by the Canopy and Soil

5.3.1 Topographic Influences on Solar Irradiance

Solar irradiance at the top of the canopy on surfaces of different aspects depended on the irradiance on the horizontal surface (Chapter 2), the altitude and azimuth of the sun, and the inclination and azimuth of the surface at the point of interest. The daily mean solar altitude was found by taking the mean of the solar altitudes calculated by List's equation (1968) at 5-min intervals throughout the day. The mean solar altitude is the same in the Northern and Southern Hemispheres in analogous months and varies between about 22° on the winter solstice and 42° on the summer solstice (Figure 5.4). At solar noon, the altitude of the sun is about 35° in winter and 81° in summer.

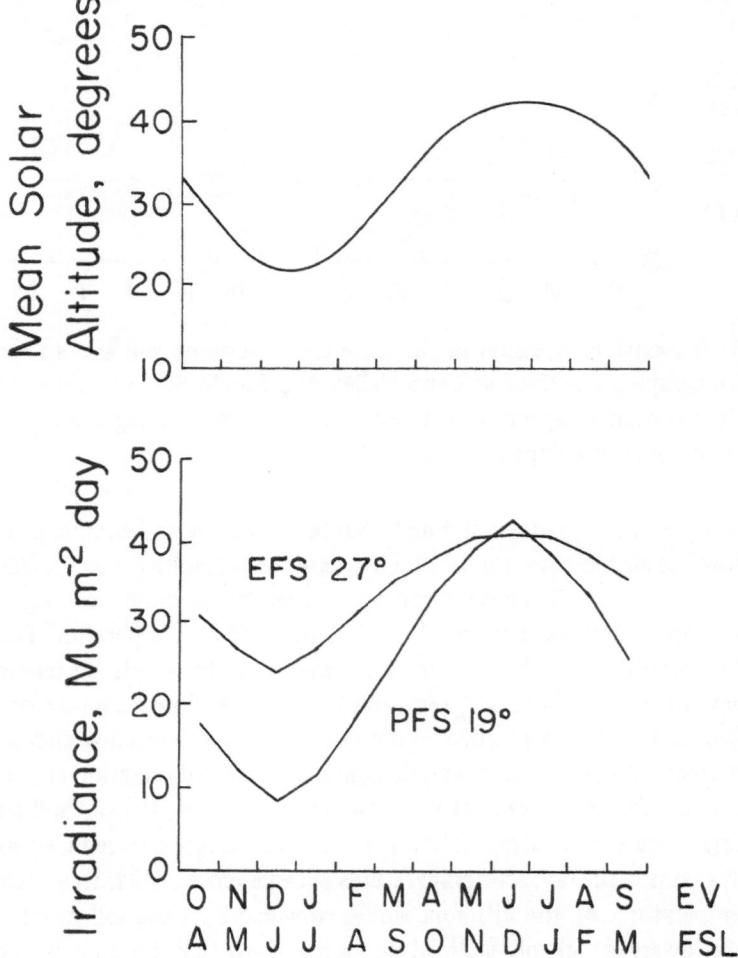

Figure 5.4. Seasonal progression of mean solar altitude at 32.5° latitude (upper graph after List, 1968) and the hypothetical irradiances on the pole-facing and equator-facing slopes at Echo Valley and Fundo Santa Laura (lower graph after Frank and Lee, 1966).

The irradiance on the pole- and equator-facing slopes was estimated from equations describing the geometrical relations between the position of the sun through the days of the year and the surface of the slope (Frank and Lee, 1966), from measurements of direct and diffuse irradiance on the ridgetop, and from measurements on each slope. As calculated from the geometrical relations, the highest total daily solar irradiance and highest noon intensity occurs when the sun's rays are nearly perpendicular to the surface of the slope (Figure 5.4), which for the equator-facing slopes is in April and September at Echo Valley and is November and March at Fundo Santa Laura. Between April and August at Echo Valley and between November and February at Fundo Santa Laura, the total daily irradiances on the pole- and equator-facing slopes are similar. During these months, the lower intensity at noon is compensated for by higher intensities near dawn and dusk because the sun rises and sets in the polar quadrants in summer. During the winter months, November to January at Echo Valley and May to July at Fundo Santa Laura, the irradiance on the pole-facing slope is only about 30% of that on the equator-facing slope. This is because in winter the sun rises and sets in the equatorial quadrants and the intensity is attenuated on the pole-facing slope because of the low solar altitudes. These lower solar intensities in winter lead to lower soil surface temperatures on bare soils, lower soil evaporation, greater soil moisture where the soil is unvegetated, cooler plant temperatures for seedlings, and lower transpiration rates for plants that are growing close to the soil surface on the pole-facing slope.

Solar irradiance on the slopes was calculated from measurements of total and diffuse irradiance on the ridgetops. Direct beam irradiance on the pole- and equator-facing slopes was calculated from Brooks (1959), Gates (1962), and Miller (1969a, b):

$$I_s = (\cos i/\cos zn)I_0 \qquad (5.1)$$

with $\cos i = \cos(zn) \cos(\alpha) + \sin(zn) \sin(\alpha) \cos(\phi - \eta)$ and where I_s is the direct beam irradiance on the slope, I_0 is the direct beam irradiance measured on the horizontal, $\cos i$ is the cosine of the angle of incidence of the direct beam to the slope, $\cos zn$ is the cosine of the zenith angle of the sun, $\cos \alpha$ is the cosine of the angle of the slope, ϕ is the azimuth of the sun, and η is the azimuth of the slope. Total solar irradiance on the slopes was the sum of the calculated direct beam irradiance and the diffuse solar irradiance. The measurements of solar irradiance on the horizontal indicated 5% to 8% more solar irradiance at Fundo Santa Laura, although this difference was within that which could be expected from experimental error. In both countries, the calculated annual solar irradiance on the pole-facing slope was 0.82 of that on the equator-facing slope. However, the irradiance on the pole-facing slope relative to the equator-facing slope varied through the season between 0.57 and 1.01 at Echo Valley and between 0.53 and 1.05 at Fundo Santa Laura with the pole-facing slope receiving less than the equator-facing slope.

The direct measurements on the slopes at Echo Valley indicated that the incoming solar irradiance, measured on a horizontal surface, was slightly less on the pole-facing slope than on the equator-facing slope because of the different horizons of the slopes. The setting of the sun behind the ridge of the pole-facing slope reduced the daily irradiance to 3% lower than that received on the equator-facing slope. At Fundo Santa

Laura, solar irradiance was also reduced by the sun setting behind a ridge on the pole-facing slope.

Thus, the annual total solar irradiance calculated from solar irradiance measured on the horizontal at Echo Valley and Fundo Santa Laura was similar and gave similar values for the slopes. Echo Valley received 6657 MJ m^{-2} yr^{-1} on the equator-facing slope, 6490 on the ridgetop, and 5527 on the pole-facing slope. Fundo Santa Laura received 6992 MJ m^{-2} yr^{-1} on the equator-facing slope, 6866 on the ridgetop, and 5736 on the pole-facing slope. The irradiance at Camp Pendleton and Mount Laguna on the horizontal (Chapter 2) was similar to the irradiance on the pole-facing slopes but was less than the irradiance on the equator-facing slopes and ridgetops at both Echo Valley and Fundo Santa Laura.

5.3.2 Interception and Absorption of Solar Irradiance

The interception of solar irradiance by the vegetation canopy was calculated by the canopy process simulator (*CAPS*, Chapter 11). Miller (1975) developed a radiation model for vegetation canopies, which was used in various vegetation types (Miller, 1975; Stoner et al., 1978a), including chaparral. The model was evaluated on *A. fasciculatum* and *C. greggii* in the chaparral and on *Baccharis linearis* and *C. odorifera* in the matorral (Lawrence, 1975). Roberts and Miller (1977) described a variation of the model to calculate irradiances on leaves in different parts of widely separated shrubs, which was tested in the Chilean shrubs *L. caustica* and *C. odorifera*. The penetration of solar irradiance through the vegetation canopy was calculated by a negative exponential equation of the form

$$I = I_0 \exp(-k_\ell c_\ell A_\ell - k_s c_s A_s) \tag{5.2}$$

where I_0 is the irradiance above the canopy; I is the irradiance below leaf and stem area indices of A_ℓ and A_s, respectively; k_ℓ and k_s are the extinction coefficients for leaves and stems, and c_ℓ and c_s are leaf and stem clustering factors to account for non-random distribution of leaves. Clustered leaves and leaves closely appressed to the stems within a stratum of the canopy cast less shade on the stratum below than do randomly dispersed leaves. For clustered leaves, the leaf cluster factor is less than 1.0, while regularly dispersed leaves have a leaf cluster factor greater than 1.0. The irradiance intercepted by the canopy is one minus the penetrated irradiance.

The clustering of leaves was measured by Lawrence (1975) in closed canopies of *A. fasciculatum* and *C. greggii* at Echo Valley and of *C. odorifera* and *B. linearis* at Fundo Santa Laura. Lawrence (1975) estimated leaf clustering factors of 0.3 to 0.7 with an average of 0.5. Roberts and Miller (1977) estimated leaf clustering factors of 1.0 to 1.3 with an average of 1.1 for widely spaced shrubs of *L. caustica* and *C. odorifera*. The different estimates for *C. odorifera*, which was measured in both studies, may be related to the scale of the clustering pattern (Greig-Smith, 1964) and different scales of measurements. The clustering factors reported by Lawrence (1975) are more appropriate to CAPS, which is a model of processes in a continuous, homogeneous canopy. Clustered leaves result in almost 100% more solar irradiance penetrating to a given level in the canopy than would have occurred with randomly distributed leaves.

Leaf and stem area indices are not easily measured in these canopies, which contained many small leaves, often closely appressed to the stem, and many small stems. The values used in this chapter were calculated from data in Krause and Kummerow (1977a) and Mooney (1977b) and from measurements on the pole- and equator-facing slopes at Echo Valley. Values of 3.5 and 1.0 for the leaf area index on the pole- and equator-facing slopes at Echo Valley were calculated from Krause and Kummerow (1977a). Mooney (1977b) gave leaf and stem area indices of 2.65 and 1.4 for Echo Valley, a leaf area index of 2.0 for Fundo Santa Laura (no stem area index was given), and leaf and stem area indices of 1.3 and 1.8 for a drought deciduous community in southern California. These area indices are full-surface areas, not intercepting areas, and are 2π times larger than the intercepting area for stems and cylindrical leaves such as those of *A. fasciculatum*. In March 1977, after a prolonged drought, Murray (*unpubl. data*) measured leaf biomasses at Echo Valley of 440 g m^{-2} on the pole-facing slope and 310 g m^{-2} on the ridgetop, which convert to leaf area indices of 1.8 and 1.2 on the two exposures, respectively, assuming a leaf specific weight of 252 g m^{-2} and adjusting for inclination (Mooney, 1977b). In May 1979, after a wet winter, the leaf area index at Echo Valley was estimated from shrub structure to be 2.2 and 0.9 on the two exposures, respectively (Stuart, *unpubl. data*). At Fundo Santa Laura, measured leaf area indices were 2.1 on the pole-facing slope and 0.8 on the ridgetop (Stuart, *unpubl. data*).

The extinction coefficient relates the leaf and stem inclinations and solar altitudes to the penetration of direct and diffuse beam solar irradiance in the canopy. Extinction coefficients for the direct solar beam were calculated by using the mean daily solar altitudes through the year and leaf inclinations of 30° and 60°. With leaves of 30°, inclination minimum values were about 0.85, a condition occurring from early March to early October in the Northern Hemisphere and indicating maximum penetration of solar irradiance. However, with leaves of 60° inclination, minimum values were about 0.95, a condition occurring near 21 June. For leaves of 30° inclination, the extinction coefficient increased to 1.35 by 21 December as the mean solar altitude decreased. For leaves with 60° inclination, the extinction coefficient varied continuously throughout the season but was nearly one from mid-April to mid-August and then increased to 2.1 on 21 December. Extinction coefficients were constant when the solar elevation was greater than the leaf angle.

With greater solar altitudes, the path length of the direct beam through the canopy becomes shorter and tends to decrease the interception of the direct beam. When the sun is high in the sky, i.e., during the summer or at midday, interception per unit leaf area is decreased by more steeply inclined leaves. When the solar altitude is relatively low, as during mornings, evenings, and winter, interception of the direct beam is increased by more steeply inclined leaves. The relations between leaf inclination and interception indicate that a canopy with horizontally inclined leaves intercepts the same fraction of the incoming direct beam from early March to early October. Interception in a canopy with steeply inclined leaves varies more throughout the year than interception in a canopy with more horizontally inclined leaves. Roberts and Miller (1977) calculated the annual course of interception of solar irradiance in *C. odorifera* and *L. caustica*. *Colliguaya odorifera* has steeply inclined leaves, while *L. caustica* has horizontally inclined leaves. *Colliguaya odorifera* intercepted more irradiance in winter and less in summer than *L. caustica*. The pattern of interception was consistent with the shallower roots, cessation of physiological activities, and higher leaf drop in summer of *C. odorifera*. The deep-rooted *L. caustica* was physiologically active throughout the year (Avila et al., 1975; Giliberto and Estay, 1978).

On either the pole- or the equator-facing slope, the interception of direct solar irradiance by a canopy with a leaf inclination of 30°C a leaf area index of 1, and leaves randomly arranged was 57% from early March to October in the Northern Hemisphere, increasing to 74% as the solar altitude decreased to its minimum on 21 December (Figure 5.5). Interception in a canopy with a leaf inclination of 60° and a leaf area index of 1 varied between 61% on 21 June and 88% on 21 December. A canopy with a leaf area index of 2 with steeply inclined leaves intercepted almost all the incoming solar irradiance throughout the winter but only 85% through the summer, while a canopy with more horizontally inclined leaves intercepted slightly less. Canopies with leaf area indices of 3 in either the Northern or Southern Hemisphere intercepted over 90% of the incoming irradiance in summer and 100% in winter. Allowing for the compensating effects of shading by stems and clustering of leaves, interception should be almost complete at Echo Valley and Fundo Santa Laura on the pole-facing slopes, nearly complete on the equator-facing slope at Fundo Santa Laura, and only partially complete on the equator-facing slope at Echo Valley and ridgetop at Fundo Santa Laura.

Compared with the direct beam, a greater fraction of the incoming diffuse solar irradiance is from near the zenith. Thus, diffuse irradiance penetrates the canopy more

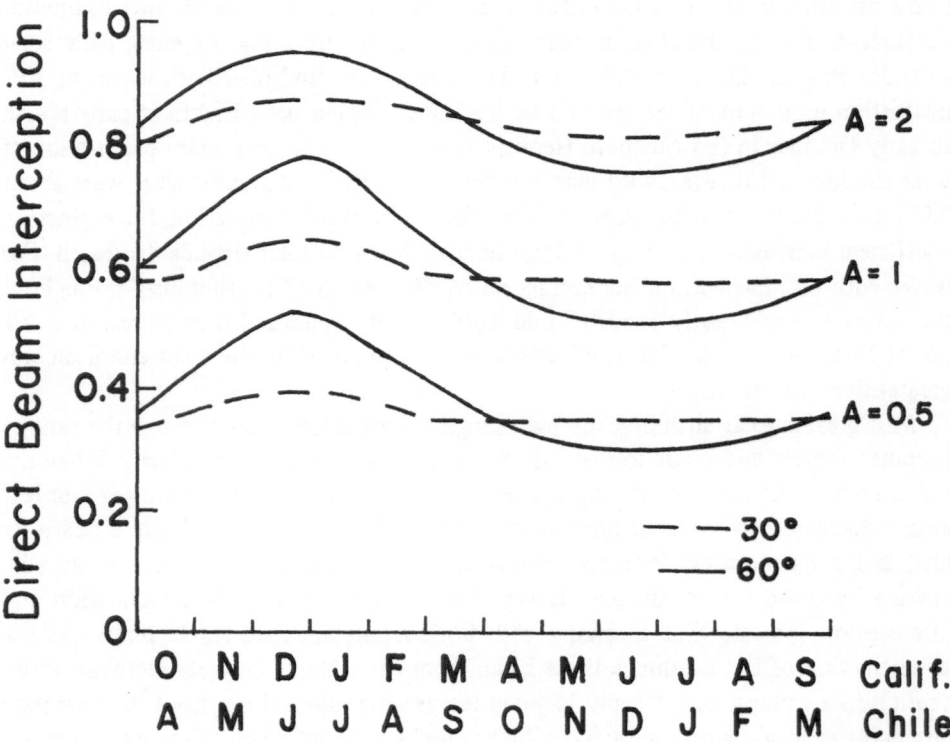

Figure 5.5. Hypothetical seasonal progression of the interception of the direct solar beam in canopies with leaves inclined 30° and 60° with foliage area indices (A) of 0.5, 1.0, and 2.0.

than direct beam irradiance. The greater degree of cloudiness at Fundo Santa Laura during the winter (Chapter 2) should result in greater penetration of diffuse solar irradiance into the canopy. The interception of diffuse irradiance was less sensitive to variations in leaf inclination than was the interception of the direct beam. In canopies with leaves inclined 30° and 60°, both with a shading area index (leaf area index plus the cross-sectional area index of stems) of 1, the interception of diffuse solar irradiance from the sky was 57% and 59%, respectively. Interception with the same leaf inclinations was about 82% with a shading area index of 2 and was about 92% with a shading area index of 3.

Some of the radiation intercepted by the canopy is reflected or transmitted to the sky and to the ground. A fraction of that reaching the ground is reflected back to the canopy. Thus, interception is only the first step in the complex process of radiation absorption by the canopy. The interception and absorption of solar irradiance in the canopies through the year was calculated with the canopy process simulator (CAPS) by using environmental conditions measured in 1973 at Echo Valley, using *C. greggii* with a leaf area index of 2 and a foliage area index (leaf area index plus intercepting stem area index) of 3.84 to represent the vegetation on the pole-facing slope, and using *A. fasciculatum* with leaf and foliage area indices of 0.52 and 1.00 to represent the vegetation on the equator-facing slope. In these simulations, *C. greggii* intercepted 94% of the direct beam in June and 97% in December, while *A. fasciculatum* intercepted between 52% in June and 62% in December. These percentages were expected on the basis of the simplified calculations described above. However, less diffuse irradiance was absorbed in the canopy in the simulation using CAPS. Of the diffuse irradiance entering the canopy from above, *C. greggii* absorbed between 62% in June and 83% in December, while *A. fasciculatum* absorbed between 10% in June and 16% in December. The low apparent absorption of the diffuse beam was largely due to the reflection of direct beam irradiance within the canopy, which was added to the diffuse beam from the sky. Of the total incoming solar irradiance, the simulations with CAPS indicate that *C. greggii* absorbs 85% to 86% and *A. fasciculatum* 46% to 51% throughout the year.

Measurements from May 1978 through April 1979 at Echo Valley indicate that on the pole-facing slope, about 69% of the total incoming solar irradiance (300 to 3000 nm) on the horizontal was absorbed in the canopy (Figure 5.6). This percentage varied little through the season (68% to 70%). About 18% was absorbed by the soil. On the equator-facing slope, about 54% was absorbed in the canopy, varying between about 48% in summer and 60% in winter. About 32% was absorbed by the soil (Miller and Poole 1981). The reduced seasonal variation of absorbed irradiance on the pole-facing slope compared to the variation on the equator-facing slope was similar to the patterns simulated. However, on the pole-facing slope, the measured amount was less than the simulated amount even though the actual foliage area index was greater than that used in the simulations. On the equator-facing slope, the measured amounts were similar to the simulated amounts.

At Fundo Santa Laura, interception and absorption of solar irradiance on the pole- and equator-facing slopes were simulated by using measured environmental conditions for 1973 and data for *L. caustica* and *C. odorifera*. Leaf and foliage area indices used were 2.0 and 3.84 for the pole-facing slope and 0.52 and 1.00 for the ridgetop. In the

P. C. Miller, E. Hajek, D. K. Poole, and S. W. Roberts

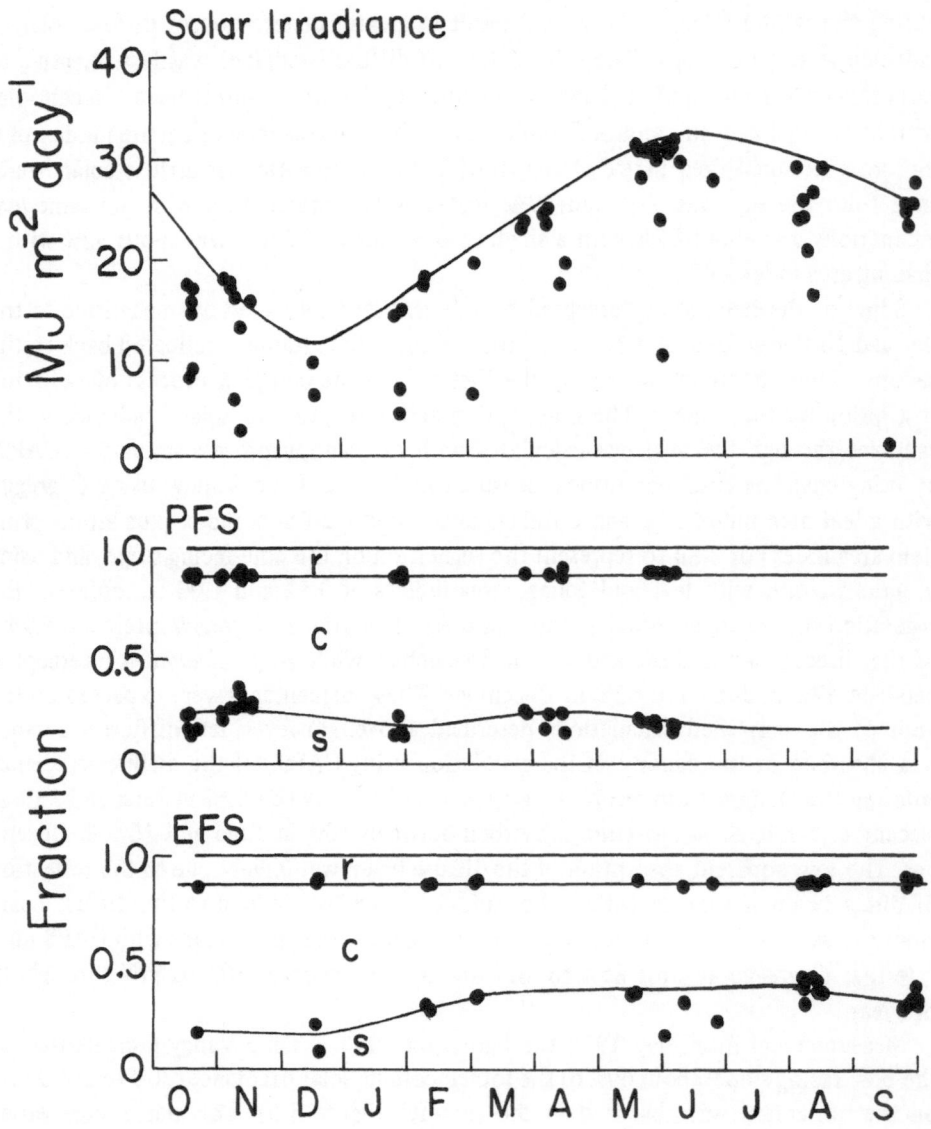

Figure 5.6. Seasonal progression of incoming solar irradiance and the fraction reflected (r), absorbed in the canopy (c), and absorbed in the ground (s) on the pole- (PFS) and equator-facing (EFS) slopes at Echo Valley in 1978-1979 (adapted from Miller and Poole 1981).

simulations, 85% to 87% of the total incoming solar irradiance was absorbed by *L. caustica* on the pole-facing slope. In measurements from October 1979 to February 1980, 75% of incoming solar irradiance on the pole-facing slope was absorbed in the canopy and 12% was absorbed at the soil surface. On the ridgetop, 39% was absorbed in the canopy and 48% in the soil, assuming 56% vegetative cover. An individual shrub on the ridgetop absorbed slightly more solar irradiance than on the pole-facing slope.

The higher absorption in the canopy on the ridgetop was because of greater reflection of solar irradiance at the soil surface.

Of the solar irradiance intercepted and absorbed by the canopy, only half is absorbed by leaves and the rest by stems. The fraction of the intercepted solar irradiance absorbed by leaves depends on leaf properties such as chlorophyll content and leaf thickness. The absorptance of chaparral leaves was not measured for either site. Shaver (1978) measured the leaf absorptances and leaf inclinations of several species of *Arctostaphylos* and concluded that leaf inclination was more related to the distribution of the species than was leaf absorption. Mooney et al. (1974) found that the incidence of pubescence, which is often vegetatively correlated with absorptance, decreased from drier to mesic areas in the chaparral zone.

5.3.3 Albedo

The albedo is the overall reflectance of the canopy and ground surface as seen from above the canopy. The albedo of the pole- and equator-facing slopes at Echo Valley and Fundo Santa Laura was measured hourly through the daylight period with upward- and downward-facing Moll-Gorzynski solarimeters oriented both horizontally and parallel to the slopes. The albedo at Echo Valley and Fundo Santa Laura with horizontal instruments was 12% to 14%, increasing about 2% as the soil dried out and became lighter. At Fundo Santa Laura, the albedo of the soil surface on the ridgetop was about 30%, while the albedo of the soil and litter on the pole-facing slope was about 11%. At Echo Valley, the albedo of the soil on the ridgetop was about 24%, while the albedo of the soil and litter on the pole-facing slope was 21%. As the soil albedo increased, irradiance absorbed in the soil decreased, and that absorbed in the canopy increased. The albedo on the pole-facing slope was 1% to 2% less than that on the equator-facing slope. This difference in albedo was due to the greater foliage area index on the pole-facing slope and the greater trapping of reflected irradiance within the canopy. At Echo Valley, the albedo with instruments oriented parallel to the slope was 22% to 25% on the equator-facing slope.

The albedo is related to the vertical profile of interception and reflection in the canopy. With steeply inclined leaves and high solar altitudes, radiation is intercepted deep within the canopy. The upward reflected fraction of this intercepted radiation may then be absorbed or reintercepted by leaves within the canopy. However, with flat, horizontal leaves and high solar altitude, interception is greater near the top of the canopy, and less reinterception and absorption occurs. Albedo is increased with leaves concentrated near the top of the canopy rather than at the bottom. Simulations with CAPS indicated that changing the distribution of leaves through the canopy may affect the albedo. The simulations indicated that the albedo for *A. fasciculatum* was 0.07 and for *C. greggii* about 0.04 to 0.05. Thus, the CAPS model underestimated the reflected irradiance and the irradiance absorbed in the canopy and overestimated the irradiance absorbed at the soil surface. This error may be due to the inclusion of the inclinations of the canopy surface in the simulations or to an erroneous treatment of reflectances within the canopy.

In summary, incoming solar irradiance is slightly greater at Fundo Santa Laura than at Echo Valley (Figure 5.7). The fraction and amount of solar radiation reflected differs more on the two slopes at each site than between research sites on the analogous slope. The fraction and amount of incoming irradiance absorbed in the canopy varies more on the two slopes at each site than between sites.

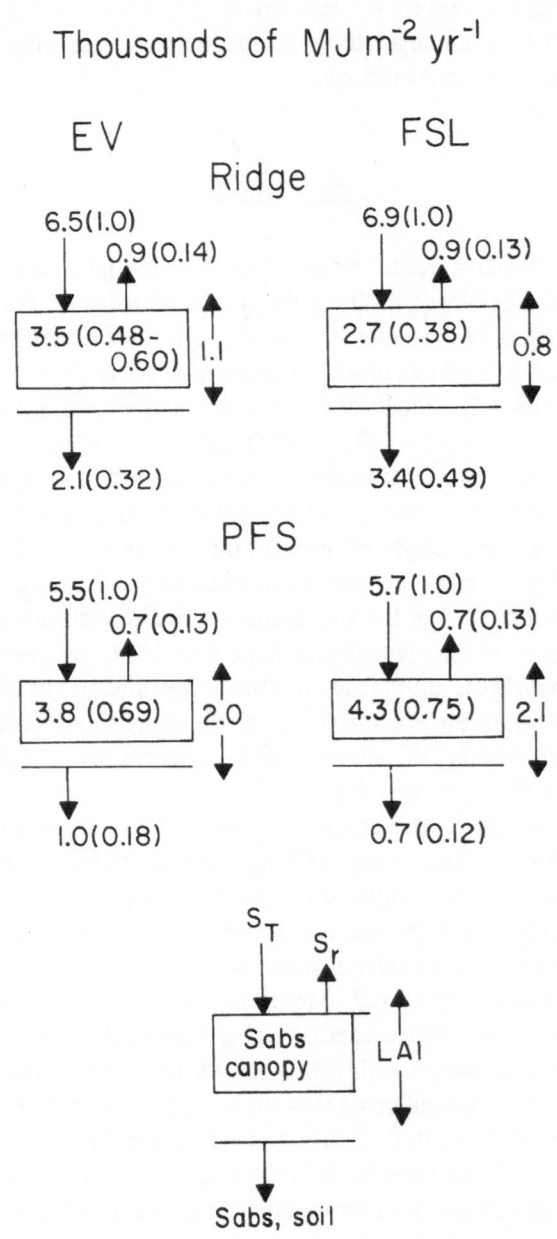

Figure 5.7. Partitioning of incoming solar. Fraction of incoming is given in parentheses.

5.4. The Partitioning of Absorbed Solar Irradiance in the Canopy

5.4.1 Infrared Irradiance

The net infrared irradiance of a leaf or of the total canopy depends on the infrared irradiance from the sky, ground, and plant parts. The infrared irradiance from the soil and plant depends on the temperature of those surfaces. Infrared irradiance from the sky in both 1972-1973 and 1978-1979 was usually 24.1 to 27.2 MJ m^{-2} day^{-1} throughout the year at Echo Valley and Fundo Santa Laura (Figure 5.8). The range in daily irradiance was 23.0 to 33.5 MJ m^{-2} day^{-1}. At Echo Valley, the highest infrared irradiance from the sky was received in August; at Fundo Santa Laura, the highest was received in January.

The loss of infrared irradiance was calculated by assuming that the canopy was radiating according to the air temperature within the canopy, and the soil was radiating according to the soil surface temperature. The contribution of each source was weighted by the canopy cover. The infrared irradiance exchange could not be measured with sufficient accuracy to distinguish differences between Echo Valley and Fundo Santa Laura. Variation between the estimates with radiometers and temperature measurements were generally greater than the variation between countries. It was clear, however, that the net infrared over the canopy was more variable at Echo Valley than at Fundo Santa Laura because of higher air temperatures at Echo Valley. The infrared loss from the canopy was greater and that from the soil was less for the canopies with higher leaf area indices. During Santa Anas, temperatures at Echo Valley were raised by the hot air and were warmer than the soil surface; thus, there was a net gain of infrared at the soil surface under these conditions. On the ridgetops, slightly more than 40% of the absorbed solar irradiance was lost by infrared irradiance, while on the pole-facing slopes 50% to 60% of the absorbed solar was lost by infrared irradiance.

5.4.2 Transpiration

Evaporating or transpiring all the annual precipitation would use 1200 and 1499 MJ m^{-2} yr^{-1} at Echo Valley and Fundo Santa Laura, respectively. The partitioning of transpiration and soil evaporation (Chapter 6) indicated that the vegetation on the slopes at Echo Valley lost more energy by transpiration than did the vegetation on the analogous slope at Fundo Santa Laura. About 11% of the solar irradiance absorbed in the canopy was lost by transpiration on the ridgetop sites at both countries; 12% to 18% was lost on the pole-facing slopes. At Echo Valley, the pole-facing slope with mixed chaparral lost about 0.3 MJ m^{-2} day^{-1} (annual average) through interception losses, 2.0 by transpiration, and 0.7 by evaporation from the soil surface. On the equator-facing slope, the chamise chaparral lost about 0.3 MJ m^{-2} day^{-1} by interception losses, 1.0 by transpiration, and 1.2 by soil evaporation. In the matorral at Fundo Santa Laura, interception losses accounted for about 0.4 MJ m^{-2} day^{-1}, transpiration for 1.3, and soil evaporation for 0.9.

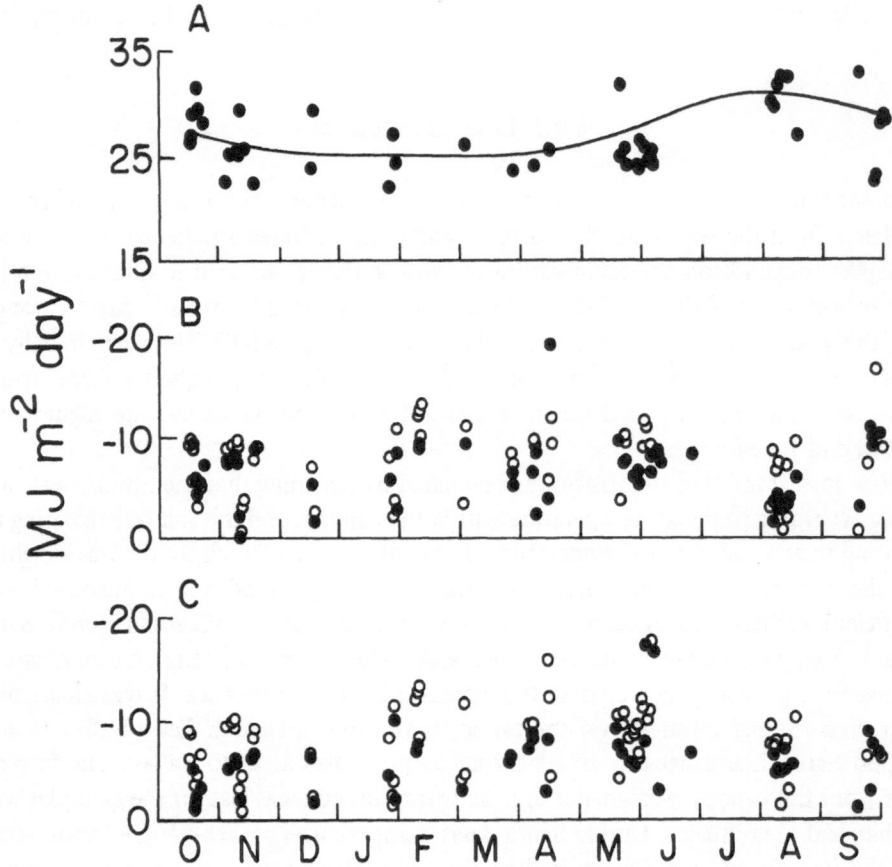

Figure 5.8. Seasonal progression at Echo Valley in 1978 and 1979 of: (A) infrared ir-radiance from the sky measured (●) and the mean (−) for the years 1971-1976 calcu-lated from

$$IR_{sky} = \sigma T_A^{\,4} \,(0.44 + 0.08 \sqrt{e}\,) \,(1 + 0.9 \;FOV)$$

where FOV = 1 − solar mean/solar max; (B) net infrared irradiance over the pole-facing slope; (C) net infrared irradiance over the equator-facing slope. In (B) and (C), net infrared irradiance is calculated from measurements of net total irradiance, total solar irradiance, and reflected solar irradiance (●), i.e., $IR_{net} = R_{net} - S_t + S_r$; and from measured sky infrared irradiance and canopy and ground temperatures (○), i.e., IR_{net} = (cover)$\sigma T_c^{\,4}$ − (1 − cover$\sigma T_g^{\,4}$) (adapted from Miller and Poole 1981).

5.4.3 Convection

Convection accounted for the difference between irradiance absorbed in the canopy and evaporation. Convection accounted for about 47% of the solar irradiance absorbed by the canopies on the ridgetops at the two sites and for about 28% of the solar irradi-ance absorbed by the canopies on the pole-facing slopes. The open canopies allow freer air movement through the canopy and greater convectional loss of heat. The greater convectional loss reduces temperature, which decreases infrared radiation loss. The tendency toward smaller leaves on the ridgetop sites further increased convectional loss. About 80% to 90% of heat lost by the canopy was from stems. In order for this convectional loss to take place, stems need be only slightly above leaf and air temperatures. In simulations using CAPS, leaf temperatures of *C. greggii* were 0.6°C below air temperature in May. Convectional heat transfer was to leaves in both December and June and was to stems in December and from stems in June. Measurements indicated leaf temperatures of *A. glauca* were 0° to 3°C below air temperature in May. These low leaf temperatures relative to air temperatures result from the steep leaf inclination and the glaucous coating in *A. glauca*. Convectional loss of energy from the soil was small on all slopes.

5.5. Conduction

The seasonal progression of heat conduction into the soil was calculated from the change in soil temperatures through the day, the soil volumetric heat capacity, and the soil thermal conductivity and was measured with heat flux plates. Both calculations and measurements indicated generally positive heat flux in the spring and negative heat flux in the fall at Echo Valley and Fundo Santa Laura. However, in the late spring, even though the temperatures in the soil profile increased, the heat content of the profile decreased because of water loss from the upper soil layers. Annually, soil heat conduction was near zero, indicating no general warming or cooling of the soil.

The disparity between annual mean air temperatures and annual mean soil temperatures apparently was due to the radiation exchange at the soil surface. The greater the irradiance at the soil surface, the greater the disparity between the temperatures.

5.6. Energy Budgets for Coastal and Montane Sites

Although all the processes of energy exchange were not measured at Camp Pendleton and Mount Laguna, these processes can be reasonably estimated from data on similar vegetation in other areas, so comparisons can be made with the energy budgets for chaparral and matorral in order to put their energy budgets in a general perspective (Table 5.1 and 5.2). At Camp Pendleton, the albedo is about 20%, based on a vegetation cover of about 50%. The solar irradiance absorbed in the canopy is about 50%, based on an interceptive foliage area index of 1.9 and a leaf-clustering factor of about 0.5. The solar irradiance absorbed by the soil is about 30%. Thus, of an average of 5600 MJ m^{-2} yr^{-1} received, 1100 is reflected, 2800 is absorbed in the canopy, and 1700 is absorbed in the ground. The net infrared loss above the canopy is about 1100 to 1900 MJ m^{-2} day^{-1}. The evaporation of all the annual precipitation consumes 500 MJ m^{-2} yr^{-1}. No subsurface drainage is expected; transpiration and evaporation from the soil surface consume all the annual precipitation. The evapotranspiration was arbitrarily divided, half to evaporation from the soil and half to transpiration. Convectional loss should account for 550 MJ m^{-2} yr^{-1} above the canopy, 510 within the canopy, and 40 at the soil surface.

Table 5.1. Annual partitioning of solar irradiance at study sites

| | Solar irradiance | | | |
	Incoming	Reflected	Absorbed in canopy	Absorbed in ground
California				
Mount Laguna, horizontal	5.9[a]	0.6 (10%)[b]	4.4 (75%)[c]	0.9 (15%)[c]
Echo Valley, pole-facing slope	5.5	0.7 (13%)	4.3 (69%)	1.0 (18%)
Echo Valley, ridgetop	6.5	0.9 (14%)	3.5 (54%)	2.1 (32%)
Camp Pendleton, horizontal	5.6[d]	1.1 (20%)[e]	2.8 (50%)[e]	1.7 (30%)[e]
Chile				
Fundo Santa Laura, pole-facing slope	5.7	0.7 (13%)	4.3 (75%)	0.7 (12%)
Fundo Santa Laura, ridgetop	6.9	0.9 (13%)	2.7 (38%)	3.3 (49%)

Note: Partitioning is based on microclimate measurements on the pole- and equator-facing slopes at Echo Valley and Fundo Santa Laura and on general properties of the vegetation and soil at Camp Pendleton and Mount Laguna. Irradiance values are expressed as thousands of megajoules per square meter per year. Percentage of incoming is given in parentheses. Solar irradiances at Mount Laguna and Camp Pendleton are from the macroclimate measurements (Chapter 2) adjusted upward to be compatible with the measurements using Moll-Gorzynski solarimeters.

[a] 0.90×6490 MJ m^{-2} yr^{-1}

[b] From values for coniferous and deciduous forests.

[c] From sunfleck measurements.

[d] 0.86×6490 MJ m^{-2} yr^{-1}.

[e] From percent cover.

Table 5.2. Partitioning of absorbed solar irradiance at four study sites

	Solar	Net infrared	Evapotranspiration	Convection
Mount Laguna, horizontal				
Above	5.3	1.9 (36%)	1.4 (26%)	2.0 (38%)
Canopy	4.4	1.8 (41%)	1.0 (23%)	1.6 (36%)
Soil	0.9	0.1 (12%)	0.4 (44%)	0.4 (44%)
Echo Valley, pole-facing slope				
Above	4.8	2.6 (54%)	1.1 (23%)	1.1 (23%)
Canopy	3.8	2.0 (53%)	0.7 (18%)	1.1 (29%)
Soil	1.0	0.6 (60%)	0.4 (40%)	0.0 (0%)
Echo Valley, ridgetop				
Above	5.6	3.0 (54%)	0.9 (16%)	1.7 (30%)
Canopy	3.5	1.5 (43%)	0.4 (11%)	1.6 (46%)
Soil	2.1	1.5 (71%)	0.5 (24%)	0.1 (5%)
Camp Pendleton, horizontal				
Above	4.2	1.5 (36%)	0.5 (12%)	2.2 (52%)
Canopy	2.6	0.7 (27%)	0.3 (12%)	1.6 (62%)
Soil	1.6	0.8 (50%)	0.2 (13%)	0.6 (38%)
Fundo Santa Laura, pole-facing slope				
Above	5.0	2.4 (48%)	0.9 (18%)	1.7 (34%)
Canopy	4.3	2.6 (60%)	0.5 (12%)	1.2 (28%)
Soil	0.7	-0.2 (29%)	0.4 (57%)	0.5 (71%)
Fundo Santa Laura, ridgetop				
Above	6.0	3.1 (52%)	0.8 (13%)	2.1 (35%)
Canopy	2.7	1.1 (41%)	0.3 (11%)	1.3 (48%)
Soil	3.3	2.0 (61%)	0.5 (15%)	0.8 (24%)

Note: Units are thousands of MJ m^{-2} yr^{-1}. Percent of absorbed is given in parentheses.

At Mount Laguna, the albedo is about 10%, based on values for forests (Sutton, 1953; List, 1968). Solar irradiance absorbed in the canopy is about 75%, based on canopy closure and distribution of light flecks on the soil surface. Solar irradiance absorbed by the soil is about 15% of incident solar irradiance. Thus, of an average of 5900 MJ m^{-2} yr^{-1}, 600 are reflected back, 4400 are absorbed in the canopy, and 900 are absorbed in the ground. The infrared loss above the canopy is about 1500 to 2300 MJ m^{-2} yr^{-1}. The evaporation of all the annual precipitation (Chapter 2) would consume 1100 MJ m^{-2} yr^{-1}, but subsurface drainage of water from the soil reduces the amount actually evaporated. The water relations of ponderosa and Jeffrey pine, measured in the San Bernadino mountains about 160 km north of Mount Laguna (Poole and Miller, *unpubl. data*), indicate annual transpiration rates equivalent to 960 MJ m^{-2} yr^{-1}. Evaporation from the soil is less than about 420 MJ m^{-2} yr^{-1}. Convectional loss above the canopy is about 2000 MJ m^{-2} yr^{-1}, including about 1500 from the canopy and 500 from the soil surface.

In terms of the partitioning of incoming solar irradiance that is reflected, absorbed in the canopy, and absorbed in the ground, the montane coniferous forest at Mount Laguna and the vegetation on the pole-facing slopes at Echo Valley and Fundo Santa Laura are similar, and the vegetation at Camp Pendleton and that on the ridgetops at Echo Valley and Fundo Santa Laura are similar (Figure 5.9). The partitioning of incoming solar irradiance is largely due to the leaf area index of the canopy, rather than species differences. The leaf area index is constrained largely by the water available for transpiration.

5.7. Effect of Vegetation Form on Microclimate, Plant Temperature, and Energy Exchange Processes

5.7.1 Influence of Shrub Form

With moderate annual precipitation concentrated in winter, there is only enough water to support, via photosynthesis and carbon use, a relatively low leaf area index, and there is a period of the year with high temperature and little evaporative cooling. The shrub characteristics common in this environment include low leaf area index, positioning of the leaves well above the soil surface, and relatively high stem area-to-leaf area ratios. With a small leaf area index and an open canopy, a large fraction of incoming solar irradiance is absorbed at the soil surface. So that this energy may be dissipated when no water is available for evaporation, the energy is lost mostly by increasing irradiance and to a lesser extent by convection. The plant absorbs some of the infrared radiation emitted by the soil surface and, because its absorptance to infrared is greater than its absorptance to solar irradiance, the total amount of absorptance by the plant increases when more solar irradiance is absorbed by the dry soil surface and then is emitted as infrared radiation. With a small leaf area index, an open canopy of scattered shrubs, rather than a continuous canopy, decreases the absorption of the outgoing infrared irradiance and is more effective in dissipating the absorbed radiation by convection. A canopy of tall plants with leaves positioned well above the soil surface can maintain relatively cool plant temperatures because the temperatures of the taller plants are coupled to the temperature of the air, rather than to that of the soil surface. The difference between air and soil surface temperatures averaged 5°C annually with small leaf area indices and was considerably greater at midday in summer. A continuous canopy with a small leaf area index intercepts more solar irradiance and more infrared irradiance from the soil surface and must reach higher temperatures to dissipate this radiation by convection into the hot air near the soil surface. In simulations

---

Figure 5.9. Partitioning of solar absorbed in the canopy and soil into infrared irradiance, evaporation, and convection. Units are thousands of megajoules per square meter per year. Fraction of solar absorbed is given in parentheses.

Thousands of MJ m^{-2} yr^{-1}

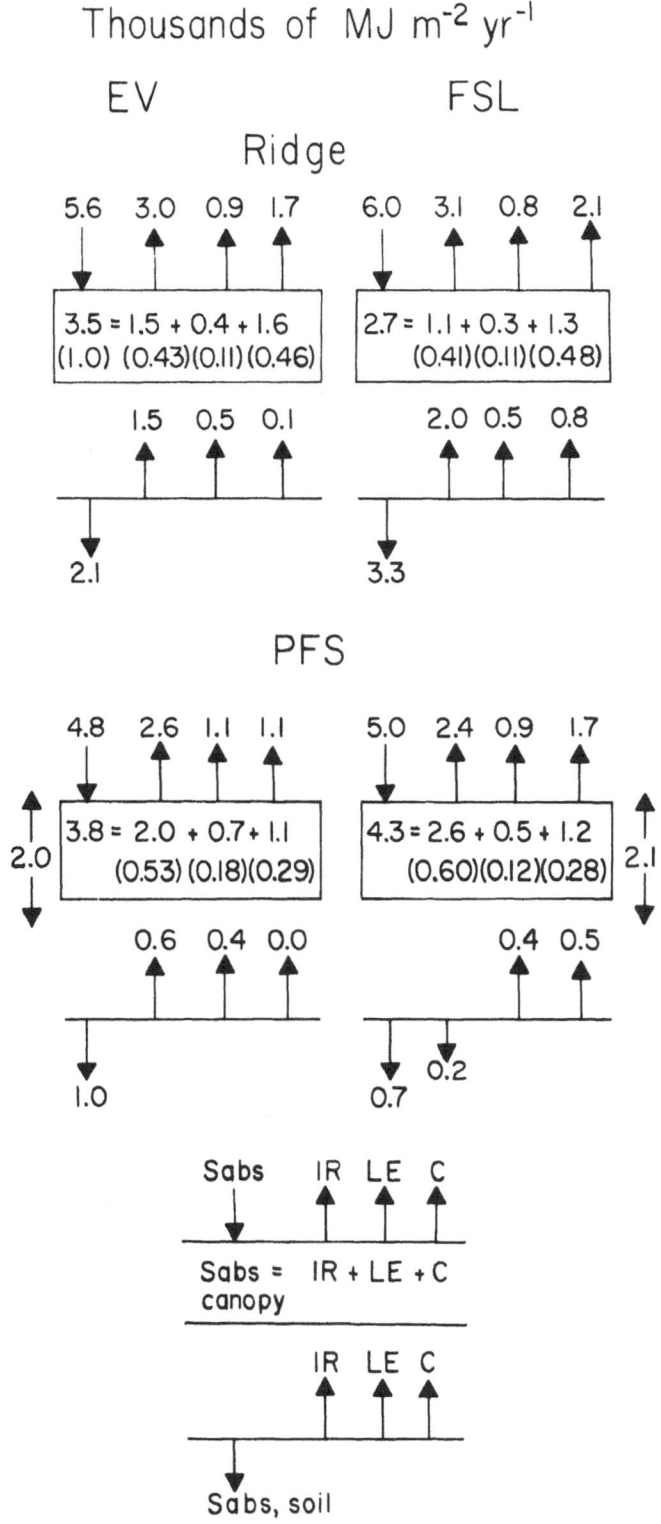

EV

FSL

Ridge

5.6 3.0 0.9 1.7   6.0 3.1 0.8 2.1

3.5 = 1.5 + 0.4 + 1.6   2.7 = 1.1 + 0.3 + 1.3
(1.0) (0.43)(0.11)(0.46)   (0.41)(0.11)(0.48)

1.5 0.5 0.1   2.0 0.5 0.8

2.1   3.3

PFS

4.8 2.6 1.1 1.1   5.0 2.4 0.9 1.7

3.8 = 2.0 + 0.7 + 1.1   4.3 = 2.6 + 0.5 + 1.2
2.0   (0.53) (0.18)(0.29)   (0.60)(0.12)(0.28) 2.1

0.6 0.4 0.0   0.4 0.5

1.0   0.7   0.2

Sabs IR LE C

Sabs = IR + LE + C
canopy

IR LE C

Sabs, soil

with *C. greggii* and *L. caustica*, as the canopy became shorter, leaf and soil surface temperatures increased 0.2° to 0.8°C at Echo Valley and decreased 0.1° to 2.1°C at Fundo Santa Laura. With a constant foliage area index, changing the vertical distribution of the foliage area changed plant temperature and transpiration. Transpiration was less with leaves concentrated at the top of the canopy than with leaves concentrated at the bottom or with leaves distributed in the canopy as they actually occur. Changing the leaf area from that which naturally occurred resulted in higher leaf temperature by 0.2° to 1.0°C and higher soil surface temperature by 0.2° to 1.9°C.

Stems absorb solar irradiance and convect this energy to the air. Although interception and dissipation of irradiance by stems was indicated in simulations to be beneficial during the late spring and summer when canopy temperatures were above the optimum for photosynthesis, during the spring interception of solar irradiance by stems decreased photosynthesis by 40%, transpiration by 15%, and the photosynthesis-to-transpiration ratio by 40%.

Presumably, the leaf area index adjusts to the irradiance and water available. An irradiance on the pole-facing slope lower than that on the equator-facing slope causes lower evaporation from the soil surface, and water becomes more available for plant growth. As the leaf area index increases, the absorption of solar irradiance in the canopy increases and transpiration increases. The leaf area on each slope increases until all available moisture is lost through transpiration or soil evaporation. On the pole-facing slope, the leaf area required to remove the available water is greater than on the equator-facing slope because of reduced irradiance on the pole-facing slope. Thus, the pole-facing slope becomes more heavily vegetated and the length of the drought becomes similar on the two slopes (Chapter 6).

5.7.2 Influence of Leaf Size and Orientation

Leaf width affects the potential for convectional and evaporational exchange of energy by affecting the resistance of the leaf air boundary layer to heat and water exchange. The interrelations between leaf width and transpirational loss are complex. As leaf width increases, leaf temperature may increase because of the increased resistance to heat and water vapor exchange. The increasing leaf temperature increases the vapor density gradient from leaf to air, tending to increase transpiration. However, increasing leaf width increases the resistance to the diffusion of water vapor from leaf to air, tending to reduce transpiration. Depending on environmental conditions, transpiration may decrease or increase as leaf width increases. The environmental conditions in the mediterranean regions are such that transpirational loss decreases as leaf width increases. With increasing solar irradiance, leaf temperatures and transpiration rates will tend to increase with leaf width (Gates, 1962, 1965, 1968). This increase may influence photosynthesis directly, depending on temperatures and the temperature response curve, and indirectly via the effect on leaf water balance and stomatal closure. Mooney et al. (1974) pointed out that wide leaves might be beneficial in winter and narrow leaves in summer, depending on prevailing temperatures and on temperature optima for photosynthesis. In simulations, Miller and Mooney (1976) showed increased production and water use efficiency with wider leaves in winter and with narrower leaves in summer. The overall effect of wider leaves was to decrease production and water-use efficiency; this effect was more pronounced at Echo Valley than at Fundo Santa Laura. Summer active plants are benefited by narrow leaves, which increase photosynthesis and water-use efficiency. The researchers' simulations indicated that more plants with narrow leaves should occur at Echo Valley than at Fundo Santa Laura. Measurements of leaf size supported these predictions (Parsons, 1973; Mooney et al., 1977). Production was more sensitive to leaf size at Echo Valley than at Fundo Santa Laura.

The effect of leaf width on convectional exchange is such that the greater the leaf width, the less the convectional exchange potential and the greater the difference between leaf and air temperatures. The narrower leaves at Echo Valley compensated for the higher annual mean temperature, so that the measured leaf temperatures at Echo Valley and Fundo Santa Laura were more similar than were the air temperatures at the two research sites. The predominance of broadleaved species on the pole-facing slope in contrast to the narrow-leaved species found on the equator-facing slopes and on the ridgetops may result partly from the need for more effective convectional exchange on equator-facing slopes because of higher surface temperatures of bare soil after a fire.

5.8. Conclusions

The air temperatures on different slopes at Echo Valley were similar to those at Fundo Santa Laura in winter but were higher in summer. At both sites, mean soil temperatures, especially in stands with low foliage area, were higher than mean air temperatures. Solar irradiance was 0% to 6% higher at Fundo Santa Laura than at Echo Valley. Solar irradiance on the pole-facing slope at Echo Valley was similar to solar irradiance on the horizontal surface at Mount Laguna and Camp Pendleton and to solar irradiance on the pole-facing slope at Fundo Santa Laura. Solar irradiance on the equator-facing slopes at Echo Valley and Fundo Santa Laura was greater than values measured at the other sites. The reflectance of the canopy and soil was similar on all slopes at Echo Valley and Fundo Santa Laura. The amount of solar irradiance absorbed in the canopy was more similar on equivalent slopes at the two sites than on the different slopes at each site. The loss of absorbed solar energy by infrared irradiance, evapotranspiration, and convection appears similar in the montane coniferous forest at Mount Laguna and the pole-facing slopes at Echo Valley and Fundo Santa Laura and similar on the equator-facing slopes at Echo Valley, Fundo Santa Laura, and Camp Pendleton.

6. Water Utilization

DENNIS K. POOLE, STEPHEN W. ROBERTS, and PHILIP C. MILLER

This chapter describes community water budgets, seasonal and annual patterns of plant water use, and the canopy and physiological controls on water use.

6.1. Introduction

Water is a major limiting resource in mediterranean regions, which are classified semi-arid. Research on water relations in mediterranean type ecosystems has focused on watershed relations and plant water relations (Rowe and Coleman, 1951; Rowe and Reiman, 1961; Leyton et al., 1967; Poole and Miller, 1975; Giliberto and Estay, 1978; Poole and Miller, 1978; Miller and Poole, 1979; Ng and Miller, 1980). Early efforts were oriented towards managing watersheds to increase water yield; later efforts were oriented toward understanding the water use by different species. Utilization of water and solar irradiance by vegetation is constrained by the processes of energy exchange in the vegetation canopy-soil system that affect transpiration and evaporation from the soil. The structure of the canopy affects the partitioning of precipitation into interception losses, soil evaporation, and transpiration and can affect the water available to the plant. The flows of water are interconnected with flows of energy (Chapter 5) because evaporation is an energy exchange process.

The water studies to date have given a general idea of the partitioning of precipitation into losses by interception, transpiration, soil evaporation, surface runoff, and subsurface drainage. The flow of water from soil to leaf is viewed as a flow of liquid water that is controlled by the water potential difference, the soil and root resistance, and the water absorbing root system. The flow of water from leaf to air is controlled by the water vapor difference, leaf conductance, and leaf air boundary layer resistance. These controls involve relations between leaf conductance and leaf water potential, leaf water potential and leaf water content, soil-root resistance and total root biomass, and absorbing root densities.

The quantities of water flowing through these paths in the system and the environmental and plant factors controlling these flows are covered in this chapter. The hypotheses are that the chaparral and matorral are similar in (1) the partitioning of precipitation into interception losses, stemflow, throughfall, runoff, evaporation from the soil, and transpiration; (2) the control of water use by the canopy and by the physiological relationships of the species; and (3) the seasonal patterns and total annual water uses.

6.2. Hydrologic Balance

6.2.1 Interception, Stem Flow, and Throughfall

The partitioning of precipitation into interception losses, stem flow, and throughfall within Californian chaparral was measured at the San Dimas Experiment Station north of San Diego. These earlier studies, conducted in mixed chaparral, showed interception losses of 9% to 20% of precipitation, stem flow of 8% to 30%, and throughfall of 62% to 81% (Hamilton and Rowe, 1949; Rowe and Colman, 1951; Hill, 1963; Corbett and Crouse, 1968).

The amounts of water lost through interception and subsequent evaporation depends on the number of storms, the intensity of each storm, and the gross morphology of the species present. Precipitation at Echo Valley and Fundo Santa Laura generally occurs in 10 to 20 storms per year. During the period of interception and throughfall measurements at these sites, 8 storms occurred at Echo Valley and 17 at Fundo Santa Laura. In two of the eight storms at Echo Valley, the precipitation was in the form of snow. One storm occurred during the summer. In the remaining five storms, the fraction of rain lost by interception had no clear relation to the total amount of rain received in the storm. The interception losses in the chaparral stands averaged 21% ± 5%. The interception losses increased in the snow storms to an average of 41% ± 8%. During the summer storm, in which 62 mm of rain fell, the interception loss was 76% to 85%. At Fundo Santa Laura, the percentage of the precipitation intercepted decreased as the storm size increased, and the percentage differed among species (Table 6.1). Data from 17 storms showed that with storms of less than 5 mm of rain, 60% to 80% of the precipitation was intercepted and evaporated by all species, except *Quillaja saponaria*, for which only 30% to 40% was evaporated. With storms of more than 25 mm precipitation, the percentage intercepted by each species was 28% ± 8%. The two shrubs with ascending branches and small leaves, *Adenostoma fasciculatum* and *Satureja gilliesii*, intercepted the smallest fraction; the two shrubs with steeply inclined leaves, *Arctostaphylos glauca* and *Colliguaya odorifera*, intercepted nearly the same fraction.

The interception of precipitation and irradiance by the overstory shrub canopy can influence community structure by affecting the distribution of understory species (Stoner et al., 1978a). In some conditions, the overstory canopy can decrease photosynthesis because of the direct effect of shading but may increase photosynthesis because of enhanced water balance of the shaded species. Production of grass is lower under shrubs than in the open with both low (<500 mm yr^{-1}) and high (>700 mm yr^{-1}) precipitation regimes due to the loss of water by interception with low precipitation

Table 6.1. Partitioning of precipitation in mixed and chamise chaparral and matorral[a]

| | Vegetation | | | Species | | | | | | | | |
| | Chaparral | | Matorral | Chaparral | | | | | | Matorral | | |
	Mixed (pole-facing slope)	Chamise (equator-facing slope)	Matorral	Adenostoma fasciculatum	Ceanothus greggii	Arctostaphylos glauca	Rhus ovata	Satureja gilliesii	Colliguaya odorifera	Quillaja saponaria	Lithraea caustica	Cryptocarya alba
Precipitation (mm yr^{-1})	475	475	590	-	-	-	-	-	-	-	-	-
Interception	0.19	0.12	0.18	0.21	0.4	0.31	-	0.00	0.28	0.18	0.18	0.45
Stemflow	0.03	0.08	0.12	0.15	0.06	0.03	0.04	0.55	0.22	0.12	0.12	0.03
Throughfall	0.78	0.80	0.70	0.64	0.53	0.66	-	0.45	0.50	0.70	0.70	0.53
Effective precipitation	0.81	0.88	0.82	-	-	-	-	-	-	-	-	-
Transpiration	0.54-0.73	0.28-0.38	0.35	0.40	0.75	0.75	0.40	0.42	0.40	-	0.45	-
Soil evaporation	0.20-0.26	0.33-0.44	0.37	0.40	0.25	0.25	0.25	0.40	0.40	-	0.25	-
Drainage	0	0.11-0.33	0.10	0.20	0.00	0.00	0.35	0.20	0.20	-	0.30	-
Surface runoff	0	0.01	-	-	-	-	-	-	-	-	-	-

[a]Note: Values are fractions of precipitation, > 25 mm at Echo Valley and Fundo Santa Laura.

and the effects of shading with high precipitation. With moderate amounts of precipitation the production of grass under the shrubs is greater than the production in the open because the suppressive effects of shading on photosynthesis are overriden as a result of the beneficial effect of shading on the moisture regime by a reduction in soil evaporation and lowering of plant temperatures. Thus, herbaceous growth is low under shrubs with normal or subnormal precipitation (Mooney, 1977b) but increases with higher precipitation.

The shrub morphology can direct stem flow toward the base of the shrub, thereby increasing the amount of soil moisture immediately available to the shrub (Specht, 1957a). At Echo Valley and Fundo Santa Laura, stem flow differed by species. At Echo Valley, *A. fasciculatum* had the highest stem flow of the Californian species, *A. glauca* the lowest. The results are consistent with the observed greater lichen cover on stems of *A. fasciculatum* than on stems of other species. At Fundo Santa Laura, the percentage of the total precipitation that appeared as stem flow increased with storm size. *Satureja gilliesii*, which has a form similar to that of *A. fasciculatum*, had the highest stem flow of the Chilean species.

The percentage of the total precipitation passing through the canopy, either unintercepted or as drip from the foliage, averaged 78% and 80% on the pole- and equator-facing slopes at Echo Valley, respectively. The trend in the differences in throughfall on the slopes was consistent with the differences in the leaf area indices on the pole- and equator-facing slopes but was not as much as expected. There was 2 to 3.5 times more leaf area on the pole-facing slope but only a 2% drop in interception. Throughfall was not measured directly at Fundo Santa Laura but could be 70% to 80% based on the species composition.

6.2.2 Drainage and Surface Runoff

Soil water potential and soil hydraulic conductivity data were used to calculate subsurface losses. The water potential curves for soils from Echo Valley and Fundo Santa Laura were measured with a pressure membrane apparatus and a soil psychrometer (Ng, 1974; Ng and Miller, 1980; Figure 6.1), by methods described by Brown (1970) and Wiebe et al. (1971). The water potentials of soils from Echo Valley decreased sharply with moisture contents less than 0.09 g cm^{-3}. The water potential curves for soils at Fundo Santa Laura differed depending on the parent rock. Water potentials of soils from the pole- and equator-facing slopes decreased sharply at about 0.07 g cm^{-3}, but those from the ridgetop decreased at about 0.18 g cm^{-3} because of the higher silt and clay content in the ridgetop soils.

Hydraulic conductivities were measured on soils from Echo Valley in order to estimate water transport toward a point of removal, i.e., plant root or site of evaporation (Ng 1974). The methods of Richards (1954) and Klute (1965) were used to measure saturated conductivity and the methods of Millington and Quirk (1959, 1960, 1961) to measure unsaturated conductivity. Hydraulic conductivity decreased markedly as soil moisture decreased below saturation and was near 0 cm day^{-1} at moisture levels below 0.15 g cm^{-3}.

Drainage, estimated from water content, hydraulic conductivity, and water potential relations, was 0 mm yr^{-1} on the pole-facing slope and 50-150 mm yr^{-1} on the equator-facing slope at Echo Valley (Ng and Miller, 1980). The greater drainage on the equa-

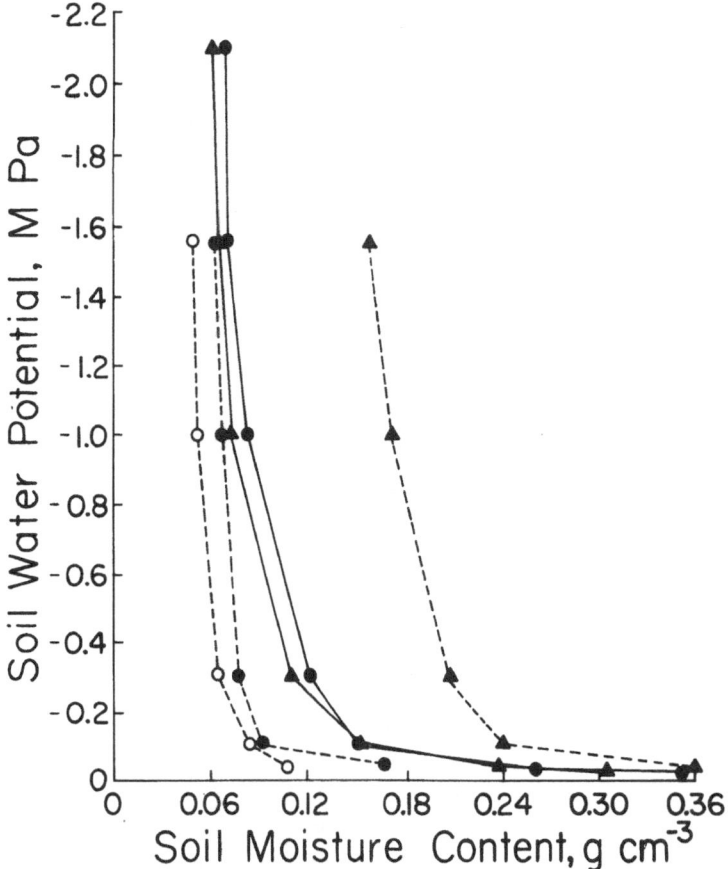

Figure 6.1. Soil water potential-water content relations for the soils at Echo Valley [pole-facing slope (−●−), ridgetop (−▲−)] and Fundo Santa Laura [pole-facing slope (--○--), equator-facing slope (--●--), ridgetop (--△--)] (after Ng, 1974; Ng and Miller, 1980; Poole, *unpubl. data*).

tor-facing slope was consistent with the lower leaf area and lower transpiration rates on that slope and with the lower levels of soil nitrate on the equator-facing slope, which may have resulted from higher leaching rates. General observations suggest that drainage losses occur consistently in the chaparral of southern California only if annual precipitation is above about 550 mm yr^{-1}. Streamflow occurred only when annual precipitation at Echo Valley was above 650 mm during three of the nine measured years. Annual precipitation during the other 6 years was well below 550 mm yr^{-1}. Surface runoff was < 1% of the precipitation on the pole-facing slope and 1% to 4% on the equator-facing slope for the average storm at Echo Valley in 1977. No runoff measurements were made at Fundo Santa Laura.

6.2.3 Evapotranspiration

Evapotranspiration, including both evaporation from the soil surface and transpiration from the vegetation, can be estimated from equations for water yield given by Shachori and Michaeli (1965) and Patric (1974) for different precipitation totals. Equations derived from those of Shachori and Michaeli, who summarized 157 studies in diverse regions, show that evapotranspiration uses all the precipitation up to 281 mm yr^{-1} in grass covered and bare areas and uses up to 398 mm yr^{-1} in woodlands (scrub and forests). Above these amounts evapotranspiration uses a fraction of the precipitation. Based on their equations, evapotranspiration (ET, in mm yr^{-1}) can be related to annual precipitation (P) as:

$$\text{grass and bare areas: if } P > 281; ET = 0.08P + 259$$
$$\text{if } P < 281; ET = P$$
$$\text{woodlands: if } P > 398; ET = 0.19P + 320$$
$$\text{if } P < 398; ET = P$$

Shachori and Michaeli concluded that woodlands could not occur with annual precipitation less than 400 mm. By means of the equations of Patric, evapotranspiration was related to annual precipitation as:

$$\text{bare areas in lysimeters: if } P > 162; \quad ET = 0.11P + 144$$
$$\text{if } P < 162; \quad ET = P$$
$$\text{grass in lysimeters: if } P > 343; \quad ET = 0.44P + 192$$
$$\text{if } P < 343; \quad ET = P$$
$$\text{wildlands (chaparral): if } P > 95.3; \quad ET = 476 \log P - 942$$
$$\text{if } P < 95.3; \quad ET = 0 \text{ (undetermined)}$$

Thus, evaporation from bare ground was calculated as about 160 mm yr^{-1} in the lysimeter data of Patric or about 280 mm yr^{-1} in the regressions of Shachori and Michaeli. Evapotranspiration from vegetated areas should be 300 to 450 mm yr^{-1}, depending on the annual precipitation and the set of equations used. Miller and Poole (1979) noted that the water loss from mixed chaparral fit the relations of Shachori and Michaeli for woodlands, while the water loss from chamise chaparral fit the relations for grass and bare areas. Miller and Poole also noted that converting chamise chaparral to grasslands might not increase water yield.

Calculations with the Mediterranean Ecosystem Simulator (Chapter 11) indicated that in both countries as foliage area increased, evaporation from the soil decreased from 200 to 250 mm yr^{-1} with bare soil to about 120 mm yr^{-1} with foliage area indices (leaf area index + stem area index) above four. Although the soil surface at Echo Valley received less infrared irradiance than did soil at Fundo Santa Laura, higher wind velocities at Fundo Santa Laura reduced surface boundary layer resistance and compensated for reduced infrared irradiance, resulting in similar evaporation rates at Echo Valley and Fundo Santa Laura. Depending on the species, transpiration increased up to 60 to 330 mm yr^{-1} as the foliage area index increased (Section on *CAPS* and *MEDECS* Simulation, below).

6.2.4 Annual Water Budgets

The available data and the simulations of plant water relations (next section) were used to estimate the annual water budgets at Echo Valley and Fundo Santa Laura (Table 6.2). With an annual precipitation of 475 mm, mixed chaparral on the pole-

Table 6.2. Annual precipitation and partitioning for 3 vegetation types under study[a]

Vegetation type	Annual precipitation	Interception	Soil evaporation	Transpiration	Subsurface drainage	Surface runoff
Mixed chaparral	475	90 (0.19)	100 (0.21)	285 (0.60)	0 (0)	0(0)
Chamise chaparral	475	57 (0.12)	184 (0.39)	154 (0.32)	15 (0.16)	5(0.01)
Matorral	590	106 (0.18)	218 (0.37)	206 (0.35)	60 (0.10)	0(0)

[a]Note: Units are mm yr^{-1}. Fraction of annual precipitation is given in parentheses.

facing slope at Echo Valley loses about 90 mm by interception, 285 mm by transpiration, and 100 mm by soil evaporation. Surface runoff and drainage were negligible. In chamise chaparral on the equator-facing slope, about 57 mm was lost via interception, 154 mm by transpiration, and about 184 mm by soil evaporation. Surface runoff accounted for about 5 mm and drainage for about 74 mm. With lower or higher precipitation drainage was decreased or increased. At Fundo Santa Laura, with an annual precipitation of about 590 mm, interception loss accounts for about 106 mm, transpiration for about 206 mm yr^{-1}, and soil evaporation for about 218 mm yr^{-1}. Runoff should be negligible, and drainage should be about 59 mm yr^{-1}. The amount of drainage predicted is consistent with observations of almost continual stream flow throughout the year at Fundo Santa Laura.

In terms of water use and water-use efficiency, matorral is between mixed and chamise chaparral, which was consistent with the moderate foliage index of matorral. The water capture efficiencies expressed as transpiration divided by precipitation were higher in mixed chaparral (0.60) and similar in chamise chaparral (0.32) and matorral (0.35). On the basis of fraction of precipitation, matorral at Fundo Santa Laura lies between mixed and chamise chaparral at Echo Valley in every process of water loss.

6.3. Control of Species Water Relations

Plant water relations of different species were compared on the basis of the relationships between leaf conductance and leaf relative water content or xylem pressure potential and between leaf relative water content and leaf water potential. The relationships were encoded in the canopy process simulator because they were viewed as the major factors controlling plant water relations (Chapter 11). The leaf water potential was divided into turgor and osmotic components.

6.3.1 Leaf Conductance

Maximum leaf conductances were between 0.15 and 1.3 cm s^{-1} for all species measured in both countries (Poole and Miller, 1975, 1978; Miller and Poole, 1979; Table 6.3). In both countries, the species with the largest leaves and most extensive root system, *Rhus ovata* and *Lithraea caustica*, had the lowest leaf conductances. The shallow-rooted species, *A. glauca* and *C. odorifera*, had the highest conductances of the sclero-

Table 6.3. Summary of water use characteristics of different chaparral and matorral shrubs

Echo Valley

Life history type	Drought semideciduous			Evergreen				
	Artemisia californica	Salvia apiana	Eriogonum fasciculatum	Arctostaphylos glauca	Ceanothus greggii	Adenostoma fasciculatum	Quercus dumosa	Rhus ovata
Maximum leaf conductance (cm s^{-1})	1.00	1.00	1.30	0.45	0.50	0.40	0.30	0.25
Minimum xylem potential during year (bars)	<−65	−40	<−65	<−65	<−65	−60	−50	−25
Xylem potential at zero conductance (bars)	<−65	ND	ND	−60	−60	−50	ND	−20
Osmotic potential at turgidity (bars)	ND	ND	ND	−30	−28	ND	ND	−30
Annual transpiration (mm yr^{-1} m^{-2} leaf)	ND	ND	ND	280-290	145-195	150-190	ND	127
Rooting depth	←———— SHALLOW			————→	DEEP ←————			————→

Fundo Santa Laura

Life history type	Drought semideciduous		Evergreen			
	Satureja gilliesii	Trevoa trinervis	Colliguaya odorifera	Cryptocarya alba	Quillaja saponaria	Lithraea caustica
Maximum leaf conductance (cm s^{-1})	0.5	0.4	0.3	0.25	0.30	0.15
Minimum xylem potential during year (bars)	−45	−60	−60	−60	−30	−35
Xylem potential at minimum conductance (bars)	ND	ND	−45	−35	−35	−30
Osmotic potential at turgidity (bars)	ND	ND	−30	−35	−25	−25
Annual transpiration (mm yr^{-1} m^{-2} leaf)	60	60	130	ND	ND	200
Rooting depth	←———— SHALLOW		————→	DEEP ←————		————→

phyllous shrubs measured. Malacophyllous shrubs and herbs had higher maximum conductances than did the sclerophyllous shrubs, in agreement with measurements by Harrison et al. (1971). In general, the maximum leaf conductances of the chaparral shrubs were higher than those of the matorral shrubs.

6.3.2 Leaf Conductance-Leaf Water Relations

The control of leaf conductance is complex, and several hypotheses concerning it have been proposed. Turner (1974), Running et al. (1975), Miller et al. (1976a), and Miller and Poole (1979) have emphasized the relation of leaf conductances to leaf water potential or leaf water content. Schulze et al. (1972a, b, 1975) emphasized the control of leaf conductance by atmospheric vapor pressure. Penning de Vries (1972a) related leaf conductance to carbon dioxide concentration inside the leaf. All workers have recognized the relation of solar irradiance to leaf conductance. The physiological mechanisms and interactions between these controlling factors are not clear (Meidner and Mansfield, 1968). At Echo Valley, leaf conductances of *A. fasciculatum* changed in a complex manner with the suggested controls and seemed to relate simply to the time since sunrise. Leaf conductances of *A. glauca, Ceanothus greggii*, and *Quercus dumosa* related more clearly to leaf water potential or atmospheric vapor pressure. In all species, the maximum leaf conductance attained during the day decreased with decreasing soil moisture and soil water potential as the season progressed.

Leaf conductances generally decreased with decreasing leaf water content and decreasing xylem pressure potential (Figures 6.2 and 6.3). Leaf conductances were at minimum values at relative water contents of 75% to 85% in the sclerophyllous, evergreen shrubs. Malacophyllous, semideciduous shrubs and herbs continued to lose water with low leaf relative water contents and low xylem pressure potentials. Of the sclerophyllous, evergreen shrubs measured, chaparral shrubs reached lower xylem pressure potentials before leaf conductances reached minimum values than did matorral shrubs. The minimum conductances were lower in the chaparral shrubs, but this difference between chaparral and matorral shrubs may be due to differences in technique.

Species of sclerophyllous, evergreen shrubs differed in the maximum conductances and the responses of conductance to decreasing water content or decreasing pressure potential. In *R. ovata* and *L. caustica*, a relative water content of 80% occurred with xylem pressure potentials of −2.0 MPa, while in *A. glauca*, it occurred at potentials of −5.0 to −6.0 MPa. Leaf conductances of *A. glauca* and *C. greggii* were similar in their responses to xylem pressure potential (Figure 6.3). However, *A. glauca* had a higher maximum leaf conductance than *C. greggii*, so that conductances of *A. glauca* decreased more rapidly with decreasing xylem pressure potential than those of *C. greggii.* Individuals within a species and morphologically similar species within a genus showed similar responses of leaf conductance to xylem pressure potential.

In *Heteromeles arbutifolia*, the response of leaf conductance to xylem pressure potential was similar at Camp Pendleton and at Echo Valley. At both sites, *H. arbutifolia* had a maximum leaf conductance of about 0.4 cm s^{-1}. Conductances decreased almost linearly as the xylem pressure potential decreased to −2.5 MPa. Below −2.5 MPa, leaf conductance decreased more rapidly until minimum values were reached at −3.5 MPa.

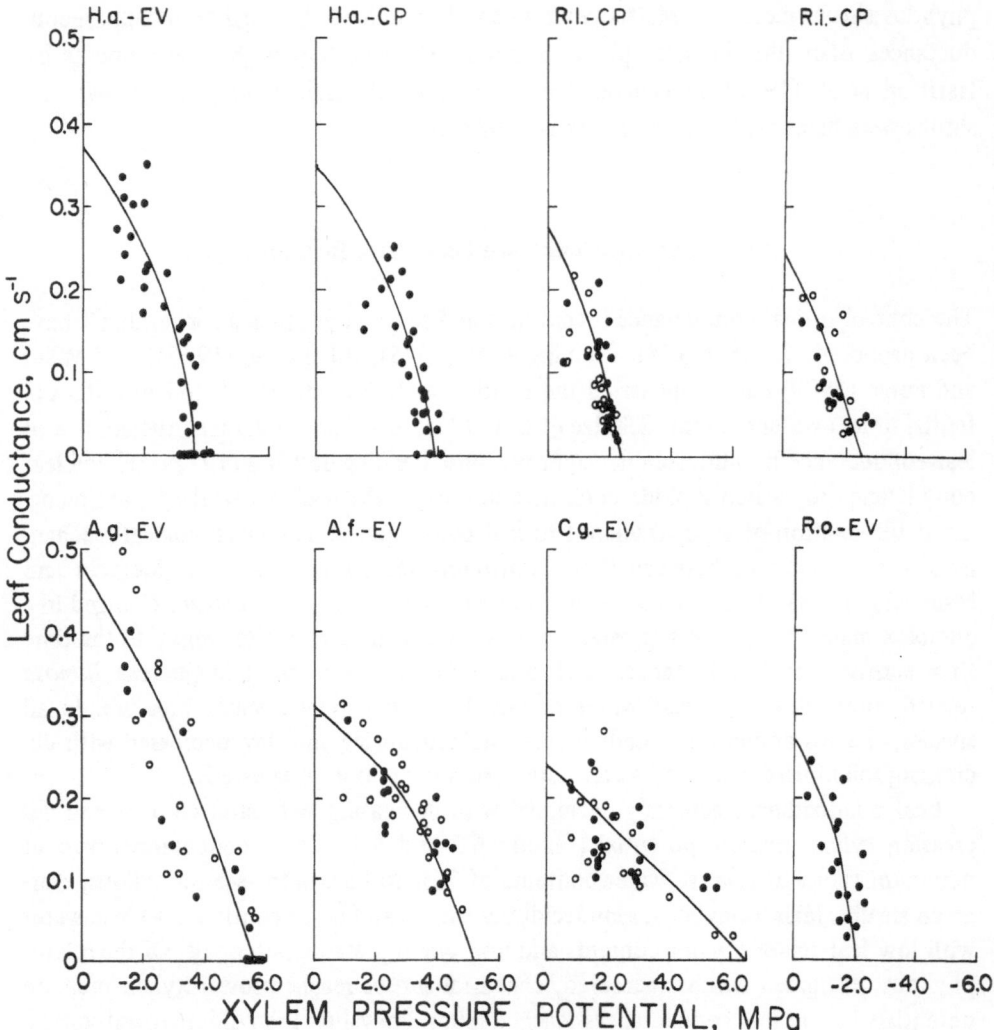

Figure 6.2. Relation between leaf conductance and xylem pressure potential in 7 Californian evergreen species of chaparral at Echo Valley (EV) and coastal sage scrub at
Camp Pendleton (CP). At any potential, the standard error for conductances is about
0.02 cm s^{-1}. Species are *Heteromeles arbutifolia* (H.a.), *Rhus laurina* (R.l.), *Rhus integrifolia* (R.i.), *Arctostaphylos glauca* (A.g.), *Adenostoma fasciculatum* (A.f.), *Ceanothus greggii* (C.g.), and *Rhus ovata* (R.o.) (from Miller, and Poole, 1979).

The responses of leaf conductances to xylem pressure potential were similar among
Rhus integrifolia, R. ovata, and *Rhus laurina*. All three species had maximum leaf conductances between 0.22 and 0.25 cm s^{-1} and had minimum leaf conductance at xylem
pressure potentials of about -2.0 MPa. Thus, the indications are that the controls on
species water relations can be generalized from specific research sites.

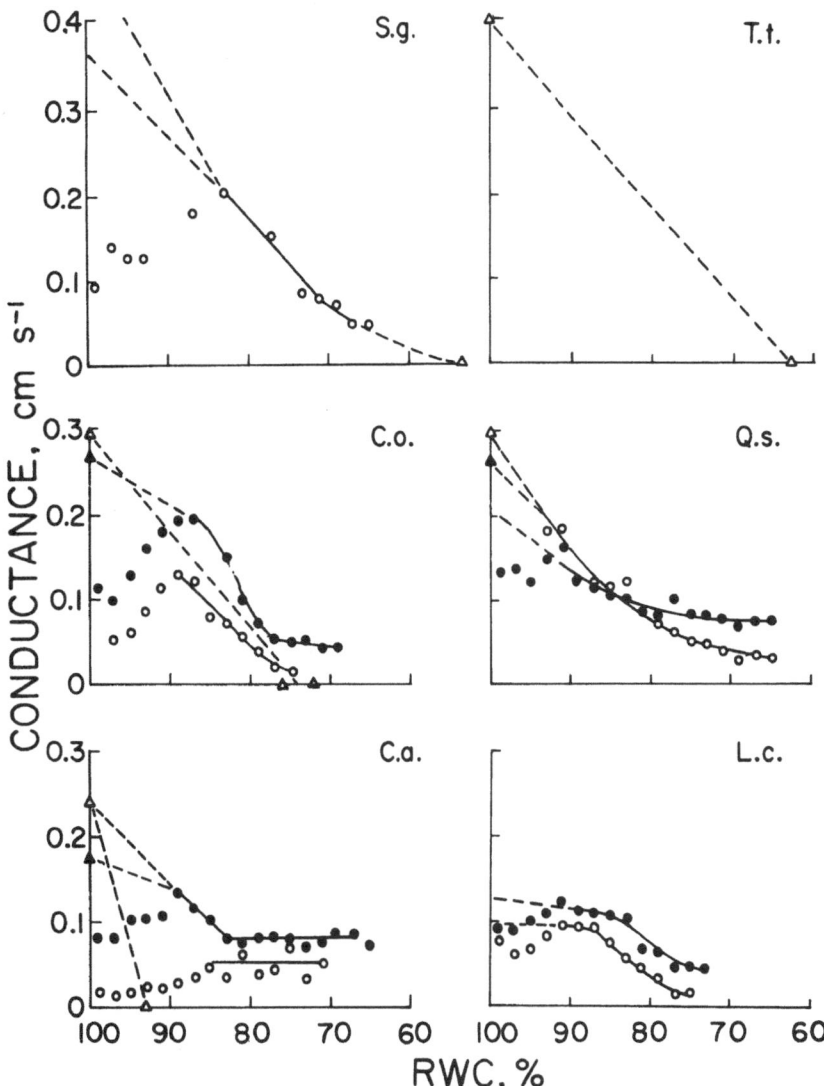

Figure 6.3. Relation of leaf conductance to leaf relative water content (RWC). Species are *Satureja gilliesii* (S.g.), *Trevoa trinervis* (T.t.), *Colliguaya odorifera* (C.o.), *Quillaja saponaria* (Q.s.), *Cryptocarya alba* (C.a.), and *Lithraea caustica* (L.c.). Each point is the mean of 8 to 10 measurements. Dashed lines are extrapolations. ▲ = January 1973; ○ = March 1975; ● = November 1975; and △ = December 1976 (after Poole and Miller, 1978).

6.3.3 Relative Water Content-Water Potential Relations

The relations between leaf water potential and leaf relative water content indicated that in the sclerophyllous shrubs, water potentials decreased dramatically as relative water content decreased below 70% to 80%. Component potentials were measured for Californian and Chilean shrubs (Poole and Miller, 1978; Poole, *unpubl. data*; Roberts, *unpubl. data*; Figures 6.4 and 6.5). Osmotic potentials for the sclerophyllous shrubs

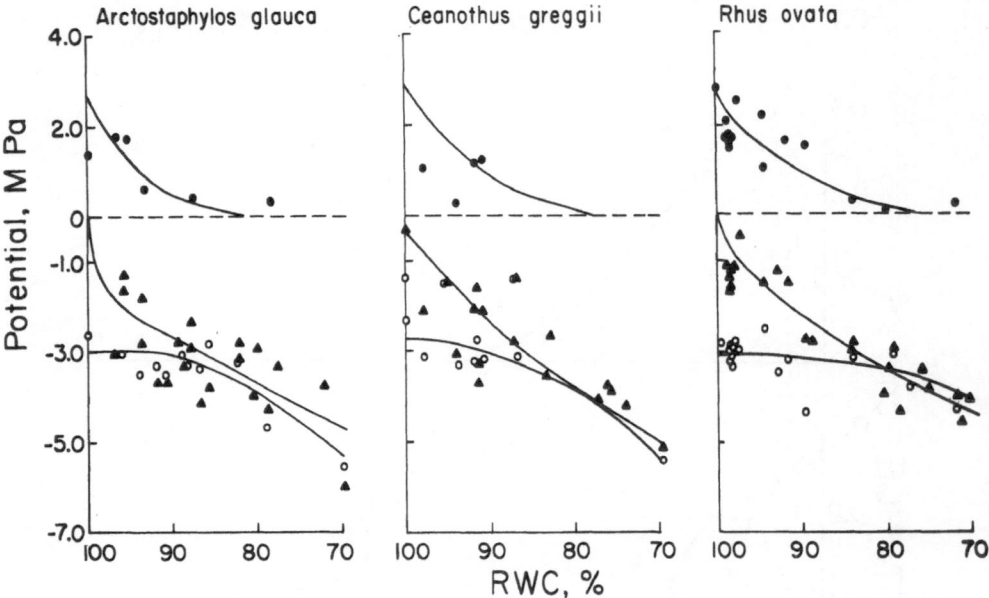

Figure 6.4. Relations of leaf (▲), osmotic (○), and turgor (●) potentials to relative water content (RWC). Each point is an individual measurement. Species are *Arctostaphylos glauca* (A.g.), *Ceanothus greggii* (C.g.), and *Rhus ovata* (R.o.).

were also estimated by using the pressure-volume technique following Tyree and Hammel (1972), Tyree et al. (1973, 1974), and Talbot et al. (1975). The water potentials did not clearly indicate a linear phase for extrapolating to zero water deficit. However, the pressure-volume method supported the independent measurements with the psychrometer. Osmotic potentials at turgidity varied from −2.5 to −3.5 MPa in the sclerophyllous shrubs, supporting the observations of Walter (1973) that shrubs in the mediterranean regions had only moderate osmotic concentrations in contrast to many drought adapted species. The component potentials have not been measured in other growth forms. Turgor pressure decreased rapidly as relative water content decreased below 80%. In all species, leaf and osmotic potentials decreased as relative water content decreased, but the rates of change differed. The shallow-rooted species in both countries had osmotic potentials below −3.0 MPa at turgidity, followed by a rapid decrease of osmotic potential at relative water contents below 94%. The deep-rooted species had osmotic potentials above −3.0 MPa at saturation, followed by a relatively constant decrease as relative water contents decreased. From saturation to 95% to 96% relative water content, all species showed little change in osmotic potential. However, this was the relative water content range where leaf water potential decreased the most because of loss of turgor. The change in osmotic potential was described by an equation in which water volume decreased while the amount of solute remained constant. Thus, no biochemical changes were suspected in the decreasing osmotic potentials as the plant dries.

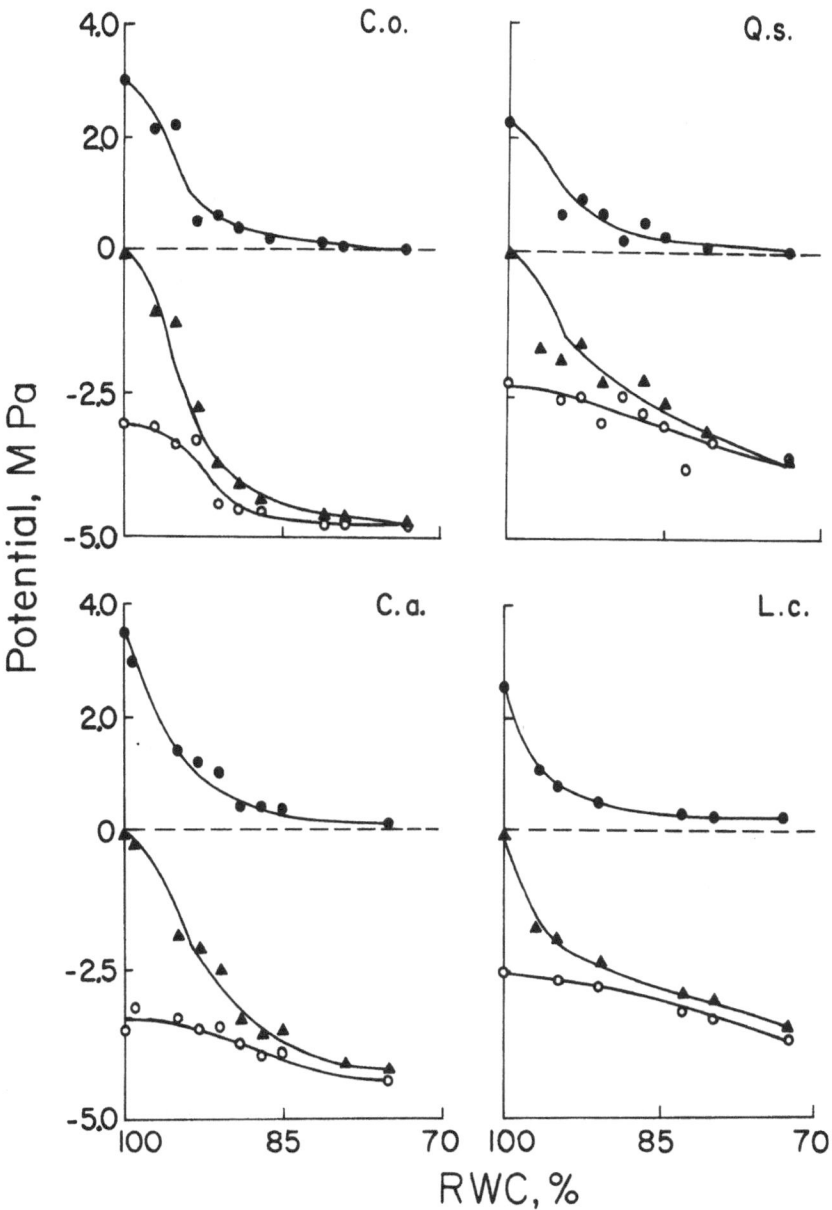

Figure 6.5. Relations of leaf (▲), osmotic (○), and turgor (●) potentials to relative water content (RWC). Each point is the mean of 3 to 5 measurements. Species are *Colliguaya odorifera* (C.o.), *Quillaja saponaria* (Q.s.), *Cryptocarya alba* (C.a.), and *Lithraea caustica* (L.c.) (from Poole and Miller, 1978).

6.3.4 Soil-Root Conductance

Soil-root conductances to water uptake, estimated from diurnal courses of plant xylem pressure potential and knowledge of the relations between leaf conductance, relative water deficit, and xylem pressure potential, varied between 1×10^{-6} cm s^{-1} MPa^{-1} during winter and spring and 0.17×10^{-6} cm s^{-1} MPa^{-1} during summer with dry soil at both Echo Valley and Fundo Santa Laura. Plant xylem pressure potentials in both countries recovered from the drought within 3 days after the first fall rains, sooner than would be expected if new root growth were necessary to provide the absorbing root tissues. The decrease in soil-root conductance as the soil dries was interpreted as caused mainly by the drying of the soil adjacent to the roots. This drying resulted in low hydraulic conductances even though the water content of the soil mass indicated high conductances and high soil water potentials. In addition, the roots in the dry soil may shrink during the summer (Huck et al., 1970). With the fall rains, roots and soil are rehydrated and absorption is again possible.

6.3.5 Correlation of Controlling Factors

The species that predominate in chaparral and matorral show a range of plant water characteristics (Table 6.3). In terms of their maximum leaf conductances, Californian species, arranged from highest (about 1.0 cm s^{-1}) to lowest (about 0.2 cm s^{-1}), were: *Artemisia californica, Salvia apiana, A. glauca, C. greggii, A. fasciculatum, Q. dumosa,* and *R. ovata*. The sequence was similar for the lowest xylem pressure potentials observed in a species, from a low of < –6.5 MPa to a high of –2.0 MPa, and for the duration of low xylem pressure potentials for a species. The sequence of species was one of increasing hydrostability, i.e., decreasing variability in pressure potentials through the year (Walter, 1973). Osmotic concentrations at turgidity were similar and moderate (–2.0 to –3.0 MPa). The depth of rooting probably also increases in the same sequence. *Artemisia californica, S. apiana, C. greggii,* and *A. glauca* are recognized as shallow rooted, while *Q. dumosa* and *R. ovata* are thought to be deep rooted (Hellmers et al., 1955b; Chapter 4). The sequences of maximum conductance, annual variations in conductance, hydrostability, and rooting depths segregate obligate seeders and sprouting shrubs. The different patterns of reproduction may reinforce the patterns of water characteristics because seedlings and resprouts have different water environments when young. Schlesinger and Gill (1980) showed lower water potentials during the summer in seedlings than in mature individuals of *Ceanothus megacarpus*. Radosevich et al. (1977) indicated that resprouts of *A. fasciculatum* generally had higher water potentials than the control vegetation following fire.

The Chilean species, arranged by maximum leaf conductance from highest (about 0.5 cm s^{-1}) to lowest (about 0.15 cm s^{-1}), were: *S. gilliesii, Trevoa trinervis, C. odorifera, Q. saponaria, Cryptocarya alba, L. caustica*. The sequence was similar for the lowest xylem pressure potentials measured but was not as consistent as with the Californian species. Root depth may be similarly arrayed (Giliberto and Estay, 1978) in the Chilean species, but the sequence is not as clearly related to reproductive patterns as with the Californian shrubs.

In general, several water related characteristics are arrayed similarly in the chaparral and matorral species studied and are mutually supportive. The maximum leaf conductances are similar to maximum conductances measured in plants in other climates, but the xylem pressure potentials are lower than those often measured in other vegetation types, even in deserts.

6.4. Diurnal and Seasonal Patterns of Plant Water Relations

6.4.1 Measured Patterns

The diurnal course of conductance varied seasonally by slope and by species (Poole and Miller, 1975; Valamanish and Roberts, *unpubl. data*). The seasonal variation was greatest in *A. glauca*, which showed the highest conductance during the drought, less in *C. greggii*, and still less in *A. fasciculatum*. The seasonal variation was more pronounced on the pole- than on the equator-facing slope. Predawn and midday xylem pressure potentials of *A. glauca, A. fasciculatum*, and *C. greggii* were lower during the summer on the pole- than on the equator-facing slope. Potentials of shallow-rooted species rose more rapidly than those of deep-rooted species after the first rains in November. During the winter, potentials of all species were high, from 0 to -2.0 MPa predawn and from -1.0 to -1.5 MPa midday. At Echo Valley the diurnal course of conductance during the winter followed the diurnal course of solar irradiance, i.e., no water stress was apparent. Conductances rose rapidly just after dawn and remained high until dusk. In late spring and summer, conductances decreased in the afternoon. High conductances occurred throughout the summer, but at the height of the summer drought conductances were high only briefly after sunrise (Poole and Miller, 1975; Valamanish and Roberts, *unpubl. data*). *Ceanothus greggii* on both pole- and equator-facing slopes showed low conductances earlier in the day as the season progressed, but the low conductances developed earlier in the day on the pole- than on the equator-facing slope, consistent with the soil drying and the earlier receipt of solar irradiance in the day (Chapter 2). *Adenostoma fasciculatum* generally developed low conductances later in the day on the pole-facing slope than on the equator-facing slope. Leaf conductances of *A. glauca* on both slopes decreased earlier in the day as the drought progressed, were low through the day during the summer, and increased after the drought ended. Leaf conductances of *A. glauca* were consistently lower on the pole- than on the equator-facing slope.

Of the species studied, *R. ovata, R. integrifolia*, and *R. laurina* generally showed the lowest leaf conductances and the greatest variations in leaf conductance throughout the day. Even at the height of the drought, these species had relatively high leaf conductances for a short time in the early morning. In earlier studies, *R. integrifolia* and *R. laurina* were found to have stem growth throughout the summer, which was attributed to their deep rooting habit (Watkins and de Forest, 1941), and were found to be conservative water users (Shapiro and de Forest, 1932). *Heteromeles arbutifolia* was the only other species that showed high conductance sometime during the day late in the drought.

The seasonal progression of daily maximum and minimum leaf conductances (Figure 6.6) showed high conductances while soil moisture was available and low conductances during the summer drought. At Camp Pendleton, leaf conductances of *H. arbutifolia* were high in the morning and low the rest of the day in June but were low throughout the day between late August and mid-November. *Rhus integrifolia* showed the greatest seasonal variation because of its relatively long period of zero leaf conductance. *Heteromeles arbutifolia* and *R. laurina* showed relatively high conductances

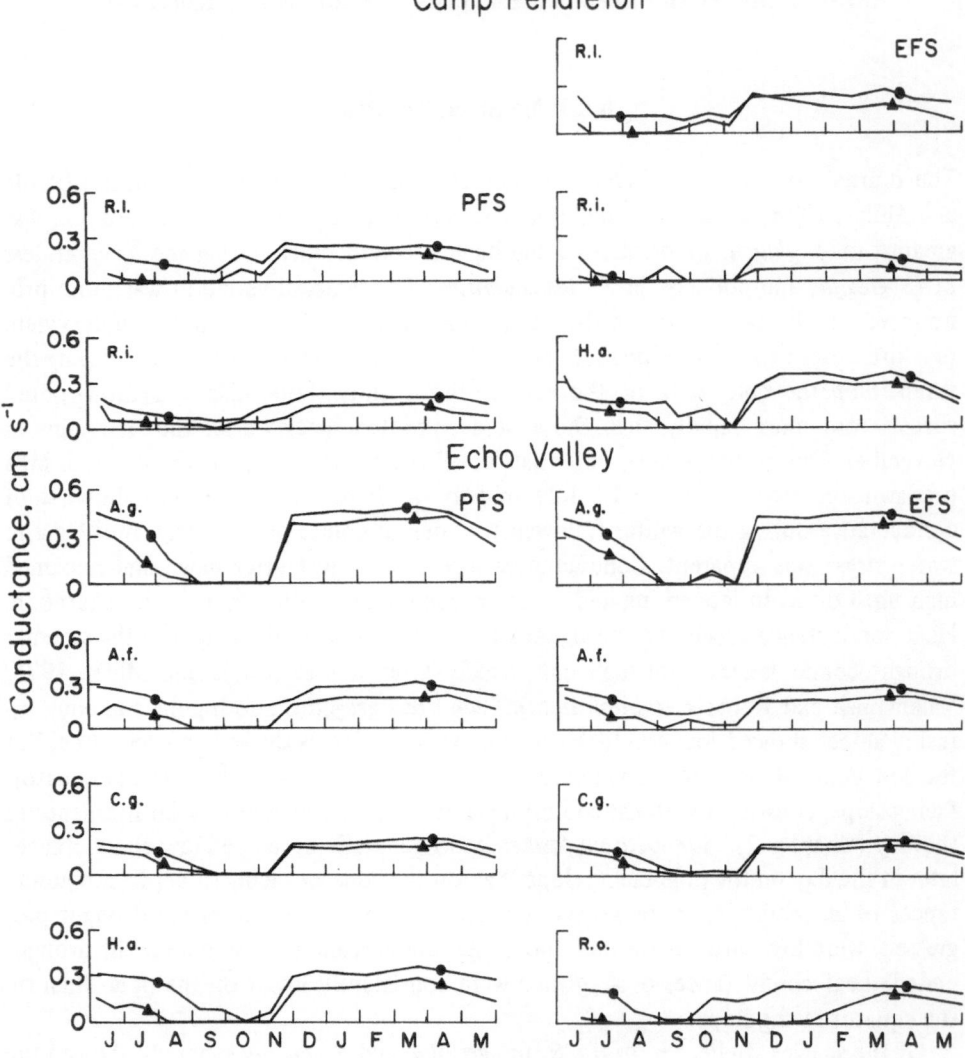

Figure 6.6. Annual course of minimum (▲) and maximum (●) leaf conductance at Camp Pendleton and Echo Valley in California on the equator- (EFS) and pole-facing (PFS) slopes. Species are *Rhus laurina* (R.l.), *Rhus integrifolia* (R.i.), *Heteromeles arbutifolia* (H.a.), *Arctostaphylos glauca* (A.g.), *Adenostoma fasciculatum* (A.f.), *Ceanothus greggii* (C.g.), and *Rhus ovata* (R.o.) (from Miller and Poole, 1979).

throughout the year, at least during early morning hours. At Echo Valley, the seasonal variation in leaf conductances was most pronounced in *A. glauca*, which showed the highest conductance during the drought. *Ceanothus greggii* showed less marked seasonality, and *A. fasciculatum* showed the least. The seasonal variation was more pronounced on the pole- than on the equator-facing slope.

Xylem pressure potentials and plant characteristics related to water use, which were measured for 15 months throughout the east-west range of sclerophyllous chaparral shrubs in San Diego County, supported the calculated wave of drying from the edges of chaparral distribution to the center of its distribution (Poole and Miller, 1981;

Chapter 2). On the coastal-facing side of the mountains, xylem pressure potentials decreased in early summer beginning at the coast and later in the summer at higher elevations (Figure 6.7). However, the lowest xylem pressure potentials were at the higher elevation sites on the desert-facing slope of the coastal mountains. Summer precipitation at the higher elevations allowed periodic recovery from the low xylem pressure potentials. Poole and Miller (1975) also showed that pressure potentials decreased earlier at Camp Pendleton than at Echo Valley. Potentials of *R. laurina* and *R. integrifolia* at Camp Pendleton decreased to minimum levels in July and August; those of *R. ovata* at Echo Valley decreased in September. Predawn potentials of *H. arbutifolia* were below −2.0 MPa for 4.5 months at Camp Pendleton and for only 2 months at Echo Valley; midday potentials were below −3.5 MPa for 3.5 months at both sites. Throughout the region *Arctostaphylos* spp. had low xylem pressure potentials, while *R. integrifolia, R. ovata*, and *R. laurina* had high xylem pressure potentials. Leaf water contents at turgidity, leaf area, and specific leaf density (g m^{-2}) were similar for a genus throughout the region indicating little morphological adaptation to the east-west aridity gradient (Table 6.4). The leaf properties varied with the age of the stand and with the water transpired per unit of leaf area. Thus, the east-west aridity gradient is masked by the increasing transpirational loss from the development of the stand as it regrows following fire.

Plants at Echo Valley showed greater diurnal variation in xylem pressure potentials than did the same species at sites with similar elevation but more coastal influence. The measurements support the notion that the Echo Valley site is relatively well protected from winds from the coast and thus, has relatively large diurnal variations in plant and environmental properties.

The seasonal progressions of predawn and midday xylem pressure potentials were measured at Camp Pendleton, Echo Valley, and Fundo Santa Laura on both pole- and equator-facing slopes (Figures 6.8 to 6.10). The deep-rooted *Q. saponaria, L. caustica*, and *Rhus* spp. showed relatively high midday potentials (−2.0 MPa) throughout the year, which were consistent with their leaf conductance-xylem pressure potential relationships. Xylem pressure potentials of soft-leaved shrubs, *A. californica, S. gilliesii*, and *T. trinervis*, decreased markedly in spring. When potentials of *A. californica* were below −6.0 MPa, the plant was leafless. The other species maintained leaves through these periods of low pressure potentials. Predawn potentials were below −6.0 MPa for 2.5 months in *A. californica*, about 2 months in *A. glauca, C. odorifera*, and *C. alba*, about 1 month in *C. greggii* and *T. trinervis*, and less than 1 month in *A. fasciculatum*.

Predawn xylem pressure potentials, which are commonly equated with soil water potentials, were 3.0 to 5.0 MPa lower than the potentials estimated from the bulk soil moisture data. At Echo Valley, predawn and midday potentials of *A. glauca, A. fasciculatum*, and *C. greggii* were lower during the summer drought on the pole- than on the equator-facing slope. Potentials of shallow-rooted species rose more rapidly than those of deep-rooted species after the first rains in November. Potentials of the *Rhus* species at Camp Pendleton and Echo Valley rose before the rains occurred in November, probably because of coastal fogs and increased air humidity. In general, the recovery of pressure potentials was synchronous in species at Camp Pendleton and Echo Valley because the winter rains are widespread. During the winter, xylem pressure potentials of all species were high, 0 to −2.0 MPa predawn and −1.0 to −1.5 MPa midday.

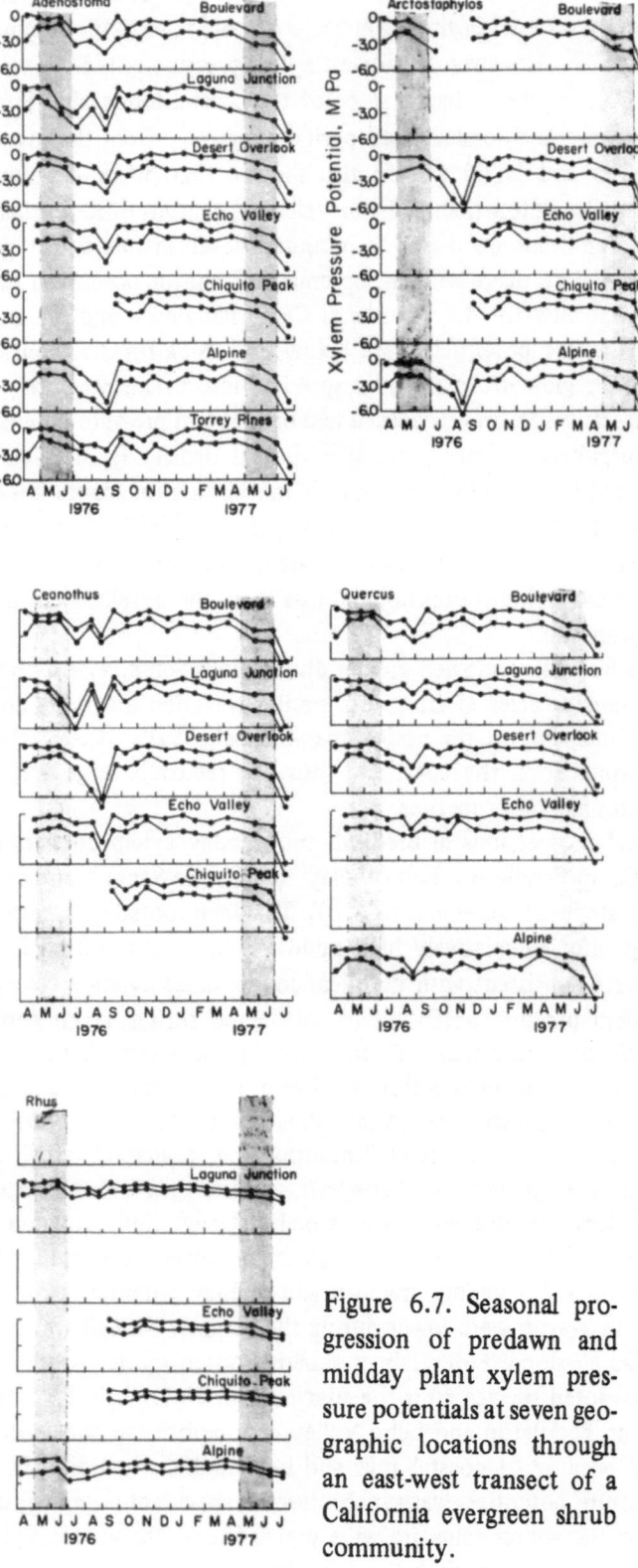

Figure 6.7. Seasonal progression of predawn and midday plant xylem pressure potentials at seven geographic locations through an east-west transect of a California evergreen shrub community.

Transpiration varied seasonally (Figure 6.11). Although leaf conductances were high in winter, transpiration was low because net radiation was low and vapor density deficits were small. During the spring, transpiration increased because of increased day-length, since leaf conductances and vapor density deficits were nearly constant. In June and July, transpiration decreased because leaf conductances decreased, even though vapor density deficits increased. *Arctostaphylos glauca* showed a complete cessation of transpiration for almost 2.5 months. *Rhus ovata, H. arbutifolia*, and *C. greggii* showed no transpiration for about 1 month. In mature shrubs, transpiration during spring in *A. glauca* was 2.0 mm day^{-1} m^{-2} of leaf area; *A. fasciculatum*, 1.1 to 1.3; *C. greggii*, 1.0; and *R. ovata*, 0.7.

Annual water use on a square meter of leaf basis was 150 to 190 mm yr^{-1} from *A. fasciculatum*, 145 to 195 mm yr^{-1} from *C. greggii*, 280 to 290 mm yr^{-1} from *A. glauca*, and 127 mm yr^{-1} from *R. ovata* (Miller and Poole, 1979). Poole and Miller (1981) argued that about 200 mm yr^{-1} must be transpired per unit leaf area to support the photosynthesis and carbon balance of the leaf. The precipitation at the edges of the chaparral distribution and the maximum leaf area indices attained in the center of the distribution support this idea.

The water capture efficiency of mature mixed chaparral is higher than that of mature chamise chaparral. With precipitation of 390 mm yr^{-1} transpiration was 390 mm yr^{-1} in mature mixed chaparral and 170 mm yr^{-1} in mature chamise chaparral (Miller and Poole, 1979). With precipitation of 300 mm yr^{-1}, 290 mm yr^{-1} and 250 mm yr^{-1} were transpired from mature mixed and mature chamise chaparral, respectively (Ng and Miller, 1980). Miller and Stoner (1979) estimated that with precipitation of 519 mm yr^{-1} a total of 430 and 210 mm yr^{-1} were transpired by mature mixed and mature chamise chaparral, respectively. These estimates and simulations with the Mediterranean Ecosystem Simulator (*MEDECS*) indicate that mature chamise chaparral, grasslands, and bare soil lose water similarly, while mature mixed chaparral loses more water.

6.4.2 Simulations with CAPS and MEDECS

The canopy process simulator (CAPS) was used to synthesize the data on the plant water relations and climate and to estimate the seasonal progression of water use. The predicted air temperatures and humidities within the canopy were close to the observed. The diurnal course of plant xylem pressure potential was used to evaluate the validity of the simulations. Transpiration increased with increasing foliage area index to 600 mm yr^{-1} for *A. glauca*, 500 mm yr^{-1} for *C. greggii*, 100 to 200 mm yr^{-1} for *R. ovata*, and 220 mm yr^{-1} for *A. fasciculatum* (Chapter 12). *Adenostoma fasciculatum* would have the highest transpiration rates at high foliage area indices but cannot tolerate the accompanying shading. Thus, its water usage was restricted by the maximum foliage areas it could maintain. The Chilean shrubs transpired 100 to 150 mm yr^{-1} at high foliage area indices in these simulations. The same soil moisture contents were used with all foliage areas. Thus, annual transpiration was overestimated because the soil did not dry faster with higher transpiration rates. The annual values represent potentials that might be achieved with high precipitation amounts.

Table 6.4. Site and plant characteristics related to water use at seven geographic locations in the sclerophyllous chaparral in California[a]

	Coastal-facing slope					Desert-facing slope	
	TP	AL	DE	EV	CU	ML	BL
Elevation (m)	134	610	915	1000	1430	1400	1150
Distance from coast (km)	1.6	47	53	56	66	66	78
Slope (%)	10	5	10	5	40	45	4
Aspect	N	N	SW	S	E	SE	NE
Precipitation (mm)	254	432	510	450	813	450	356
Soil water holding capacity (mm)	102	25-40	-	25-50	50-75	50-75	100-125
Age of stand (yr)	-	49	54	28	28	8	34
Leaf size (cm^2)							
Arctostaphylos	-	4.3 ± 0.3	4.6 ± 0.3	5.4 ± 0.2	4.4 ± 0.3	1.6 ± 0.0	1.8 ± 0.1
Ceanothus	0.7 ± 0.0	-	1.2 ± 0.1	0.8 ± 0.0	1.3 ± 0.1	0.8 ± 0.0	1.1 ± 0.0
Rhus	17.0 ± 1.5	20.2 ± 0.7	-	13.3 ± 0.5	-	15.4 ± 1.0	-
Quercus	0.8 ± 0.0	1.7 ± 0.1	1.4 ± 0.1	1.5 ± 0.1	2.6 ± 0.2	1.3 ± 0.1	1.6 ± 0.1
Specific leaf area (g m^{-2})							
Arctostaphylos	-	314 ± 5	262 ± 10	188 ± 4	260 ± 8	108 ± 2	281 ± 13
Ceanothus	308 ± 9	-	374 ± 9	368 ± 11	367 ± 5	319 ± 5	407 ± 9
Rhus	196 ± 4	347 ± 6	-	288 ± 8	-	322 ± 2	-
Quercus	199 ± 6	190 ± 5	172 ± 3	180 ± 2	166 ± 5	178 ± 3	181 ± 9
Water content per leaf area (g m^{-2})							
Arctostaphylos	-	278 ± 5	267 ± 8	195 ± 6	284 ± 9	224 ± 4	262 ± 14
Ceanothus	360 ± 8	-	357 ± 7	415 ± 25	357 ± 8	364 ± 8	384 ± 17
Rhus	270 ± 5	444 ± 8	-	337 ± 11	-	425 ± 5	-
Quercus	187 ± 6	152 ± 3	147 ± 5	160 ± 2	141 ± 4	169 ± 2	189 ± 16

Site characteristics

Water content at turgidity (g g^{-1})

Arctostaphylos	-	0.471 ± 0.003	0.506 ± 0.003	0.510 ± 0.002	0.523 ± 0.001	0.518 ± 0.03	0.489 ± 0.005
Ceanothus	0.540 ± 0.001	-	0.490 ± 0.002	0.525 ± 0.001	0.492 ± 0.005	0.532 ± 0.003	0.482 ± 0.001
Rhus	0.576 ± 0.007	0.563 ± 0.004	-	0.538 ± 0.001	-	0.569 ± 0.002	-
Quercus	0.487 ± 0.006	0.44 ± 0.005	0.461 ± 0.006	0.469 ± 0.003	0.460 ± 0.005	0.488 ± 0.003	0.502 ± 0.002

From Poole and Miller (1981).

[a]Note: *Quercus dumosa* occurred at all sites. For other genera, similar species had to be substituted. Thus, at Torrey Pines (TP), *C. verruco-sus* and *R. laurina* were measured: no *Arctostaphylos* species occurred there. At Alpine (AL), Descanso (DE), and Mount Laguna (ML), *A. glandulosa* was measured. No *Ceanothus* species were present at the Alpine site. At the Cuyamaca (CU) and Boulevard (BL) sites, *A. pungens* was measured. At Echo Valley (EV), *A. glauca* was measured. Otherwise, *R. ovata* and *C. greggii* were measured.

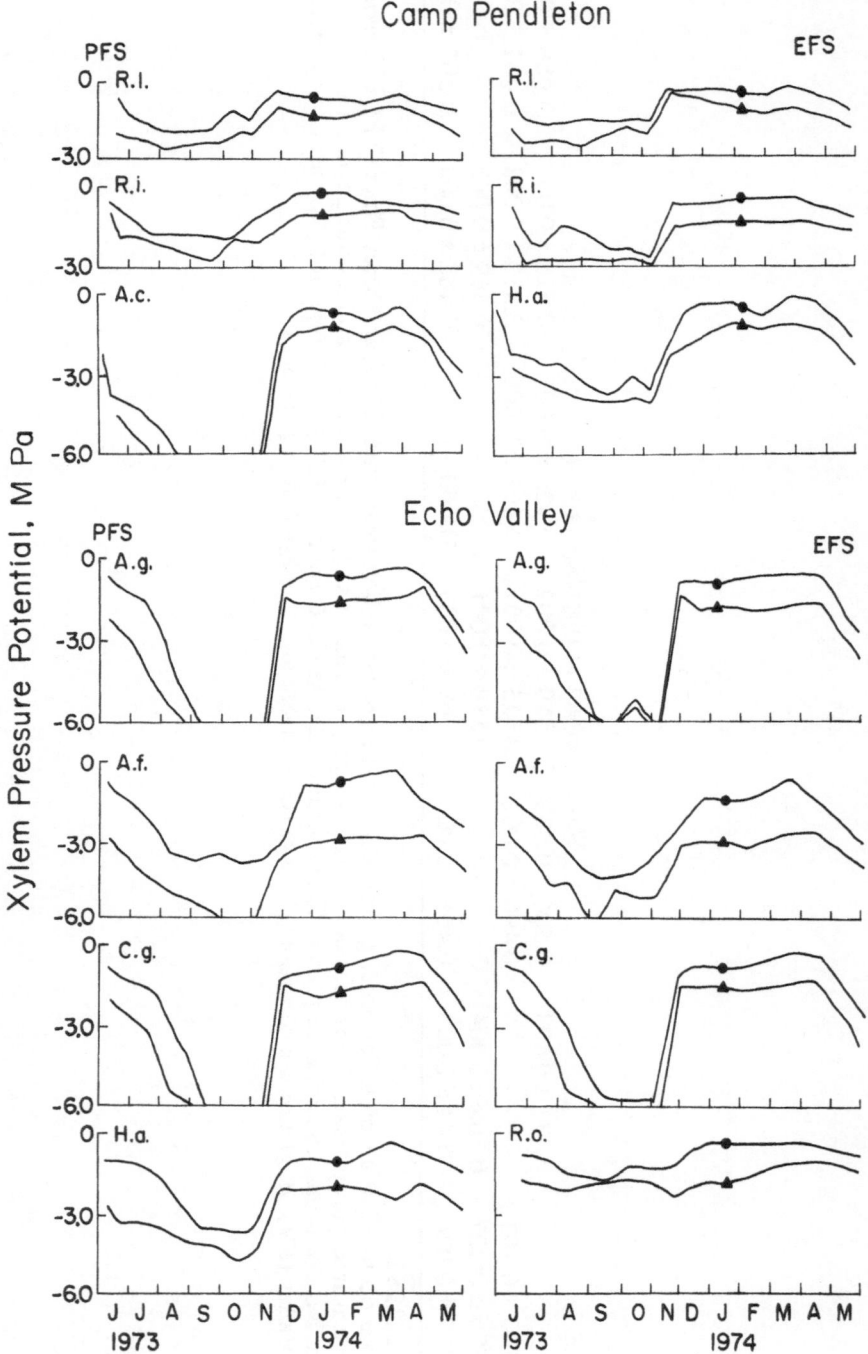

Figure 6.8. Annual course of predawn (●) and midday (▲) xylem pressure potentials at Camp Pendleton on the pole- (PFS) and equator-facing (EFS) slopes for *Rhus laurina* (R.l.), *Rhus integrifolia* (R.i.), *Artemisia californica* (A.c.), and *Heteromeles arbutifolia* (H.a.); and at Echo Valley on the pole- (PFS) and equator-facing (EFS) slopes for *Arctostaphylos glauca* (A.g.), *Adenostoma fasciculatum* (A.f.), *Ceanothus greggii* (C.g.), *Heteromeles arbutifolia* (H.a.), and *Rhus ovata* (R.o.) (from Miller and Poole, 1979).

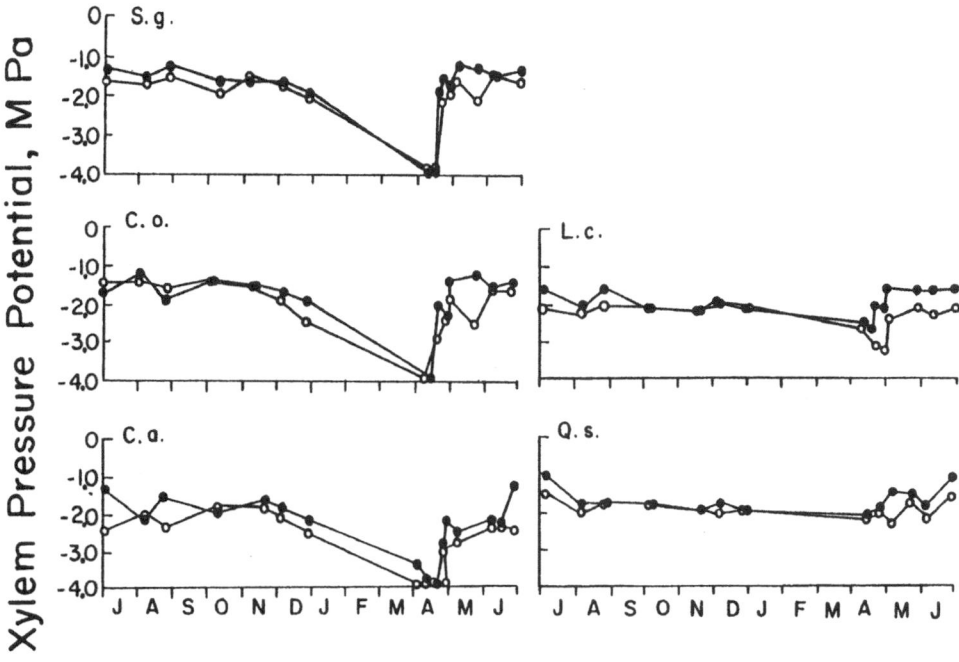

Figure 6.9. Seasonal courses of midday xylem pressure potentials on the pole- (−●−) and equator-facing (−○−) slopes at Fundo Santa Laura for *Satureja gilliesii* (S.g.), *Colliguaya odorifera* (C.o.), *Lithraea caustica* (L.c.), *Cryptocarya alba* (C.a.), and *Quillaja saponaria* (Q.s.) (from Giliberto and Estay, 1978).

Simulations with MEDECS allowed soil moisture to vary with the transpirational loss and foliage area. With increasing foliage area, transpiration increased to 330 mm yr^{-1} with *A. glauca*, 300 mm yr^{-1} with *C. greggii*, 150 mm yr^{-1} with *R. ovata*, and 160 mm yr^{-1} with *A. fasciculatum*. The Chilean shrubs transpired less. *Lithraea caustica* transpired up to 200 mm yr^{-1}, *C. odorifera* up to 130 mm yr^{-1}, and *S. gilliesii* and *T. trinervis* up to 60 mm yr^{-1}. When conditions were similar, the annual transpiration loss calculated with MEDECS increased with precipitation similarly to the annual transpiration calculated with CAPS. Thus, the simulations indicate that water use by the Chilean shrubs is generally lower than water use by the Californian shrubs. The notable exception is *R. ovata*, which is not abundant in Californian chaparral.

6.4.3 Similarity of Water Use

In simulations and measurements, the highest rates of transpiration (mm day^{-1}) occurred in spring in both countries when soil moisture was available, and daylength, solar irradiance, and temperatures were increasing. Transpiration was low during the winter because of low incoming irradiance and during the summer because of the lack of water. In the winter, transpiration was higher at Echo Valley than at Fundo Santa Laura because of the higher seasonal solar irradiance at Echo Valley. The annual transpirational water use was similar between chamise chaparral at Echo Valley and mator-

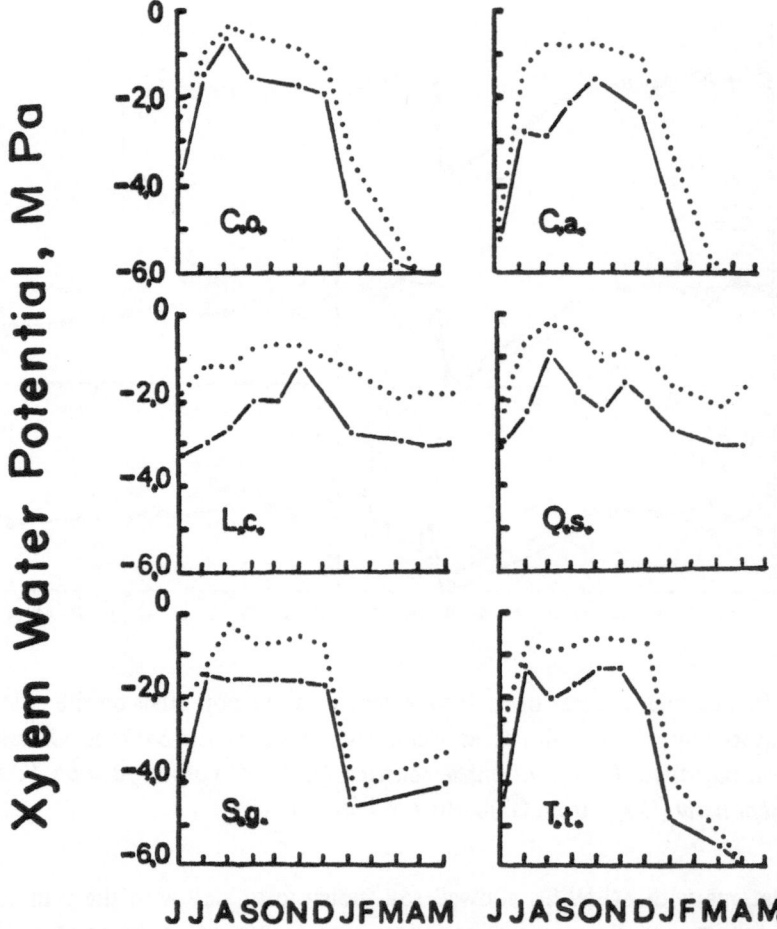

Figure 6.10. Seasonal course of predawn ($\cdots$) and midday ($-\cdot-$) xylem pressure potentials of six Chilean species: *Colliguaya odorifera* (C.o.), *Cryptocarya alba* (C.a.), *Lithraea caustica* (L.c.), *Quillaja saponaria* (Q.s.), *Satureja gilliesii* (S.g.), and *Trevoa trinervis* (T.t.) (Martinez, *unpubl. data*).

ral at Fundo Santa Laura, and both were less than the water use in mixed chaparral at Echo Valley. The pattern of plant water use at Echo Valley and Fundo Santa Laura is related to their mediterranean type climates, but this water use pattern is not necessarily unique to these climates. In the Great Plains grasslands, the mature grasses show the highest transpiration in the spring, using moisture that has been stored in the soil during winter and early spring snow melt. In most other climatic types, the seasonal progression of transpiration differs from that in the mediterranean type climates.

6.5. Environment and Plant Influences on Water Capture Efficiency

The water capture efficiency appears to be constrained by annual precipitation through its effects on the potential plant biomass and foliage area index. The water capture efficiency, calculated as transpiration divided by precipitation, in mediterranean climates is directly related to the foliage area index. With increasing foliage area indices, soil evaporation decreases while transpiration and interception increases. Because inter-

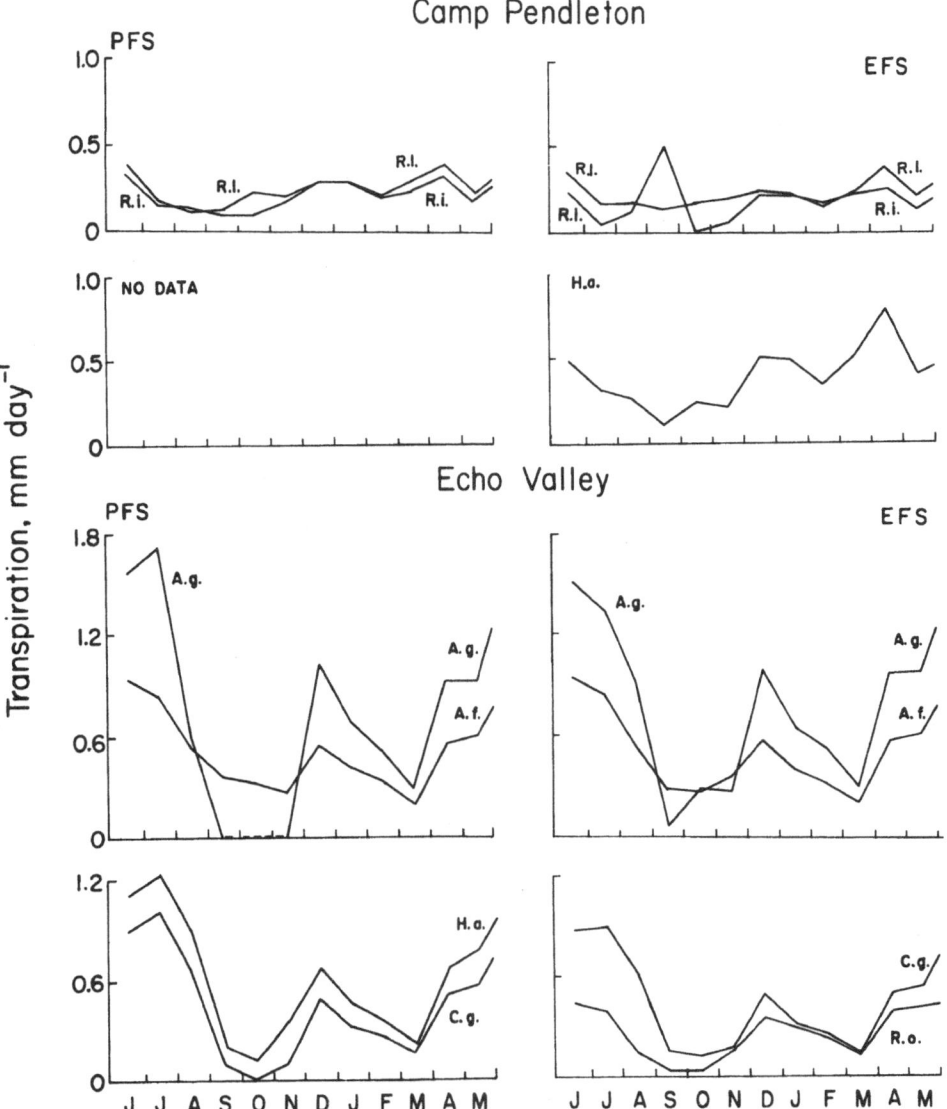

Figure 6.11. Annual course of mean monthly transpiration at Camp Pendleton and Echo Valley on the pole- (PFS) and equator-facing (EFS) slopes. Species are *Rhus laurina* (R.l.), *Rhus integrifolia* (R.i.), and *Heteromeles arbutifolia* (H.a.) at Camp Pendleton; and *Arctostaphylos glauca* (A.g.), *Adenostoma fasciculatum* (A.f.), *Ceanothus greggii* (C.g.), *Heteromeles arbutifolia* (H.a.), and *Rhus ovata* (R.o.) at Echo Valley (from Miller and Poole, 1979).

ception losses are less than soil evaporation losses, the vegetation becomes more efficient at using water as the foliage area index increases. The foliage area index cannot increase indefinitely but reaches a maximum at which photosynthesis just sustains tissue maintenance and replacement costs. At these high foliage areas, transpiration reaches a plateau even though precipitation increases, and in this range of precipitation, water capture efficiency decreases. The maximum foliage area that can be maintained is constrained by the amount of photosynthate fixed, which is related to the water transpired, the ratio of photosynthesis to transpiration, the costs of replacing leaves and stems, the turnover rates of leaves and stems, and the maintenance costs.

At the lower precipitation amounts ($<$ 400 mm), only low foliage areas can be maintained ($<$ 2.0), and soil evaporation is high, giving a low water capture efficiency. In the simulations, as the foliage area index increased from 0 to 2, soil evaporation was reduced to two-thirds. From field observation and theory, an annual precipitation of 400 to 450 mm yr^{-1} is required to maintain a foliage area index of 2. Thus, with precipitation below about 400 mm yr^{-1}, water capture efficiency can be expected to decrease rapidly with decreasing precipitation. Above 450 mm yr^{-1}, the foliage area that can be maintained is limited by the light relations of the shrubs and the leaf drop during the summer drought. Transpiration is limited by foliage area, and the water capture efficiency again decreases as precipitation increases above 450 mm. The lower water capture efficiency is accompanied by increased drainage. The low water capture efficiency of chamise chaparral, at low elevations and drier sites, and the higher capture efficiency of mixed chaparral, at midelevations and more mesic sites, is consistent with the precipitation constraints on water capture efficiency. The low water capture efficiency of matorral is caused by the low foliage area, higher precipitation, and low net radiation in winter.

Other plant characteristics should correlate with the precipitation gradient. At low precipitation ($<$ 400 mm yr^{-1}) and low foliage area index, plants compete with soil evaporation for water. Transpiration and water capture efficiency increase with high leaf conductances and high leaf area to dry weight ratios, which increase soil shading. Because the soil drought is lengthened by soil evaporation and high conductances, drought deciduous species are favored. According to leaf energy budget theory (Gates, 1965), water is conserved by steeply inclined leaves, narrow leaves, and high leaf reflectances. Thus, in areas with low precipitation, the water capture efficiency is increased with leaves with high leaf area to dry weight ratios, steep inclinations, narrow widths, and high reflectances. At higher precipitation levels ($>$ 550 mm yr^{-1}), water is lost by drainage; transpiration is increased with higher leaf conductances; the length of the drought is short. Under these conditions, evergreen species are favored unless restricted by cold temperatures. Leaf width, inclination, and color can be more variable without affecting water loss. At intermediate precipitation levels (400 to 550 mm yr^{-1}), the composition of the vegetation changes the length of the soil drought by including species with different leaf conductances and different leaf area indices. The water capture efficiency is controlled by the vegetation composition and can be relatively high. The mixed chaparral usually occurs in the intermediate precipitation range and has a high water capture efficiency and moderate leaf conductances.

Ng and Miller (1980) postulated that shrubs with leaves able to survive through a drought could support higher leaf area indices and higher leaf conductances than shrubs that could not endure a drought. The increased transpiration from the higher leaf area and conductance would increase the length of the drought. In a competitive situation, a species captures and uses more water by using water lavishly when it is available, rather than by conservatively using water. Such a pattern occurs in *A. glauca* and *C. greggii*, which occur together and are lavish water users. This pattern contrasts with the water-conserving pattern postulated by Specht (1972a, b).

Specht (1972a, b) suggested that evergreen species in a mediterranean climate have a pattern of moderate water use throughout the year, so soil moisture is not depleted until just prior to the onset of the wet season following the summer drought. A pattern of moderate water use would allow sufficient water throughout the drought period for photosynthesis to offset maintenance respiration and for growth. Specht's suggestion is consistent with the pattern of water use on the equator-facing slope at Echo Valley (Ng and Miller, 1980), which was dominated by a single species, *A. fasciculatum*, with the pattern of water use in *Q. dumosa* and *R. ovata*, and with the pattern of water use in the Chilean shrubs. *Rhus ovata* also has a high temperature requirement for growth, similar to Australian shrubs (Specht 1969a, b); the Chilean shrubs appear to have low temperature requirements for growth. Specht's suggestion, however, is not consistent with the pattern of soil moisture on the pole-facing slope which is dominated by several species, including *A. glauca* and *C. greggii*, that are also evergreen shrubs. Thus, the patterns of water use must be viewed in the context of various physiological characteristics and the flora history of the area. The Australian mediterranean vegetation has no cool temperate species, but such species are present in chaparral and matorral because of migrations of the flora during the Pleistocene.

6.6. Conclusions

Transpiration and water capture efficiency in mixed chaparral were higher than in the matorral, but those of the chamise chaparral were similar to those of matorral. The controls on water use differ between chaparral and matorral, mainly because of the higher maximum leaf conductances measured in the chaparral. The relative water contents at which minimum leaf conductance occurs were similar among sclerophyllous shrubs and among malacophyllous shrubs in both California and Chile but were dissimilar between the two shrub types within a country. The seasonal patterns of water use were similar within a shrub type in each country but differed between shrub types. The water capture efficiency at Fundo Santa Laura was reduced by low maximum leaf conductances, low net irradiance in winter, and higher precipitation.

7. Energy and Carbon Acquisition

WALTER C. OECHEL, WILLIAM LAWRENCE, JAMIL MUSTAFA, and JOSÉ MARTÍNEZ

This chapter describes the patterns of energy and carbon acquisition in plants growing at Echo Valley in southern California and Fundo Santa Laura in central Chile. The pattern of temperature and light dependency of photosynthesis is presented and evaluated, and geographic variability is discussed. Resource-use efficiency in important species is compared. Also, the relationship between carbon acquisition and species abundance in the vegetation is considered.

7.1. Introduction

Patterns of resource-use efficiency are affected by the rates of physiological processes and the benefit to the plant derived from these processes. The calculation of resource use efficiency requires a knowledge of key physiological processes that affect the cost and benefit of the resource use. Many studies of adaptive responses of photosynthesis to the environment have assumed that an increase in carbon assimilation carries with it an adaptive advantage to the individuals displaying the greater uptake rates (Mooney, 1972; Mooney et al., 1975; Solbrig and Orians, 1977). These studies on adaptive response also assumed that an increase in available carbon may be used to increase the success and fitness of the population. Within the plant community, there may be several carbon uptake and allocation patterns possessed by different plant species. The combination of these species possessing differing carbon-use efficiencies may result in maximized total resource use and minimized overlap in resource use.

Similarities in the physiological processes and in resource-use efficiency associated with carbon uptake and energy acquisition between plants from Echo Valley in southern California and Fundo Santa Laura in central Chile are considered in this chapter. Physiological processes showing major effects on productivity, carbon uptake, and resource use were compared in the field. Assuming that maximal carbon uptake is strongly selected, the initial hypothesis was that similar climatic patterns at Echo

Valley and Fundo Santa Laura resulted in similar patterns and efficiencies of energy capture by major species in the respective vegetations of the two countries. A comparison of plant cover with carbon acquisition rates determined the extent of correlation between carbon uptake rates and success in the vegetation.

Photosynthetic response surfaces representing the magnitude of net photosynthetic rates at various light and temperature levels may be constructed. The shape and magnitude of such response surfaces are known to vary seasonally. This phenomenon may result from acclimation of photosynthesis (Strain, 1969; Smith and Hadley, 1974; Hicklenton and Oechel, 1976; Oechel, 1976), leaf age effects (Hardwick et al., 1968; Ludlow and Wilson, 1971; Syvertsen and Cunningham, 1977), end product inhibition (Neales and Incoll, 1968; Cunningham and Syvertsen, 1977), or a combination of these factors. Seasonal changes in the rate of leaf respiration may also have marked effects on the shape of the photosynthetic response surface (Sveinbjörnsson, 1979; Sveinbjörnsson and Oechel, *unpubl. data*). Photosynthetic capacity varies for plants of diverse origin and growth form (Larcher, 1975). Total carbon uptake is affected not only by the maximum rate of photosynthesis, but also by the temperature, light, and moisture sensitivity of photosynthesis. Differences in the shape of the photosynthetic response surface with respect to major environmental factors could have a marked effect on total carbon uptake. Mooney et al. (1975) found that evergreen shrubs photosynthesized all year, with the major limitations to photosynthesis in the chaparral of southern California at a given nutrient availability being (in order of decreasing importance) water, photoperiod, and temperature. Little evidence for temperature acclimation of photosynthesis was found in the chaparral or matorral (Harrison, 1971; Mooney et al., 1977). Photosynthetic response surfaces have been characterized previously for relatively few species and growth forms in mediterranean areas (Mooney and Dunn, 1970b; Harrison et al., 1971; Morrow and Mooney, 1974; Mooney et al., 1977; Oechel and Lawrence, 1979).

The distribution of the coastal sage and chaparral growth forms along the aridity gradient from the coast to the mountains may reflect their adaptive carbon-gaining strategies (Mooney et al., 1977). Mooney and Dunn (1970a, b) considered the moisture regimes that would favor deciduous or evergreen species. The assumption was that the vegetational composition should reflect the relative competence of the respective growth forms in carbon accumulation (Mooney and Dunn, 1970a, b; Harrison et al., 1971; Mooney et al., 1975; Miller, 1979). Solbrig and Orians (1977) suggested that the percent distribution of evergreen and deciduous plants in the desert was affected by the relative length of the drought period. The species that were most productive in a given moisture region would prevail. These studies assumed that species abundance in the extant vegetation reflects the photosynthetic competence of the respective species. This reasoning indicates that in the chaparral, the evergreen shrub form is restricted at the coast, at high elevations, and at the desert transition by low carbon uptake rates. At the coast, this limitation has been assumed to be due to insufficient moisture to support the evergreen growth form in competition with drought deciduous species (Miller and Mooney, 1976; Miller, 1979).

The photosynthetic and respiratory response surfaces were determined for *Rhus ovata, Ceanothus greggii, Adenostoma fasciculatum*, and *Arctostaphylos glauca* at Echo Valley, and *Satureja gilliesii, Colliguaya odorifera, Trevoa trinervis*, and *Lithraea caustica* at Fundo Santa Laura. Measurements were made in late spring–May in California and November and early December in Chile–to represent periods of near-peak productivity (Gigon, 1979). Response patterns for the selected species at Echo Valley and Fundo Santa Laura were determined by using a transportable, six-cuvette infrared carbon dioxide analysis system to ensure sufficient replication of samples and species. The system employed Peltier-effect cooled chambers and an ADC Mark II[©] infrared gas analyzer with a computer-based data acquisition system (Oechel and Lawrence, 1979). Six terminal branches, generally two on each of three plants, were monitored for one week. One branch was held under ambient temperature and light conditions and provided an estimate of in situ diurnal photosynthesis, while in the other five cuvettes, the branches received ambient light under a variety of constant temperature regimes. A constant temperature was randomly assigned to each cuvette and generally changed every 24 h. During the measurement period, each of the five chambers experienced temperatures, generally at 5° intervals, from 5° to 40°C. The resulting data allowed the calculated response surface for each species to be plotted in a three-dimensional representation to facilitate comparisons between species at Echo Valley and Fundo Santa Laura over the range of temperature and light conditions measured. Curvilinear regressions of photosynthesis on temperature for various light classes were performed. These regressions were then used to calculate key values on the photosynthetic response surface to compare species at each research site in the two countries. The parameters compared were temperature optimum (i.e., the temperature at which maximum photosynthesis occurs), percent maximum photosynthesis at 5° and 40°C, and the temperature range that yields 85% of maximum photosynthesis.

Simulations using the Canopy Process Simulator (CAPS) (Chapter 11) were made to test the degree to which carbon uptake was maximized and to predict the effect of a change in the position of the temperature optimum for net photosynthetic carbon dioxide incorporation on daily carbon uptake.

The relationship between growth form and carbon uptake rate was determined through extensive measurements of carbon dioxide uptake at Echo Valley and Fundo Santa Laura using carbon-14 dioxide labeling techniques (Tieszen et al., 1974; Mustafa, 1979). The growth forms measured at Echo Valley were annual herbs, drought semideciduous shrubs, evergreen sclerophyllous shrubs, and succulents. In addition to these growth forms, perennial herbs, drought deciduous shrubs, photosynthetic stem shrubs, and evergreen phreatophytes were measured at Fundo Santa Laura.

A T-shaped experimental design was utilized to determine the relative variation in response pattern between Echo Valley and Fundo Santa Laura compared to the variation within southern California (Chapter 1). Major growth forms were characterized along an elevational transect from 110 m to 1400 m running from the coast at Torrey Pines State Park to the desert scrub transition at Boulevard (Figure 3.1). Results from this transect were compared to differences between growth forms at Echo Valley and Fundo Santa Laura. Productivity of major species along the transect was compared to the cover data collected by Steward and Webber (Chapter 3) in order to determine the relationship between photosynthetic capacity and cover in the chaparral.

7.2. Patterns of Temperature and Light Dependency

All species showed a generalist response to temperature. That is, species were not photosynthetically specialized for maximum photosynthetic rate over a narrow temperature range. Rather, high photosynthetic rates were maintained over a wide range in temperature, and temperature suppression of photosynthesis at the extremes of temperatures measured was surprisingly low.

At full sunlight, *A. fasciculatum* incorporated carbon dioxide at a rate of at least 85% of the maximum photosynthetic rate over a temperature range of more than 27°C; these high rates of photosynthesis occurred at temperatures from < 5° to > 30°C. At high light intensities (e.g., 1800 μE m^{-2} s^{-1}), *C. greggii* photosynthesized at a rate greater than 85% of maximum photosynthesis over a temperature range of $\geqslant$ 25°C. Under similar conditions, *R. ovata* and *A. glauca* had slightly smaller temperature ranges, 22° and 18°C, respectively, which yielded 85% or more of the maximum carbon uptake rate. The same pattern also seemed to hold at lower light intensities.

This broad response pattern appeared well adapted to the wide temperature range experienced by the vegetation in the chaparral and matorral where both freezing conditions and extreme heat may be experienced during the growing season, and positive photosynthesis is maintained through the year by the evergreen species (Harrison, 1971; Mooney, 1977a; Mustafa, 1978; Gigon, 1979). Except for important differences in maximum carbon uptake rates, visual inspection of the response surfaces indicates no major difference in the pattern of carbon uptake with respect to temperature and light intensity between the Californian and Chilean species measured. This impression is validated in comparisons between aspects of the response surfaces for species, which are made in the following sections.

Similar calculations were made from the data of Mooney et al. (1975) for photosynthesis in *Heteromeles arbutifolia*. Under saturating light, *H. arbutifolia* also showed broad temperature response patterns. The data indicated that at peak season, 85% of maximum photosynthesis was maintained over a temperature range exceeding 20°C. However, as the summer drought increased, the temperature effect on assimilation increased. By early September, 85% or greater of maximum photosynthesis occurred over a temperature range of only 11°C. During the summer, the rate of photosynthesis at 30°C dropped to 42% of that at optimum temperature. These results suggest that, in some species, production may be relatively insensitive to temperature during certain parts of the year but may be very sensitive at other seasons.

7.2.1 Temperature Optima

At saturating light intensities, the temperature optima for photosynthesis generally occurred at temperatures between 15° and 31°C (Figures 7.1 and 7.2). For *Arbutus menziesii* and *H. arbutifolia*, optimum temperature values were at the low end of the above range (15° to 20°C during February) (Morrow, 1971). Although Dunn (1975) reported values for *R. ovata* as low as 5°C or less, his values generally ranged from 15° to 25°C. *Adenostoma fasciculatum* had the highest temperature optima measured and *C. greggii* the lowest. *Rhus ovata* and *A. glauca* had temperature optima similar to that of *C. greggii*. In most cases, temperature optima were affected little by light intensity and at most radiation levels were within $\pm 2°C$ of the value at 1800 μE m^{-2} s^{-1}. *Adenostoma fasciculatum* displayed the highest optimum temperature of all species observed. While the broadness of the response surfaces decreased the importance of the location of the optimum temperature somewhat, the high temperature optimum of *A. fasciculatum* may reflect adaptation to the warm, exposed, equator-facing slopes it frequently occupies. The temperature optima for photosynthesis in the Chilean species measured were similar and occurred between 18° and 23°C. The narrower range of temperature optima observed at Fundo Santa Laura was perhaps a result of the more maritime climate at Fundo Santa Laura compared to the climate at Echo Valley.

7.2.2 Maximum Carbon Uptake Rates

To understand the basis for similarity or dissimilarity in community resource use, comparisons were made in physiological responses between species at Echo Valley and Fundo Santa Laura. The species chosen for intensive analysis of the photosynthetic systems were important by virtue of cover, biomass, and frequency of occurrence or because they represented an identifiable growth pattern with equivalences in both countries. The pairs of Californian and Chilean species used were generally those identified as analogous by Thrower and Bradbury (1977) except that instead of *Ceanothus leucodermis*, *Ceanothus greggii* was paired with *T. trinervis*, because *C. greggii* was more abundant at Echo Valley. These analogous species were initially identified on the basis of growth form and morphology; the similarity in form between the analogous pairs is often striking. The level of similarity between analogous pairs in phenology, anatomy, and chemical composition has also been investigated (Thrower and Bradbury, 1977). The species were combined into paired samples, and the species within the pairs were compared with respect to pertinent aspects of their photosynthetic response to temperature. In selecting the species to form the pairs, consideration was given to the importance of each species in the vegetation (e.g., abundance and cover), as well as to its morphology, habitat, and physiological characteristics (e.g., nitrogen fixation). The pairs selected were *C. greggii* and *T. trinervis*, both of which are reported nitrogen fixers and members of the Rhamnaceae; *A. glauca* and *C. odorifera*, which are evergreen broad sclerophylls; *A. fasciculatum* and *S. gilliesii*, which are narrow-leaved shrubs; and *R. ovata* and *L. caustica*, which show low leaf conductances to water vapor (Poole and Miller, 1975) and late season growth patterns (Thrower and Bradbury, 1977), are members of the Anacardiaceae, and are evergreen broad sclerophylls (Figure 7.3). Certain important differences, however, exist within these species pairs. Bark photosynthesis is more pronounced in *T. trinervis* than in *C. greggii*. *Trevoa trinervis* and *S. gilliesii* are drought deciduous and drought semideciduous malacophylls, respectively, and *C. odorifera* is an evergreen sclerophyll but drops some of its leaves during the summer after they are formed, while all of the principal Californian species studied were evergreen. The results showed that, even though *T. trinervis* and *S. gilliesii* differed from their paired species by being drought deciduous, the similarity in photosynthetic performance within representative pairs was marked. Despite large differences in the low and high temperature response of *L. caustica* and *R. ovata*, the paired

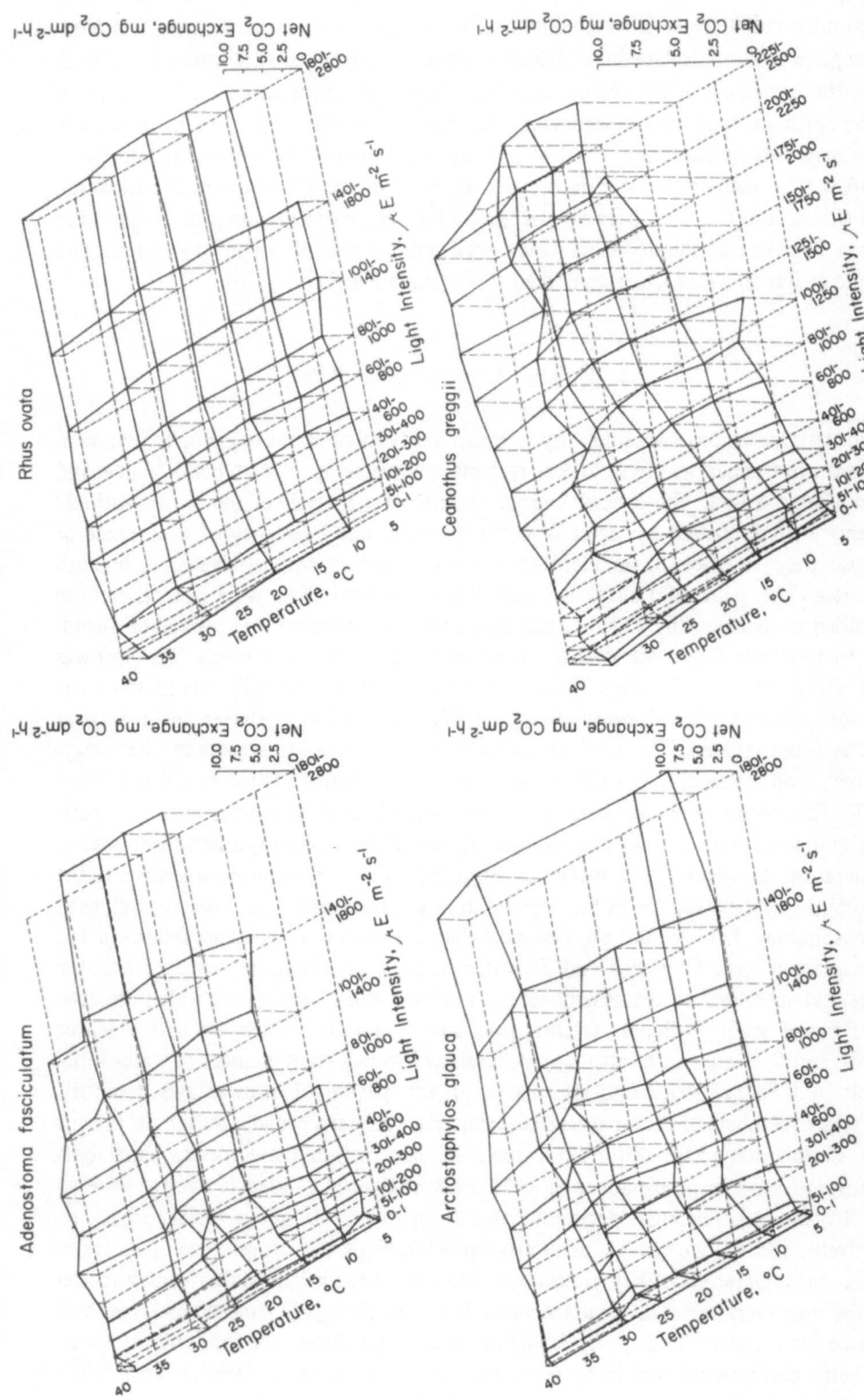

Figure 7.1. Photosynthetic response surfaces as a function of light and temperature. (a) Response surfaces were determined in May and early June 1976 at Echo Valley for *Adenostoma Fasciculatum*, *Rhus ovata*, *Arctostaphylos glauca*, and *Ceanothus greggii*; and

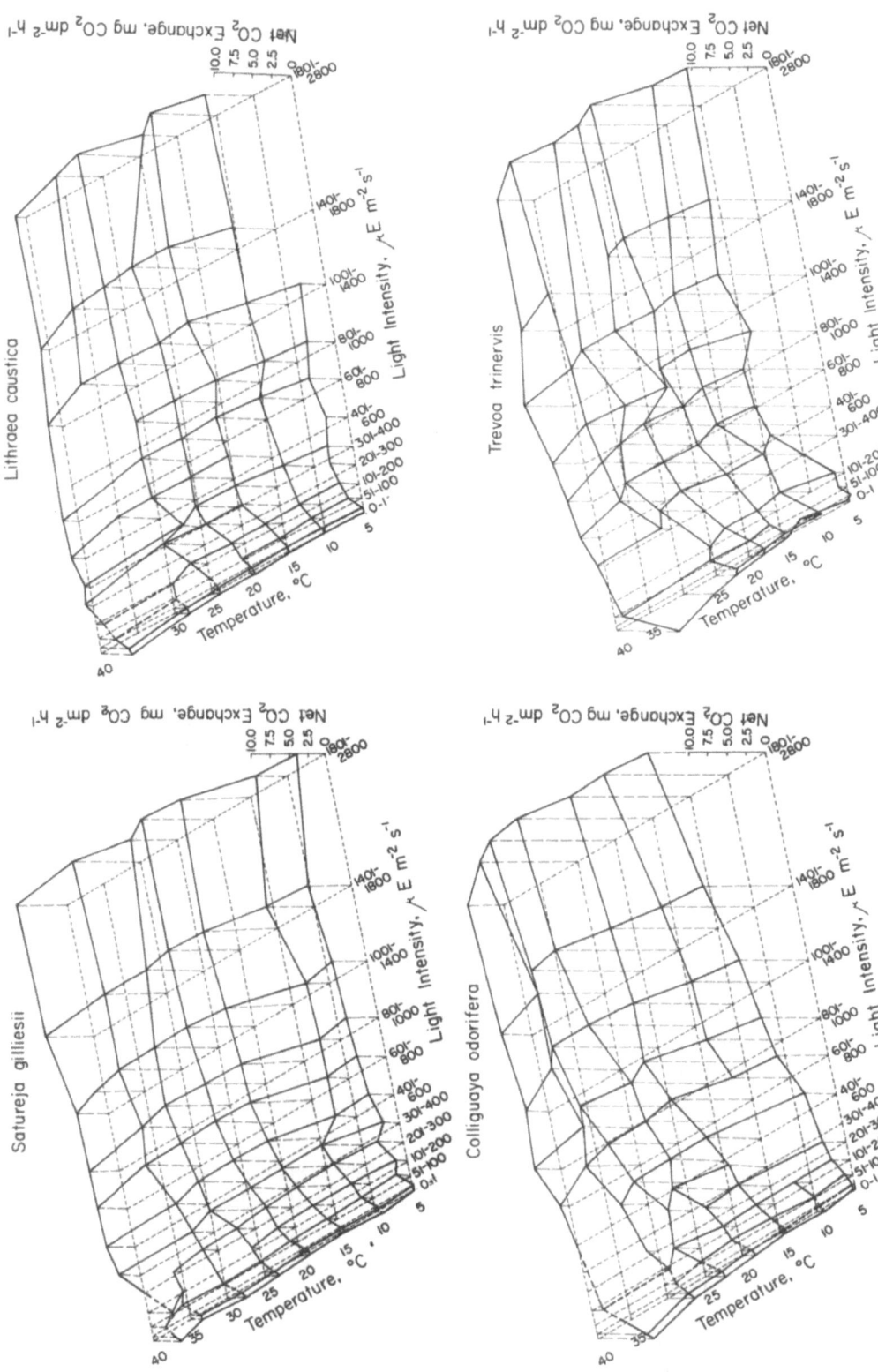

(b) November to early December 1976 at Fundo Santa Laura for *Satureja gilliesii, Lithraea caustica, Colliguaya odorifera*, and *Trevoa trinervis*.

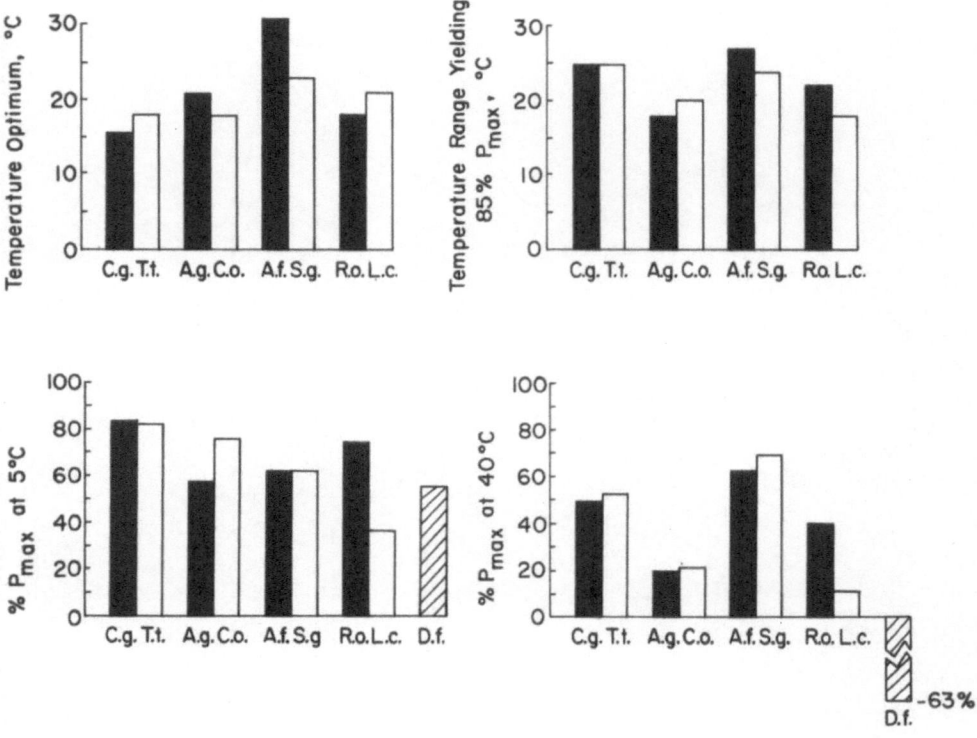

Figure 7.2. Temperature responses of photosynthesis (P_{max}) for *Ceanothus greggii* (C.g.), *Arctostaphylos glauca* (A.g.), *Adenostoma fasciculatum* (A.f.), and *Rhus ovata* (R.o.) measured at Echo Valley; and *Trevoa trinervis* (T.t.), *Colliguaya odorifera* (C.o.), *Satureja gilliesii* (S.g.), and *Lithraea caustica* (L.c.) measured at Fundo Santa Laura during the spring and early summer (May-June and November-December 1976, respectively) under saturating light intensity (1401 to 1800 $\mu E\ m^{-2}\ s^{-1}$). Comparisons are made to *Dupontia fischeri* (D.f.), an arctic graminoid (Tieszen, 1973).

t-test for means showed no overall statistical difference at the 0.05 probability level in the relative 5° or 40°C photosynthetic responses between the paired Californian and Chilean species studied. However, the Californian plants showed a broader temperature range in which at least 85% of the maximum photosynthesis rate occurred and showed significantly higher temperature optima ($P < 0.05$, paired t-test). This may have reflected the warmer temperatures and the greater temperature variability at Echo Valley (Miller et al., 1977; Chapter 2).

    The Chilean species showed significant and consistently higher maximum photosynthetic rates than did their paired Californian counterparts ($P < 0.01$, paired t-test). The mean rate for the Californian species was 8.45 mg CO_2 dm^{-2} h^{-1}. The equivalent value for the Chilean species was 11.1 mg CO_2 dm^{-2} h^{-1}. The vegetation in central Chile has more drought deciduous shrub elements than does the vegetation in southern California. This is reflected in the paired Chilean species and may contribute to the higher mean photosynthetic rates in these species. However, all Chilean species showed higher photosynthetic rates than their paired counterparts, including the evergreens *L. caustica* and *C. odorifera*. Because these values represent measurements taken once during the spring, the observed pattern may shift somewhat when viewed on an annual basis.

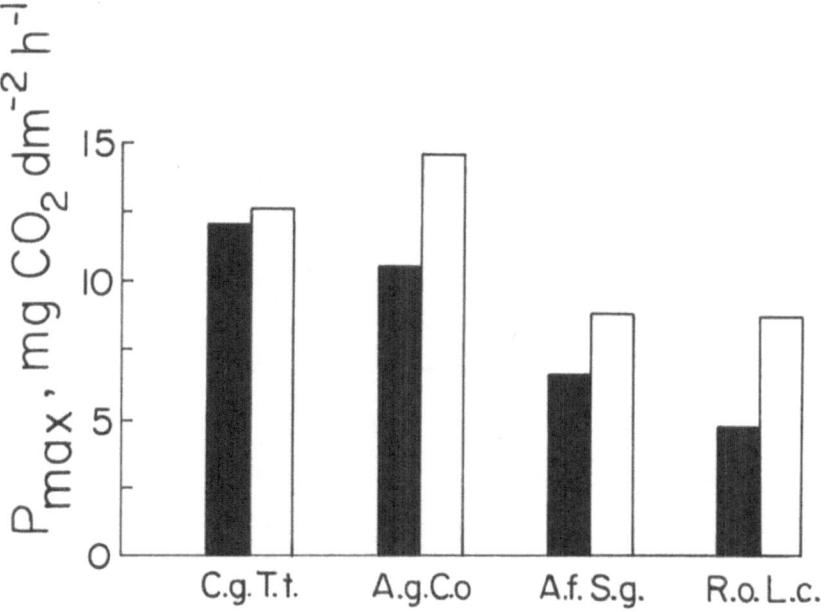

Figure 7.3. Maximum spring photosynthetic rates (P_{max}) for *Ceanothus greggii* (C.g.), *Arctostaphylos glauca* (A.g.), *Adenostoma fasciculatum* (A.f.), and *Rhus ovata* (R.o.) at Echo Valley in May 1976; and for *Trevoa trinervis* (T.t.), *Colliguaya odorifera* (C.o.), *Satureja gilliesii* (S.g.), and *Lithraea caustica* (L.c.) at Fundo Santa Laura in November 1976 as determined with infrared gas analysis at saturating sunlight (1401 to 1800 μE m^{-2} s^{-1}).

The maximal carbon uptake rates were positively correlated with observed conductances to water vapor for all species except the two nitrogen-fixing species, *C. greggii* and *T. trinervis*. In *C. greggii* and *T. trinervis*, photosynthetic rates were higher than would be predicted from their conductances. *Ceanothus greggii* and *T. trinervis* show relatively high tissue nitrogen levels, possibly as the result of significant nitrogen fixation occurring at some point in their life histories (Chapter 9), which may result in higher RuBP carboxylase levels and lower cellular resistances to carbon dioxide flux. This would result in higher photosynthetic rates at any given conductance.

Although the maximal carbon uptake rates reported here for chaparral species at Echo Valley were higher than some reported values for these species (Mooney and Dunn, 1970a, b; Harrison et al., 1971), they were generally bracketed by the existing literature values. Dunn (1975) reported a range of maximal carbon uptake values for *R. ovata* of about 2.0 mg CO_2 dm^{-2} h^{-1} in October 1966 to 15.4 mg CO_2 dm^{-2} h^{-1} during August 1967.

The Californian and Chilean species measured showed remarkable similarity in their temperature dependency of photosynthesis; their temperature optima and temperature response curves were indistinguishable from one another, especially in contrast to the patterns exhibited by species from other biomes or ecosystems (Figure 7.4). In species from other ecosystems, the temperature dependency of photosynthesis was more pro-

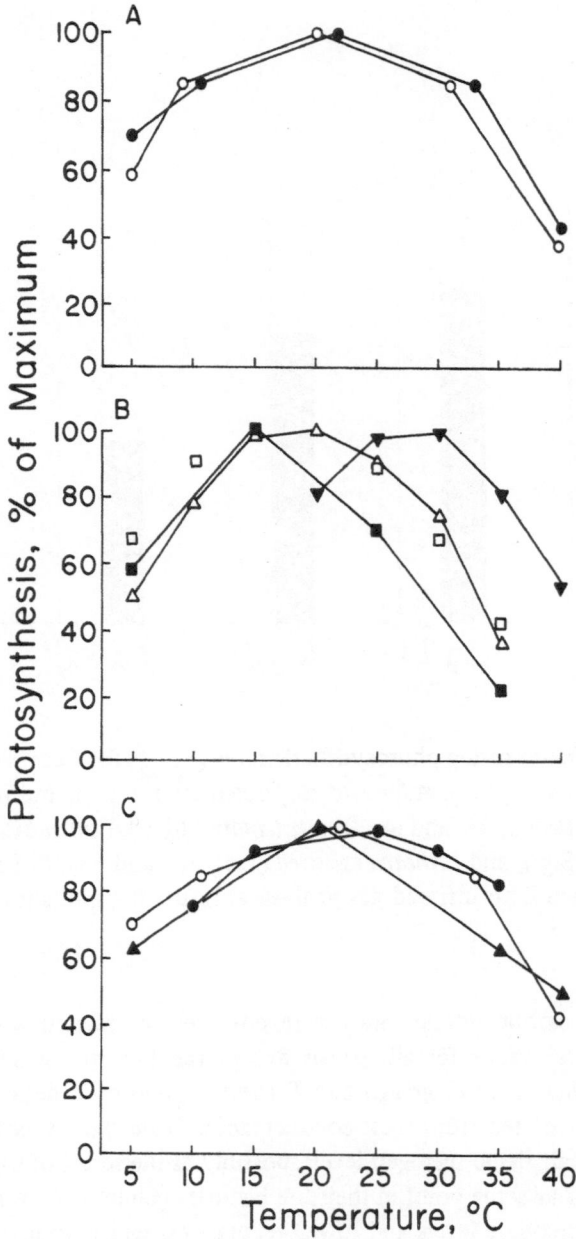

Figure 7.4. (A) Average photosynthetic response of four chaparral species (●): *Ceano-thus greggii, Arctostaphylos glauca, Adenostoma fasciculatum,* and *Rhus ovata*; and four matorral species (○): *Trevoa trinervis, Colliguaya odorifera, Satureja gilliesii,* and *Lithraea caustica* to temperature at saturating light intensity (1401 to 1800 $\mu E\ m^{-2}\ s^{-1}$) (derived from Figure 7.1). (B) Average photosynthetic response of representative species from montane, *Picea abies* (△) (Pisek and Winkler, 1959); desert, *Hammada scoparia* (▼) (Schulze et al., 1972a); boreal, *Picea mariana* (□) (Vowinckle et al., 1975); and tundra, *Dupontia fischeri* (■) (Tieszen, 1973) habitats. (C) Comparison of average photosynthetic response of chaparral (○) to data for *Rhus ovata* from August 1967 (●) (Dunn, 1975) and to *Pinus ponderosa* (▲) from the San Bernardino Mountains of southern California (May response; Coyne, *unpubl. data*).

nounced and the position of these species along the temperature axis was determined in response to environmental conditions (Larcher, 1975). Arctic vascular plants were photosynthetically very competent at moderately low temperatures of 5°C but were photosynthetically less competent at temperatures greater than 35°C (Billings, 1973; Skre, 1972, 1975; Tieszen, 1973; Limbach and Oechel, *unpubl. data*). The opposite is often true of desert species (e.g., Schulze et al., 1972a). Northern temperate species such as *Picea abies* (Pisek and Winkler, 1959) and *Picea mariana* (Vowinckle et al., 1975) may be intermediate. However, broad response surfaces for photosynthesis have been reported for *Larrea divaricata*, an evergreen desert shrub photosynthetically active throughout the year (Oechel et al., 1972). Dunn (1975) found that *R. ovata* had patterns that were both temperature sensitive and insensitive, but the generalized pattern was similar to that reported here. Ponderosa pine at an elevation of about 3100 m also had similar temperature dependencies to those reported here. Coyne (*pers. comm.*) reported photosynthetic maxima in May of 8 mg CO_2 dm^{-2} h^{-1} at 20°C in *Pinus ponderosa*. A photosynthetic rate of 63% of the maximum occurred at 5°C and a rate of 50% of the maximum at 40°C. Similar patterns of photosynthetic insensitivity to low temperature may be common in evergreen species capable of year-round photosynthesis.

7.3. Simulated Temperature Sensitivity

The broad, generalized temperature response pattern of photosynthesis in the eight species studied indicates that energy capture in these species should not be affected by modest temperature changes due to global temperature change or by changes due to temperature differences likely to occur over fairly large elevational or regional ranges. CAPS was used to predict the effect on seasonal carbon dioxide incorporation of a ± 5°C change from recorded climatic values for Echo Valley. A ± 5°C temperature change corresponds to an elevational displacement of more than ± 800 m or to moving from near sea level at the coast to 1800 m at the chaparral/montane forest transition (Miller et al., 1977). The simulations suggest that this magnitude of temperature change has relatively little effect on seasonal carbon dioxide incorporation (Figure 7.5). Species spanning this elevational range, such as *A. fasciculatum*, should show little temperature limitation of photosynthesis anywhere along their range of distribution in San Diego County. Seasonal temperature changes caused by alterations in atmospheric levels of carbon dioxide or aerosols, or by climatic shifts are, therefore, predicted to have had only minor impacts on seasonal carbon dioxide incorporation of chaparral species with broad temperature response patterns similar to those reported here.

7.4. Simulated Fit of the Photosynthetic Temperature Optima

The position of the temperature optimum for photosynthesis can significantly affect carbon uptake rates and resource-use efficiency. For example, an inefficient or non-optimal placement of the temperature at which maximum photosynthesis occurs can decrease photosynthesis by a few percent (Miller et al., 1976a) or by more than 25% of that theoretically possible (Oechel et al., 1975). Broad photosynthetic response patterns decrease the dependency of photosynthesis on the placement of the temperature

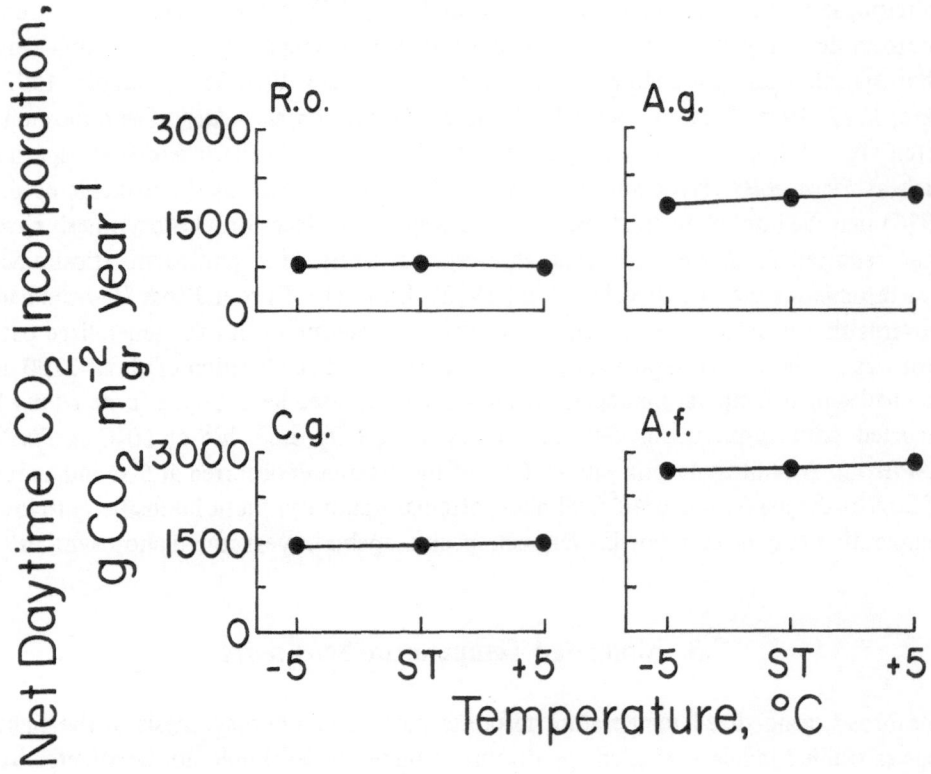

Figure 7.5. The effects on simulated seasonal (daytime) photosynthesis incorporated in grams CO_2 per square meter of ground per year of an increase or decrease in temperature of 5°C from observed standard temperatures (ST) in 1973-1974 in *Rhus ovata* (R.o.), *Arctostaphylos glauca* (A.g.), *Ceanothus greggii* (C.g.), and *Adenostoma fasciculatum* (A.f.) with a leaf area index of 1 and a stem area index of 0.9.

optima. A CAPS simulation was used to determine the fit of the temperature optima and the effect of moving the position of the temperature optima on daily carbon uptake rates under spring conditions.

The fit of the extant temperature optima were analyzed in *C. greggii, A. glauca, A. fasciculatum,* and *R. ovata* in southern California, and *T. trinervis, C. odorifera, S. gilliesii,* and *L. caustica* in central Chile. In these simulations, the shapes of the response surfaces were held constant, but optimum temperatures for photosynthesis were varied from 5° to 30°C. In general, the placement of the temperature optima had little effect on daily spring carbon dioxide incorporation rates in Californian species (Figure 7.6). However, *A. glauca* was particularly sensitive to the placement of temperature optima below 25°C because of the combination of high frequent ambient temperatures and markedly lower photosynthetic rates at temperatures above the optimum. As the temperature optima decrease, ambient temperatures are more frequently above the optimum, thus lowering photosynthetic yield. The Chilean species measured are more sensitive to the location of the temperature optima than are the Californian species. This partially results from the narrower range of temperatures experienced at Fundo Santa Laura due to the greater maritime influence at this site.

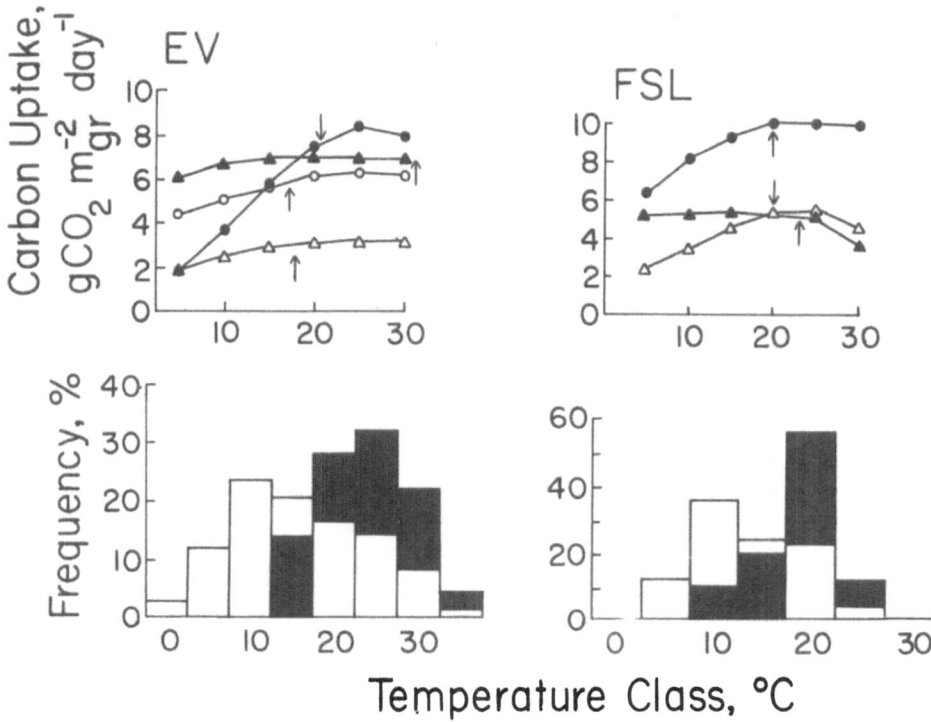

Figure 7.6. The simulated effect of variations in the placement of temperature optima from 5° to 30°C on daily spring carbon uptake in *Arctostaphylos glauca* (●), *Adenostoma fasciculatum* (▲), *Ceanothus greggii* (○), and *Rhus ovata* (△) at Echo Valley (EV); and *Colliguaya odorifera* (●), *Lithraea caustica* (△), and *Satureja gilliesii* (▲) at Fundo Santa Laura (FSL). Arrows indicate the observed temperature optimum for each species. Vertical bars indicate the midday (■) and 24-hour (□) temperature frequencies for the period indicated. Physiological data are for 1976, climatological data are for May 1973 in California and November 1973 in Chile.

All eight species measured had a temperature optimum that would produce 90% or greater of the maximal uptake rate under the extant environment. At Fundo Santa Laura, the temperature optima were found to be similar to the most frequently occurring temperatures. At Echo Valley, this was not necessarily the case. The more narrow temperature distribution in central Chile apparently results in stronger selective pressure on the placement of the temperature optima because the location of the temperature optimum has a greater effect on carbon uptake in Chilean species than in Californian species. This in turn results in similar temperature optima for the Chilean species.

7.5. Quantum Efficiency at Light Saturation

The photosynthetic rates of Chilean species, which are generally higher than those of the Californian species, result in higher resource-use efficiency of light in photosynthesis at Fundo Santa Laura (Table 7.1). Percent quantum efficiency, as used here, is the Einsteins of radiation (6.03×10^{23} photons, 400 to 700 nm) intercepted by the leaf divided by the moles of carbon dioxide fixed times 100. The average quantum effi-

Table 7.1. Quantum yields (mol CO_2 E^{-1}) for chaparral species at Echo Valley and matorral species at Fundo Santa Laura[a]

Species	P_{max} (mg CO_2 dm^{-2} h)	$\dot{P}_{max}$ (μmol CO_2 m^{-2} s^{-1})	Percent quantum efficiency (mol $CO_2 \times 1/E^{-1} \times 100$)
	Echo Valley		
Ceanothus greggii	12.0	7.57	0.47
Arctostaphylos glauca	10.1	6.37	0.40
Rhus ovata	4.7	2.97	0.16
Adenostoma fasciculatum	6.6	4.17	0.26
Mean	8.4	5.27	0.32
	Fundo Santa Laura		
Colliguaya odorifera	14.2	8.96	0.56
Satureja gilliesii	8.4	5.30	0.33
Trevoa trinervis	12.0	7.57	0.47
Lithraea caustica	8.4	5.30	0.33
Mean	10.8	6.78	0.42

[a]Calculated from maximum photosynthesis (P_{max}) values at a quantum flux of 1600 μE m^{-2} s^{-1}.

ciency at light saturation at Echo Valley for *C. greggii*, *A. glauca*, *R. ovata*, and *A. fasciculatum* was 0.32%. The average quantum efficiency was 0.42% for *C. odorifera*, *S. gilliesii*, *T. trinervis*, and *L. caustica* at Fundo Santa Laura, an increase of 31% over the value for Echo Valley.

7.6. Growth Form Effects on Photosynthetic Rate

Photosynthetic capacity varies for plants not only of diverse origin (Cooper, 1975), but also of diverse growth forms (Larcher, 1975). Drought deciduous plants in the Californian coastal sage scrub community have higher capacities than do the more sclerophyllous evergreens of the chaparral (Harrison et al., 1971).

Paeonia californica, an herbaceous perennial, is active in the winter and spring. The photosynthetic tissue does not tolerate drought and dies back with the onset of summer drought, leaving the belowground structure, including the large fleshy root, to persist until the next growing season. This species also had a wide temperature range for photosynthesis, and 85% or more of maximum photosynthesis was maintained over a 21°C range. The temperature optimum was at 25°C. It is interesting that this species, whose photosynthetic tissue does not experience the warmest portions of the year, also displays a wide temperature response.

A photosynthetic rate of 52.5 mg CO_2 dm^{-2} h^{-1} was measured in *P. californica* at its temperature optimum of 25°C. This unusually high rate reflects the mesomorphic leaves and high leaf conductances of this species. If maintained for a photosynthetic period of approximately 6 months, these rates would result in significantly higher carbon uptake rates than those of evergreen sclerophyllous shrubs that comprise the dominate growth form in the chaparral.

The patterns of maximum carbon incorporation displayed by representative species from several growth forms at Echo Valley (Table 7.2) and Fundo Santa Laura (Table 7.3) were determined using carbon-14 dioxide techniques (Oechel and Mustafa, 1979) in order to assess the effect of growth form on production and to interpret the photosynthetic responses of the predominant species relative to other elements in the vegetation. In southern California, five growth forms, annual herbs, perennial herbs, evergreen sclerophyllous shrubs, drought semideciduous shrubs, and succulents, were sampled during late May and early June 1976. The drought semideciduous form was also sampled during May 1976 at lower elevations near Torrey Pines State Park (110 m) and Alpine (610 m). The maximum carbon uptake rates were calculated on a dry weight and leaf area basis. At Fundo Santa Laura, three additional growth forms, drought deciduous shrubs, photosynthetic stem shrubs, and evergreen phreatophytes, were measured monthly from June 1976 to July 1977.

Considered on a dry weight basis, the maximum carbon uptake rates decreased in southern California as follows: annual herbs > perennial herbs > drought deciduous shrubs > evergreen sclerophyllous shrubs > succulents (Figure 7.7). On a leaf-area basis, the sequence was slightly altered as follows: drought deciduous shrubs > perennial herbs > annual herbs > evergreen sclerophyllous shrubs > succulents. However, on a leaf-area basis, no significant difference occurred in mean maximum carbon uptake rates among the annuals, perennials, and the drought deciduous shrubs or between evergreen and succulent forms (Duncan's new multiple range test, $P = 0.005$). Thus, on

Table 7.2. Maximum photosynthesis rates determined by $^{14}CO_2$ and infrared gas analysis (IRGA) techniques and leaf density for chaparral species along an elevational transect in San Diego County

Growth form and species	Elevation (m)	Sampling date (Day/Mo/Yr)	Leaf density (g dry wt dm^{-2})	Maximum photosynthesis rates		
				(mg CO_2 g^{-1} dry wt h^{-1}) $^{14}CO_2$	(mg CO_2 dm^{-2} h^{-1}) $^{14}CO_2$	IRGA
Annual herbs						
Clarkia purpurea	1000	04/01/76	0.94	9.29	8.77	
Mimulus breuipes	1000	04/06/76	0.94	9.31	8.78	
Vicia spp.	1000	30/05/76	0.47	32.62	15.41	
Mean				17.07	10.99	
Perennial herbs and low shrubs						
Castilleja foliolosa	1000	30/05/76	–	12.5	–	
Eriophyllum confertiflorum	1150	10/06/76	–	13.4	–	
Lupinus excubitus	1000	29/05/76	0.94	13.6	12.86	
Paeonia californica	1000	04/06/76	0.63	12.6	7.95	
Penstemon centranthifolius	1000	29/05/76	0.94	17.6	16.62	
Penstemon spectabilis	1000	29/05/76	0.94	7.8	7.30	
Mean				12.9	11.18	
Drought semideciduous shrubs						
Salvia apiana	610	11/06/76	1.46	14.0	13.1	
Salvia mellifera	85	12/06/76	1.56	6.6	8.3	
Trichostema parishii	1000	03/06/76	1.58	9.0	15.2	
Mean				9.9	12.2	
Evergreen sclerophyllous shrubs						
Adenostoma fasciculatum						
Equator-facing slope	1000	03/06/76	1.17	3.6	4.3	
Equator-facing slope	1000	17/05-01/06/76				6.6
Pole-facing slope	1000	05/06/76	1.17	3.0	3.5	

Arctostaphylos glauca						
Equator-facing slope	1000	03/06/76	2.08	4.1	9.0	
Equator-facing slope	1000	03-14/06/76				10.5
Pole-facing slope	1000	05/06/76	2.18	3.5	7.6	
Ceanothus greggii						
Equator-facing slope	1000	03/06/76	3.13	3.3	10.4	
Equator-facing slope	1000	16-20/06/76				7.6
Ceanothus leucodermis	1000	05/06/76	0.94	7.6	7.2	
Ceanothus verrucosus	120	12/06/76	3.13	2.6	8.3	
Cercocarpus betuloides	1150	10/06/76	1.56	4.3	6.8	
Quercus dumosa	1000	05/06/76	1.56	4.0	6.2	
Rhus integrifolia	85	12/06/76	3.13	1.4	4.3	
Rhus ovata						
Equator-facing slope	1000	03/06/76	3.13	1.1	3.5	
Equator-facing slope	1000	22/27/06/76				4.7
Mean				3.5	6.5	7.4
Succulent						
Yucca whipplei	1000	01/06/76	2.5	2.3	5.8	

Table 7.3. Annual Gross photosynthesis and assimilation rates of species occurring in matorral vegetation at Fundo Santa Laura calculated from $^{14}CO_2$ measurements

Species	Growth form[a]	Gross photosynthesis (g CO_2 dm^{-2} yr^{-1})	Assimilation (g organic matter m^{-2} yr^{-1})
Quillaja saponaria	ES	22.3	1,373.5
Colliguaya odorifera	ES	24.8	1,525.8
Lithraea caustica	ES	16.4	1,012.5
Cryptocarya alba	ES	13.2	813.6
Mean		18.7	1,181.3
Trichocline aurea	PH	8.7	537.9
Pasithea coerulea	PH	4.1	253.2
Mean		6.4±	395.5
Satureja gilliesii	DD	16.5	1,012.5
Trevoa trinervis	DD	25.2	1,547.7
Mean		20.9	1,280.1
Trevoa trinervis	PS	8.0	493.7
Retanilla ephedra	PS	12.6	772.4
Mean		10.3	633.0
Puya berteroniana	S	2.5	155.3
Trichocereus chilensis	S	2.4	148.8
Mean		2.5	152.0
Drimys winteri	EP	6.5	398.8
Myrceugenella chequen	EP	12.4	762.8
Mean		9.5	580.8
Avena fatua	AH	15.6	958.9
Madia sativa	AH	12.4	762.8
Mean		14.0	860.9

[a] ES = evergreen schlerophyllous shrubs; PH = perennial herbs; DD = drought deciduous and semideciduous shrubs; PS = photosynthetic stem shrubs; S = succulents; EP = evergreen phreatophytes; and AH = annual herbs.

area basis, the five growth forms could be divided into two major groups: (1) the malacophyllous group, which included annual herbs, perennial herbs, and drought semideciduous shrubs; and (2) the sclerophyllous group, which included evergreen shrubs and succulents. Species in the malacophyllous group had mesophytic leaves, a low leaf-density thickness (specific leaf weight) ranging from 0.78 to 1.53 g dry weight dm^{-2} (Figure 7.8), and photosynthesized at high rates. Species in the sclerophyllous group had a high leaf-density thickness and photosynthesized all year at lower rates than did the first group. Carbon uptake rate was sampled in annuals during June 1976 when most of their leaves were fully mature, which may have contributed to the lower carbon uptake rates measured compared to rates reported for annual herbs in the desert (Mooney et al., 1976).

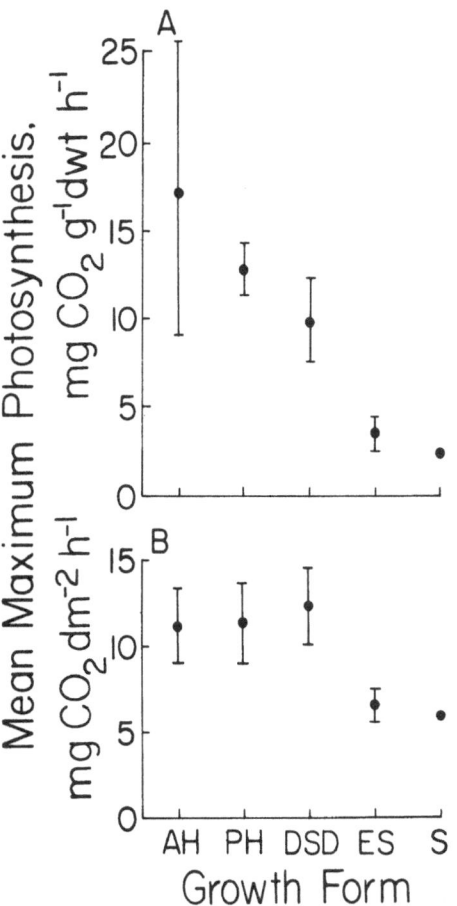

Figure 7.7. Mean maximum photosynthetic rates in southern California chaparral of five growth forms on a dry-weight basis (A) and on a unit-leaf area basis (B). Bars represent 1 SE resulting from the ranges in mean values displayed by the species comprising a growth form: annual herbs (AH), perennial herbs (PM), drought semideciduous shrubs (DSD), evergreen shrubs (ES), and succulents (S).

The only succulent species included in this study was *Yucca whipplei*, a leaf succulent. Photosynthetic maxima for *Y. whipplei* were obtained during daylight hours using carbon-14 dioxide techniques. Because crassulacean acid metabolism (CAM) has been reported in other members of the genera (e.g., Kluge and Ting, 1979), diurnal gas exchange measurements were made near the end of the drought period in October 1979 to determine if there was substantial nighttime assimilation of carbon dioxide in *Y. whipplei*, which would increase its maximum calculated rate of carbon dioxide uptake. No evidence of CAM activity was found; rather there was the diurnal pattern of carbon dioxide assimilation typical of C-3 plant species. Therefore, while facultative CAM activity may exist, *Y. whipplei* is not an obligate CAM species.

Denser leaves, those with greater dry weight per unit area, generally have lower carbon uptake rates on a weight basis. Sclerophytic leaves conserve water more efficiently than malacophyllous leaves; however, regulation of water loss results in regulation of gas exchange due to partial or complete stomatal closure. The evergreen and

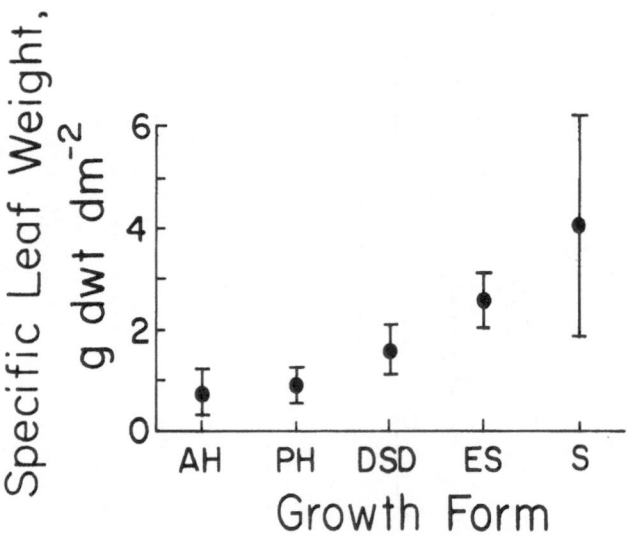

Figure 7.8. Mean specific leaf weight of the various growth forms in southern California chaparral. Growth forms are: annual herbs (AH), perennial herbs (PH), drought semideciduous shrubs (DSD), evergreen shrubs (ES), and succulents (S). Bars represent 1 SD of the mean and are affected by the variation in measurements within and between species sampled in each growth form.

succulent forms are able to reduce water loss in part by their higher resistance to water vapor loss, but this occurs at the expense of high carbon dioxide gas exchange rates. Malacophyllous leaves are shed as a means of avoiding drought while allowing high rates of photosynthesis. Malacophyllous leaves lose water rapidly through transpiration if water is available because they have no extraordinary means for water control (Harrison et al., 1971; Gigon, 1979; Miller and Poole, 1979).

Although chaparral species may be ranked by growth form in terms of their maximum field carbon uptake rates on dry-weight basis, the range of variability among species is such that knowing a plant's growth form is not sufficient to predict, a priori, whether it will have had a higher or lower carbon uptake rate than species in other growth forms. For example, the evergreen shrub *Ceanothus verrucosus* had a similar carbon uptake rate to the drought semideciduous shrub *Salvia mellifera* when measured at the same time and location. When comparing carbon uptake rates on unit-leaf area basis, only two groups exist: the malacophyll form, consisting of annuals, perennial herbs, and drought deciduous shrubs, that have leaf longevities of less than one year; and the sclerophytic evergreen form, consisting of evergreen shrubs and succulents that have leaf longevities of a year or greater. Thus, the evergreen versus deciduous or sclerophyll versus malacophyll categories were more important than other growth form classifications when predicting photosynthesis on a unit-leaf area basis.

Chilean species showed relationships between growth form and carbon uptake rate (Figure 7.9) similar to that described above for Californian species. Deciduous leaves generally had much higher carbon uptake rates than did evergreen leaf tissues. However, the overall pattern of photosynthesis (area basis) by growth form was drought deciduous ≥ annual herbs > perennial herbs ≥ photosynthetic stem shrubs ≥ evergreen

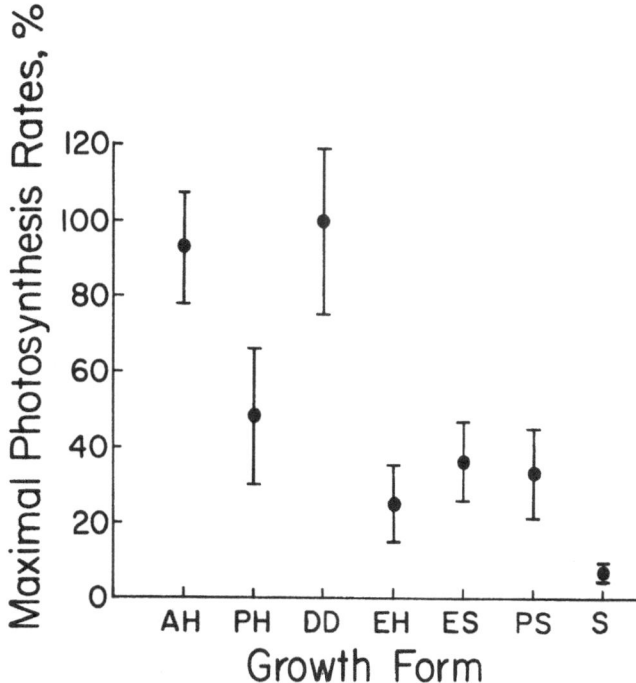

Figure 7.9. Maximal photosynthetic rates as a percent of the mean for the most productive growth form, the drought deciduous form. Relative mean rates for the growth form (based on mg CO_2 dm^{-2} h^{-1}) and the highest and lowest rates observed for species within each growth form (bars) are presented. The species included in each growth form are listed in Table 7.3: annual herbs (AH), perennial herbs (PH), drought deciduous and drought semideciduous shrubs (DD), evergreen phreatophytes (EP), evergreen sclerophyllous shrubs (ES), photosynthetic stem shrubs (PS), and succulents (S).

phreatophytes > succulents. It was not predicted that the maximum carbon uptake rates of the photosynthetic stem shrub growth form would be higher than the evergreen phreatophytes and of similar magnitude to the evergreen sclerophyllous shrubs. Also, maximum photosynthetic rates of the perennial herbs (geophytes) and evergreen phreatophytes measured were lower than were predicted given the mesomorphic leaf in the former case and the increased water supply in the latter case. Because these measurements were made monthly over the course of the year and because a large number of measurements were made, the relative rates are believed to be representative of the species measured. However, in many growth form categories, only two species were represented, so the results may be biased by the species selected.

7.7. Relation Between Photosynthetic Competence and Cover

An attempt was made to determine to what extent the success of a plant correlated with the rate of carbon accumulation. While many criteria for success are possible, the parameter chosen in this study to indicate the level of success in the vegetation was the percent cover of the species. This was felt to be a good indication of the relative performance of species in the community and an indication of one available resource (light) captured by the species. Other indicators, such as seed production or successful reproduction, were outside the scope of this study.

To determine the relationship between total seasonal carbon uptake and species success, seasonal carbon uptake was measured using carbon-14 dioxide techniques and was compared to percent cover data as presented in Chapter 3. It was reasonable to expect that, when widely different growth forms are contrasted, the importance of carbon uptake rate may be overridden by other factors such as height, competitive interactions for sunlight, growth rate, and requirements for germination and establishment. Therefore, one growth form, the evergreen sclerophyllous shrub, was selected to test the hypothesis that species with higher realized seasonal carbon uptake capacities would have a higher percent cover in the chaparral community. Because the basic growth form and adaptive strategies are similar for all species in the growth form, energy- and carbon-capturing abilities of a species would have a major effect on species success.

Carbon uptake was measured in southern California at six locations along a transect from near sea level at Torrey Pines (110 m) to 1400 m on the west-facing slope of the Laguna mountains (Figure 3.1). Climatic conditions and the vegetation varied along the transect (Parsons and Moldenke, 1975; Miller, 1979; Chapter 3). Measurements were made from May through August 1976, months that included periods of peak productivity and summer drought (Oechel and Mustafa, 1979).

Seasonal carbon incorporation was compared to vegetation data collected at the six sites where carbon uptake was measured. At each site, percent cover was determined from four line-intercept transects of 20 m each in the immediate area of the carbon uptake measurements. A more general relationship was determined by comparing the carbon uptake rate of a species at each transect site to the generalized cover of the species throughout San Diego County at similar elevations. The cover values used were those of Steward and Webber (Chapter 3), who determined the percent cover for 10 species along a complex elevational gradient. Their data were for 130 stands chiefly in San Diego County. To avoid rain shadow effects, only photosynthetic data from coastal areas and the western slopes of Mount Laguna were compared to the elevational ordinations of Steward and Webber (Figure 7.10). All carbon uptake determinations were made under natural conditions of temperature and light with carbon-14 dioxide techniques. Maximum carbon uptake values were derived from the highest values for the day. Measurements were generally made from sunrise until 1400 Pacific Standard Time.

When total seasonal carbon uptake from May to August at each transect site was compared to the cover at the specific study site, little correlation was found between these two variables within or among species. *Adenostoma fasciculatum*, in particular, showed large changes in cover for relatively small changes in carbon uptake. Other species showed variation in carbon uptake accompanied by little corresponding change in percent cover.

When total seasonal carbon uptake for an elevation was compared to the generalized vegetation for that elevation, a strong pattern among species emerged (Figure 7.10. The species that were the most photosynthetically active were more abundant in the chaparral vegetation as a whole. Yet, this strong positive relationship between carbon uptake rates and percent cover was not found at small-scale resolution. Many interrelated environmental and physiological factors that influence the success of an individual plant may mask or alter the relationship between photosynthesis and the percent

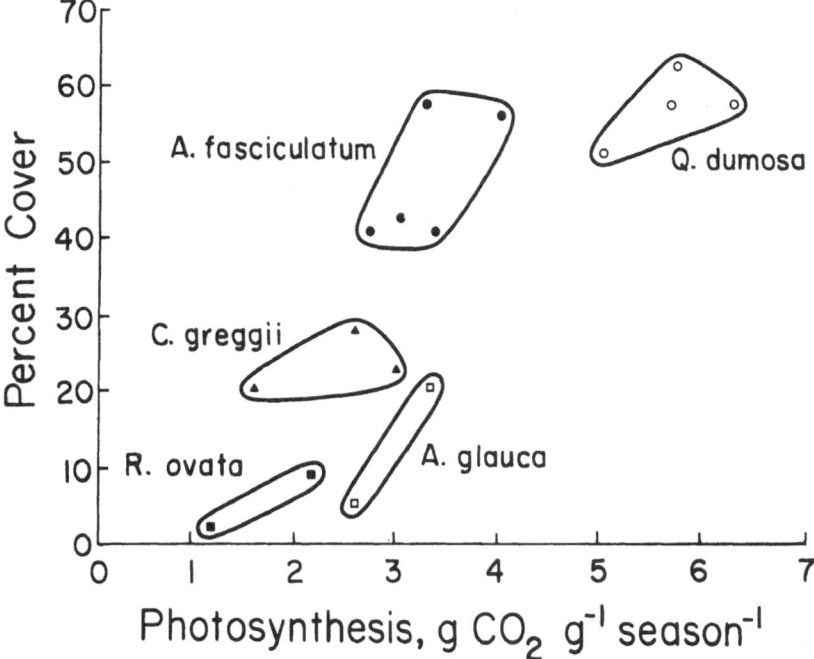

Figure 7.10. The relationship between total seasonal photosynthesis (g CO_2 g^{-1}, May-August 1976) and species success (percent cover) at various elevations from 100 to 1450 m. Each point represents seasonal photosynthetic uptake at a particular elevation and the cover at that elevation on western mountain slopes throught San Diego County, California.

cover estimate. Once a plant becomes established, its carbon uptake rate is determined to a large degree by the physiology of the plant and by the extant environment. Past history, such as fire history, grazing, available seed source, and climate following the disturbance, may explain much of the current distribution of species.

A main limitation of carbon uptake in the chaparral, especially at lower elevations, is the lack of moisture (Mooney et al., 1975; Miller, 1978). Mooney et al. (1975) indicated that the evergreen growth form is disadvantaged at the coast and low elevations because of the long drought period that reduces seasonal carbon uptake. In general, evergreen shrubs should show the highest uptake at midelevations and low uptake at elevational extremes along the coast, high in the mountains, and at the desert transition, which are limits of the chaparral vegetation type.

Over the transect measured, there was less variation in carbon uptake than was expected from the major change in climate observed over the elevational gradient measured. Only at Boulevard did *Quercus dumosa* and *C. greggii* show marked ($> 40\%$) depression of carbon uptake during the drought period. Throughout the transect, *A. fasciculatum* always varied by less than 32% of the maximum carbon uptake rates (Figure 7.11). No marked depression in carbon uptake was noted at the edges of the distribution of the chaparral vegetation, and evergreen shrubs appeared to be less disadvantaged at the coast than predicted by earlier papers. Except for *C. greggii* at Boulevard, no major decrease in carbon uptake occurred in association with the border

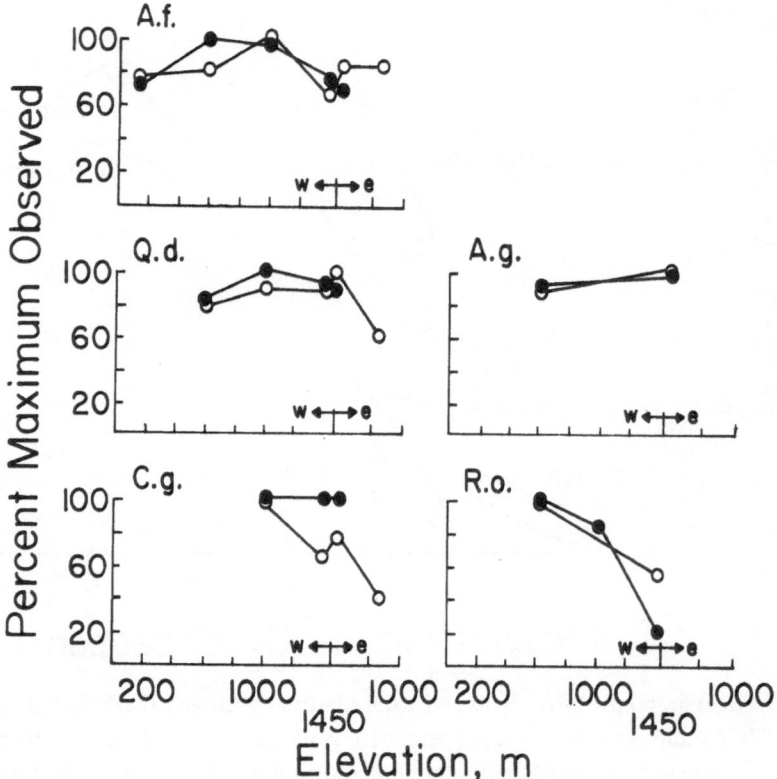

Figure 7.11. Percent of the maximum observed cover (●) and total seasonal photosynthesis (○) across an elevational gradient in southern California from May-August 1976 for four evergreen species: *Adenostoma fasciculatum* (A.f.), *Quercus dumosa* (Q.d.), *Arctostaphylos glauca* (A.g.), *Ceanothus greggii* (C.g.), and *Rhus ovata* (R.o.).

of the vegetation type or with the limits of the distribution of a species. It has been shown that species adjust through acclimation, ecotypic differentiation, and morphological adaptations to environmental differences associated with separate geographic locations (e.g., Billings et al., 1971; Smith and Hadley, 1974). These processes may have contributed to the relatively constant rate of uptake with elevation in the species measured.

7.8. Seasonal Patterns of Carbon Uptake

7.8.1 Chaparral Species

The literature provides an unclear picture of the level of suppression of carbon uptake that results from summer drought. Several authors predicted a marked summer reduction in carbon uptake at the coast in southern California caused by a lengthy summer drought (Mooney et al., 1975; Miller and Mooney, 1976). Mooney et al. (1975) estimated a summer depression in carbon uptake of almost 70% when compared with the spring maxima.

Photosynthetic measurements made along the transect with carbon-14 dioxide revealed that both evergreen and drought deciduous species showed positive daily carbon uptake throughout the study period (Figure 7.12). In general, the seasonal carbon uptake rates declined at all elevations as the summer drought progressed from May through August 1976. All plants showed the lowest water stress in May with water stress generally increasing through the summer. Daily carbon uptake rates correlated with dawn water potential and were reduced when dawn water potentials decreased in July and August.

Quercus dumosa, Cercocarpus betuloides, and *A. fasciculatum* showed less depression in uptake rates, based on dry weight, during the summer season than did the other species measured. *Quercus dumosa* has the deepest and most extensive root system of the major chaparral species (Hellmers et al., 1955b). Its extensive root system may enable *Q. dumosa* to exploit deep soil water and to maintain relatively high carbon uptake rates. *Cercocarpus betuloides* had high carbon uptake rates at Boulevard, which borders on the desert and was the most xeric location studied. Based on root measurements of two young shrubs, Hellmers et al. (1955 b) suggested that *C. betuloides* has a shallow root system. However, his data may not represent the general rooting habit of *C. betuloides.*

Ceanothus greggii and *A. glauca* are relatively shallow-rooted shrubs. When water was available early in the season, both species photosynthesized at high rates. As water became depleted from the upper soil levels in summer, these species showed a greater water stress and more reduced carbon uptake rates than did the deep-rooted shrubs. *Ceanothus verrucosus*, a coastal evergreen species, had a pattern of carbon uptake and water potential similar to *C. greggii*, which occurs at higher elevations.

Where maximum rooting depth is not limited by a shallow soil profile, *A. fasciculatum* had a deep root system, which was less extensive than that of *Q. dumosa* (Hellmers et al., 1955b; Kummerow et al., 1977). *Adenostoma fasciculatum* had intermediate rates of carbon uptake and water potentials (Mustafa, 1978) and showed only moderate depression of photosynthesis during drought. The greatest drought effect on carbon uptake in *A. fasciculatum* was at the coast (Figure 7.13).

Of the species measured, *R. ovata* and *R. integrifolia* had relatively low daily carbon uptake rates, the highest water potential values, and seasonally the least variable water potentials. The water potential data for *R. ovata* and *R. integrifolia* agreed with those of Poole and Miller (1975), who reported that these *Rhus* species showed high water potentials and suggested that these species were restricted to rock outcrops where soil moisture was available. They also reported that *R. ovata* had the most responsive stomata at water potentials of about −2.0 MPa. The low and variable carbon uptake rates of *R. ovata* and *R. integrifolia* measured in the present study were presumably caused by low carbon dioxide and water vapor conduction at relatively high water potentials and by early stomatal closure when water potentials approached and exceeded −2.0 MPa (Poole and Miller, 1975; Mustafa, 1978). Marked variability of carbon fixation in *R. ovata* was noted by Mooney and Dunn (1970a, b) when carbon fixation measured during a dry year was less than half that measured during a wet year.

Relative to the photosynthetic rates at other transect sites measured, carbon uptake rates of *A. fasciculatum* and *C. greggii* were depressed at the Sunrise Highway site (1400 m) where both species underwent greater water stress. Shrubs at the Sunrise Highway site were regenerating following fire in 1970. The shrubs were small and

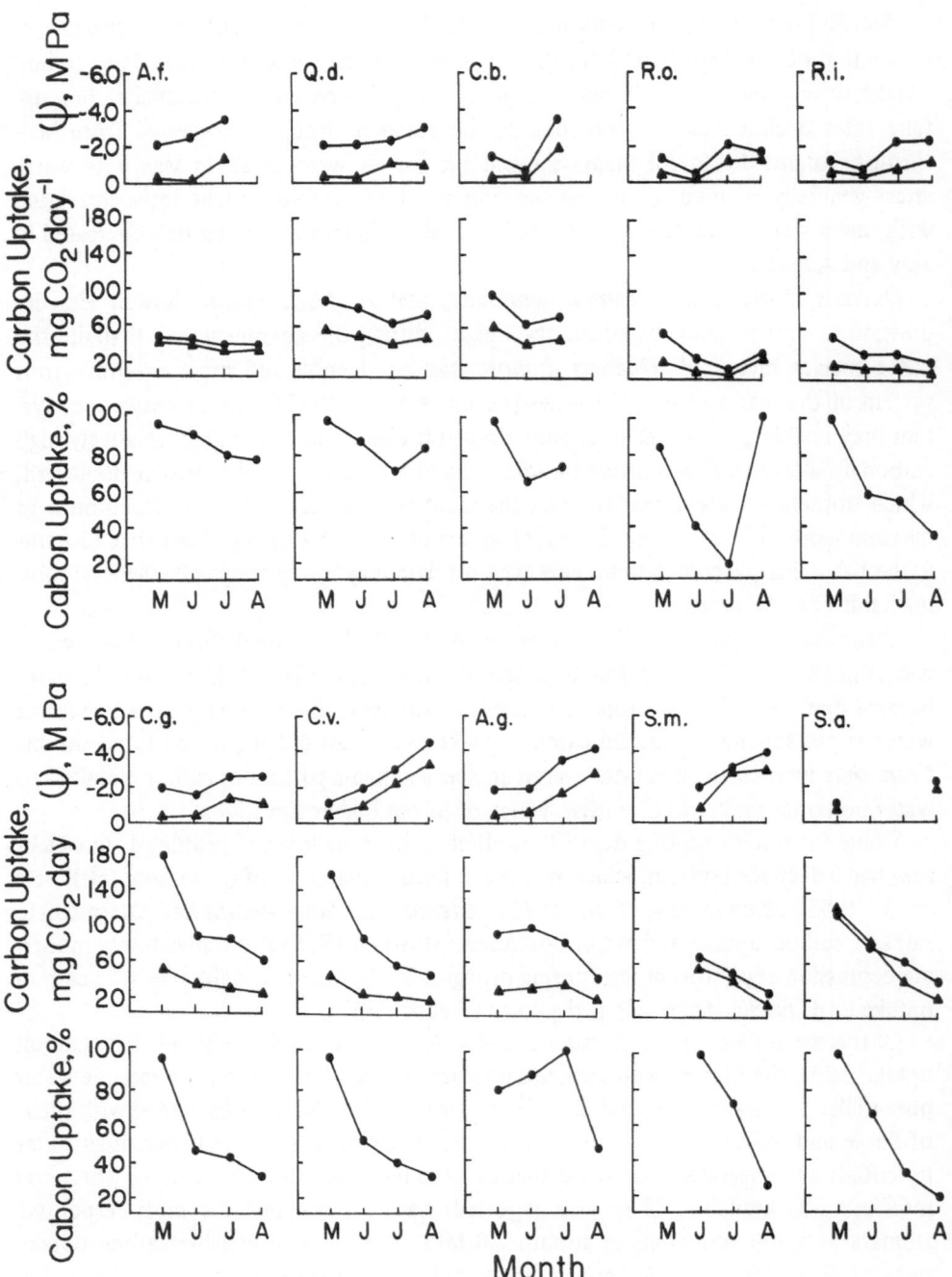

Figure 7.12. Seasonal patterns of dawn (▲) and midday (●) water potentials in bars, carbon uptake in milligrams of CO_2 per day per gram dry weight of leaf (▲) and per square decimeter of leaf (●), and the relative carbon uptake as a percent of the maximum monthly rate for *Adenostoma fasciculatum* (A.f.), *Quercus dumosa* (Q.d.), *Cercocarpus betuloides* (C.b.), *Rhus ovata* (R.o.), *Rhus integrifolia* (R.i.), *Ceanothus greggii* (C.g.), *Ceanothus verrucosus* (C.v.), *Arctostaphylos glauca* (A.g.), *Salvia mellifera* (S.m.), and *Salvia apiana* (S.a.).

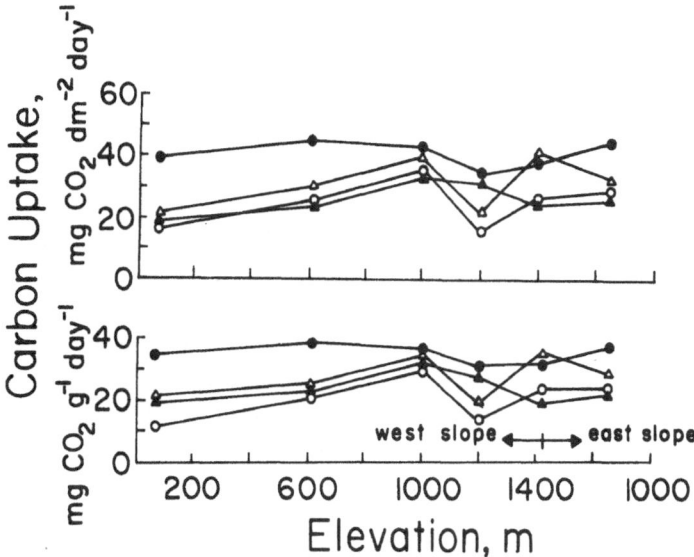

Figure 7.13. Seasonal daily carbon uptake of *Adenostoma fasciculatum* at six transect elevations in San Diego County, California in May (●), June (△), July (▲), and August (○) 1976.

widely spaced on a steep slope. Species at this location were differentially affected by water stress according to their rooting habit. Because of its deep, extensive root system, *Q. dumosa* maintained high water potentials and high carbon uptake rates (Mustafa, 1978). In contrast, *C. greggii* developed severe stress. Schlesinger and Gill (1980) show much greater midsummer water stress in 6-year-old stands of *Ceanothus megacarpus* than in 22-year-old stands and link these results to the time needed for this species to develop an extensive root system. Previous workers (Christensen and Muller, 1975) have also indicated that because of decreased water penetration and increased evaporation from the exposed soil surface, the moisture content of burned sites may be less than that of unburned mature chaparral.

The evergreen shrubs were divided into three groups based on carbon uptake rates and water potential values. The first group was composed of *Q. dumosa, C. betuloides*, and *A. fasciculatum*, species with deep root systems that had high carbon uptake and showed relatively low water stress when measured with the Scholander pressure chamber (Oechel and Mustafa, 1979). Except for *C. betuloides*, these species were sampled at most elevations (1000 to 1414 m) along the transect. The second group of species, *C. greggii, C. verrucosus*, and *A. glauca*, showed generally less carbon uptake and greater water stress during the drought period and had a shallower root system. The highest daily carbon uptake for *C. greggii*, the only species in the group sampled over a wide elevational gradient, occurred at a midelevation (1000 m). Unfortunately, incomplete occurrence of these species at all elevations made generalizations difficult. The third group, composed of *R. integrifolia* and *R. ovata*, showed little water stress compared to the other evergreen shrubs, but their carbon uptake rates were generally low and variable. The highest carbon uptake rates for this group occurred at low elevations (610 m) (Mustafa, 1978). Over the elevational ranges of the species studied, the shallow-rooted, evergreen shrubs may be better adapted to the lower elevations of the chaparral. Because lower elevations receive less precipitation, which consists of smaller amounts of precipitation per storm, a shallow root system is a better adaptation to utilize the light rains that may not penetrate deeply into the soil. Higher elevations receive more total precipitation and more precipitation per storm. At these elevations, there is greater penetration, and a deep root system is an advantage in reaching deeper soil water. However, the deeply rooted *A. fasciculatum* is one of the few evergreen species that occurred in moderate abundance near the coast. Other adaptations, including the narrow-leaved form that helps reduce water loss, may be important in favoring the distribution of *A. fasciculatum* over such a wide environmental range. Evergreen species are photosynthetically well adapted at all elevations within the chaparral of southern California. However, at the species level, elevation may be an important factor in segregating species according to their rooting habit.

The data presented here indicate that the lower daily carbon uptake rates of evergreen species were compensated for by a reduced sensitivity of carbon uptake to drought. Throughout the study period, the daily carbon uptake rates of *C. verrucosus* remained higher on a leaf-area basis than did those of the drought semideciduous shrub, *S. mellifera* (Figure 7.14). Differences between the data presented here and those of Mooney et al. (1975) may be due to a more pronounced drought during the year of the study of Mooney and co-workers. Parsons and Moldenke (1975) found that evergreen species were as abundant as drought deciduous species at the coast in southern California. Steward and Webber (Chapter 3) found that at sea level, sclerophylls were about 60% as abundant as malacophylls on a percent-cover basis, but at 300 m elevation, sclerophylls were more abundant than malacophylls. Except for a generally very narrow band along the coast, little evidence exists for the competitive superiority of drought deciduous plants near the coast.

The seasonal carbon uptake rates of two drought semideciduous species, *Salvia mellifera* and *Salvia apiana*, which were sampled at low elevations, were compared with carbon uptake rates of evergreens at low elevations and with other growth forms sampled at Echo Valley. Daily carbon uptakes of the drought deciduous species decreased more rapidly with the progression of drought than did those of the evergreen species. Both species of *Salvia* shed most of their leaves in the summer but remained photosynthetically active throughout the study period. *Salvia mellifera*, a coastal species sampled at Torrey Pines, underwent the greatest water stress in August (-4.0 MPa) when its carbon uptake rates were reduced by 75% of the June value. Because no data were available for May, the actual percentage rates, relative to maximum values, would probably be lower. *Salvia apiana*, sampled at Alpine, reached carbon uptake rates of 123 mg CO_2 g^{-1} day^{-1} in May. These rates were reduced by 83% in August as the plant became water stressed.

The drought semideciduous *Salvia* species, while demonstrating high carbon uptake rates when well hydrated, were more sensitive to water stress than were the evergreen species measured. On one hand, regression data indicated that no net carbon dioxide incorporation occurred in these two *Salvia* species when dawn water potentials were less than -3.5 MPa. *Arctostaphylos glauca* and *A. fasciculatum*, on the other hand, showed significant daily carbon dioxide incorporation at dawn water potentials of less than -3.5 MPa. Extrapolation of the regression of carbon dioxide uptake on maximum

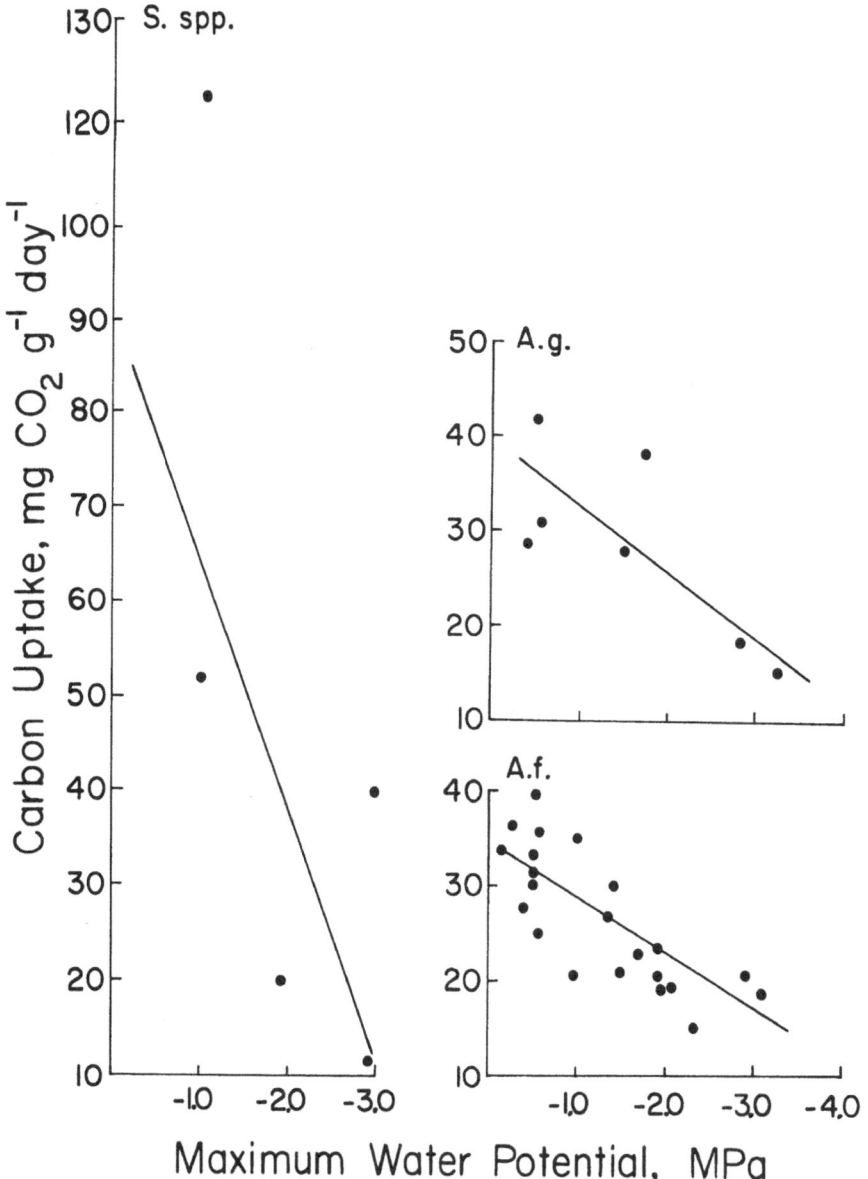

Figure 7.14. Relationship between maximum (dawn) water potential and seasonal daily carbon uptake of *Salvia apiana* and *Salvia mellifera* (S. spp.), *Arctostaphylos glauca* (A.g.), and *Adenostoma fasciculatum* (A.f.) in San Diego County, California.

water potential indicated than in *A. glauca* and *A. fasciculatum*, carbon dioxide uptake may persist down to dawn water stresses of −5.0 to −5.5 MPa, respectively.

7.8.2 Matorral Species

Maximal photosynthetic yield under saturating light intensity indicates the level of resource-use efficiency under saturated light conditions. On an area basis, species from central Chile tended to have much higher carbon uptake rates and maximal resource-use efficiencies with respect to light than did species from southern California. Carbon-14 dioxide techniques indicated that evergreen shrub growth form at Fundo Santa Laura ranged from a mean carbon uptake rate of about 3.5 to 19 mg CO_2 dm^{-2} h^{-1}; at Echo Valley the range was about 3.5 to 10.4 mg CO_2 dm^{-2} h^{-1} (Mustafa, 1978). The range in maximum photosynthetic rates for deciduous shrubs was 40 to 66 mg CO_2 dm^{-2} h^{-1} at Fundo Santa Laura versus a range of 8 to 15 mg CO_2 dm^{-2} h^{-1} at Echo Valley. In Chile, photosynthetic stem species showed carbon uptake rates almost equivalent to those of evergreen sclerophylls and greater than those of the average evergreen phreatophytes. This photosynthetic stem growth form is largely lacking in the Californian chaparral.

Evergreen sclerophylls at Fundo Santa Laura showed positive carbon uptake rates through the year (Figure 7.15). *Quillaja saponaria* showed the smallest depression, which resulted in minimal values of 40% or more of its maximal uptake rate. The other evergreen sclerophyllous species were depressed to around 15% of their maximal rate during the height of the drought. Evergreen species at Fundo Santa Laura peaked in photosynthetic activity in September or October. Drought deciduous and photosynthetic stem species peaked in photosynthetic activity between August and October. Herbs peaked later, with photosynthetic maxima in November or December. Peak photosynthetic activity occurred earlier in shrubs and allowed a competitive advantage for water use in these species. The annual herbs measured did not reach maximal activity until November.

The period of photosynthetic activity was continuous throughout the year for evergreen hygrophilous and sclerophyllous species and for photosynthetic stem species, although considerable seasonal variation in rate existed. Drought deciduous shrubs had about a 6.5-month period of photosynthetic activity (Figure 7.16) and also showed pronounced variation in carbon uptake rates over this period. Annual and perennial herbs were limited to about 4.5 and 5 months of photosynthetic activity, respectively. At Fundo Santa Laura, the total seasonal carbon uptake per unit leaf area of drought deciduous and evergreen species was similar.

Evergreen phreatophytes, because of low photosynthetic rates or depression of photosynthesis during the drought period or both, were markedly less productive than were evergreen sclerophyllous shrubs. Annual herbs, because of very high photosynthetic rates over a brief period, had yearly carbon uptake totals approaching those of the drought deciduous and evergreen sclerophyllous species. Drought deciduous species at Fundo Santa Laura lost their foliage and, therefore, their photosynthetic capacity more quickly and more completely than did deciduous species at Echo Valley (Figures 7.12 and 7.15). Photosynthesis ceased in *S. gilliesii* by the end of December and in *T. trinervis* by the end of January. In contrast, *Salvia mellifera* and *S. apiana* retained leaves and remained photosynthetically active in August at Echo Valley; this period corresponds climatically and phenologically to about February in central Chile (Miller et al., 1977; J. Kummerow, *pers. comm.*). The seasonal decrease in carbon uptake in drought semideciduous and evergreen species was more similar at Echo Valley than was the decrease in these growth forms at Fundo Santa Laura. Leaf drop in response to drought was more complete at Fundo Santa Laura than at Echo Valley. These factors may result in significant differences in resource-use efficiency by the respective growth forms in southern California and central Chile.

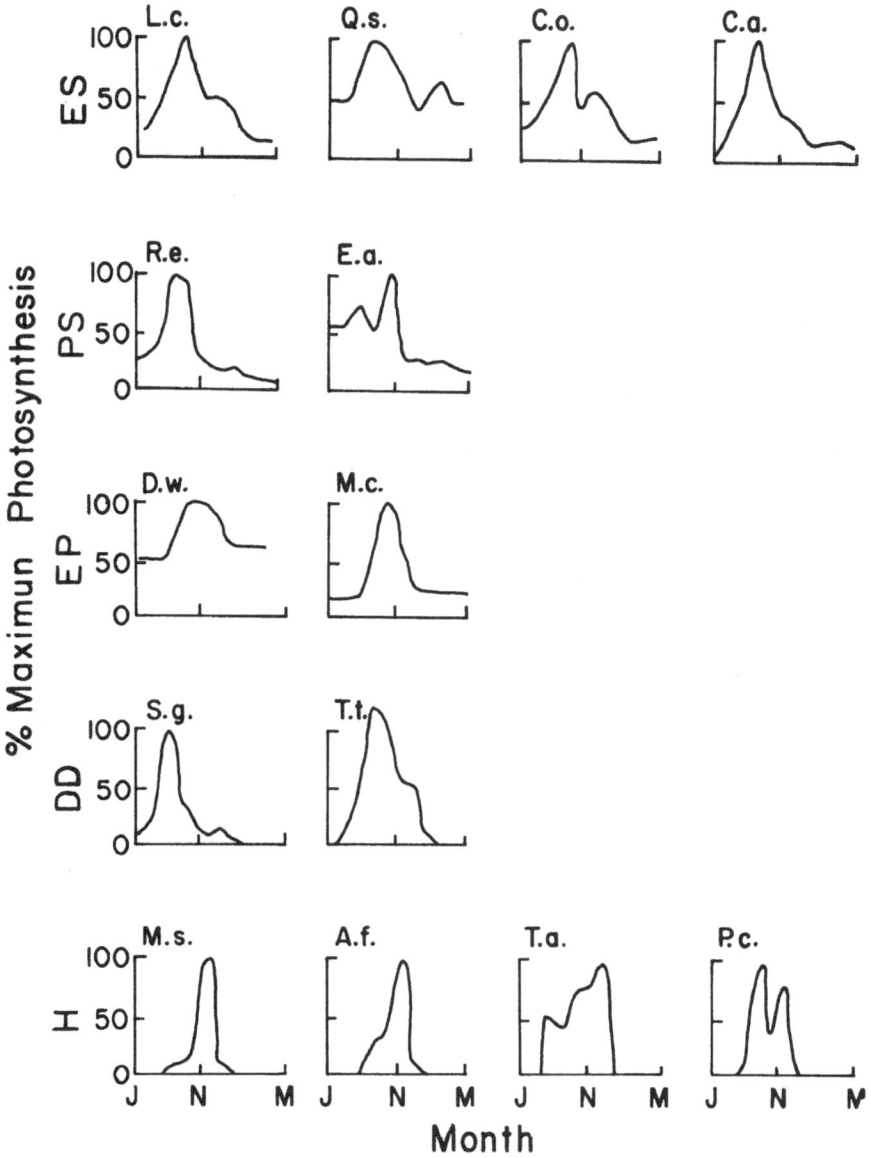

Figure 7.15. Relative seasonal pattern of maximal photosynthetic rates from June 1976 to May 1977 for *Lithraea caustica* (L.c.), *Quillaja saponaria* (Q.s.), *Colliguaya odorifera* (C.o.), *Cryptocarya alba* (C.a.), *Retanilla ephedra* (R.e.), *Ephedra andina* (E.a.), *Drimys winteri* (D.w.), *Myrceugenella chequen* (M.c.), *Satureja gilliesii* (S.g.), *Trevoa trinervis* (T.t.), *Madia sativa* (M.s.), *Avena fatua* (A.f.), *Trichocline aurea* (T.a.), and *Pasithea coerulea* (P.c.) representing five growth forms: herbaceous (H), drought deciduous and drought semideciduous shrubs (DD), evergreen phreatophytes (EP), photosynthetic stem shrubs (PS), and evergreen sclerophyllous shrubs (ES) at Fundo Santa Laura.

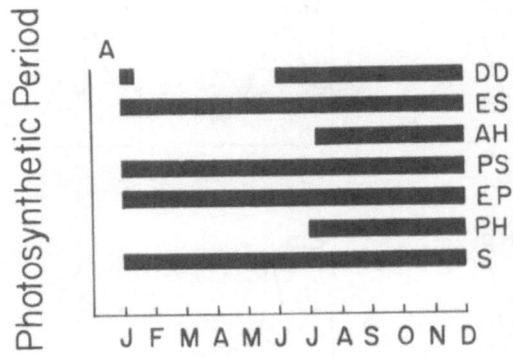

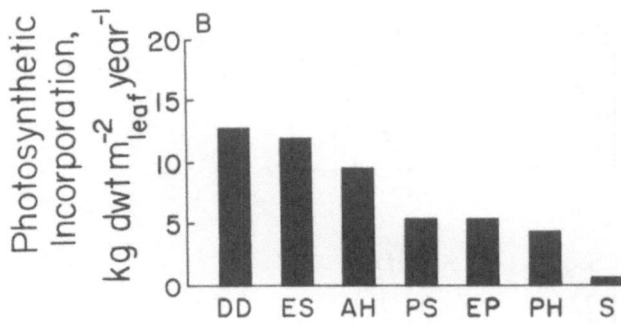

Figure 7.16. Duration of the photosynthetic period (A) and annual incorporation (dry matter equivalent) per square meter of leaf area (B) in Chilean growth forms: drought deciduous and drought semideciduous shrubs (DD), evergreen sclerophyllous shrubs (ES), annual herbs (AH), photosynthetic stem shrubs (PS), evergreen phreatophytes (EP), perennial herbs (PH), and succulents (S).

7.9. Conclusions

The evergreen, mediterranean shrub species studied at Echo Valley and Fundo Santa Laura were very similar in their patterns of photosynthetic response and in their efficiencies of resources capture and use. The photosynthetic response surfaces were relatively temperature insensitive, and temperature optima for photosynthesis are near the theoretical values that yield maximal carbon uptake. Temperature limitation on photosynthesis is minimal.

Although growth forms differed in carbon uptake rates, the differences were less when photosynthesis rates were expressed on a surface-area basis than when expressed on a dry-weight basis which included the effects of leaf-density differences among growth forms. On a leaf-area basis, basically two plant groups exist: those with a leaf duration of less than one year, which have higher carbon uptake rates, and those with a leaf duration of one year or longer, which have lower carbon uptake rates. In California, the photosynthetic rates of the second group are about 55% of those of the first group.

In southern California, evergreen species with higher carbon uptake rates appear to have a competitive advantage over evergreen species with lower carbon uptake rates. This advantage is reflected by the fact that species with the highest carbon uptake rates also display the greatest cover. These relationships may hold only when species of similar growth forms are compared.

Data on carbon uptake rates do not support the idea that the transition from chaparral to coastal sage scrub in southern California is caused by drought effects on production as was previously suspected. Most evergreen species maintain substantial late summer carbon uptake rates, which are not markedly reduced at the coast when compared to the reduction that occurs in other areas.

In southern California, relatively small differences in carbon uptake rates are observed across large elevational and climatic gradients. Similar carbon uptake rates may be caused by marked homeostatic control resulting from acclimation or ecotypic differentiation or both.

8. Carbon Allocation and Utilization

Walter C. Oechel and William Lawrence

This chapter describes the patterns of carbon metabolism and utilization in plants from southern California and central Chile. Rates of dark respiration and total carbon cost of leaves, stems, and roots are presented and compared. Growth and maintenance respiration rates for roots are presented, as is the effect of light on net carbon dioxide evoluation by stems. Allocation of carbon to various organic chemical fractions in leaves and stems is analyzed.

8.1. Introduction

A plant's productivity and success are due to its patterns of carbon utilization, as well as its capacity for carbon assimilation. A small change in carbon allocation from leaves to stems or roots may have a dramatic impact on net carbon uptake, nutrient uptake, and water relations (Monsi and Murata, 1970; Miller and Tieszen, 1972; Wareing and Patrick, 1975). Respiration rates, the size and strength of carbon sinks, and carbon allocation patterns may have major influences on plant productivity (Evans, 1975a, b) and ecosystem dynamics. Potential points of impact include competition, recovery from fire or grazing, and resource-use efficiency. Growth, maintenance respiration, and carbohydrate storage are major carbon sinks and must be accurately described to understand the carbon balance of woody shrubs and to avoid large errors in the computation of the carbon budget (Syvertsen and Cunningham, 1977). Respiration may utilize a major portion of the carbon assimilated in photosynthesis. For example, in 16 forest types surveyed by Kira (1975), 32% to 77% of the carbon assimilated was lost in dark respiration.

Despite the importance of respiration as a major carbon sink, there is little published information on the respiration rates for chaparral or matorral species, especially of nonphotosynthetic tissues. Even less is known of the carbon allocated to the two components of total respiration: maintenance respiration and growth respiration. In herbaceous agricultural plants, maintenance respiration is about 1% to 4% per day of the carbon present in the biomass (McCree, 1974). Even in agricultural plants, only limited experimental data exist on the magnitude of growth respiration (Challa, 1976). The patterns of respiration, carbon allocation, and carbon utilization and the extent of convergence in these patterns in major shrub species in southern California and central Chile are analyzed and evaluated in this chapter.

The level of functional similarity in carbon utilization between mediterranean species in southern California and central Chile is investigated in this chapter. Respiration rates and temperature dependency of respiration are examined to determine if functional convergence in metabolic patterns is indicated within analogous species pairs in southern California and central Chile. Patterns of allocation of carbon to organic chemical constituents are examined to ascertain similarities and differences between the species groups. Overall allocation of carbon to growth and respiration of various plant structures is compared, and similarities in carbon allocation to the production of diverse organic constituents in leaves and stems are analyzed.

8.2. Carbon Losses Through Respiration

8.2.1 Leaf Dark Respiration

The difference between carbon uptake in photosynthesis and carbon loss in dark respiration is an indication of the net carbon production by a leaf. Leaf dark respiration is, therefore, a major determinant of the amount of carbon exported from a leaf and made available to the rest of the plant. In mature tissues of determinant growth, dark respiration is composed solely of maintenance respiration; in growing tissues, dark respiration includes both maintenance and growth respiration.

Leaf dark respiration rates were measured for *Adenostoma fasciculatum, Arctostaphylos glauca, Ceanothus greggii,* and *Rhus ovata* at Echo Valley, California and *Satureja gilliesii, Colliguaya odorifera, Trevoa trinervis,* and *Lithraea caustica* at Fundo Santa Laura, Chile by using the infrared gas analysis system described by Oechel and Lawrence (1979). Photosynthetic rates for these species are reported in Chapter 7. Both photosynthetic and leaf dark respiration rates were measured between 5° and 40°C on plants growing in the field. Dark respiration rates were measured throughout the natural night period, so that any effect of decreasing carbohydrate levels on dark respiration rates was included in the net response (Challa, 1976). Five plants were exposed to varying experimental temperatures, while one plant experienced ambient temperature for the entire night period. The following night, different experimental temperatures were used for each of the five plants.

While the pattern of photosynthesis with respect to temperature was found to be similar in the Californian and Chilean species examined in Chapter 7, leaf dark respiration rates were more variable. Respiration in almost all species measured was more temperature sensitive than was photosynthesis (e.g., compare Figures 7.1 and 8.1).

8.2.1.1 Leaf Respiration Rates. The mean leaf dark respiration rate at 20°C for Californian species at Echo Valley was 0.62 mg CO_2 g^{-1} dry weight h^{-1} (0.95 ± 0.46 mg

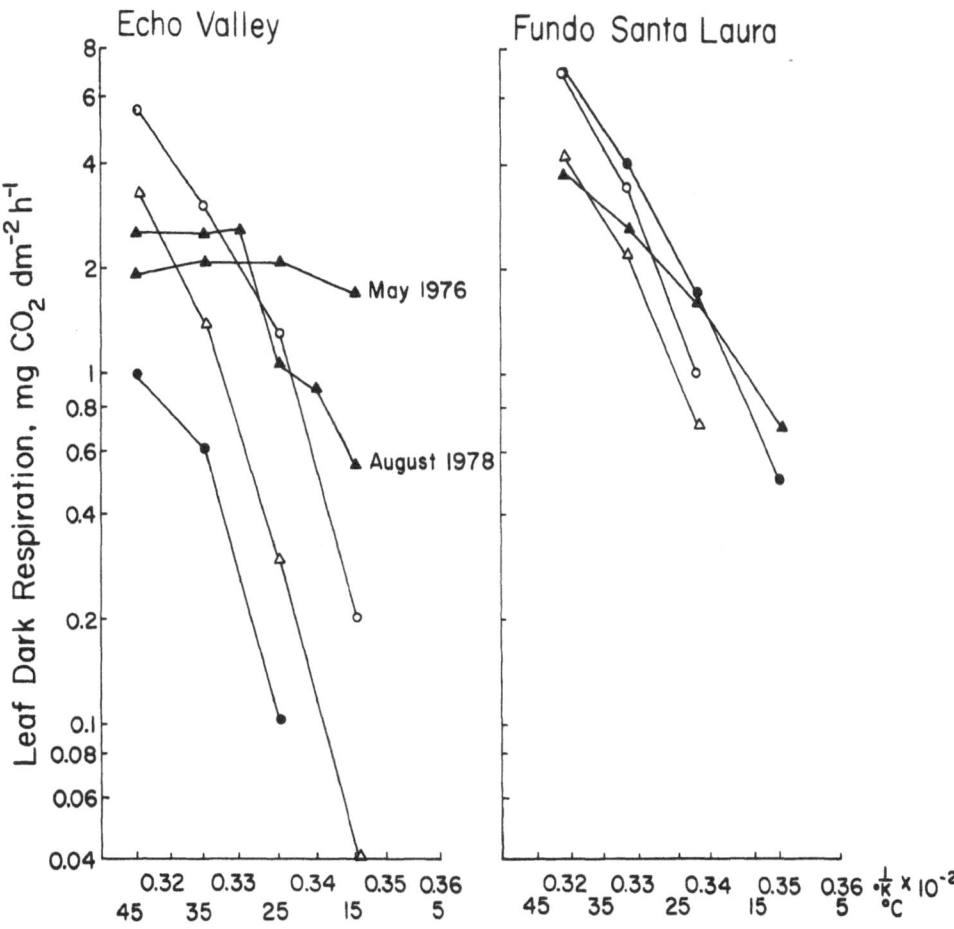

Figure 8.1. Arrhenius plot of the log of leaf dark respiration versus the reciprocal of the absolute temperature for: *Adenostoma fasciculatum* (▲), *Rhus ovata* (△), *Ceanothus greggii* (○), and *Arctostaphylos glauca* (●) in May-June 1976 at Echo Valley and for: *Trevoa trinervis* (○), *Satureja gilliesii* (▲), *Colliguaya odorifera* (●), and *Lithraea caustica* (△) in November-December 1976 at Fundo Santa Laura. August 1978 data for *Adenostoma fasciculatum* (▲) are also presented.

CO_2 dm^{-2} h^{-1}), which is lower than the 0.91 mg CO_2 g^{-1} dry weight h^{-1} (1.25 ± 0.24 mg CO_2 dm^{-2} h^{-1}) for Chilean species at Fundo Santa Laura. Leaf dark respiration rates for Californian and Chilean species were similar to the 1.0 mg CO_2 g^{-1} dry weight h^{-1} measured for sun leaves of evergreen broadleaved trees and the 0.7 mg CO_2 g^{-1} dry weight h^{-1} measured for coniferous trees. The chaparral and matorral shrubs had dark respiration rates considerably below the 3 to 4 mg CO_2 g^{-1} dry weight h^{-1} measured for winter deciduous foliage trees and the 5 to 8 mg CO_2 g^{-1} dry weight h^{-1} measured for wild herbaceous and crop plants (Larcher, 1975).

Species at Fundo Santa Laura generally had higher leaf dark respiration rates than did their paired Californian counterparts at Echo Valley. In separate paired *t*-tests of respiration rates at 30°C and of respiration rates between 20° and 40°C, the Chilean species had consistently higher dark respiration rates ($P \leqslant 0.01$) than the Californian species. The difference tended to become more pronounced at higher temperatures.

The two-way analysis of variance indicated a similar pattern (Table 8.1). Leaf dark respiration rates of the four species pairs were not statistically different one from another, but there was a statistical difference between the respiration rates for Californian and Chilean species.

The concept of convergent, functionally analogous species pairs between the mediterranean vegetation of Echo Valley and Fundo Santa Laura was not supported by leaf dark respiration rates. Leaf respiration rates for Chilean species were generally higher than were those of the Californian species, and respiration rates were less variable among the Chilean species measured. The paired species showed no consistent pattern of similarity with respect to dark respiration rates. In the dark at 10°C, both *C. greggii* and *R. ovata* respire only about 1% of the maximum photosynthetic rate that occurs at the optimal temperature; however, this percentage increased with increased temperature. At 40°C, *C. greggii* consumed 46% and *R. ovata* 69% of maximum photosynthesis in respiration. Most of the nitrogen in a leaf is generally contained in protein. Because dark respiration rates are closely correlated with leaf protein turnover rates (Evans, 1975b), the high respiration rates in *C. greggii* may result from high levels of leaf protein (see Table 8.7). The relative temperature insensitivity in *A. fasciculatum* resulted in dark respiration rates at all temperatures of about 0.25 of the maximum photosynthesis rate at the temperature optimum and light saturation for photosynthesis.

Respiration rate, as a percent of the maximal rate of photosynthesis, was generally higher in the Chilean species measured. Between temperatures of 20° and 40°C, where a complete data set is available, the percentage of the maximal rate of photosynthesis consumed in dark respiration was significantly higher in Chilean species (paired *t*-test, $P \leqslant 0.01$) (Table 8.2). The cost of leaf maintenance was greater in the Chilean species relative to the maximum photosynthetic return, which may result in a depressed resource-use efficiency in the Chilean species. The reduced variability in respiratory rate among Chilean species may indicate greater selective pressure on respiratory rate of plants in central Chile compared to those in southern California. Also indicated is minimal convergence in respiratory patterns between the species measured.

8.2.1.2 Q_{10} Values. Q_{10} values indicate the fractional increase in dark respiration rate for a 10°C increase in temperature. Higher Q_{10} values, therefore, indicate a greater rate of increase in respiration with temperature. Dark respiration rates may acclimate to changes in temperature, which under certain circumstances may appear to modify the Q_{10} response. The response of respiration to temperature is not necessarily uniform over a large temperature gradient. The Q_{10} may be different at lower temperatures than at higher temperatures, indicating a change in temperature sensitivity and a possible switch in enzyme systems. Q_{10}, as used here, refers to the temperature sensitivity of total dark respiratory activity, which is a combination of numerous independent processes with potentially differing temperature dependencies.

Q_{10} values for leaf respiration decreased in the mediterranean shrubs measured, as temperatures increased from 10° to 40°C. Values were generally greater than 2 at temperatures below 20°C and less than 2 at temperatures above 20°C. The mean value for the Californian species, *C. greggii*, *A. fasciculatum*, *R. ovata*, and *A. glauca*, ranged from 5.4 at 10° to 20°C to 1.6 at 30° to 40°C (Table 8.2). The Chilean species, *C. odorifera*, *S. gilliesii*, *L. caustica*, and *T. trinervis*, had lower Q_{10} values at lower tempera-

Table 8.1. Summary of statistical differences among species pairs, and between Californian and Chilean groups of species[a]

	Among species pairs	Between Californian and Chilean groups of species	Direction of difference
Q_{10} of leaf respiration (10° to 40°C)	+	−	Echo Valley < Fundo Santa Laura
Leaf dark respiration (20° to 40°C)	−	+	Echo Valley < Fundo Santa Laura
Leaf respiration (20° to 40°C)/P_{max}	−	+	Echo Valley < Fundo Santa Laura
Q_{10} of stem respiration	−	+	Echo Valley < Fundo Santa Laura
Stem respiration	+	−	Echo Valley < Fundo Santa Laura

[a] Statistical difference at $P \leqslant 0.05$ (+) and no statistical difference at $P = 0.05$ (−); P_{max} is maximum photosynthetic rate at saturating light intensity and optimal temperature.

Table 8.2. Key parameters of dark respiration (R_s) of leaves in selected shrub species from Echo Valley and Fundo Santa Laura[a]

Temperature (°C)	Ceanothus greggii	Arctostaphylos glauca	Rhus ovata	Adenostoma fasciculatum	$\overline{X} \pm SE$	Trevoa trinervis	Colliguaya odorifera	Lithraea caustica	Satureja gilliesii	$\overline{X} \pm SE$
	Californian species					Chilean species				
					Respiration ($mg\ CO_2\ dm^{-2}\ h^{-1}$)					
10	0.2	-[b]	0.04	1.7[c]	0.6 ± 0.4	-	0.5	-	0.7	0.60 ± 0.1
20	1.3	0.1	0.30	2.1	0.9 ± 0.5	1.0	1.7	0.7	1.6	1.25 ± 0.2
30	3.0	0.6	1.40	2.1	1.8 ± 0.5	3.4	4.0	2.1	2.6	3.00 ± 0.8
40	5.5	1.0	3.30	1.8	2.9 ± 1.0	7.0	7.0	4.2	3.7	5.40 ± 0.9
					Q_{10}					
10-20	7.4	-	7.7	1.2	5.4 ± 2.1	-	3.4	-	2.3	2.8 ± 0.6
20-30	2.4	5.4	4.4	1.0	3.3 ± 1.0	3.4	2.4	3.0	1.6	2.6 ± 0.4
30-40	1.8	1.5	2.3	0.8	1.6 ± 0.6	2.1	1.8	2.0	1.4	1.8 ± 0.1
					$(R_s/P_{max})(100)$					
10	1	-	1	21	7.7 ± 6.7	-	3	-	8	5.5 ± 2.5
20	11	1	7	25	11.0 ± 5.1	8	12	8	18	11.5 ± 2.4
30	25	6	30	25	21.5 ± 5.3	27	28	24	30	27.2 ± 1.2
40	46	10	69	21	36.5 ± 13.2	56	48	49	43	49.0 ± 2.7

[a] Respiration values presented are those from the regression relationship fitted to the data points. Maximal photosynthetic rate (P_{max}) was determined under saturating light intensities at optimal temperatures (Chapter 7).
[b] Rate not detectable with system.
[c] From direct measurements rather than regression analysis.

tures, with a mean value of 2.8 at 10° to 20°C. The mean Q_{10} for Chilean species was 1.8 at 30° to 40°C. Leaf dark respiration rates in the Chilean species tended to be less sensitive to temperature changes than those in the Californian species. Between 20° and 40°C, the mean Q_{10} value was 2.1 in the Chilean species and 2.4 in the Californian species (Figure 8.1). This lower temperature sensitivity of respiration in the Chilean species measured was surprising, given the greater uniformity in temperature at Fundo Santa Laura compared to Echo Valley (Miller et al., 1977; Chapters 2 and 7). The more variable climate at Echo Valley might be expected to produce a greater homeostasis with respect to temperature fluctuations. However, at Echo Valley, short-term temperature fluctuations, especially below 30°C, have a large effect on dark respiration rates in the species measured.

Ceanothus greggii and *R. ovata* had similar responses of respiration to temperature, although leaf respiration rates in *C. greggii* were somewhat higher. Respiration was strongly temperature dependent in both species, especially at lower temperatures where the Q_{10} was 7.4 to 7.7 from 10° to 20°C (Figure 8.1; Table 8.2). Despite substantial Q_{10} values of up to 5.4 between 20° and 30°C, *A. glauca* had the lowest leaf dark respiration rates at each measured temperature. *Arctostaphylos glauca* also had the lowest respiratory cost relative to photosynthetic maxima of all species tested at all temperatures.

Of the Californian and Chilean species measured, *A. fasciculatum* had the least temperature sensitive leaf respiration rates. In May 1976, the Q_{10} for leaf respiration rate in *A. fasciculatum* was less than 1.25 between 10° and 40°C. At 10°C, the respiration rate of *A. fasciculatum* was much greater than any other species, 1.7 mg CO_2 dm^{-2} h^{-1}, but at 40°C, *A. fasciculatum* had the second lowest respiration rate following that of *A. glauca*. Because the observed flat response of respiration to temperature is uncommon, the relationship was again determined for *A. fasciculatum* in August 1978. Leaf dark respiration rates at temperatures below 25°C in August showed normal temperature sensitivity. Between 25° and 40°C, there was again a very temperature insensitive pattern in leaf dark respiration with Q_{10} values of around 1.0. Because the leaf dark respiration rate was insensitive to higher temperatures, high night temperature should not result in a major decrease in leaf productivity in *A. fasciculatum* due to increased dark carbon dioxide loss from the leaves. Its low Q_{10} and relatively low respiration rates at high temperatures may partially explain the success of *A. fasciculatum* in exposed, open, south-facing slopes where high temperatures are common.

The same pairs of Californian and Chilean species were used to analyze the pattern of respiration in California and Chile as were used to compare photosynthetic response (Chapter 7). These were *C. greggii* and *T. trinervis*, *A. glauca* and *C. odorifera*, *R. ovata* and *L. caustica*, and *A. fasciculatum* and *S. gilliesii.*

The two-way analysis of variance was used to test for differences among the four species pairs and for differences between the groups of Californian and Chilean species (Table 8.1). The interaction term between species pairs and Californian and Chilean data sets was often significant. Despite this interaction, the probabilities for differences can still be compared (Sokal and Rohlf, 1969). The results of the two-way analysis of variance showed a statistical difference in Q_{10} among the four species pairs studied ($P = 0.02$). The four species pairs were functionally different one from another with respect to Q_{10} values between 10° and 40°C for mean leaf respiration rates. However, taken as a group, Q_{10} values of the Californian and Chilean species were not statistically different, a circumstance suggesting analogous response patterns to temperature of the plants to their respective environments.

8.2.2 Stem Respiration

Carbon lost through stem respiration is required for growth, maintenance, and normal functioning of stem tissues, including the phloem, xylem, cortex, and periderm. Despite low stem respiration rates, the large mass of stems in mediterranean shrubs may result in a significant respiratory loss of carbon. The large cost of stem maintenance implies considerable selective pressure to maximize return for carbon lost or to minimize carbon lost. The large expenditure of carbon to the stems in relation to the total carbon budget indicates the relative importance of stem structures in the various species. Minimal information of the carbon lost through stem respiration in mediterranean shrubs is available in the literature. Respiration rates are higher in growing or otherwise metabolically active tissues, and major decreases in respiration are normally associated with decreases in metabolic activity. However, stem photosynthesis may reincorporate respired carbon dioxide, thereby reducing carbon cost of maintaining stem activity without reducing the rate of stem metabolism.

8.2.2.1 Rates and Temperature Dependence of Stem Respiration. Of the species measured at Echo Valley, *A. glauca* showed the highest stem respiration rates at temperatures of 15°C and higher (Figure 8.2; Table 8.3). *Adenostoma fasciculatum* showed the lowest respiration rates at all temperatures. *Quercus dumosa, C. greggii*, and *R. ovata* were intermediate and similar in respiration rates over most of the temperature range measured. Between 5° and 25°C, *A. fasciculatum* had the lowest Q_{10} values observed. *Arctostaphylos glauca* showed the highest Q_{10} values between 5° and 15°C. Except for *A. glauca*, the lowest Q_{10} values were observed at the lowest temperatures measured, 5° to 15°C. The average Q_{10} values for all Californian species was 1.6 for temperatures from 5° to 25°C, which indicated moderate temperature insensitivity of stem respiration. Stem respiration, like leaf respiration, was generally much more temperature sensitive than was photosynthesis (Chapter 7).

Of the species measured at Fundo Santa Laura, *T. trinervis* showed the highest stem respiration rate at all temperatures. *Colliguaya odorifera* and *L. caustica* showed intermediate rates of stem respiration. *Satureja gilliesii* had the lowest rate of stem respiration. Between 5° and 25°C, the average Q_{10} was 2.2 for all species and ranged from 1.8 to 1.9 for *L. caustica* and *C. odorifera* to 2.4 to 2.6 for *T. trinervis* and *S. gilliesii*.

The slope of the line in an arrhenius plot indicates the energy of activation for the process considered. Arrhenius plots of stem respiration indicate that the energy of activation decreased below 25°C for *A. fasciculatum* and below 15°C for the Chilean species and the other Californian species measured except *A. glauca*, which showed an increase in the energy of activation below 15°C. Higher activation energies result in greater temperature sensitivity of the reaction rates considered. A change in the energy of activation with temperature indicates a likely switch in enzyme systems and the presence of isozymes. The lower energies of activation observed at the lower temperatures act to maintain higher respiration rates, and therefore, higher metabolic activity at these temperatures.

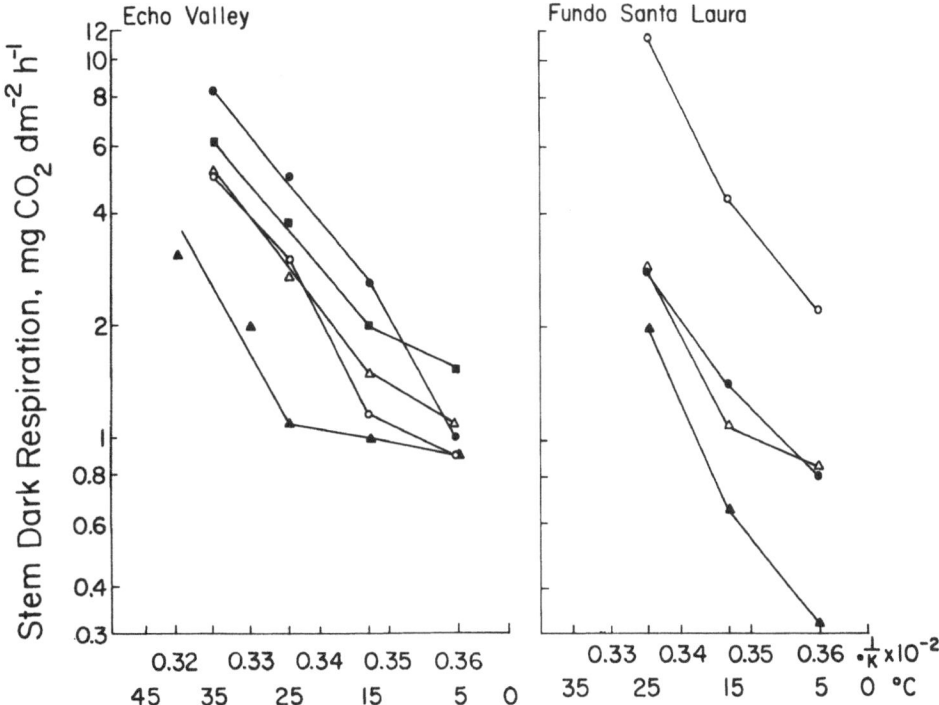

Figure 8.2. Arrhenius plots of stem dark respiration for: *Arctostaphylos glauca* (●), *Quercus dumosa* (■), *Rhus ovata* (△), *Ceanothus greggii* (○), and *Adenostoma fascicula-tum* (▲) measured in May-June 1976 at Echo Valley and for: *Colliguaya odorifera* (●), *Lithraea caustica* (△), *Trevoa trinervis* (○), and *Satureja gilliesii* (▲) measured in November-December 1979 at Fundo Santa Laura. Rates are for largest stem diameter classes measured for each species, which ranged from 4-6 mm to 10 mm for Chilean species.

A two-way analysis of variance was used to test the similarity of the stem respiration rates among each of the four species pairs and between the results for the groups of Californian and Chilean species. *Quercus dumosa*, not part of the original paired sample design, was not included in these statistical tests. Stem respiration rates between 5° and 25°C were used and temperature was entered as a covariant. These results indicate that the stem respiration rates varied significantly among the four species pairs ($P \leqslant 0.05$) (Table 8.1). However, at $P = 0.05$, there was no statistical difference in the stem respiration rates between the Californian and Chilean species. The variation in rates between the pairs of species was greater than the variation in the rates of stem respiration between the Californian and Chilean groups of species. While the mean stem respiration rate of the four Chilean species at 25°C was 58% higher than the mean rate of the four Californian species, no difference was found in the stem respiration rates of Californian and Chilean species at this temperature by using the *t*-test for sample means. The Q_{10} from 5° to 25°C was 29% higher for the Chilean species than for the Californian species.

The two-way analysis of variance showed no significant differences in Q_{10} values among pairs of species (at $P = 0.05$), but there was a significant difference between the Q_{10} values for stem respiration between the groups of Californian and Chilean species

Table 8.3. Key parameters of stem respiration in shrub species at Echo Valley and Fundo Santa Laura

Species	Temperature (°C)								Q_{10}			
	5	10	15	20	25	30	35	40	5-15	15-25	25-35[a]	5-25
	Respiration rate (mg dm^{-2} h^{-1}) ± SE											
					Echo Valley							
Ceanothus greggii	0.9 ± 0.1	-	1.3 ± 0.2	-	3.0 ± 0.3	-	5.0 ± 0.3	-	1.4	2.3	1.7	1.8
Arctostaphylos glauca	1.0 ± 0.3	-	2.6 ± 0.7	-	5.5 ± 1.4	-	8.2 ± 1.9	-	2.6	2.1	1.5	2.3
Rhus ovata	1.2 ± 0.2	-	1.5 ± 0.4	-	2.7 ± 0.6	-	5.2 ± 1.0	-	1.2	1.8	1.9	1.5
Adenostoma fasciculatum	0.9 ± 0.3	0.8 ± 3	1.0 ± 0.4	0.9 ± 0.3	1.2 ± 0.3	2.0 ± 0.4	-	3.1 ± 0.7	1.0	1.2	1.6	1.2
Quercus dumosa	1.6 ± 0.2	-	2.0 ± 0.3	-	2.8 ± 0.4	-	6.1 ± 0.9	-	1.2	1.4	1.6	1.3
Mean ± SE	1.1 ± 0.1	-	1.7 ± 0.3	-	3.2 ± 0.7	-	5.4 ± 0.9	-	1.5 ± 0.3	1.8 ± 0.2	1.7 ± 0.1	1.6 ± 0.2
					Fundo Santa Laura							
Trevoa trinervis	2.1 ± 0.3	-	4.4 ± 0.4	-	11.8 ± 1.1	-	-	-	2.1	2.7	-	2.4
Colliguaya odorifera	0.8 ± 0.5	-	1.4 ± 0.8	-	2.9 ± 1.7	-	-	-	1.8	2.1	-	1.9
Lithraea caustica	0.9 ± 0.1	-	1.2 ± 0.1	-	2.9 ± 0.4	-	-	-	1.3	2.4	-	1.8
Satureja gilliesii	0.3 ± 0.01	-	0.7 ± 0.1	-	2.0 ± 0.05	-	-	-	2.3	2.9	-	2.6
Mean ± SE	1.0 ± 0.4	-	1.9 ± 0.8	-	4.9 ± 2.3	-	-	-	1.9 ± 0.2	2.5 ± 0.2	-	2.2 ± 0.2

[a] 30° to 40°C for *A. fasciculatum*.

$(P \leqslant 0.05)$. Californian species have a statistically lower Q_{10} for stem respiration, which may be in response to the more variable temperature regime of Echo Valley (Chapter 7).

8.2.2.2 Effect of Stem Size on Respiration Rates. As stem diameter increased from 1 to 15 mm, stem respiration on a dry-weight basis decreased (Figure 8.3). This decrease was presumably because larger diameter stem classes of tissue had a higher surface-to-volume ratio and contained a larger fraction of metabolically inactive structural materials, which reduced the stem respiration rate on a mass basis. The most dramatic decrease in respiration rate on a dry-weight basis occurred as stems increased in diameter to 5 mm in most of the Californian species and to 3 to 5 mm in the Chilean species measured. Above 5 mm, increasing stem diameter resulted in a more gradual decrease in stem respiration rate.

Stem respiration on a surface-area basis did not show the consistent decrease in rate with increasing diameter that was shown by stem respiration expressed on a dry-weight basis in either the Californian or Chilean species. *Trevoa trinervis* showed a marked increase in stem respiration rate per unit area with increasing stem diameter (Figure 8.4). The three other Chilean species studied, *C. odorifera*, *L. caustica*, and *S. gilliesii*, showed little change in stem respiration rate per unit area with increasing stem diameter. In these species, the cambial and phloem activities appear to remain more or less constant on an area basis over the stem diameter size classes measured.

For the Californian species measured at Echo Valley, the stem respiration rate on a stem-area basis was more variable than that on a dry-weight basis, possibly because of difficulties in accurately estimating stem area. Over the range of stem areas measured, there was little consistent change in stem respiration rate on an area basis with varying stem diameter. The lower sensitivity of respiration on an area basis to diameter

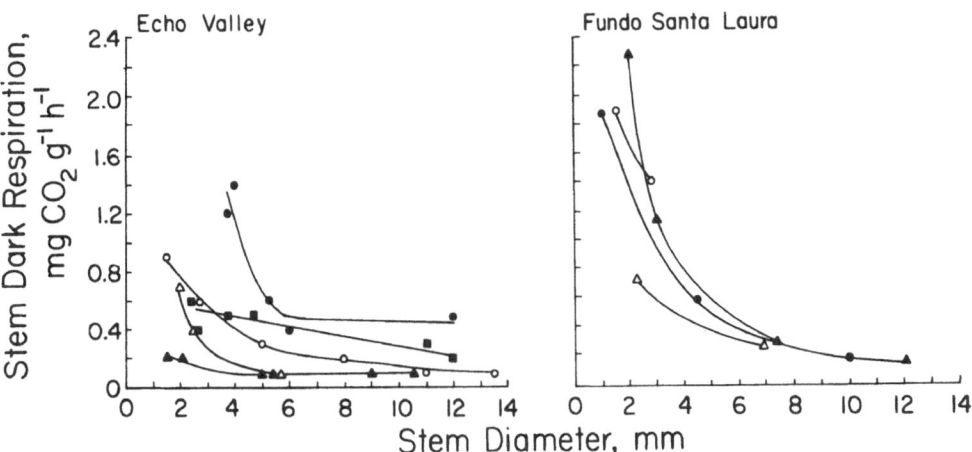

Figure 8.3. Relationship between stem diameter and stem dark respiration rate at 25°C for: *Arctostaphylos glauca* (●), *Quercus dumosa* (■), *Rhus ovata* (△), *Ceanothus greggii* (○), and *Adenostoma fasciculatum* (▲) measured in May-June at Echo Valley and for: *Colliguaya odorifera* (●), *Lithraea caustica* (△), *Trevoa trinervis* (○), and *Satureja gilliesii* (▲) measured in November-December 1979 at Fundo Santa Laura. Rates are for largest stem diameter classes measured for each species.

changes compared to respiration on a unit stem-weight basis is consistent with the results of Larcher (1975). The rates reported here for a stem diameter of 1 cm ranged from 0.10 to 0.44 mg CO_2 g^{-1} dry weight h^{-1} for the Californian species measured. Chilean species had stem respiration rates from 0.14 to 0.27 mg CO_2 g^{-1} dry weight h^{-1} for stems 0.72 to 1.2 cm in diameter. These stem respiration rates were generally higher than those reported for *Larix decidua* (Larcher, 1975), in which 1-cm stems at 20°C respired at about 0.08 mg CO_2 g^{-1} dry weight h^{-1}.

8.2.2.3 Stem Respiration in the Light. Respiratory loss of carbon dioxide in plants relative to carbon incorporation is potentially very high. However, stem photosynthesis (e.g., Adams et al., 1967) recycles respired carbon dioxide and reduces the net loss of carbon and energy from the plants. In most chaparral species, ambient midday light intensities are sufficient to support bark photosynthesis. This situation significantly reduces the carbon dioxide loss from the stems, especially at moderate and warm temperatures.

At 25°C, a frequent midday spring temperature in the chaparral, there was generally a marked reduction of net carbon dioxide loss from stems in the light in the species measured (Figure 8.5). This suppression undoubtedly resulted from photosynthetic fixation of respired carbon dioxide. In the light, net carbon dioxide flux from the stems of *C. greggii* and *A. glauca* was only 17% to 18% of that under otherwise similar conditions in the dark. In the light, *R. ovata* and *Q. dumosa* stems respired at 63% to

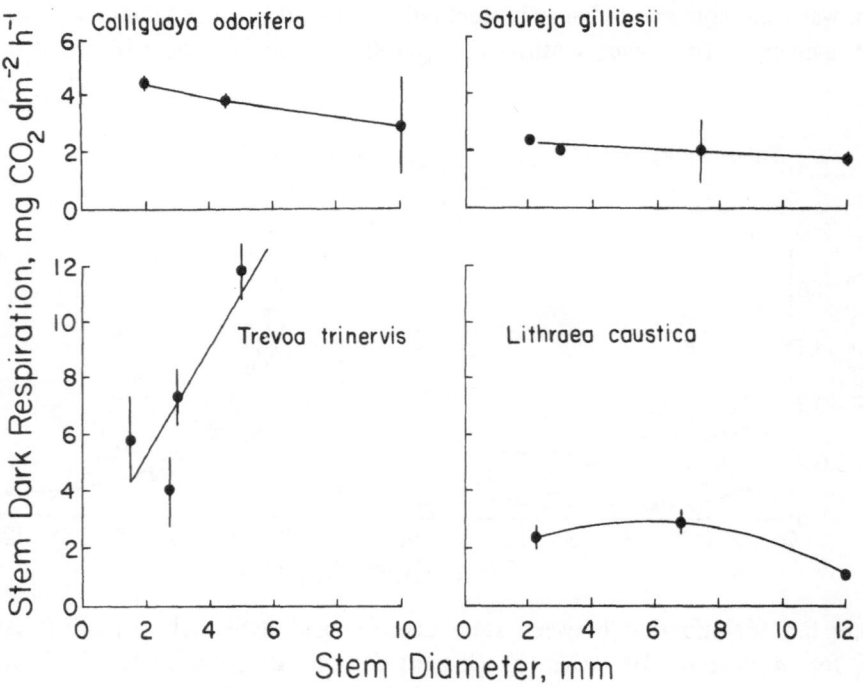

Figure 8.4. Relationship between stem diameter and stem dark respiration at 25°C on an area basis in four shrub species measured in May-June 1979 at Fundo Santa Laura. Circles and bars represent mean values ± 1 SE.

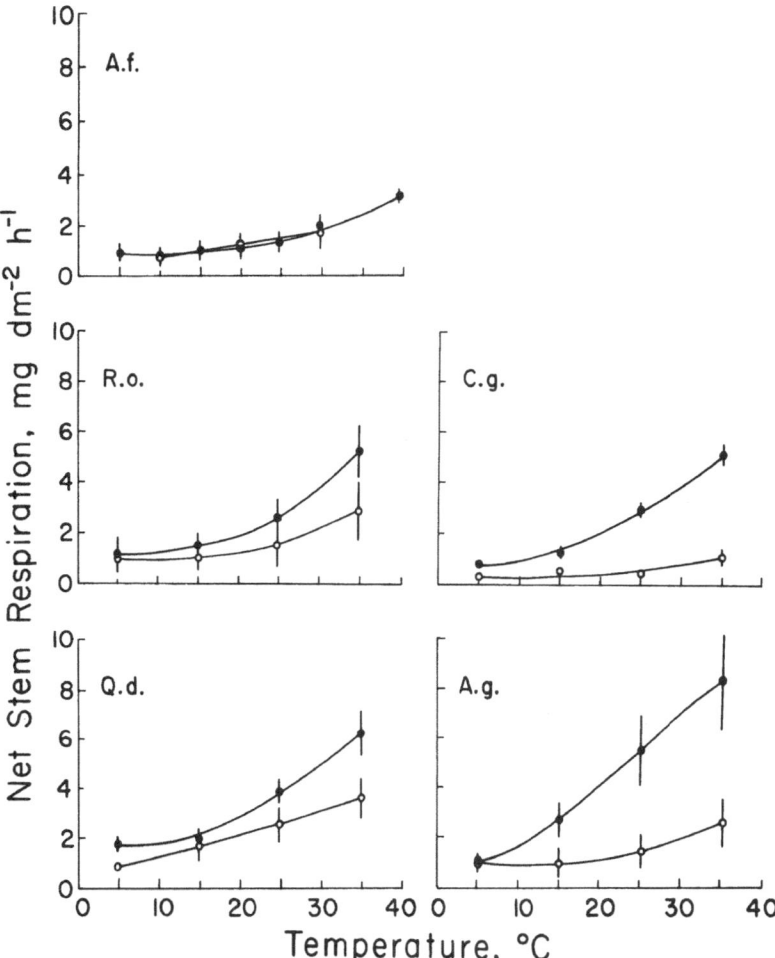

Figure 8.5. Net stem respiration in the light (o) and in the dark (●) as a function of temperature for five Californian evergreen species: *Adenostoma fasciculatum* (A.f.), *Rhus ovata* (R.o.), *Ceanothus greggii* (C.g.), *Quercus dumosa* (Q.d.), and *Arctostaphylos glauca* (A.g.). Light measurements were made using normal stem angles at midday (1000 to 1400 h) under ambient illumination ($> 1800\ \mu E\ m^{-2}$ PhAR). Bars represent mean values ± 1 SE.

64% of the dark rate. This suppression of net carbon dioxide loss results in substantial savings to the carbon economy of the plants. Only in the case of *A. fasciculatum* was there little or no apparent bark photosynthesis in the light.

Chilean species showed an even greater suppression of stem respiration in the light than did the Californian species measured (Figure 8.6). At least some stem size classes of *S. gilliesii*, *L. caustica*, and *C. odorifera* showed net positive photosynthetic uptake of carbon dioxide under the same measurement conditions as were used for the Californian species. *Trevoa trinervis* showed positive photosynthesis in the light for most temperatures and size classes observed.

While there was no consistent change in stem respiration rate with size on surface-area basis, the rate of photosynthetic reincorporation of carbon dioxide respired by

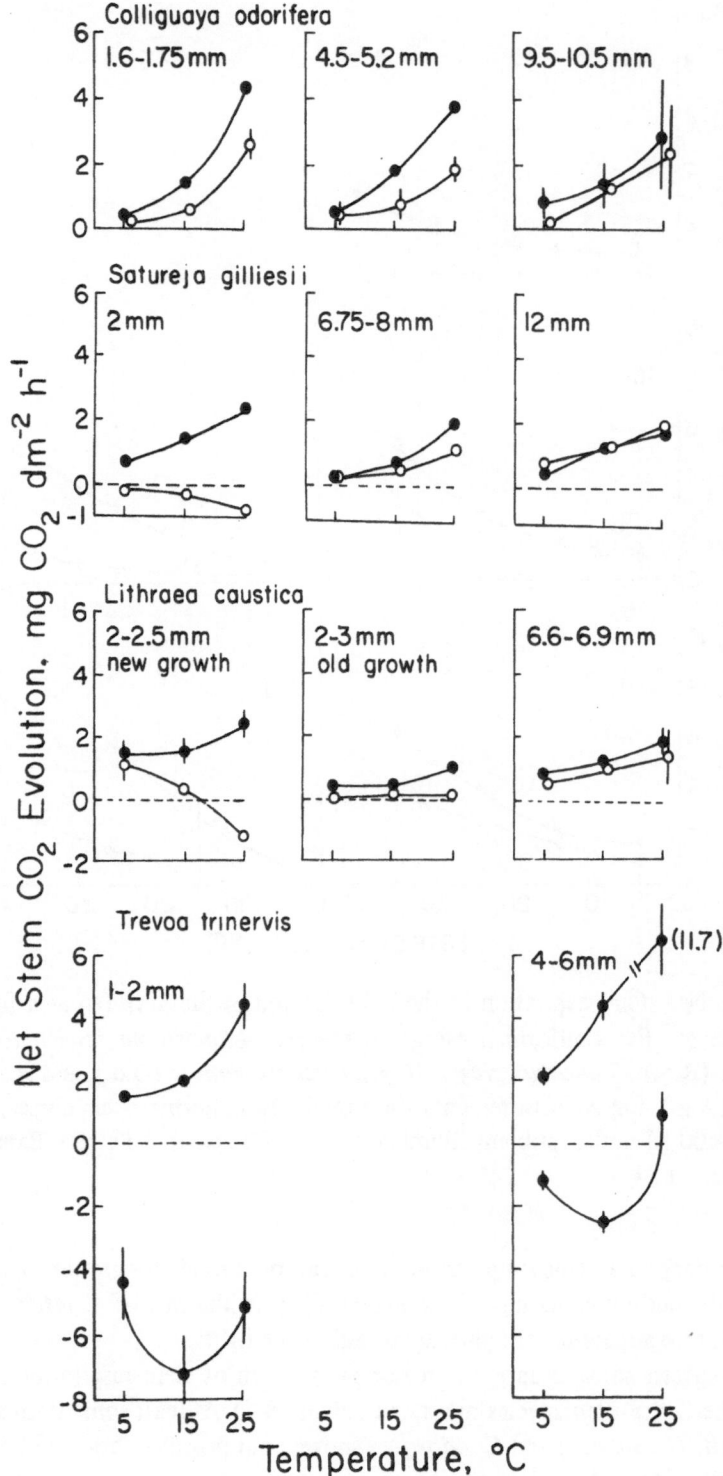

Figure 8.6. Effect of temperature on net CO_2 evolution by stems in the light (o) ($\geq$ 1500 μE m^{-2} s^{-1}) and in the dark (●) for various stem sizes of four Chilean species at Fundo Santa Laura, Chile. Negative values indicate positive net photosynthesis. Bars represent mean values ± 1 SE.

the stem decreased with increasing stem size and age. Two-millimeter stems of *S. gilliesii* showed net positive photosynthesis in full sunlight at 25°C, which was a 135% decrease in the stem dark respiration rate, 3-mm stems showed an 80% decrease in stem dark respiration rate in the light, 6.75-to-8-mm stems showed at 40% decrease, and 12-mm stems showed no major difference between stem respiration rates in the light or in the dark. Age also appeared to have an effect. Current-year, 2-to-2.5-mm stems of *L. caustica* showed positive photosynthesis at 25°C in the light, a 146% decrease in the net stem respiration rate. One-year-old stems with similar diameters showed only an 86% decrease in net stem respiration rate in the light. *Trevoa trinervis* was the only species examined that showed positive net stem photosynthesis over a wide range of stem sizes and ages.

At 25°C, stem photosynthesis in *T. trinervis* was not light saturated at 850 μE m^{-2} s^{-1} (Figure 8.7). Light compensation appeared to occur at about 160 μE m^{-2} s^{-1}. Because of the spread between light intensities examined, additional data are required to identify the precise light compensation point for *T. trinervis* stems.

The net carbon budget of mediterranean shrubs from Echo Valley and Fundo Santa Laura was affected to a large extent by the photosynthetic capabilities of stems of various age and size classes, as well as by the distribution of these stem age classes in the plant canopies. Smaller stem diameter classes had higher respiration rates on a per-gram basis, but also experienced greater apparent photosynthesis in the light.

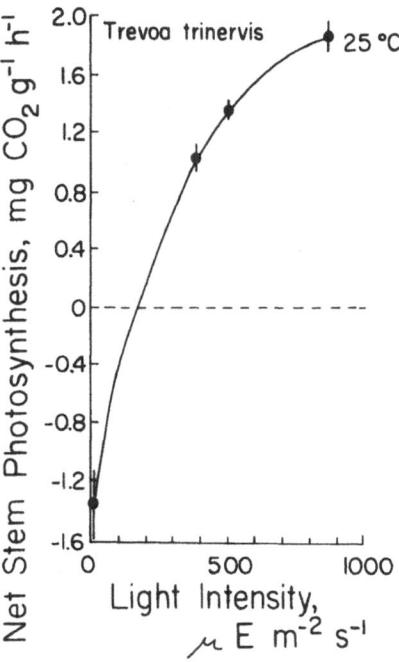

Figure 8.7. Effect of light intensity on net photosynthesis of 2.5-to-3.0-mm stems of *Trevoa trinervis* at 25°C at Fundo Santa Laura, Chile, November-December 1979. Bars represent mean values ± 1 SE.

8.2.3 Root Respiration

Root respiration potentially represents a major loss of carbon in plants (Billings et al., 1973; Caldwell et al., 1977; Hansen and Jensen, 1977). Values for root respiration are of major importance in evaluating carbon costs in chaparral and matorral shrubs and in comparing the costs and benefits associated with diverse rooting habits. Root growth and maintenance respiration rates affect the seasonal carbon costs of roots during times of growth and dormancy.

Allocation of carbon to root systems benefits the plant in terms of nutrient and water uptake and support. However, roots can form a sizable sink for carbon. Excessive allocation of carbon to roots can decrease the formation of other plant parts, such as leaves, and can depress overall plant production. A critical balance exists between the allocation of carbon to belowground versus aboveground plant parts in order to optimize productivity and resource use and to ensure survival. While there has been recent interest in root biomass, distribution, and production in chaparral and matorral ecosystems (Chapter 4), there is little previously published information on root metabolic rates and root turnover rates in chaparral and matorral ecosystems. Recent reviews on mediterranean ecosystems have not included data on the magnitude of root respiration, the temperature dependence of total root respiration, or the relative magnitude of root growth and maintenance respiration (e.g., di Castri and Mooney, 1973; Mooney, 1977a). Because roots function as support and storage organs and as the interface for water and nutrient uptake between the aboveground plant and the soil, allocation of carbon to roots influences their role in nutrient uptake and support and thereby affects the pattern of resource utilization within chaparral and matorral ecosystems.

Two separate experiments were performed to determine the rate of root respiration in mediterranean shrubs. In the first experiment, root respiration was determined as carbon dioxide efflux from the rooting medium of *Arctostaphylos pungens*, *R. ovata*, and *Yucca whipplei*, which were grown in washed sand in 10-liter pots (Evans, 1972; Ledig et al., 1976), and from the roots of *Ceanothus leucodermis* and *Quillaja saponaria*, which were grown in hydroponics.

Plants were established and grown under controlled environments in chambers. Throughout the establishment and measurement phase, potted plants were watered well with distilled water. Approximately every two weeks, each pot received an excess volume of 0.10 strength Hoagland's solution. Before root respiration was measured, the plants were transferred to growth chambers at constant temperatures of 10°, 20°, and 30°C and a 16-hour photoperiod for acclimation during periods of at least two weeks. The pots were then sealed by placing Plexiglas© over the tops of the pots through which the shoots protruded. The Plexiglas© contained inlet and outlet ports for the air stream and a port for watering the pots. The bottoms of the pots were sealed prior to measurement. Total carbon dioxide evolution from the sand was measured in the growth chamber at the acclimation temperature by sampling the carbon dioxide content of the returning air stream from the pots with infrared gas analysis for a period of at least 24 hours. Carbon dioxide evolution was measured in *C. leucodermis* and *Q. saponaria* by passing air through a sealed root chamber containing a nutrient solution in which the roots were submerged. Maintenance respiration was determined after plant carbohydrate reserves were lowered by dark starvation over several days (Baker et al., 1972; McCree, 1974). Following carbon dioxide flux determinations, the roots were removed from the pots and their dry weights and surface areas determined. Sand blanks and potting medium remaining after the roots were extracted were checked for carbon dioxide evolution. Both the blanks and the residual sand yielded negligible values of carbon dioxide evolution.

Results of the first experiment indicate that at 20°C total root respiration rates for sand culture-grown plants ranged from 0.32 mg CO_2 g^{-1} h^{-1} in *Y. whipplei* to 1.12 mg CO_2 g^{-1} h^{-1} in *R. ovata* (Figure 8.8; Table 8.4). The roots of chaparral species respired at rates similar to those of woody roots from other biomes. These values are in the same range as rates found in other woody root systems when the same method of measurement was used, i.e., infrared gas analyses to measure carbon dioxide efflux from attached roots (Huck et al., 1962; Bate and Canvin, 1972). The root respiration rates reported here were an order of magnitude lower than those reported for *Senecio* spp. (Lambers and Steingröver, 1978). The differences may have been due to the ages of root systems used, the use of hydroponics versus sand culture measurement medium, or species and growth form differences, or a combination of these factors. On an average, chaparral and matorral plants grown and measured in hydroponics had much higher root respiration rates than did those grown in sand culture. The higher root respiration rates may be the result of changed oxygen availability or nutrient levels in hydroponics versus sand culture and raise doubts concerning the ability to extrapolate from hydroponic experiments to field situations. Moreover, the sample size of plants used in the hydroponic experiments was small, and definite conclusions concerning differences between root respiration rates using the two measurement mediums require a systematic study with a larger sample size.

The mean maintenance respiration of the root mass of all species measured was about 74% of the total root respiration at 20°C (Table 8.4). These values showed little variability among species. For the chaparral plants under near optimal conditions, the carbon consumed to maintain the root systems was about three times greater than that

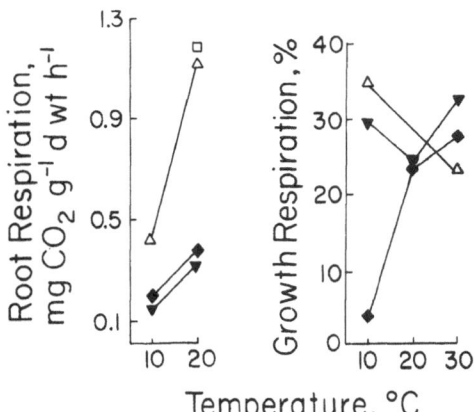

Figure 8.8. Temperature dependence of root respiration and growth respiration as a percent of total root respiration for *Arctostaphylos pungens* (♦), *Rhus ovata* (△), *Yucca whipplei* (▼), and *Quillaja saponaria* (□).

Table 8.4. Total root (Rs_{TOT}), root maintenance (Rs_m), and growth (Rs_g) respiration rates (mg CO_2 g^{-1} dry wt h^{-1}) at 10° and 20°C

Species	Respiration component	Root temperature 10°C		Root temperature 20°C	
Arctostaphylos pungens (8)[a]	Rs_{TOT}	0.19		0.38	
	Rs_m	0.19	(97%)[c]	0.29	(76%)
	Rs_g[b]	0.01		0.09	
Rhus ovata (9)	Rs_{TOT}	0.42		1.12	
	Rs_m	0.27	(65%)	-	
	Rs_g	0.18		-	
Yucca whipplei (5)	Rs_{TOT}	0.14		0.32	
	Rs_m	0.10	(70%)	0.24	(75%)
	Rs_g	0.04		0.08	
Quillaja saponaria[d] (2)	Rs_{TOT}	-		1.21	
	Rs_m	-		0.66	(55%)
	Rs_g	-		0.55	
Ceanothus leucodermis[d] (1)	Rs_{TOT}	-		0.73	
	Rs_m	-		0.59	(88%)
	Rs_g	-		0.14	

[a] Numbers in parentheses represent the number of individuals measured for that species.
[b] Growth respiration is calculated as $Rs_{TOT} - Rs_m$.
[c] Percentages in brackets indicate proportion of maintenance respiration to total respiration as a function of dry weight.
[d] Hydroponically grown plants.

allocated to growth of new root tissue. However, plant root growth is quite plastic. In general, root growth is favored under nutrient-limiting conditions, and the ratio of growth to maintenance respiration may increase under more limiting conditions where the percent allocation of carbon to root systems is increased and root growth is favored.

Rhus ovata and *A. pungens* have finely divided root systems and large root surface areas. *Rhus ovata*, an evergreen species that actively photosynthesizes throughout the year, maintained the highest root respiration rate on a dry-weight basis at all temperatures (Figure 8.8). *Rhus ovata* appears to tolerate the summer drought, in part, by maintaining extensive root systems and is often associated with rock outcrops and granite boulders. At least partially offsetting the gain in primary productivity due to an extensive root system is the increased cost of producing and maintaining a root system with a large surface area and extensive biomass. *Yucca whipplei*, whose roots were much thicker, fleshier, and had fewer side branches than those of the other species measured, exhibited the lowest rates of respiration on a dry-weight basis. *Yucca whipplei* may tolerate low rates of root activity through its ability to control water loss effectively and to utilize water stored in tissues while maintaining positive photosynthesis through crassulacean acid metabolism, although the latter has not yet been demonstrated (Chapter 7).

The total root respiration rates at 10° and 20°C were 0.14 and 0.32 for *Y. whipplei*, 0.19 and 0.38 for *A. pungens*, and 0.42 and 1.12 for *R. ovata* (Table 8.4). The two hydroponically grown plants had higher than average rates at 20°C, 1.21 mg g^{-1} h^{-1} for *Q. saponaria* and 0.73 for *C. leucodermis*. There was little variability in percent of total root respiration allocated to root growth respiration at 20°C; generally, this value was 24% to 25% of total root respiration for the sand culture-grown plants *A. pungens* and *Y. whipplei*. These species made similar energy allocations to root growth and maintenance respiration under similar nutrient and thermal regimes. The hydroponically grown plants were more variable, with percent growth respiration ranging from 45% of total root respiration for *Q. saponaria* to 22% of total root respiration for *C. leucodermis* (Table 8.4). While these values represent differences between resource allocation in *Q. saponaria* and *C. leucodermis*, the values may not be strictly comparable to those obtained from sand-grown plants.

Root growth respiration at 30°C ranged from 24% to 37% of total root respiration in all species. At 10°C, the species measured showed the greatest variability in the percent of total root respiration allocated to root growth respiration; values ranged from 30% to 35% in *R. ovata* and *Y. whipplei* to below 3% in *A. pungens* and *C. greggii*. The low growth respiration in *R. ovata* and *Y. whipplei* indicated little root growth in these two species at 10°C.

In all species measured in the first experiment, total root respiration rates decreased between 20° and 30°C. In *A. pungens*, the Q_{10} between 10° and 20°C was 2.0 and the Q_{10} between 20° and 30°C was 0.64. Between 20° and 30°C, total root respiration rate was depressed 5% in *C. greggii* and 50% in *R. ovata* (Oechel, *unpubl. data*). Because the plants were maintained at the measurement temperatures for a minimum of two weeks, the lower total root respiration rates at 30°C may reflect acclimation to these conditions. The measured respiration rates at 30°C would probably have been higher if the plants had been acclimated to lower temperatures and only briefly exposed to 30°C. A decrease in respiration rate at higher temperatures following acclimation has been reported for leaf dark respiration rates. Over an 8°C temperature range, plants acclimated to warm temperatures had lower leaf respiration rates than did cold-acclimated plants when leaf respiration rates were compared at the same measurement temperature (Smith and Hadley, 1974; Larcher, 1975). In other words, each root treatment may show a typical Q_{10} response if temperatures were changed too rapidly for acclimation. However, the pattern following acclimation may result in decreased total root respiration rates at 30°C when compared to 20°C.

Because 30°C is a moderate temperature, which does not damage the aboveground chaparral plant parts, and because microclimate data from Echo Valley indicated surface soil temperatures of 50°C in the summer, the observed decrease in total root respiration rates at 30°C was not thought to be the result of heat damage to plant tissue. It is possible that the decrease in root respiration at high temperatures represents a mechanism in chaparral roots that allows the plants to avoid potentially high carbon loss rates. However, elevated temperatures may have induced some level of dormancy or root senescence. Soil temperatures of 30°C and higher are likely to be associated with seasons of the year and soil profiles characterized by reduced available moisture for plant uptake. Decreasing total root respiration under these conditions would serve to conserve carbon in situations of low potential water uptake.

In the second experiment, root maintenance and root growth respiration were estimated in *Ceanothus crassifolius*. The plants were grown in the University of California, Riverside greenhouse for about nine months in 10-liter pots containing washed #16 silica sand before being transferred to San Diego State University for the root respiration measurements. At San Diego State University, the pots were placed in modified, temperature-controlled Styrofoam© ice chests. Plants received excess Hoagland's solution five times per day during the first month, three times per day during the second month, and twice per day until the initiation of the gas exchange measurements. One-quarter strength Hoagland's solution was used except that nitrogen was supplied at rates of 1, 5, 10, 15, and 20 mM NO_3^-N. Root maintenance plus growth respiration rates were determined using shoots of plants growing in the light. Measurements of temperature dependence of root respiration were completed as quickly as possible, within a few hours of the plant reaching thermal equilibrium. Measurements from all levels of nitrogen were combined for the root respiration results presented here.

In this experiment, which was designed to test short-term response to temperature change, *C. crassifolius* demonstrated temperature-dependency curves typical for rapid temperature change, which contrasted to the results from the first experiment where two-week acclimation periods between temperature changes were allowed. In *C. crassifolius*, total root respiration rates increased with increasing temperatures between 5° and 35°C (Figure 8.9). Total root respiration rate at 20°C was about 1.5 mg CO_2 g^{-1} h^{-1}. The Q_{10} between 5° and 35°C was about 1.9 and decreased with temperature from a Q_{10} of 2.0 between 5° and 15°C to a Q_{10} of 1.8 between 25° and 35°C. The respiration rates reported under rapid temperature change are higher than those for all other species measured in the first experiment. The pattern of temperature dependency of respiration is also much different than those reported from the first experiment. The decrease in total root respiration rate above 20°C and the low Q_{10} values between 10° and 20°C in the first experiment may reflect acclimation of the plants to the measurement temperatures.

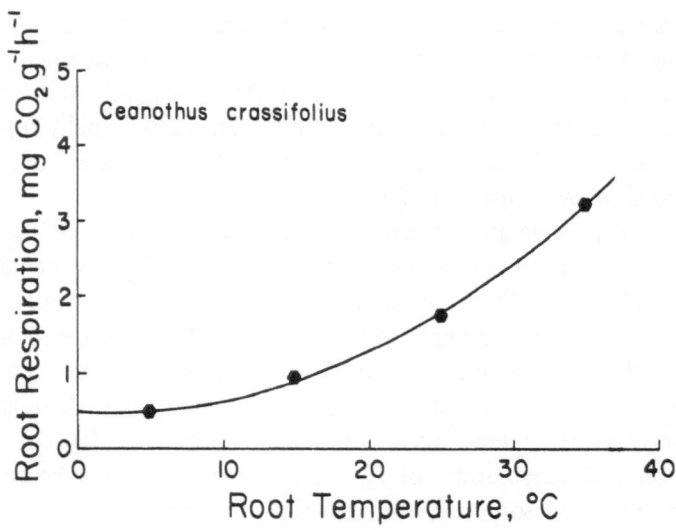

Figure 8.9. Effect of temperature on the rate of total root respiration of *Ceanothus crassifolius*. The line presented is for root respiration with the shoots in the light. The line is described by $y = 0.478 - 0.0117X + 0.0026X^2$, $r^2 = 0.97$.

Further data are required to determine the relationships between root metabolic activity and variables of soil moisture content and temperature in chaparral and matorral species. Additional data are also required for root respiration of herbs and grasses, which probably exhibit different rates and patterns of temperature dependence than do the woody shrubs. However, it is apparent from these two experiments that the experimental design may have marked impact on the results and the conclusions drawn from research on carbon dioxide gas exchange patterns. Short-term laboratory responses to temperature may not accurately reflect the effect of temperature on carbon metabolism in the field under conditions of gradual temperature change, where acclimatization may be a major factor affecting plant response patterns.

8.3. Carbon-14 Dioxide Utilization

Carbon, once incorporated within the plant, may be differentially allocated and utilized. Carbon-14 dioxide labeling allows comparison of allocation patterns and the analysis of differing adaptive patterns among various mediterranean plant species.

Individual shrubs less than 45 cm in height of *A. glauca, A. fasciculatum, Salvia apiana, Paeonia californica*, and *C. greggii* were labeled in the spring and early summer of 1976 by enclosing the entire aboveground portion of each shrub in a transparent plastic tent. Depending on plant size, 1 to 2 mCi of carbon-14 dioxide was introduced into the tent. The plant remained covered for about 4 h to allow maximal incorporation of the label. Labeling was done early in the morning to ensure adequate light and to minimize temperature buildup in the tent. Early morning labeling maximized the number of daylight hours available for biochemical fixation and allocation of the incorporated carbon-14 dioxide, thereby reducing the initial respiration of fixed carbon-14 dioxide and maximizing carbon retention. Air was circulated within the labeling tent with a fan, and the tents were removed before air temperature inside reached 45°C. Material from these shrubs was used both for analysis of carbon allocation and for determining root turnover rates (Figure 8.10).

Leaf, stem, and root material from the labeled plants was sampled after 1 to 2 days, 2 weeks, and in monthly or bimonthly intervals for 12 months to determine the distribution, use, and retention of photosynthate. During the four-month period following carbon-14 dioxide labeling, the greatest concentration of the labeled photosynthate remained in the leaves, the lowest concentration was found in the roots, and the stems were intermediate. The relative specific activity of carbon-14 dioxide in current-season roots, stem bark, and leaves of *A. glauca* 4 months after labeling was 1.0:1.2:2.5, respectively. Total label in the leaves decreased rapidly within the first week and more slowly thereafter. Young *A. glauca* leaves lost about 74% of the original label within one week and 94% of the total label after 4 months. In the other species sampled, the amount of labeled photosynthate lost from the leaves in the first week ranged from 30% to 91%. The percent of total label partitioned to the roots varied by species. After one week, *A. glauca* had distributed about 4% of the total label to the roots, while *A. fasciculatum* directed about 37% of the total label to the root compartment. In all species, the greatest carbon-14 activity in the roots was in the smallest root size classes. Activity was generally the highest in the roots of all size classes located in the 20-to-40-cm soil horizon.

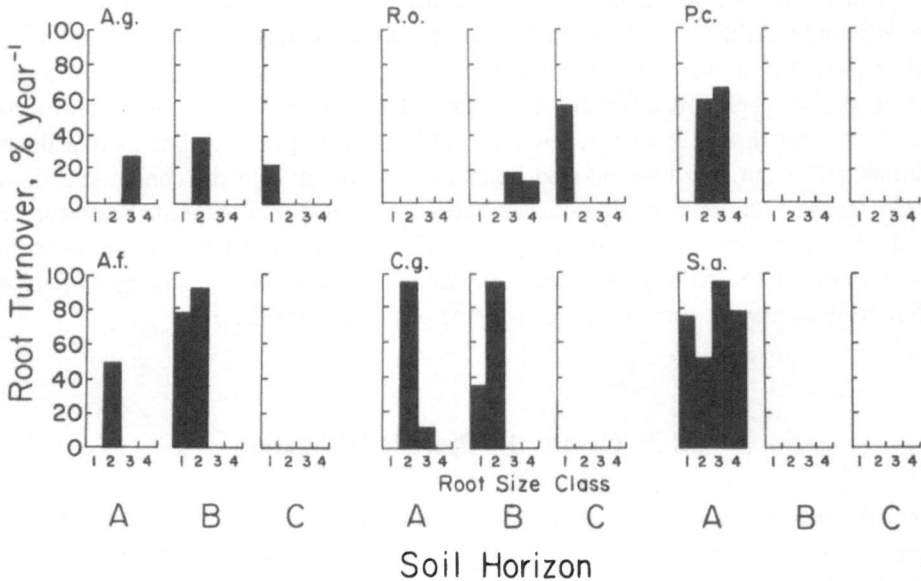

Figure 8.10. Percent annual root turnover in *Arctostaphylos glauca* (A.g.), *Rhus ovata* (R.o.), *Paeonia californica* (P.c.), *Adenostoma fasciculatum* (A.f.), *Ceanothus greggii* (C.g.), and *Salvia apiana* (S.a.) from June 1976 to May 1977 in California. A is the 0-to-20-cm soil horizon, B is 20 to 40 cm, and C is 40 to 60 cm. The root size classes are: 1, to < 0.25 cm; 2, 0.25 to 0.5 cm; 3, 0.5 to 1.0 cm; and 4, to > 1.0 cm.

8.4. Root Turnover

In general, root maintenance respiration in conjunction with root turnover accounts for the major cost of the root system. Seasonal root dynamics indicate the temporal sink that roots provide for photosynthates. High fractional turnover rates of the root systems have been reported from such diverse habitats as arctic tundra (Shaver and Billings, 1975), short grass prairie (Coleman, 1976), and cold desert (Caldwell et al., 1977). Caldwell et al. calculated a large cost in respiration and turnover associated with the belowground compartment. Root turnover was approximately three times the annual shoot productivity in both *Atriplex*- and *Ceratoides*-dominated communities.

Little knowledge exists of in situ root growth and turnover in the Californian mediterranean shrub vegetation. The four principal evergreen shrub species, *A. fasciculatum*, *R. ovata*, *A. glauca*, and *C. greggii*; a drought semideciduous subshrub, *S. apiana*; and two herbaceous species, *P. californica* and *Stipa coronata*, were labeled at Echo Valley in May be means of a carbon-14 dioxide labeling technique to determine the turnover of structural root constituents. Root samples were collected after one week and then at increasingly longer intervals for 16 months. When combined with root biomass, these data yield an estimate of root production (Caldwell and Camp, 1974).

Initial results indicated relatively low root turnover rates in *A. glauca* when compared to other species (Figure 8.10). This corresponded with the low carbon allocation to the roots in *A. glauca*. *Salvia apiana* and *P. californica* displayed high turnover rates in all root size classes including those greater than 1 cm. This pattern may reflect the

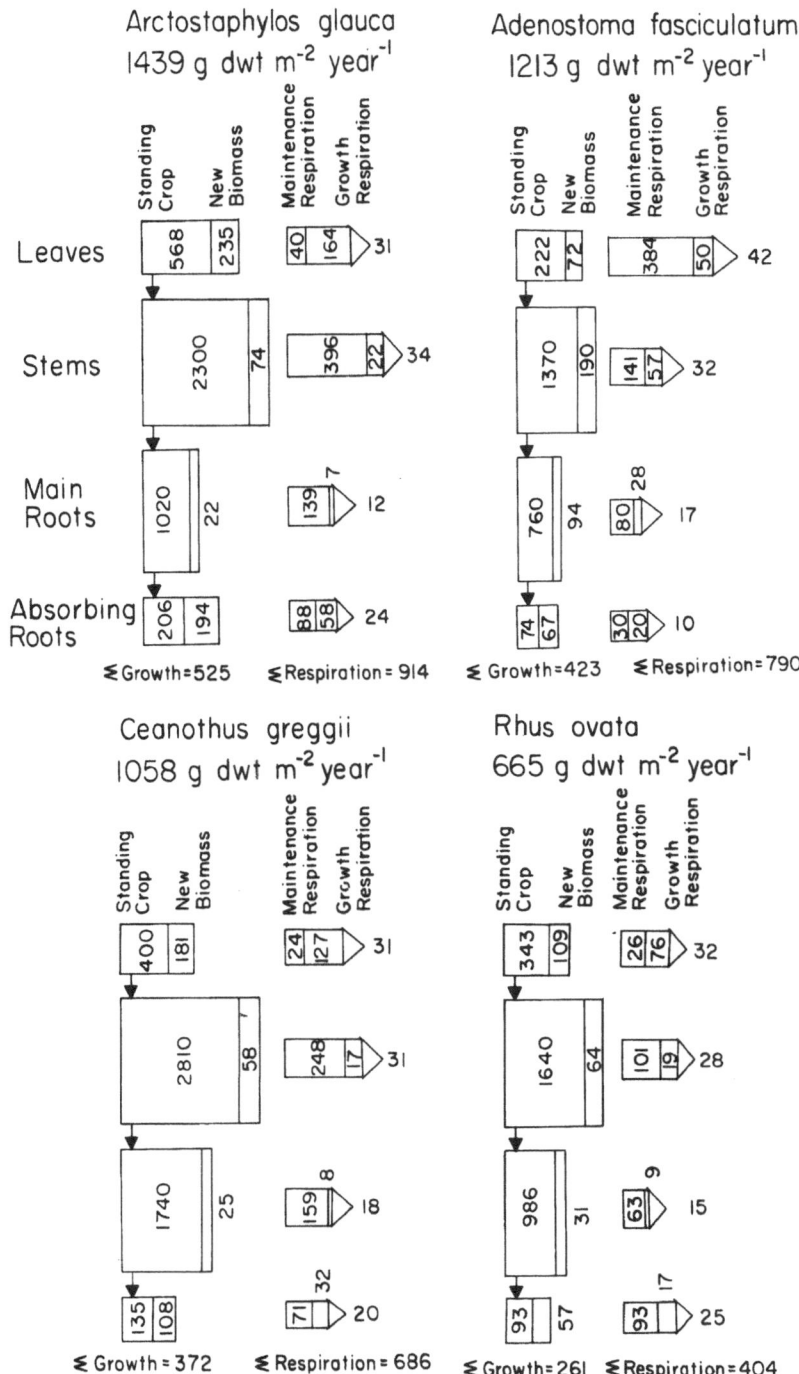

Figure 8.12. Standing crop and allocation to new biomass, maintenance respiration, and growth respiration in four Californian species at Echo Valley. The values presented represent field observations of growth and biomass, field and laboratory measurements of CO_2 flux, and calculations and extrapolations by means of the CAPS and MEDECS models (Mooney et al., 1977; Chapters 4 and 7). Percentages shown do not necessarily total 100% because of rounding error.

8.12). Despite this range, all species showed similar allocation patterns. Growth utilized about one-third of the available carbon; growth and maintenance respiration, about two-thirds. There were differences, however, in the way that carbon was allocated to new growth and biomass between the species, which suggested possible differences in strategies of resource use. *Adenostoma fasciculatum* showed a larger allocation of carbon to new stem biomass than did the other species studied. Of all Californian species studied, *A. glauca* invested the greatest percentage of available carbon to root growth.

The general pattern of carbon use was similar for the Chilean species studied. New biomass utilized between 37% of the carbon expenditures in *C. odorifera* and 45% in *L. caustica*. With the exception of those of *L. caustica*, the stems of the Chilean species utilized between 24% and 37% of the carbon expenditures for growth and respiration. The bulk of this carbon was used for respiration. *Lithraea caustica* utilized only 11% of its carbon budget in stem respiration and growth. The carbon costs of root growth and maintenance ranged from a low of 22% of the carbon expended in *S. gilliesii* and *T. trinervis* to a high of 42% in *C. odorifera*. Absorbing root growth was a major carbon cost. Absorbing root biomass turnover required from 9% to 22% of the total

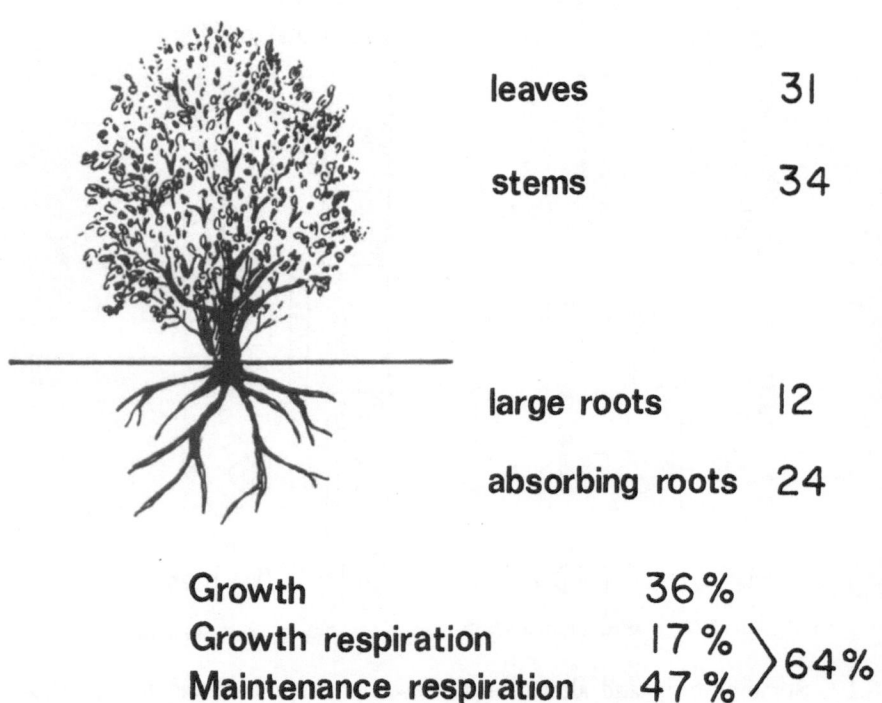

Arctostaphylos glauca   Carbon Allocation (% of Total)

leaves   31

stems   34

large roots   12

absorbing roots   24

Growth   36%
Growth respiration   17% ⟩ 64%
Maintenance respiration   47%

Figure 8.11. Estimated carbon allocation to growth and respiration of various structures in *Arctostaphylos glauca*. Values were calculated from field and laboratory experiments, field observations, and calculations and extrapolations by means of the CAPS and MEDECS models (Mooney et al., 1977; Chapters 4 and 7).

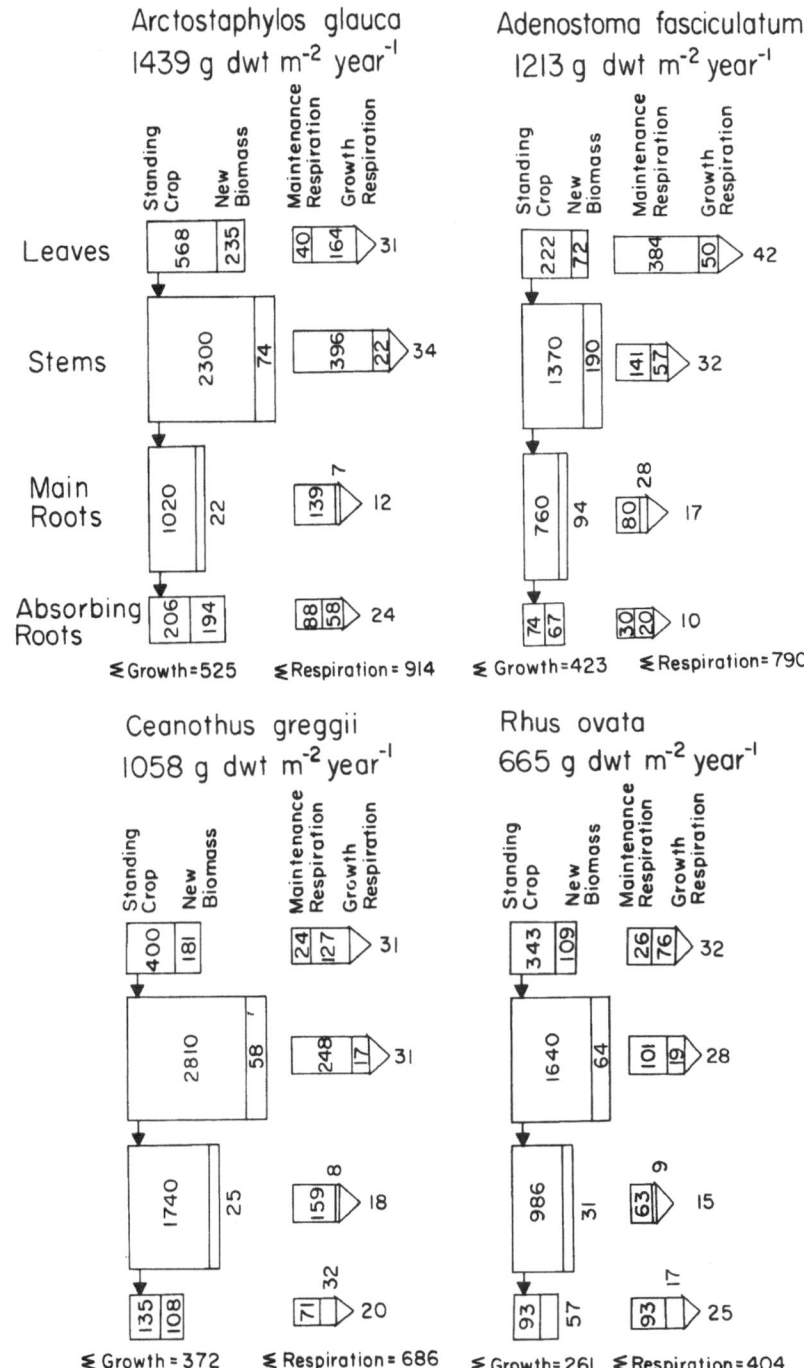

Figure 8.12. Standing crop and allocation to new biomass, maintenance respiration, and growth respiration in four Californian species at Echo Valley. The values presented represent field observations of growth and biomass, field and laboratory measurements of CO_2 flux, and calculations and extrapolations by means of the CAPS and MEDECS models (Mooney et al., 1977; Chapters 4 and 7). Percentages shown do not necessarily total 100% because of rounding error.

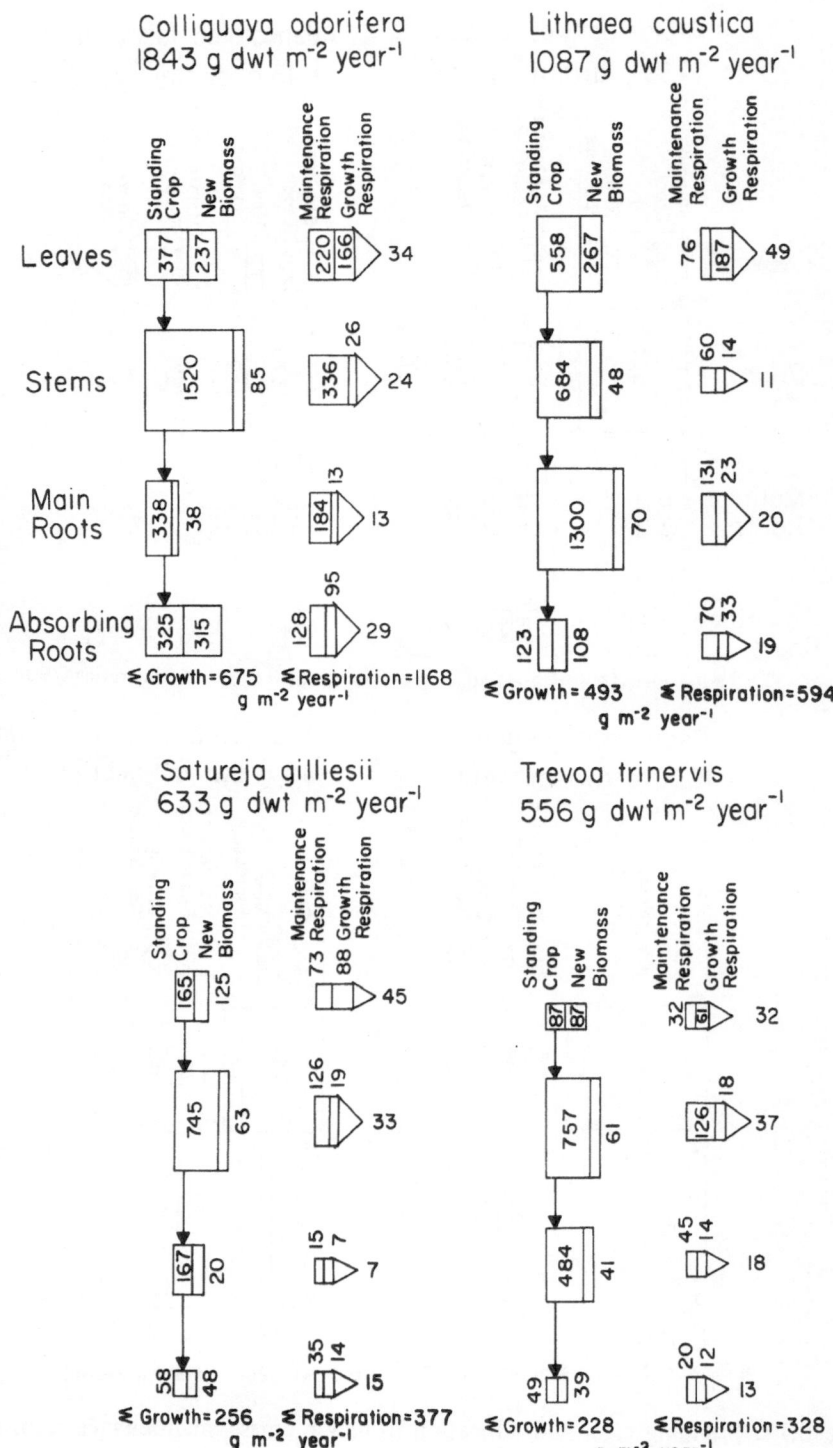

Figure 8.13. Standing crop and allocation to new biomass, maintenance respiration, and growth respiration in four Chilean species at Fundo Santa Laura. The values presented field observations of growth and biomass, field and laboratory measurements of CO_2 flux, and calculations and extrapolations by means of the CAPS and MEDECS models (Mooney et al., 1977; Chapters 4 and 7). Percentages shown do not necessarily total 100% because of rounding error.

carbon utilized for biomass replacement and concomitant growth respiration. Recent estimates for root production (Chapters 4 and 8) indicate that actual fine root production and maintenance costs may be even higher than those calculated here.

The apportioning of carbon between growth and respiration processes and among the various plant structures was similar in the Californian and Chilean species studied. In the Californian species studied, 36% of the carbon utilized went into biomass increase; in the Chilean species, this value was about 40%.

The Californian species utilized an average of 34% of their carbon budget for leaf growth and maintenance; the Chilean species utilized 40%. *Ceanothus greggii*, an evergreen, and *T. trinervis*, a deciduous species, invested similar percentages of their carbon budget for leaf growth and maintenance, 33% and 32%, respectively. The Californian species invested 30% of the carbon utilized in stem maintenance and growth; the Chilean species invested 26%. *Colliguaya odorifera* and especially *L. caustica* spent less on stem construction and maintenance than did *A. glauca* and *R. ovata*.

Californian species utilized about 35% of their growth and maintenance carbon for the main and absorbing roots, compared to 31% utilized for the same purpose by Chilean species. Except for *A. fasciculatum* and *L. caustica*, most species annually invest more carbon in the fine root system than in the main root system. Overall, Californian species utilize 16% of the available carbon in main roots and 20% in the absorbing roots. Chilean species utilized 12% of the carbon for the major roots and 20% for the fine roots.

8.6. Simulated Temperature Sensitivity of Carbon Metabolism

As discussed in Chapter 7, an annual climatic increase or decrease in temperature of 5°C would have little effect on the photosynthetic rates of the chaparral species studied (see Figure 7.6). However, because of the greater sensitivity of leaf, stem, and root respiration to temperature, an increase or decrease of 5°C in global temperature is predicted to have major effects on the carbon balance of chaparral species. This prediction was tested by using a MEDECS simulation of the effect of a 5°C temperature change on carbon flux, growth, carbohydrate reserves, and photosynthesis in *A. glauca*.

The MEDECS simulation indicated that, while photosynthesis was little affected by a simulated ± 5°C change in annual temperature, maintenance respiration increased by 31% or decreased by 17% over the temperature range (Table 8.6). The simulation of growth respiration was more complicated than was the simulation of photosynthesis because the effect of temperature on growth respiration was calculated through its effect on the rate of growth for various plant structures. Overall, a decrease in temperature of 5°C reduced growth respiration by 45% by severely limiting total growth. A 5°C decrease in temperature resulted in a 56% increase in net carbon dioxide flux because of the depressed maintenance and growth respiration. An increase in temperature of 5°C decreased net carbon dioxide incorporation for the plant by 21%, from 672 g m^{-2} yr^{-1} to 528 g m^{-2} yr^{-1}. In the simulations, actual growth was depressed by either an increase or a decrease in temperature. The standard year resulted in a reduction in carbohydrate reserves of 82 g m^{-2} yr^{-1}. In the scenario of decreased temperature, the net effect of decreased temperature and respiration was to yield a large surplus in carbohydrate reserves of 588 g m^{-2} yr^{-1}. A 5°C increase in temperature decreased carbohydrate

Table 8.6. Simulated response in *Arctostaphylos glauca* of growth, respiration, photosynthesis, and carbohydrate reserves to a 5°C increase or a decrease in temperature[a]

	Standard	−5°C	+5°C
Net photosynthesis	2230	2144 (−4%)	2273 (+2%)
Maintenance respiration	1000	834 (−17%)	1310 (+31%)
Growth respiration	478	262 (−45%)	435 (−9%)
Net incorporation	672	1048 (+56%)	528 (−21%)
Growth			
Leaves	571	341	481
Stems	57.9	42.4	77.6
Roots	125.8	76.6	147.6
Total	755	460 (−39%)	706 (−6%)
Net change in reserves	−83	+588	−178

[a] All units are carbon dioxide equivalents, g CO_2 m^{-2} yr^{-1}, and based on 16 stems per square meter of ground. The climate of the standard year is that of 1973-1974.

reserves 178 g m^{-2} yr^{-1}, which was an additional loss of 95 g m^{-2} yr^{-1} over that which occurred in the standard year. The calculated loss in reserves of 83 g m^{-2} yr^{-1} resulted from the conditions selected for the standard year 1972-1973. This suppression may reflect the normal year-to-year variability in augmentation to or loss from the reserve pool.

MEDECS indicated that respiration and growth processes are much more temperature sensitive than is photosynthesis in the plants simulated, but dynamic changes within the plant could modify these results. For example, carbohydrate buildup is known to be capable of enhancing respiration rates, suppressing photosynthesis rates, and increasing growth rates (Challa, 1976; Cunningham and Syvertsen, 1977). These factors might act to moderate the accumulation of carbohydrate under a depressed temperature regime to less than the predicted 588 g m^{-2} yr^{-1}.

Feedback relationships and acclimation potentials for photosynthesis, respiration, and growth are poorly known and are worthy of further study and quantification. Such information is needed before the effect of global temperature change on plant growth, carbon balance, and carbohydrate reserve levels can be predicted accurately.

8.7. Organic Nutrient Levels

Mooney and Chu (1974) determined the seasonal variations in the use of carbon-14-labeled photosynthate for the growth of stems, leaves, roots, and reproductive parts in *Heteromeles arbutifolia*, an evergreen sclerophyll shrub of the chaparral. They suggested that not all growth functions are met simultaneously even with a year-round carbon gain, because of heavy allocation demands by different plant parts in different seasons. Certain aspects of plant metabolism may be favored at various times, a trend that is even more pronounced in deciduous species. For example, the annual carbohydrate cycles of the evergreen oak *Quercus agrifolia* showed much less seasonal variation than did *Aesculus californica*, a drought deciduous species. Species with shorter seasons of carbon gain may have greater seasonal variation in both absolute and relative carbohydrate supply among growing points (Mooney and Hays, 1973). Seasonal changes in plant carbohydrate levels are expected to affect potential growth rates. Jones and Laude (1960) and Laude and Jones (1961) found that seasonal changes in reserve carbohydrate levels were correlated with resprouting ability of *A. fasciculatum* following fire or other disturbance. Little is known about specific relationships between carbohydrate supply and the percent success in resprouting and the effects of carbohydrate supply on the rates of stem elongation or leaf expansion at a particular time of the year. In the study reported here, the organic composition of leaves and stems of species from California and Chile were compared. Organic composition was analyzed by tissue age and season for the general organic chemical classes particularly important to plant structure and metabolism. The data were collected to test for underlying similarity or differences in the absolute or temporal distribution of resources to organic chemical constituents in species from California and Chile.

8.7.1 Methods

Plant tissue was collected from September 1975 to December 1976 at Echo Valley and Fundo Santa Laura. Leaves and stems of *C. greggii, A. glauca, A. fasciculatum,* and *R. ovata* were sampled from pole- and equator-facing slopes. However, little difference in the content of organic constituents was detected in samples from two slopes and only results from the equator-facing slope at Echo Valley are presented here. Leaves and stems of *T. trinervis, C. odorifera, Q. saponaria, Cryptocarya alba, Kageneckia oblonga, S. gilliesii,* and *L. caustica* were collected from a single site at Fundo Santa Laura in 1975-1976. Only October and November samplings for 1975 in Chile are reported here.

Leaf and stem tissue samples were divided into three age classes: Age One leaves and stems or current tissue, which were produced in the 1975 growing season, March to May 1975 in California and October 1975 to January 1976 in Chile; Age 2 or 1-year-old tissue, produced in the 1974 growing season; and Age 3 or 2-year-old tissue, produced in 1973 or before. Because *S. gilliesii* and *T. trinervis* are drought deciduous species and lack leaf material in the older age classes, average values of tissue contents in 1- and 2-year-old leaves of Chilean species were generally not compared to values for Californian species. The two-way analysis of variance was used to compare the various organic fractions measured for the four species pairs and groups of Californian and Chilean species. Probabilities and interaction terms are given in Table 8.7.

The samples were dried at 75° to 80°C immediately after they were collected. The samples were analyzed for total nonstructural carbohydrate, crude fat, crude protein, hemicellulose, lignin, cellulose, acid detergent fiber, and neutral detergent fiber (cell wall) content at the Agricultural Experiment Station, University of Alaska, Palmer, Alaska. Nonstructural carbohydrate and crude fat contents were determined by the methods described in Chapter 9.

Because nitrogen represents about 16% of the weight of protein, percent crude protein was calculated by multiplying the percent inorganic nitrogen contents in leaves and stems (Chapter 9) by the factor 6.25, resulting in crude protein as a percentage of dry weight. This method is a reasonable approximation of the true protein level because proteins generally account for 75% to 80% of the leaf nitrogen (Lyttleton, 1973).

Analyses for hemicellulose, lignin, cellulose, fiber, and cell wall followed the methods of Goering and Van Soest (1970). Acid detergent fiber, after correcting

Table 8.7. Tail probabilities of a two-way analysis of variance for differences between species pairs and the groups of Californian and Chilean species in chemical composition of leaves and stems[a]

Variable	Species pairs	Groups of Species	Species pair-Groups of Species interaction
Percent total nonstructural carbohydrate			
Leaves	0.111	0.000	0.209
Stems	0.001	0.000	0.000
Percent fat			
Leaves	0.001	0.000	0.004
Stems	0.000	0.000	0.000
Percent crude protein			
Leaves	0.000	0.001	0.001
Stems	0.038	0.088	0.490
Percent hemicellulose			
Leaves	0.012	0.536	0.447
Stems	0.810	0.088	0.490
Percent cellulose			
Leaves	0.000	0.000	0.000
Stems	0.000	0.236	0.014
Percent lignin			
Leaves	0.023	0.059	0.020
Stems	0.055	0.064	0.136
Percent acid detergent fiber			
Leaves	0.000	0.006	0.000
Stems	0.000	0.435	0.001
Percent cell wall			
Leaves	0.000	0.073	0.001
Sclerophyll index			
Leaves	0.040	0.148	0.003
Stems	0.001	0.004	0.001

[a] Data are from Tables 8.6 to 8.12, 9.10, and 9.11.

for the acid insoluble ash, contains cellulose, lignin, and cutin. The cell wall fraction was determined by neutral detergent fiber technique and consisted of total fiber, including hemicellulose, lignin, and cellulose. Subtracting the acid detergent fiber from the cell wall fraction yielded the hemicellulose fraction. Lignin, cellulose, and insoluble ash were determined by the permanganate technique.

The sclerophyll index used was modified from the index proposed by Loveless (1962) and used by Mooney et al. (1977) and Cromack and Monk (1975), among others. In Loveless' index, the formula: (% crude fiber/% crude protein) X 100, considers the amount of cell wall material relative to the protein level, which gives a good indication of the amount of metabolically active material relative to structural material.

The crude fiber fraction reported by Loveless (1962) was less than the acid detergent fiber or true fiber in the samples because of the partial digestion of cellulose and lignin that occurred when the crude fiber analysis methods of Horwitz (1960) were used. Cromack and Monk (1975) and Mooney et al. (1977) multiplied their true fiber data by 0.64 to approximate the crude fiber data used by Loveless. Therefore, the sclerophyll indices reported by Loveless (1962), Cromack and Monk (1975), and Mooney et al. (1977) were 36% lower than they would have been if the true fiber values had been used in the calculations. In the study reported here, sclerophyll values were calculated with the true acid detergent fiber values, and earlier published values based on crude fiber or on conversions from the fiber to crude fiber equivalents were corrected by multiplying by 1.56. The corrections return the data of Cromack and Monk (1975) and Mooney et al. (1977) to the level they originally measured and corrects Loveless (1962) to values approximating more recent extraction techniques (Cromack, *pers. comm.*).

8.7.2 Total Nonstructural Carbohydrates

The total nonstructural carbohydrate fraction is mainly comprised of soluble sugars, starches, and/or fructosans. Nonstructural carbohydrates are relatively inexpensive to produce; Penning de Vries (1972b) has calculated that only 1.1 g glucose is required to produce 1 g sucrose and that 1.2 g glucose is utilized in the production of 1 g starch. Starches are readily converted to sugars and form an important energy and carbon source within the plant. As indicated earlier, sugar levels may enhance or suppress photosynthetic, respiratory, and growth rates within the plant.

One of the largest differences in organic composition of leaves and stems between the groups of Californian and Chilean species was in total nonstructural carbohydrate levels. Total nonstructural carbohydrate content was markedly lower in Chilean species compared to Californian species (Figure 8.14; see also Table 9.10). A two-way analysis of variance showed no statistical difference among the leaf total nonstructural carbohydrate contents of the analogous species pairs ($P = 0.05$) (Table 8.7). However, the analogous pairs were significantly different with respect to stem carbohydrate levels ($P \leqslant 0.01$). There was also significant differences in total nonstructural carbohydrate levels of leaves and stems between the groups of Californian and Chilean species ($P \leqslant 0.01$). These data indicated a lack of functional convergence in extant carbohydrate levels between the two sites studied. Tissue total nonstructural carbohydrate levels are discussed in more detail in Chapter 9.

8.7.3 Crude Fats

The crude fat extraction includes lipids, resins, and waxes. Lipids are important constituents of cell membranes. Fifty percent of the chloroplast lamellae may be lipids. Membrane lipids are primarily phospholipids or glycolipids. Lipid contents of the leaves of most plants are generally between 3% and 10%. Lipids, which may provide important energy storage functions especially in seeds, are extremely expensive to pro-

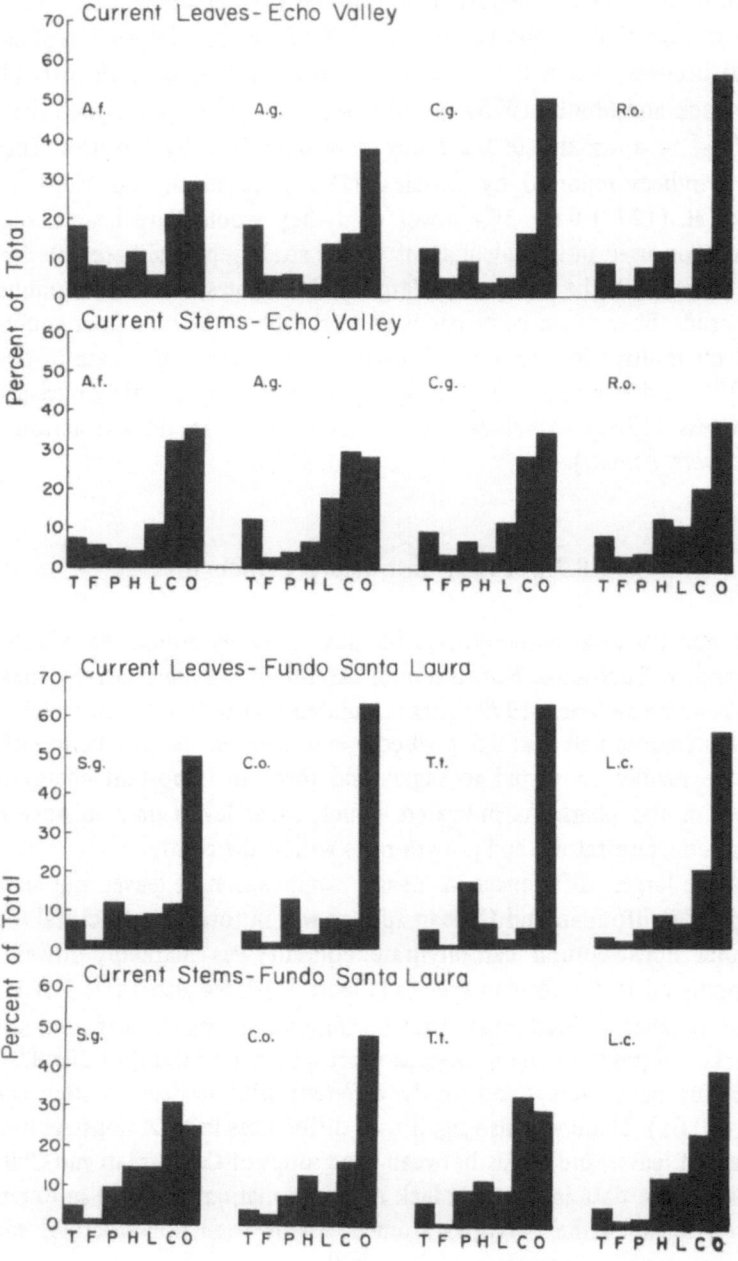

Figure 8.14. Percent composition of major cell constituents and residual fraction in current-year leaves and stems of species collected from equator-facing slopes at Echo Valley in the 1975-1976 season and collected at Fundo Santa Laura in 1975. Species samples at Echo Valley are: *Adenostoma fasciculatum* (A.f.), *Arctostaphylos glauca* (A.g.), *Ceanothus greggii* (C.g.), and *Rhus ovata* (R.o.). Species sampled at Fundo Santa Laura are: *Satureja gilliesii* (S.g.), *Colliguaya odorifera* (C.o.), *Trevoa trinervis* (T.t.), and *Lithraea caustica* (L.c.). Letters below bars represent: total nonstructural carbohydrate (T), crude fats (F), crude protein (P), hemicellulose (H), lignin (L), cellulose (C), and other (O).

duce. Penning de Vries (1975) calculated that the production of 1 g lipid requires 3.0 g glucose.

Similar to the results for total nonstructural carbohydrate, crude fat contents are generally much lower in leaves and stems of the Chilean species sampled than they are in the Californian species sampled (Figure 8.14; see also Table 9.11). The two-way analysis of variance indicates significant differences in leaf and stem crude fat contents between the groups of Californian and Chilean species ($P \leqslant 0.01$) (Table 8.7). There are also significant differences among the four species pairs tested ($P \leqslant 0.01$). As in the case of total nonstructural carbohydrate, there appears to be no convergence in the pattern of fat content between the species sampled in California and Chile. Crude fat content of species at Echo Valley and Fundo Santa Laura is discussed in more detail in Chapter 9.

8.7.4 Crude Protein

Up to 50% of the soluble leaf protein may be fraction one protein. This protein fraction is thought to be primarily RuBP carboxylase, which is of central importance to photosynthetic carbon dioxide incorporation. The protein content in cell walls is small, and the bulk of the protein is found in cytoplasm.

Because enzymes form the major protein component in leaves, protein content is important in regulating metabolic rates of plants. Protein levels also affect the food value of plants as forage. Proteins are expensive to produce; between 1.7 and 2.5 g glucose are required for the production of 1 g protein, depending on whether ammonia or nitrate is the substrate (Penning de Vries, 1975).

The mean crude protein content of current-year leaves from the four Californian species was 7.4 ± 0.7% (Table 8.8). Mean crude protein content of similar aged leaf tissue was 12.0 ± 1.8% for all Chilean species (Table 8.9). Calculated crude protein content decreased with leaf age in all species. In the Californian species, average crude protein contents dropped from 7.4% in current-year tissue to 4.5% in 2-year-old tissue. Leaf crude protein content in the Chilean species dropped from 12% in current-year tissue to 6.5% in 2-year-old tissue. These decreases correlate with decreases in leaf photosynthetic rates and leaf dark respiration rates, which occur with increasing leaf age.

A comparison of the four sets of species pairs, *A. glauca* and *C. odorifera*, *A. fasciculatum* and *S. gilliesii*, *C. greggii* and *T. trinervis*, and *R. ovata* and *L. caustica*, showed that crude protein levels in current-year leaf tissue differed significantly (two-way analysis of variance at $P \leqslant 0.01$) among the pairs. The group of Californian species considered as a population differed significantly ($P \leqslant 0.05$) from the Chilean species in crude protein levels of current-year leaves. Of the four principal species studied in Chile, only *L. caustica* had crude protein levels less than 10% in current-year leaf tissue. None of the calculated crude protein levels in current-year leaf tissue for the Californian species were as high as 10%. *Trevoa trinervis* and *C. greggii*, both potential nitrogen fixers, showed crude protein contents higher than average for their respective vegetations.

The calculated average crude protein content in stems also decreased markedly with age. Crude protein content in the Californian species decreased from 5.2% in current-year stems to 3.7% in 2-year-old stems. Chilean shrubs decreased from an average pro-

Table 8.8. Percent crude protein in leaves and stems from shrubs on equator-facing slopes at Echo Valley, California[a]

	Age 1			Age 2						Age 3				
	Summer 06/27/76	Fall 09/30/75	Seasonal X̄ ± SE	Fall 09/30/75	Late fall 11/14/75	Spring 03/19/76	Summer 06/27/76	Fall 09/30/76	Seasonal X̄ ± SE	Fall 09/30/75	Late fall 11/14/75	Summer 06/27/76	Fall 09/30/76	Seasonal X̄ ± SE
Leaves														
Adenostoma fasciculatum	6.9	7.5	7.2 ± 0.3	5.0	5.6	7.5	5.6	6.2	6.0 ± 0.4	4.4	-	-	-	4.4
Arctostaphylos glauca	6.2	5.6	5.9 ± 0.4	6.2	5.0	5.0	6.2	5.6	5.6 ± 0.3	4.4	3.8	5.0	-	4.4 ± 0.4
Ceanothus greggii	10.0	8.3	9.1 ± 0.8	7.5	8.3	9.4	7.5	8.8	5.0 ± 0.4	5.0	3.8	5.0	-	4.6 ± 0.6
Rhus ovata	8.8	6.2	7.5 ± 1.3	8.3	5.6	6.2	6.2	5.6	6.4 ± 0.5	5.0	3.8	5.0	-	4.4 ± 0.6
X̄ ± SE of X̄ for above species	8.0 ± 0.8	6.9 ± 0.6	7.4 ± 0.7	6.8 ± 0.7	6.1 ± 0.7	7.0 ± 0.9	6.4 ± 0.4	6.6 ± 0.8	5.8 ± 0.3	4.7 ± 0.2	3.8 ± 0.0	5.0 ± 0.0		4.5 ± 0.1
X̄ ± SE of X̄ for all species	7.4 ± 0.6					6.6 ± 0.2					4.5 ± 0.4			
Stems														
Adenostoma fasciculatum	5.0	4.4	4.7 ± 0.3	4.4	4.4	3.8	3.8	4.4	4.2 ± 0.1	5.0	3.8	3.8	5.0	4.4 ± 0.4
Arctostaphylos glauca	4.4	3.7	4.0 ± 0.4	3.1	2.5	2.5	3.8	3.1	3.0 ± 0.2	2.5	2.5	3.1	3.1	2.8 ± 0.2
Ceanothus greggii	7.5	6.9	7.2 ± 0.3	5.6	6.2	6.2	6.2	5.6	6.0 ± 0.1	4.4	4.4	5.0	5.0	4.7 ± 0.2
Rhus ovata	5.6	3.7	4.7 ± 1.0	5.6	2.5	3.1	3.8	3.1	3.6 ± 0.5	2.5	1.9	3.1	3.1	2.6 ± 0.3
X̄ ± SE of X̄ for above species	5.6 ± 0.6	4.7 ± 0.8	5.2 ± 0.7	4.7 ± 0.6	3.9 ± 0.9	3.9 ± 0.8	4.4 ± 0.6	4.0 ± 0.6	4.2 ± 0.6					3.6 ± 0.5
X̄ ± SE of X̄ for all species	5.2 ± 0.5					4.2 ± 0.2						3.7 ± 0.2		

[a]Collected from September 1975 to September 1976. Leaf and stem age classes 1, 2, and 3 refer to tissue produced in the spring of 1976 (current), 1975 (1-year-old), or 1974 (2-years-old), respectively.

Table 8.9. Percent crude protein, percent hemicellulose, and percent lignin in leaves and stems from shrubs at Fundo Santa Laura, Chile[a]

	Percent crude protein			Percent hemicellulose			Percent lignin		
	Age 1	Age 2	Age 3	Age 1	Age 2	Age 3	Age 1	Age 2	Age 3
Leaves									
Satureja gilliesii	11.9	-	-	8.6	-	-	9.0	-	-
Colliguaya odorifera	12.5	6.9	-	3.7	-	-	7.6	6.8	5.2
Trevoa trinervis	16.9	-	-	6.6	-	-	2.4	-	-
Lithraea caustica	5.0	3.8	-	-	8.6	6.5	6.4	9.0	9.1
Quillaja saponaria	17.5	6.9	5.6	3.0	2.4	3.7	3.4	4.1	4.0
Cryptocarya alba	6.9	5.0	6.2	20.3	16.5	17.0	8.9	8.2	7.7
Kageneckia oblonga	13.1	12.0	7.5	6.2	10.8	8.1	7.0	5.5	6.2
$\overline{X}$ ± SE for all species	12.0 ± 1.8	6.9 ± 1.4	6.5 ± 0.6	8.1 ± 2.6	9.6 ± 2.9	8.8 ± 2.9	6.4 ± 1.0	6.7 ± 0.8	6.7 ± 1.1
Stems									
Satureja gilliesii	6.9	9.4	6.9	24.4	12.0	16.4	14.4	17.2	17.3
Colliguaya odorifera	13.8	7.5	4.4	12.6	10.2	16.2	8.6	15.2	16.8
Trevoa trinervis	18.1	9.4	3.8	11.7	13.9	18.2	9.2	12.0	16.2
Lithraea caustica	3.8	2.5	2.5	12.6	15.3	22.1	14.2	14.5	15.2
Quillaja saponaria	11.9	5.0	1.9	9.0	8.2	17.4	10.0	14.3	14.7
Cryptocarya alba	4.4	3.8	4.4	11.8	16.8	21.4	10.0	11.2	13.7
Kageneckia oblonga	13.1	5.6	7.5	13.6	17.8	25.2	9.1	12.2	13.6
$\overline{X}$ ± SE for all species	10.3 ± 2.0	6.2 ± 1.0	4.5 ± 0.8	12.2 ± 0.6	13.5 ± 1.3	19.1 ± 1.4	10.8 ± 0.9	13.8 ± 0.8	1.54 ± 0.6

[a]Collected in October and November 1975. Leaf and stem age classes 1, 2, and 3 refer to tissue produced in the spring of 1975 (current), 1974 (1-year-old), and 1973 (2-years-old), respectively.

tein content of 10.3% in current-year stems to 4.5% in 2-year-old stems. *Trevoa trinervis*, which exhibits substantial bark photosynthesis in young tissue, had very high stem crude protein values, especially in current-year stems where the crude protein was 18%, which was somewhat higher than the value for its leaves. The crude protein level dropped to 9.4% and 3.8% in 1- and 2-year-old *T. trinervis* stems, which correlates with decreasing respiratory and photosynthetic rates per gram dry material in older age classes. Similar relationships of crude protein content with age hold for the other Chilean and Californian species measured.

With the two-way analysis of variance, both the species pairs (at $P \leqslant 0.05$) and the groups of Californian and Chilean species (at $P \leqslant 0.01$) showed statistical difference in the crude protein content in stems (Table 8.7). No convergence was found in the levels of crude protein content in leaf or stem tissue in the Californian and Chilean species studied.

8.7.5 Hemicellulose

Hemicelluloses form a heterogeneous group of poorly defined pentosans and hexosans, which are constituents of primary and secondary plant cell walls. Hemicelluloses may be hydrolyzed and utilized as a carbon source under some conditions (Terent'ev and Petrovich, 1968; Reid and Wilkie, 1969). However, when lignins are abundant in secondary cell walls, they may interfere with the enzymatic hydrolysis of hemicelluloses. Primary plant cell walls generally have more hemicellulose relative to cellulose than do secondary cell walls. Similarly, stems have lower hemicellulose contents relative to cellulose than do leaves (Bailey, 1973).

Current and 1-year-old leaf tissue of the Californian and Chilean species averaged 6% to 7% and 8% to 10% hemicellulose, respectively (Tables 8.9 and 8.10). There was no major effect of leaf age on hemicellulose content in current through 2-year-old leaf tissue. There was more variability in percent hemicellulose between species in California or Chile than among leaf age classes of all species combined. In Californian species, current-year leaf tissue ranged from an average value of about 3% for *A. glauca* to 11% for *R. ovata*. The range in Chilean species was from 3% for *Q. saponaria* to 20% for *C. alba*.

Current-year leaf tissue in both *R. ovata* and *L. caustica* averaged 11% hemicellulose. Percent hemicellulose in current-year leaf tissue showed only a 0% to 3% variation between species in the other three species pairs. However, the two-way analysis of variance indicated that the four species pairs were significantly different from one another in leaf hemicellulose content ($P \leqslant 0.01$), but there was no significant difference between the group of Californian species and the group of Chilean species (at $P = 0.05$).

Stems generally increased in hemicellulose content with age. The mean for Californian species ranged from an average of 7% hemicellulose in current-year stems to 18% in 2-year-old stems. Chilean species ranged from an average of 12% hemicellulose in current-year and 1-year-old stems to 19% in 2-year-old stems. Stems varied less in hemicellulose content among species than did leaves. Stem hemicellulose content was significantly lower in Californian species than in Chilean species and was not significantly different among the four species pairs. These data indicate a similarity between plants in Chile and California in leaf hemicellulose levels but not in stem hemicellulose content.

Table 8.10. Percent hemicellulose in leaves and stems from shrubs on equator-facing slopes at Echo Valley, California collected from November 1975 to September 1976. Leaf and stem age classes, 1, 2, and 3 refer to tissue produced in the spring of 1976 (current), 1975 (1-year-old), or 1974 (2-years-old), respectively.

	Age 1				Age 2					Age 3			
	Late fall 11/14/75	Summer 06/27/76	Fall 09/30/76	Seasonal $\overline{X} \pm SE$	Late fall 11/14/75	Spring 03/19/76	Summer 06/27/76	Fall 09/30/76	Seasonal $\overline{X} \pm SE$	Late fall 11/14/75	Summer 06/27/76	Fall 09/30/76	Seasonal $\overline{X} \pm SE$
Leaves													
Adenostoma fasciculatum	5.2	8.9	2.6	5.7 ± 3.1	11.3	8.4	12.4	5.0	9.3 ± 1.7				
Arctostaphylos glauca	5.2	0.8	3.9	3.3 ± 1.3	1.3	4.8			2.7 ± 1.3				
Ceanothus greggii	2.3	5.4	3.3	3.7 ± 0.9		2.1	0.5		1.3 ± 0.8				
Rhus ovata	4.0	12.5	17.6	11.4 ± 4.0	0.6	18.4	15.1	6.1	10.0 ± 4.1	1.8	9.3		5.6 ± 3.8
$\overline{X} \pm SE$ of seasonal $\overline{X}$ for above species	3.8 ± 0.8	6.4 ± 2.5	6.8 ± 3.6	6.0 ± 1.9	14.4 ± 3.5	8.4 ± 9.3	9.3 ± 4.5	5.5 ± 0.5	5.9 ± 2.3				5.6 ± 3.8
$\overline{X} \pm SE$ of $\overline{X}$ for all species	5.9 ± 1.0				6.9 ± 1.2								
Stems													
Adenostoma fasciculatum	1.4	9.4	1.8	4.2 ± 2.6	6.6	15.2	5.2	8.5	8.9 ± 2.2	16.8	17.5	23.1	19.1 ± 2.0
Arctostaphylos glauca	3.2	7.9	9.0	6.7 ± 1.8	14.1	14.1	12.5	18.0	14.7 ± 1.2	20.9	22.3	21.3	21.5 ± 0.4
Ceanothus greggii	5.2	3.1	4.9	4.4 ± 0.7	8.7	4.7	4.5	10.4	7.1 ± 1.5	24.9	13.1	18.8	18.9 ± 3.4
Rhus ovata	13.0	16.2	10.2	13.1 ± 1.7	11.2	11.6	12.2	13.9	12.2 ± 0.6	13.2	15.1	13.5	13.9 ± 0.6
$\overline{X} \pm SE$ of seasonal $\overline{X}$ for above species	5.7 ± 2.5	9.1 ± 2.7	6.5 ± 1.9	7.1 ± 2.1	10.1 ± 1.6	11.4 ± 2.4	8.6 ± 2.2	12.7 ± 2.1	10.7 ± 1.7	19.0 ± 2.5	17.0 ± 2.0	19.2 ± 2.1	18.4 ± 5.8
$\overline{X} \pm SE$ of $\overline{X}$ for all species	7.1 ± 1.0				10.7 ± 0.9					18.4 ± 0.7			

8.7.6 Lignin

Lignin encrusts cellulose and secondary cell walls, giving strength to plant structures and forming the second most abundant compound, following cellulose, in most wood. Lignin is usually associated with plant supporting or conducting tissues and is one of the most difficult plant products to decompose because it resists enzymatic digestion. Lignin content in litter tends to be inversely proportional to decomposition rates in the soil (Alexander, 1961). Lignin forms covalent bonds to plant polysaccharides, which prevent swelling of plant fibers after ingestion by animals. This fact slows the breakdown by rumen of cecum microorganisms found in the alimentary tracts of domesticated animals. The crosslinks formed between lignin and cellulose inhibit digestion close to the crosslinks by the enzyme cellulase. Lignin is a fairly expensive product to produce. Penning de Vries (1975) calculated that it requires 2.15 g glucose to produce 1 g lignin compared to the 1.2 g glucose required to produce 1 g cellulose or starch. Because of its abundance and cost to produce, lignin represents a sizable carbon sink in woody plants.

Lignin percentages for leaves in Californian species averaged 6% for current-year tissue, 7% for 1-year-old tissue, and 8% for 2-year-old leaf tissue (Table 8.11). *Arctostaphylos glauca* had the highest leaf lignin content, 11.0 ± 3.0 in current-year tissue, and *R. ovata* had the lowest lignin content, 2.1 ± 0.4. While the percent lignin in current-year tissue generally increased rapidly from the fall until the following summer, there was no strong pattern of varying lignin content with season in the older age classes.

The values for lignin for the leaves of measured Californian and Chilean shrubs generally fall in the range of lignin reported by Cromack and Monk (1975) for the leaves of deciduous and evergreen tree species from Coweeta, North Carolina, where the range for tissue at mid-growing season was from 3.5% lignin in dogwood to 16.4% in white pine. However, in the case of white pine, senescent tissue ranged up to 31.0% lignin.

The four California-Chile species pairs were significantly different in leaf lignin content (two-way analysis of variance, $P \leqslant 0.05$), but there was no significant difference in leaf lignin content between the groups of Californian and Chilean species (at $P = 0.05$), indicating convergence with respect to this leaf constituent (Tables 8.9 and 8.11).

As expected, stems showed higher lignin contents than did the leaves due to the greater percentage of woody tissues in stems. Stem age had little effect on lignin content. The average lignin content for all Californian species ranged from 11.6% ± 1.5% in current-year stems to 12.3% ± 1.1% in 2-year-old stems. Stem lignin content in current-year tissue of principal Chilean species was similar to that in the Californian species but tended to be slightly higher in the older age classes. *Satureja gilliesii* had the highest stem lignin content with stem lignin averaging 16.3% for all age classes.

Stem lignin content was not statistically different among the four species pairs or between groups of species in California and Chile (at $P = 0.05$, two-way analysis of variance).

Table 8.11. Percent lignin in leaves and stems from shrubs on equator-facing slopes at Echo Valley, California[a]

	Age 1				Age 2					Age 3			
	Late fall 11/14/75	Summer 06/27/76	Fall 09/30/76	Seasonal $\overline{X} \pm SE$	Late fall 11/14/75	Spring 03/19/76	Summer 06/27/76	Fall 09/30/76	Seasonal $\overline{X} \pm SE$	Late fall 11/14/75	Summer 06/27/76	Fall 09/30/76	Seasonal $\overline{X} \pm SE$
Leaves													
Adenostoma fasciculatum	-	4.6	6.8	5.7 ± 1.1	7.7	7.7	5.7	6.7	7.0 ± 0.5	-	-	-	-
Arctostaphylos glauca	5.0	14.5	13.4	11.0 ± 3.0	6.7	6.8	12.8	13.0	9.8 ± 1.8	10.9	14.3	-	12.6 ± 1.7
Ceanothus greggii	3.7	5.6	5.0	4.7 ± 0.6	-	8.3	6.4	6.9	7.2 ± 0.6	-	-	-	-
Rhus ovata	2.6	1.2	2.5	2.1 ± 0.4	4.4	2.0	2.2	3.2	3.0 ± 0.6	3.7	3.3	-	3.5 ± 0.2
$\overline{X} \pm SE$ for above species	3.8 ± 0.7	6.5 ± 2.8	6.9 ± 2.3	5.9 ± 1.6	6.2 ± 1.0	6.2 ± 1.4	6.8 ± 2.2	7.4 ± 2.0	6.7 ± 1.4	7.3 ± 3.6	8.8 ± 5.5		8.0 ± 4.6
$\overline{X} \pm SE$ of $\overline{X}$ for all species		5.7 ± 1.0					6.7 ± 0.3				8.0 ± 0.8		
Stems													
Adenostoma fasciculatum	15.6	10.6	10.4	12.2 ± 1.0	14.4	12.4	17.5	13.2	14.4 ± 1.1	12.8	16.9	10.2	13.3 ± 1.9
Arctostaphylos glauca	13.4	15.9	16.7	15.3 ± 1.0	11.0	10.6	17.1	13.5	13.1 ± 1.5	12.7	10.4	8.7	10.6 ± 1.2
Ceanothus greggii	9.1	11.0	11.7	10.6 ± 0.8	10.9	13.4	12.9	11.3	12.1 ± 0.6	15.1	17.7	12.2	15.0 ± 1.6
Rhus ovata	9.6	4.0	11.2	8.3 ± 2.2	7.9	5.5	10.7	12.5	9.1 ± 1.5	9.7	10.6	10.9	10.2 ± 0.4
$\overline{X} \pm SE$ for above species	11.9 ± 1.6	10.4 ± 2.4	12.5 ± 1.4	11.6 ± 1.5	11.0 ± 1.3	10.5 ± 1.8	14.6 ± 1.6	12.6 ± 0.5	12.2 ± 1.1	12.6 ± 1.1	13.9 ± 2.0	10.5 ± 0.7	12.3 ± 1.1
$\overline{X} \pm SE$ of $\overline{X}$ for all species		11.6 ± 0.6					12.2 ± 0.8				12.3 ± 1.0		

[a] Collected from November 1975 to September 1976. Leaf and stem age classes 1, 2, and 3 refer to tissue produced in the spring of 1976 (current), 1975 (1-year-old), or 1974 (2-years-old), respectively.

8.7.7 Cellulose

Cellulose is important in contributing strength to primary and secondary cell walls. A linear polymer of D glucose, cellulose is simple in structure and is relatively inexpensive to produce, requiring only 1.2 g glucose per gram of cellulose produced (Penning de Vries, 1975). Cellulose is easily broken down sequentially by exoenzymes or randomly by endoenzymes. However, encrustation by lignins greatly slows decomposition in the soil or digestion in the gut of herbivorous animals.

Leaf cellulose content was similar in all species studied. Cellulose content of current-year leaves for all Californian species averaged 16%, which decreased to 15% in 2-year-old leaves (Table 8.12). Average cellulose content of current-year leaves of the Chilean species was 11% (Table 8.13). The two-way analysis of variance indicated difference in leaf cellulose content among the four species pairs ($P \leqslant 0.01$) as well as between groups of Californian and Chilean species ($P \leqslant 0.01$). No convergence is, therefore, indicated in allocation patterns of cellulose in the leaves.

Cellulose ranged from an average of 28% in current-year stems for all Californian species to 38% in 2-year-old stems. These results were similar for the principal Chilean species, which ranged from 26% cellulose in current-year stems to 36% in 2-year-old stems. Stem cellulose content showed statistical differences among the four species pairs ($P \leqslant 0.01$), but cellulose contents of stems showed no statistical difference between Californian and Chilean species (at $P = 0.05$). Despite the tendency to greater cellulose content in older age classes of tissues, little consistent seasonal variation in stem cellulose content was apparent in the Californian species.

8.7.8 Acid Detergent Fiber

Acid detergent fiber content (cellulose and lignin) was higher in current leaves of Californian species than in current leaves of Chilean species. The average for current leaves measured in California was 24% compared to 15% measured in Chile (Tables 8.13 and 8.14). Acid detergent fiber content showed little variation with leaf age in the Californian species. However, the Chilean evergreen species *C. odorifera* and *L. caustica* showed increases in acid detergent fiber as leaves aged. Current-year and 1-year-old stems in the Californian species showed similar acid detergent fiber content to the Chilean species of the same age.

The two-way analysis of variance indicated significant differences in percent leaf acid detergent fiber among the four species pairs and between the groups of Californian and Chilean species (at $P \leqslant 0.01$).

Stem percent acid detergent fiber was found to be significantly different among the four species pairs ($P \leqslant 0.01$) but not between the groups of species measured in California and Chile (at $P = 0.05$).

8.7.9 Cell Wall

The cell wall fraction includes all major structural components, hemicellulose, cellulose, and lignin. The cell wall component of current-year leaves averaged about 26% to 29% for Californian and principal Chilean species measured (Tables 8.15 and 8.16).

Table 8.12. Percent cellulose in leaves and stems from shrubs on equator-facing slopes at Echo Valley, California[a]

	Age 1				Age 2					Age 3			
	Late fall 11/14/75	Summer 06/27/76	Fall 09/30/76	Seasonal X̄ ± SE	Late fall 11/14/75	Spring 03/19/76	Summer 06/27/76	Fall 09/30/76	Seasonal X̄ ± SE	Late fall 11/14/75	Summer 06/27/76	Fall 09/30/76	Seasonal X̄ ± SE
Leaves													
Adenostoma fasciculatum	-	22.5	18.9	20.7 ± 0.82	20.8	23.1	19.7	19.4	20.8 ± 0.8	-	-	-	-
Arctostaphylos glauca	16.7	19.6	14.5	16.0 ± 0.75	14.2	13.8	15.2	13.9	14.3 ± 0.3	17.0	14.9	-	16.0 ± 1.1
Ceanothus greggii	16.1	16.8	15.2	16.0 ± 0.46	-	17.1	14.7	15.0	15.6 ± 0.8	-	-	-	-
Rhus ovata	-	11.2	13.0	12.2 ± 0.40	13.7	16.3	14.9	13.0	14.5 ± 0.8	13.7	13.8	-	13.8 ± 0.0
X̄ ± SE of X̄ for above species	16.4 ± 0.3	17.5 ± 2.4	15.6 ± 1.1	16.2 ± 1.8	16.2 ± 2.3	17.6 ± 2.0	16.1 ± 1.2	15.3 ± 1.4	14.9 ± 1.1	15.4 ± 1.7	14.4 ± 0.6		14.9 ± 1.1
X̄ ± SE of X̄ for all species		16.5 ± 0.6				16.3 ± 0.5					14.9 ± 0.5		
Stems													
Adenostoma fasciculatum	38.1	27.8	30.3	32.1 ± 3.1	44.6	40.0	32.7	39.6	39.2 ± 2.4	47.7	36.7	38.2	40.9 ± 3.5
Arctostaphylos glauca	33.9	28.5	25.3	29.2 ± 2.5	38.0	28.1	29.0	29.3	31.1 ± 2.3	44.5	38.6	33.8	39.0 ± 3.1
Ceanothus greggii	31.1	26.4	28.5	28.7 ± 1.4	42.7	31.2	30.3	36.2	35.1 ± 2.8	43.4	35.5	40.7	39.9 ± 2.3
Rhus ovata	26.5	15.9	20.3	20.9 ± 3.1	35.2	29.3	23.0	25.5	28.2 ± 2.6	39.1	28.6	24.9	30.9 ± 4.2
X̄ ± SE of X̄ for above species	32.4 ± 2.4	24.6 ± 3.0	26.1 ± 2.2	27.7 ± 2.4	40.1 ± 2.2	32.2 ± 2.7	28.8 ± 2.1	32.6 ± 3.2	33.4 ± 2.4	43.7 ± 1.8	34.8 ± 2.2	34.4 ± 3.5	37.7 ± 2.3
X̄ ± SE of X̄ for all species		27.7 ± 2.4				33.4 ± 2.4					37.6 ± 3.0		

[a]Collected from November 1975 to September 1976. Leaf and stem age classes 1, 2, and 3 refer to tissue produced in the spring of 1976 (current), 1975 (1-year-old), or 1974 (2-years-old), respectively.

Table 8.13. Percent cellulose and percent acid detergent fiber (lignin and cellulose) in leaves and stems from shrubs at Fundo Santa Laura, Chile[a]

	Percent cellulose			Percent acid detergent fiber		
	Age 1	Age 2	Age 3	Age 1	Age 2	Age 3
Leaves						
Satureja gilliesii	11.6	-	-	20.7	-	-
Colliguaya odorifera	7.4	10.8	9.4	10.4	18.4	-
Trevoa trinervis	5.4	-	-	8.4	-	-
Lithraea caustica	20.0	21.2	19.5	19.4	31.4	29.4
Quillaja saponaria	7.2	15.0	12.7	-	-	-
Cryptocarya alba	21.0	16.8	17.9	-	-	-
Kageneckia oblonga	14.4	16.8	14.7	-	-	-
X̄ ± SE for all species	12.4 ± 2.4	16.1 ± 1.7	16.2 ± 1.5	14.7 ± 3.11	24.9 ± 6.5	29.4
Stems						
Satureja gilliesii	14.4	17.2	17.3	46.5	61.2	61.6
Colliguaya odorifera	16.2	26.3	35.5	25.5	43.3	53.2
Trevoa trinervis	33.1	35.3	36.8	43.0	48.6	53.7
Lithraea caustica	23.9	28.3	30.6	39.2	44.8	46.4
Quillaja saponaria	18.8	27.2	33.4	-	-	-
Cryptocarya alba	21.6	26.9	34.3	-	-	-
Kageneckia oblonga	19.6	24.3	31.5	-	-	-
X̄ ± SE for all species	23.4 ± 2.4	30.0 ± 2.4	34.9 ± 1.4	35.9 ± 5.3	45.6 ± 1.6	53.9 ± 4.4

[a] Collected in October and November 1975. Leaf and stem age classes 1, 2, and 3 refer to tissue produced in the spring of 1975 (current), 1974 (1-year-old), and 1973 (2-years-old), respectively.

Table 8.14. Percent acid detergent fiber (lignin and cellulose) in leaves and stems from shrubs on equator-facing slopes at Echo Valley, California[a]

	Age 1				Age 2					Age 3			
	Late fall 11/14/75	Summer 06/27/76	Fall 09/30/76	Seasonal $\bar{X}$ ± SE	Late fall 11/14/75	Spring 03/19/76	Summer 06/27/76	Fall 09/30/76	Seasonal $\bar{X}$ ± SE	Late fall 11/14/75	Summer 06/27/76	Fall 09/30/76	Seasonal $\bar{X}$ ± SE
Leaves													
Adenostoma fasciculatum		27.2	26.4	26.8 ± 0.40	28.8	30.9	25.6	26.7	28.0 ± 1.2	-	-	-	-
Arctostaphylos glauca	21.6	34.2	30.5	28.8 ± 3.7	21.2	20.2	28.1	28.1	24.4 ± 2.1	28.6	29.4	-	29.0 ± 0.4
Ceanothus greggii	20.1	22.4	20.7	21.2 ± 0.58		25.2	21.4	22.4	23.0 ± 1.1	-	-	-	-
Rhus ovata	20.6		17.0	18.8 ± 1.8	18.1	18.8	17.5	16.6	17.8 ± 0.46	17.4	18.1	-	17.8 ± 0.4
$\bar{X}$ ± SE of $\bar{X}$ for above species	20.8 ± 0.44	27.9 ± 3.4	23.6 ± 3.0	23.9 ± 2.3	22.7 ± 3.2	23.8 ± 2.7	23.2 ± 2.3	23.4 ± 2.6	23.3 ± 2.1	23.0 ± 5.7	23.8 ± 5.7		23.4 ± 5.6
$\bar{X}$ ± SE of $\bar{X}$ for all species		24.1 ± 2.1					23.3 ± 0.23				23.4 ± 0.4		
Stems													
Adenostoma fasciculatum	55.6	38.4	51.1	48.4 ± 5.2	59.6	53.2	50.6	56.9	55.1 ± 2.0	63.5	54.6	55.8	58.0 ± 2.8
Arctostaphylos glauca	48.7	44.6	43.2	45.5 ± 1.6	49.7	39.4	46.1	45.8	45.0 ± 2.0	57.3	49.1	52.4	53.0 ± 2.4
Ceanothus greggii	40.3	37.4	40.5	39.4 ± 1.0	53.7	44.9	43.9	47.9	47.6 ± 2.2	59.0	53.3	53.9	55.4 ± 1.8
Rhus ovata	36.7	20.1	33.2	30.0 ± 5.0	43.6	39.2	34.6	39.3	39.2 ± 1.8	49.9	39.9	37.2	42.3 ± 3.9
$\bar{X}$ ± SE of $\bar{X}$ for all species	45.3 ± 4.2	35.1 ± 5.2	42.0 ± 3.7	40.8 ± 4.1	51.6 ± 6.4	44.2 ± 3.3	43.8 ± 3.4	47.5 ± 3.6	46.7 ± 3.3	57.4 ± 2.8	49.2 ± 3.3	49.8 ± 4.3	52.7 ± 2.7
$\bar{X}$ ± SE of $\bar{X}$ for all species		40.8 ± 3.0					46.8 ± 1.8				52.1 ± 3.4		

[a] Collected from November 1975 to September 1976. Leaf and stem age classes 1, 2, and 3 refer to tissue produced in the spring of 1976 (current), 1975 (1-year-old), or 1974 (2-years-old), respectively.

Table 8.15. Percent cell wall in leaves and stems from shrubs on equator-facing slopes at Echo Valley, California[a]

	Age 1				Age 2					Age 3			
	Late fall 11/14/75	Summer 06/27/75	Fall 09/30/75	Seasonal $\bar{X} \pm$ SE	Late fall 11/14/76	Spring 03/19/76	Summer 06/27/76	Fall 09/30/76	Seasonal $\bar{X} \pm$ SE	Late fall 11/14/76	Summer 06/27/76	Fall 09/30/76	Seasonal $\bar{X} \pm$ SE
Leaves													
Adenostoma fasciculatum	-	36.1	29.0	32.5 ± 3.6	32.1	31.5	32.1	31.7	31.8 ± 0.15	-	-	-	
Arctostaphylos glauca	26.8	35.0	34.4	32.1 ± 2.6	22.3	25.0	22.4	24.9	23.8 ± 0.6	22.7	20.5	-	21.6 ± 1.1
Ceanothus greggii	20.1	22.4	20.7	21.1 ± 0.7	-	25.2	21.4	22.4	23.0 ± 1.2	-	-	-	
Rhus ovata	24.6	36.8	34.6	32 ± 3.8	18.7	37.2	32.6	22.7	27.8 ± 4.3	19.2	27.6	-	23.3 ± 4.1
$\bar{X} \pm$ SE of $\bar{X}$ for above species	23.8 ± 2.0	32.6 ± 3.4	29.7 ± 3.8	29.4 ± 2.3	27.3 ± 3.9	29.7 ± 2.9	27.2 ± 3.0	25.4 ± 2.2	26.6 ± 2.0	21.0 ± 1.8	24.0 ± 3.5		22.0 ± 0.85
$\bar{X} \pm$ SE of $\bar{X}$ for all species	28.7 ± 2.6				27.4 ± 0.9					22.5 ± 1.5			
Stems													
Adenostoma fasciculatum	57.0	47.8	52.4	52.6 ± 2.7	67.8	68.4	56.2	65.4	64.5 ± 2.8	80.3	72.1	78.9	77.1 ± 2.5
Arctostaphylos glauca	51.9	52.5	52.2	52.2 ± 0.17	63.8	53.5	58.6	63.8	59.9 ± 2.5	78.2	71.4	73.7	74.4 ± 2.0
Ceanothus greggii	40.3	37.4	40.5	39.4 ± 1	53.7	44.9	43.9	47.9	47.6 ± 2.2	57.0	53.3	53.9	55.4 ± 1.8
Rhus ovata	49.7	36.3	43.4	43.1 ± 3.9	54.8	50.8	46.8	53.2	51.1 ± 1.7	63.1	55.0	50.7	56.3 ± 3.6
$\bar{X} \pm$ SE of $\bar{X}$ for above species	49.7 ± 3.5	43.5 ± 4.0	47.2 ± 3.1	46.8 ± 3.3	60.0 ± 3.5	55.4 ± 4.8	53.4 ± 3.6	57.6 ± 4.2	55.8 ± 3.9	70.1 ± 5.3	63.0 ± 5.1	64.3 ± 7.0	65.8 ± 5.8
$\bar{X} \pm$ SE of $\bar{X}$ for all species	46.8 ± 1.8				56.6 ± 1.4					65.8 ± 2.2			

[a]Collected from November 1975 to September 1976. Leaf and stem age classes 1, 2, and 3 refer to tissue produced in the spring of 1976 (current), 1975 (1-year-old), or 1974 (2-years-old), respectively.

Table 8.16. Percent cell wall and sclerophyll index in leaves and stems from shrubs at Fundo Santa Laura, Chile[a]

	Percent cell wall			Sclerophyll index[b]		
	Age 1	Age 2	Age 3	Age 1	Age 2	Age 3
Leaves						
Satureja gilliesii	29.4	-	-	174	-	-
Colliguaya odorifera	20.8	15.4	-	83.2	267	-
Trevoa trinervis	15.3	-	-	49.7	-	-
Lithraea caustica	37.6	40.0	35.9	530	836	-
Quillaja saponaria	13.8	21.7	21.1	61.7	279	310
Cryptocarya alba	50.7	44.2	44.0	441	553	431
Kageneckia oblonga	28.8	34.1	29.8	173	194	289
$\overline{X}$ ± SE for all species	28.1 ± 4.9	31.1 ± 5.4	32.7 ± 4.8	216 ± 73	426 ± 119	343 ± 45
Stems						
Satureja gilliesii	61.0	73.3	78.1	677	652	896
Colliguaya odorifera	33.1	53.4	69.4	186	577	1214
Trevoa trinervis	54.6	54.0	72.0	237	518	1432
Lithraea caustica	56.8	59.0	68.4	1046	1750	1854
Quillaja saponaria	32.8	51.0	65.6	254	1143	1098
Cryptocarya alba	43.8	55.6	69.9	730	1036	1106
Kageneckia oblonga	43.2	54.1	71.1	223	646	611
$\overline{X}$ ± SE for all species	46.5 ± 4.3	57.2 ± 3.8	70.6 ± 1.5	479 ± 128	903 ± 167	1173 ± 149

[a] Collected in October and November 1975. Leaf and stem age classes 1, 2, and 3 refer to tissue produced in the spring of 1975 (current), 1974 (1-year-old), and 1973 (2-years-old), respectively.

[b] Sclerophyll index = (% acid detergent fiber/% crude protein) × 100 (Loveless, 1962). Percent acid detergent fiber and percent crude protein are from Tables 8.13-8.14 and 8.8-8.9, respectively.

Values diverged in 1- and 2-year-old leaf tissue in part because of missing sample values and the absence of leaf material from the deciduous Chilean species.

Average cell wall content in stem tissue was very similar in all three age classes of the species measured at Echo Valley and Fundo Santa Laura. Cell wall content averaged 46% to 47% for current-year stems, 57% for 1-year-old stems, and 66% to 71% for 2-year-old stems. The two-way analysis of variance indicated a significant difference in leaf cell wall content among the four species pairs ($P \geqslant 0.01$), but no significant difference was found in the cell wall content between the groups of Californian and Chilean species (at $P = 0.05$) (Table 8.7). Convergence in amount of carbon allocated to cell wall in the Californian and Chilean species measured is, therefore, indicated.

8.7.10 Sclerophyll Index

Although sclerophyll index varied among species, sclerophyll index was quite high for the leaves of most Californian and Chilean species measured. Current-year leaves in chaparral species had an average sclerophyll index of 386 ± 78 when the mean of the seasonal mean of each species was taken (Table 8.17). The four principal matorral species had a lower average sclerophyll index of 209 ± 110 (Table 8.16). The larger standard error of the mean in the Chilean data represents the larger variability in sclerophyll index among matorral species compared to the range in values found among chaparral species. The range of values in California was 236 to 548 for seasonal means of current leaves. In Chile, current leaves ranged from a seasonal mean of 50 for *T. trinervis* to 530 for *L. caustica*. *Lithraea caustica* showed the highest sclerophyll index of any species with a sclerophyll index of 836 in 1- and 2-year-old leaves. The value for *L. caustica* leaves corresponded closely to the sclerophyll index of 805 reported by Mooney et al. (1977) after conversion of Mooney's value to a true fiber basis.

While not significant at $P = 0.05$ (two-way analysis of variance) (Table 8.7), the average sclerophyll index for Californian species was consistently higher than the average of all Chilean species. Because phosphorus tends to be more limiting in Chile than in California (Chapter 9), the sclerophyll indices reported here contradict the hypothesis of Loveless (1962) that sclerophyll indices above 150 to 230 indicate a vegetation adapted to growing on phosphorus-deficient soils and that phosphorus is a major cause of sclerophylly. In fact, in current-year tissue of the Chilean species, only *L. caustica* exceeds the level established by Loveless to indicate sclerophylly. If low phosphorus availability was a major cause of sclerophylly, Chilean species, on the average, should have higher sclerophyll indices than Californian species, but they do not. Sclerophylly in chaparral and matorral species may be more of a response to moisture stress than to low phosphorus availability.

As pointed out by Mooney et al. (1977), many of the Californian and Chilean species have sclerophyll indices that equal or surpass the highest values given by Loveless for a variety of plants. Of the 89 sclerophyll indices reported by Loveless for species from diverse locations and habitats, 80 species have a corrected sclerophyll index of 251 or less. The highest reported value is 534 for *Mangifera indices* from India. Cromack and Monk (1975) report sclerophyll values for nine tree species at Coweeta, North Carolina that range from a corrected value of 203 for white oak to 456 for white pine at mid-growing season. The sclerophyll index reported for senescent tissue was up to three times higher.

Table 8.17. Sclerophyll index in leaves and stems from shrubs on equator-facing slopes at Echo Valley, California[a]

	Age 1			Age 2					Age 3			
	Summer 06/27/76	Fall	Seasonal X̄ ± SE	Late fall 11/14/75	Spring 03/19/76	Summer 06/27/76	Fall 09/30/76	Seasonal X̄ ± SE	Late fall 11/14/75	Summer 06/27/76	Fall 09/30/76	Seasonal X̄ ± SE
Leaves												
Adenostoma fasciculatum	394	352	373 ± 21	514	412	457	431	453 ± 22	-	-	-	-
Arctostaphylos glauca	551	545	548 ± 3	424	404	453	502	446 ± 21	753	588	-	670 ± 83
Ceanothus greggii	224	249	236 ± 13	-	268	285	255	269 ± 9	458	362	-	410 ± 48
Rhus ovata	-	274	274	323	303	282	296	301 ± 9	-	-	-	-
X̄ ± SE of X̄ for above species			386 ± 78					367 ± 24				540 ± 130
X̄ ± SE for all species	390 ± 94	355 ± 55		420 ± 56	347 ± 36	369 ± 50	371 ± 58		605 ± 149	475 ± 114		
X̄ ± SE of X̄ for all species		372 ± 18				379 ± 15				540 ± 156		
Stems												
Adenostoma fasciculatum	768	1161	964 ± 198	1355	1400	1332	1293	1345 ± 22	1671	1436	1116	1408 ± 161
Arctostaphylos glauca	1014	1168	1091 ± 78	1988	1576	1213	1477	1567 ± 160	2292	1552	1690	1845 ± 196
Ceanothus greggii	499	587	543 ± 44	866	724	708	855	788 ± 42	1340	1070	1080	1163 ± 88
Rhus ovata	359	897	628 ± 270	1744	1264	911	1268	1296 ± 170	2626	1287	1200	1704 ± 462
X̄ ± SE of X̄ for above species			806 ± 131					1249 ± 165				1530 ± 152
X̄ ± SE for all species	660 ± 145	953 ± 137		1488 ± 245	1241 ± 184	1041 ± 142	1223 ± 131		1982 ± 292	1336 ± 104	1272 ± 142	
X̄ ± SE of X̄ for all species		806 ± 148				1248 ± 92				1530 ± 231		

[a] Collected from November 1975 to September 1976. Leaf and stem age classes 1, 2, and 3 refer to tissue produced in the spring of 1976 (current), 1975 (1-year-old), or 1974 (2-years-old), respectively. Sclerophyll index = (% acid detergent fiber/% crude protein) X 100 (Loveless, 1962). Percent acid detergent fiber and percent crude protein are from Tables 8.13 and 8.7, respectively.

Sclerophyll indices, especially of senescent material, are a good predictor of decomposition potential of plant material (Cromack and Monk, 1975). High sclerophyll indices might, therefore, be expected to result in reduced decomposition and nutrient cycling rates in the chaparral and matorral vegetation.

Sclerophyll indices of 1-year-old chaparral leaves varied somewhat by season and were lowest in spring and highest in late fall. Mature tissue appeared to decrease in sclerophylly from fall to spring and then increase until the subsequent fall. Overall, sclerophyll indices tended to increase as the tissue aged. Several species with low sclerophyll indices in the first year showed major increases in subsequent years. For example, *Q. saponaria* had a sclerophyll index of 62 in the current-year leaves, which increased by 450% to 279 in 1-year-old leaves. Five of eight Chilean species had sclerophyll indices of less than 180 in current-year leaves, but no species had a sclerophyll index below 180 in older leaves. This change in sclerophyll index may be an adaptation allowing the leaf to survive the summer drought or to endure for several years.

8.7.11 Relative Organic Composition

The chemical procedures routinely used may account for 95% of the leaf fraction in herbaceous agricultural species. However, the recovery of identified constituents was significantly lower for the Californian and Chilean sclerophylls analyzed in this study (Figure 8.14). A major portion of the leaf was unaccounted for after total nonstructural carbohydrate, crude fat, crude protein, hemicellulose, lignin, and cellulose were measured. The unspecified fraction was generally larger in the Chilean leaf fractions than in the Californian leaf fractions.

The chemical procedures applied were generally those developed for agricultural species. Chemical fractions not normally abundant in agricultural species, such as phenolics and alkaloids, were not measured and, in some cases, might interfere with the extracts of other compounds. Phenolics average about 10% of the dry weight of chaparral evergreen leaves (P. Rundel, *pers. comm.*). Pectins may contribute 4% to 15% the dry weight of aerial parts in herbage species (Bailey, 1973). Organic acids can form another sizable pool, which was not measured by the present techniques. In addition to the existence of fractions not sampled, errors may be encountered in the analysis of wildland plants due to the effects of tannins, resins, and other compounds or the extraction procedures used. More work is required to determine chemical fractions of chaparral and matorral species and the appropriate extraction techniques for wildland species in general.

There are interesting differences in the relative distribution of organic compounds in leaves and stems among the Californian and Chilean speices measured. Chilean species had extracted fat levels as low as 1%. Chilean species also show lower total nonstructural carbohydrate levels. As discussed in Chapter 9, the lower fat and total nonstructural carbohydrate levels may result from greater available inorganic nutrient levels in Chile, resulting in utilization of carbon pools for growth. Chilean species may also have higher secondary compound concentrations (Montenegro et al., *in press*; P. Rundel, *pers. comm.*), which account for part of the undetermined fraction and which interfere with the extraction procedures for other chemical components, causing decreased recovery of the affected organic constituents. Total recovery of organic constituents of the species measured was generally higher in stems than in leaves. The higher recovery rate may be due to the greater percentage of cell wall fractions in stems. These cell wall fractions appear to be more completely recovered in chemical analyses than are many of the other organic constituents. The higher recovery rates of cell wall fractions may reduce the unspecified organic fraction in stems. Lower concentrations in stems of secondary compounds, such as tannins, which interfere with extraction procedures, may also result in greater recovery of organic compounds and reduce the unspecified organic fraction. Older age classes of stems showed an even higher recovery of specified chemical fractions than did current-year stems.

The causes of the differences in overall chemical composition are intriguing. Development of improved chemical analysis techniques may yield interesting differences in the chemical composition including secondary compounds in Californian and Chilean evergreen sclerophyll species.

8.8. Conclusions

There was a tendency for leaf and stem respiration rates and Q_{10} values to be lower in the Californian group of species than in the Chilean group. This contributed to a significantly lower ratio of leaf respiration rates relative to maximum photosynthetic rates in Californian species compared to Chilean species.

Leaf respiration rates were generally more variable among the chaparral species measured than among the matorral species. Thus, in the more uniform temperature environment at Fundo Santa Laura, there was less diversity in respiration rates among species than in the more variable temperature environment at Echo Valley. Leaf dark respiration rates for the group of Californian species were significantly lower than were the rates for the Chilean species (two-way analysis of variance). Q_{10} values for leaf dark respiration were also somewhat lower for the Californian species, but the difference was not statistically significant. The same trend was found for stem respiration, but the statistical significance was reversed; Q_{10} values were significantly lower in California than in Chile, while mean stem respiration rates were lower in California but were not statistically significantly different. Of five parameters tested—leaf and stem dark respiration rates, leaf and stem Q_{10} values, and the ratio of dark respiration to maximum photosynthetic rates—only leaf Q_{10} for dark respiration and stem dark respiration rates were not statistically different between the groups of species in California and Chile. These data, therefore, do not support convergence in respiratory pattern with respect to temperature in California and Chile. However, there may be other aspects of respiration not investigated that are more similar between the Californian and Chilean groups of species.

Calculated stem respiration represented between 15% and 29% of the carbon utilized each year in the chaparral and matorral shrubs studied. Stem respiration decreased on a dry-weight basis with increasing stem size. During the daylight period, chaparral species reincorporated up to 85% of the carbon dioxide respired from the stems. Stem photosynthesis significantly reduces the cost of stem maintenance in these species. Calculated root respiration accounted for 11% to 29% of the annual carbon utilization in chaparral and matorral species. Growth respiration was 25% to 35% of total root respiration between 20° and 30°C.

MEDECS simulations indicated that above- and belowground growth utilized 34% to 38% of the carbon available to *C. greggii, A. fasciculatum, A. glauca*, and *R. ovata.* Respiration utilized 72% to 66% of the available photosynthate. Chilean species allocated between 37% and 45% of the total carbon utilized for the production of new biomass, a somewhat higher allocation to growth than in the Californian species. Calculated carbon utilization for growth and respiration in *A. glauca* was almost evenly divided between the leaves (31%), stems (33%), and roots (36%). These ratios generally hold for the Californian species. On the average, Chilean species allocate similar percentages to roots, but allocation in Chilean species was higher to leaves and lower to stems than in the case of the Californian species. Sizable variation exists in reported fine root biomass and root production values. Uncertainty exists, therefore, as to actual fine root biomass, production, and respiration values. Fine roots represent a major carbon sink within the plant. Further research is needed to calculate accurately the carbon allocation to the root systems for growth and respiration and to assess the validity of the MEDECS calculations of allocation to fine roots presented here. Another area of uncertainty is in secondary growth rates of roots and stems. Further research is needed to quantify and validate these rates.

Simulations with MEDECS showed that respiration and growth rates were much more sensitive to an increase or a decrease in temperature than were photosynthetic rates. A 5°C decrease in temperature was calculated to decrease maintenance respiration by 17% and growth respiration by 45% while only suppressing photosynthesis by 4%. However, acclimation to altered thermal environments may dampen some of the calculated response to temperature changes.

Chemical constituents varied between the pairs of Californian and Chilean species. Overall, the analogous species pairs tended to be different from each other with respect to both leaf and stem organic composition. For six out of seven chemical leaf fractions, the species pairs were significantly different. For four of six chemical stem fractions, the species pairs were significantly different. These data indicate the wide variation in the range of responses possible among species from the same continent.

The groups of species from California and Chile tended to show nonconvergent patterns for most organic constituents tested. For five of the eight leaf chemical fractions analyzed, the groups of Californian and Chilean species were found to be different. Stems were more similar in chemical composition between continents, and five of the seven chemical constituents analyzed showed no significant difference between groups of Californian and Chilean species. Further evidence of a lack of convergence in leaf and stem chemical constituents was demonstrated by canonical correlation.

While there is little evidence for convergence in leaf chemistry between plants from California and Chile, stems are more similar. This may be because stem structures in woody shrubs are more conservative, with less latitude for variation or adaptive modification between Californian and Chilean species.

The striking pattern of convergence in carbon assimilation patterns between species groups from California and Chile (Chapter 7) was not found with respect to patterns of carbon allocation. Both absolute respiration rates and temperature sensitivities of respiration were different between the Californian and Chilean species. Further, the allocation patterns of carbon to chemical fractions in the stems and especially the leaves were quite different.

The lower fat and nonstructural carbohydrate levels in the Chilean species may reflect greater inorganic nutrient availability in Chile (Chapter 9), resulting in use of storage materials for growth. However, the large unidentified other fraction represents chemical compounds not included in the chemical fractions analyzed or extracted by traditional techniques for obtaining these fractions.

The unidentified fraction is larger in Chilean plants than in the Californian group of species. This may be due, in part, to the presence of larger amounts of secondary compounds in Chilean species. These compounds might also interfere with the extraction procedures for other compounds, including total nonstructural carbohydrates, thereby reducing apparent levels of certain of the specific compounds and increasing the apparent size of the unidentified fraction even further. These observed differences in chemical composition of tissues, respiration rates, and temperature sensitivity of respiration would be expected to affect the patterns of resource-use efficiency between the Californian and Chilean species. The direction and magnitude of the effect depends on the impact of the parameter on resource capture and on the efficiency of resource capture. These aspects are analyzed in Chapter 12.

9. Mineral Nutrient and Nonstructural Carbon Utilization

GAIUS R. SHAVER

This chapter describes and compares the seasonal course of mineral nutrients, total nonstructural carbohydrate, and crude fat in plant tissues from Echo Valley and Fundo Santa Laura.

9.1. Introduction

9.1.1 Background

Low nutrient availability is an important influence on the structure and distribution of mediterranean type vegetation, particularly in South Africa (Cody and Mooney, 1978) and in south and southwest Australia (Specht, 1973). Nutrient limitation of primary production was demonstrated by fertilization experiments in California (Hellmers et al., 1955a; J. Kummerow, *unpubl. data*) and Australia (Beadle, 1954, 1962, 1966; Specht, 1963; Specht and Groves, 1966). However, most interpretations of controls on primary production and growth form composition in mediterranean type vegetation in California and Chile have emphasized control by the interaction between temperature and water availability through the year (Specht, 1969a, b; Mooney et al., 1970; Mooney and Dunn, 1970b; Miller and Mooney, 1976; Parsons, 1976b; Miller, 1979; Chapters 2, 5, and 7). Relatively little is known of the mechanisms whereby plants of the Californian chaparral and Chilean matorral have adapted to nutritional factors in their environment.

The major objective of this chapter is to describe and compare the seasonal course of mineral nutrient content of the dominant shrub species of Echo Valley in southern California and Fundo Santa Laura in central Chile. Mooney and Rundel (1979) showed that *Adenostoma fasciculatum* at Echo Valley took up much of its nitrogen and phosphorus during the winter, when the shrubs were not growing, and suggested that this pattern was possible because evergreen leaves provided a sink for nutrients during non-growth periods. The leaves acted as a storage pool that could be used later when growth demands are high. Mooney and Rundel felt that storage of nutrients in evergreen leaves reduced leaching losses from the ecosystem. No comparable data are available for other species in southern California or central Chile, although a similar pattern was de-scribed in Australia (Specht, 1973). Similar patterns in other species would be evidence for a functional convergence in ecosystem nutrient-cycling processes and in plant adap-tation to the generally low nutrient status in mediterranean regions.

Mineral nutrient availability, in general, and that of phosphorus, in particular, was lower at Echo Valley than at Fundo Santa Laura, although nitrate-nitrogen in the soil was higher at Echo Valley than at Fundo Santa Laura (Miller et al., 1977). Mooney et al. (1977) found higher average leaf nitrogen contents at Echo Valley and higher leaf phosphorus contents at Fundo Santa Laura, although the nitrogen and phosphorus percentages of the species overlapped, particularly that of nitrogen. The sclerophyll index for leaves was high at both sites. The somewhat less fertile soils and shorter growing period at Echo Valley (Miller et al., 1977; Cody and Mooney, 1978) suggested that selection for nutrient uptake during nongrowth periods and nutrient storage in evergreen leaves would be greater at Echo Valley than at Fundo Santa Laura. If this were true, one might expect nutrient uptake during the winter at Echo Valley, as shown by Mooney and Rundel (1979), but not necessarily at Fundo Santa Laura, i.e., nutrient uptake and cycling patterns in the two sites would not be similar.

A second objective of this chapter is to describe and compare the seasonal patterns of total nonstructural carbohydrate and crude fat accumulation in the dominant shrub species of Echo Valley and Fundo Santa Laura. Such data were available for *Heteromeles arbutifolia* (Mooney and Chu, 1974) grown in pots in California, but no data were available for any Chilean species. Mooney and Chu (1974) concluded that season-al variation in mobile carbon pool sizes within the plant resulted from an inability to meet all carbon needs simultaneously throughout the year. As a result, carbohydrate reserves were lowest in late spring-early summer, after depletion due to rapid spring growth. The highest carbohydrate levels were found in late winter, presumably result-ing from photosynthate accumulation during a time when temperatures were too low for growth but not too low for photosynthesis. Thus, the seasonal course of total non-structural carbohydrate in *H. arbutifolia* was similar to that for nitrogen and phos-phorus in *A. fasciculatum* (Mooney and Rundel, 1979). The similarity in seasonal pat-terns of total nonstructural carbohydrate concentrations in several species of shrubs from southern California and central Chile would support a hypothesis of functional similarity in carbon accumulation and use, analogous to the predicted similarity in mineral nutrition.

9.1.2 Methods

Plant tissue collections were made over a period of 16 months, from September 1975 to December 1976, at both Echo Valley and Fundo Santa Laura. At Echo Valley, leaves and stems of *Ceanothus greggii*, *Arctostaphylos glauca*, *A. fasciculatum*, and *Rhus ovata* from pole- and equator-facing slopes were sampled and analyzed separately, while root samples were collected from both slopes and lumped. Only results for leaf and stem nutrient contents on the equator-facing slope at Echo Valley are presented. Seasonal patterns were essentially identical on the two slopes, with no consistent differences in nutrient content. At Fundo Santa Laura, samples of leaves, stems, and roots of *Trevoa trinervis, Colliguaya odorifera, Quillaja saponaria, Cryptocarya alba, Kageneckia oblonga, Satureja gilliesii,* and *Lithraea caustica* were collected from a single, relatively flat area.

Leaf and stem tissue samples were further subdivided according to age. Three age classes were recognized: Age 1, leaf and stem tissue newly produced in the 1975 growing season (March to May in California, October to January 1976 in Chile); Age 2, tissue produced in the 1974 growing season; and Age 3, tissue produced in 1973 or before. Thus, the nutrient status of yearly age classes of leaves and stems was followed throughout the sampling period.

Root tissue samples were subdivided into two classes: large roots 1 to 3 cm in diameter and fine roots 2 to 5 mm in diameter. The larger size class was considered representative of roots greater than 3 years old, and the smaller size class was considered to include only roots less than 3 years old. Only results for the older, larger roots are presented.

All samples were analyzed for total nitrogen, phosphorus, potassium, calcium, and magnesium content by the Agricultural Experiment Station, University of Alaska. Drying was done immediately after collection at 75° to 80°C, before shipment to Alaska. Nitrogen and phosphorus were analyzed colorimetrically with a Technicon© autoanalyzer following acid digestion. The same digest was used in determination of potassium, calcium, and magnesium content by atomic absorption spectrophotometry. Nonstructural carbohydrate content was determined for the same samples by enzymatic digestion (Smith, 1969) followed by a colorimetric analysis based on reaction of sugars with alkaline potassium ferricyanide (Technicon© method #280-73A). Crude fat, including all fats, waxes, and oils, was determined by the method of Randall (1974).

9.2. Seasonal Patterns of Plant Nutrient Content

9.2.1 Nitrogen

The nitrogen content of samples collected at Echo Valley showed consistent overall decreases from age class to age class in leaf and stem tissue (Table 9.1). There was considerable overlap in percent nitrogen between age classes due to seasonal variation. Roots showed a tendency toward increased nitrogen at the end of the sampling period (Table 9.2). *Ceanothus greggii* consistently had the highest percent nitrogen in all tissues and age classes, as might be expected from its nitrogen-fixing ability. *Arctostaphylos glauca* tissues were usually lowest in percent nitrogen, with *R. ovata* and *A. fasciculatum* intermediate.

In the samples from Fundo Santa Laura, the same decrease in nitrogen content with tissue age class was apparent, with a particularly marked decrease within the youngest leaf and stem age classes (Table 9.3). The observed decrease in nitrogen content in youngest tissues in Fundo Santa Laura probably was greater than that observed at Echo Valley because in Chile the sample collection started with immature leaves and

Table 9.1. Nitrogen percentages in leaves and stems of Californian shrubs[a]

Species	Age 1		Age 2					Age 3			
	Summer 26 Jun	Fall 30 Sep	Fall 20 Sep	Late fall 14 Nov	Spring 19 Mar	Summer 27 June	Fall 30 Sep	Fall 20 Sep	Late fall 14 Nov	Summer 27 Jun	Fall 30 Sep
Leaves											
Adenostoma fasciculatum	1.1	1.2	0.8	0.9	1.2	0.9	1.0	0.7			
Arctostaphylos glauca	1.0	0.9	1.0	0.8	0.8	1.0	0.9	0.7	0.6	0.8	
Ceanothus greggii	1.6	1.3	1.2	1.3	1.5	1.2	1.4				
Rhus ovata	1.4	1.0	1.3	0.9	1.0	1.0	0.9	0.8	0.6	0.8	
Mean	1.3	1.1	1.1	1.0	1.1	1.0	1.0	0.7	0.6	0.8	
Stems											
Adenostoma fasciculatum	0.8	0.7	0.7	0.7	0.6	0.6	0.7	0.8	0.6	0.6	0.8
Arctostaphylos glauca	0.7	0.6	0.5	0.4	0.4	0.6	0.5	0.4	0.4	0.5	0.5
Ceanothus greggii	1.2	1.1	0.9	1.0	1.0	1.0	0.9	0.7	0.7	0.8	0.8
Rhus ovata	0.9	0.6	0.9	0.4	0.5	0.6	0.5	0.4	0.3	0.5	0.5
Mean	0.9	0.8	0.8	0.6	0.6	0.7	0.7	0.6	0.5	0.6	0.6

[a]Samples were collected between September 1975 and September 1976. Tissues were produced during the spring; "Age 1" denotes 1976, "Age 2" denotes 1975, and "Age 3" denotes 1974.

Table 9.2. Nitrogen and phosphorus percentages in roots more than three years old, or 1 to 3 cm in diameter[a]

Percent nitrogen

Species	Late fall 17 Nov	Winter 17 Jan	Spring 19 Mar	Spring 13-27 May	Summer 20 Jul	Fall 30 Sep-5 Oct	Winter 12 Dec
California							
Adenostoma fasciculatum	0.3	0.5	0.3	0.3	0.6	0.8	0.7
Arctostaphylos glauca	0.2	0.2	0.2	0.3	0.3	0.5	0.5
Ceanothus greggii	0.6	0.5	0.5	0.5	0.6	0.8	0.9
Rhus ovata	0.2	0.3	0.3	0.3	0.6	0.7	0.6
Mean	0.3	0.3	0.3	0.3	0.5	0.7	0.7

Percent phosphorus

Species	Late fall 17 Nov	Winter 17 Jan	Spring 19 Mar	Spring 13-27 May	Summer 20 Jul	Fall 30 Sep-5 Oct	Winter 12 Dec
California							
Adenostoma fasciculatum	0.06	0.06	0.12	0.10	0.31	0.13	0.18
Arctostaphylos glauca	0.02		0.12	0.10	0.19	0.08	0.17
Ceanothus greggii	0.03		0.10	0.10	0.18	0.09	0.17
Rhus ovata	0.04	0.10	0.13	0.17	0.26	0.19	0.23
Mean	0.04	0.08	0.12	0.12	0.24	0.12	0.19

Percent nitrogen

Species	Spring 10-28 Nov	Summer 7-25 Jan	Fall 4-19 Mar	Winter 5-26 May	Winter 6-10 Jul	Spring 21 Sep
Chile						
Colliguaya odorifera	0.7	0.7	0.5	0.8	1.0	1.1
Lithraea caustica	0.4	0.2	0.8	0.5	0.4	0.6
Quillaja saponaria	0.6	0.6	0.6	0.5	0.9	0.8
Trevoa trinervis	1.1	0.8	1.0	1.3	0.9	0.7
Cryptocarya alba	0.5	0.6				
Kageneckia oblonga	0.8	0.9				
Satureja gilliesii	1.0	0.9				
Mean	0.8	0.7	0.7	0.8	0.8	0.8

Percent phosphorus

Species	Spring 10-28 Nov	Summer 7-25 Jan	Fall 4-19 Mar	Winter 5-26 May	Winter 6-10 Jul	Spring 21 Sep
Chile						
Colliguaya odorifera	0.11	0.19	0.12	0.15	0.15	
Lithraea caustica	0.10	0.08	0.11	0.12	0.13	0.17
Quillaja saponaria	0.13	0.07	0.17	0.18	0.14	0.15
Trevoa trinervis	0.15	0.16	0.16	0.18	0.13	0.15
Cryptocarya alba	0.13	0.13				
Kageneckia oblonga	0.15	0.10				
Satureja gilliesii	0.18	0.16				
Mean	0.14	0.13	0.14	0.16	0.14	0.16

[a]Samples were collected from November 1975 to December 1976.

Table 9.3. Nitrogen and phosphorus percentages in leaves and stems of Chilean shrubs[a]

| Species | Percent nitrogen | | | | | | Percent phosphorus | | | | | |
| | Age 1 | | | Age 2 | | Age 3 | Age 1 | | | Age 2 | | Age 3 |
	Spring 24-29 Oct	Late fall 5-26 May	Spring 21 Sep	Spring 24-29 Oct	Late fall 5-26 May	Spring 24-29 Oct	Spring 24-29 Oct	Late fall 5-26 May	Spring 21 Sep	Spring 24-29 Oct	Late fall 5-26 May	Spring 24-29 Oct
Leaves												
Colliguaya odorifera	2.0	1.3	1.3	1.1			0.27	0.18	0.19	0.21		
Lithraea caustica	0.8	1.0	0.8	0.6			0.15	0.18	0.15	0.19		0.14
Quillaja saponaria	1.8	1.2	0.8	1.1		0.9	0.35	0.15	0.15	0.17		0.14
Trevoa trinervis	2.7						0.20					
Cryptocarya alba	1.1			0.8		1.0	0.22			0.18		0.21
Kageneckia oblonga	2.1			2.4		1.2	0.36			0.29		0.24
Satureja gilliesii	1.9						0.26					
Mean	1.8	1.2	1.0	1.2		1.0	0.26	0.17	0.16	0.21		0.20
Stems												
Colliguaya odorifera	2.2	1.2	1.0	1.2	1.0	0.7	0.26	0.21	0.18	0.15	0.19	0.11
Lithraea caustica	0.6	0.6	0.5	0.4	0.5	0.4	0.14	0.12	0.12	0.11	0.12	0.08
Quillaja saponaria	1.9	1.1	0.7	0.8	1.0	0.3	0.26	0.18	0.14	0.17	0.17	0.15
Trevoa trinervis	2.9	1.7	1.4	1.5	1.6	0.6	0.25	0.16	0.15	0.15	0.15	0.13
Cryptocarya alba	0.7			0.6		0.7	0.15		0.17		0.13	
Kageneckia oblonga	2.1			0.9		1.2	0.35		0.28		0.26	
Satureja gilliesii	1.1			1.5		1.1	0.27		0.24		0.20	
Mean	1.7	1.2	0.9	1.0	1.0	0.7	0.24	0.17	0.15	0.18	0.16	0.15

[a]Samples were collected between October 1975 and September 1976. Tissues were produced during the spring; "Age 1" denotes 1975, "Age 2" denotes 1974, "Age 3" denotes 1973.

stems, which were still expanding in October to January. The young tissues collected from Echo Valley were at least 3 months old at the time of their first collection in late June. *Trevoa trinervis*, also a nitrogen-fixer, consistently had high percent nitrogen, while leaf and stem tissue of *L. caustica* had the lowest percent nitrogen content among Chilean species measured.

The range of variation among species in percent nitrogen was less at Echo Valley than at Fundo Santa Laura, particularly in older leaves, stems, and roots (Tables 9.1-9.3). Although comparison of results from the two sites was confounded by a lack of precise correlation between tissue age classes and developmental status, percent nitrogen in older leaves and stems from Echo Valley averaged 0.3% to 0.5% lower than in the corresponding tissues from Fundo Santa Laura (Tables 9.1 and 9.3). Mooney et al. (1977) reported that percent nitrogen in old leaves and in leaves overall was higher at Echo Valley than at Fundo Santa Laura.

The percent nitrogen at the start of the sampling period averaged 0.3% in old roots in species measured at Echo Valley versus 0.7% in old roots measured at Fundo Santa Laura (Table 9.2). By the end of sampling, the mean percent nitrogen increased to 0.6% in old roots collected at Echo Valley versus 0.8% in roots from Fundo Santa Laura.

At Echo Valley, the seasonal pattern of the percent nitrogen in new and mature leaves of *A. fasciculatum* was very similar to that reported by Mooney and Rundel (1979) for *A. fasciculatum* at the same site during the period 1972-1974 (Table 9.1). In 1976, the percent nitrogen in new leaves of *A. fasciculatum* was higher than in 1973, 1.1% to 1.2%, versus about 0.8%, but showed the same slight increase in early fall, relative to the summer value. The percent nitrogen in mature leaves increased from September to March, then decreased during the period of rapid growth in spring-early summer, and then increased again in late summer-fall. The range in nitrogen concentration in mature leaves was 0.7% to 1.2% versus about 0.6% to 1.4% reported by Mooney and Rundel (1979). *Ceanothus greggii* showed the same seasonal trend. However, mature leaves of *A. glauca* and *R. ovata* exhibited seasonal trends in percent nitrogen that were almost the inverse of those in *A. fasciculatum* and *C. greggii*.

The percent nitrogen in mature stems of *R. ovata* and *A. glauca* followed a pattern similar to that of their leaves, i.e., a decrease in fall-early winter followed by an increase in the spring (Table 9.1). In mature *A. fasciculatum* stems, percent nitrogen was highest in September of both 1975 and 1976 and lowest in winter, spring, and midsummer.

At Fundo Santa Laura, only 1-year-old stems of *T. trinervis* showed a marked increase in percent nitrogen in the nongrowing season (Table 9.3), but 1-year-old leaves and 2-year-old stems of all Chilean species increased significantly in percent nitrogen during the growing season. The percent nitrogen in current-year leaves and stems also increased in several Chilean species from October to November, but in all species there was a decline in percent nitrogen from winter to spring. This pattern is the reverse of that observed in *A. fasciculatum* and *C. greggii* leaves at Echo Valley (Table 9.3).

9.2.2 Phosphorus

Phosphorus concentrations in leaves, stems, and roots in the Californian plants measured (Tables 9.2 and 9.4) showed much greater seasonal variation but a more uniform pattern among species than was observed for nitrogen concentrations. Relative differences between species at any one sample date also were less for phosphorus than for

Table 9.4. Phosphorus percentages in leaves and stems of Californian shrubs[a]

Species	Age 1		Age 2					Age 3			
	Summer 27 Jun	Fall 30 Sep	Fall 20 Sep	Late fall 14 Nov	Spring 19 Mar	Summer 27 Jun	Fall 30 Sep	Fall 30 Sep	Late fall 14 Nov	Summer 27 Jun	Fall 20 Sep
Leaves											
Adenostoma fasciculatum	0.20	0.16	0.07	0.14	0.17	0.17	0.16	0.07			
Arctostaphylos glauca	0.22	0.10	0.07	0.09	0.14	0.22	0.11	0.04	0.08	0.27	
Ceanothus greggii	0.21	0.09	0.08	0.09	0.14	0.19	0.07				
Rhus ovata	0.76	0.12	0.14	0.10	0.14	0.23	0.10	0.07	0.10	0.25	
Mean	0.25	0.12	0.09	0.11	0.15	0.20	0.11	0.06	0.09	0.26	
Stems											
Adenostoma fasciculatum	0.10	0.11	0.10	0.10	0.12	0.18	0.12	0.11	0.09	0.17	0.09
Arctostaphylos glauca	0.26	0.12	0.06	0.08	0.12	0.25	0.08	0.06	0.06	0.23	0.07
Ceanothus greggii	0.22	0.10	0.08	0.11	0.12	0.19	0.07	0.08	0.09	0.18	0.05
Rhus ovata	0.30	0.13	0.17	0.10	0.14	0.23	0.11	0.06	0.06	0.21	0.09
Mean	0.24	0.12	0.10	0.10	0.13	0.21	0.10	0.08	0.08	0.20	0.08

[a]Samples were collected between September 1975 and September 1976. Tissues were produced during the spring; "Age 1" denotes 1976, "Age 2" denotes 1975, "Age 3" denotes 1974.

nitrogen concentrations. Seasonal variation of phosphorus concentrations within age classes was greater than the variation between age classes.

At Fundo Santa Laura, phosphorus concentrations varied most among species from October to January, the period of most rapid growth (Tables 9.2 and 9.3). At other times of the year, the phosphorus concentrations were more similar among species and showed little seasonal variation. The lack of seasonal variation was most evident in roots (Table 9.2).

The percent phosphorus was lower in all species and all tissues measured from Echo Valley than in samples from Fundo Santa Laura for all periods sampled except in late June 1976 and except for root samples collected at Echo Valley in late December 1976. Mooney et al. (1977) also found lower average leaf phosphorus concentrations a at Echo Valley. However, their mean phosphorus percentages for California in June 1973 were 0.13% and 0.06% for new and old leaves, respectively, in contrast with 0.25% and 0.20% measured in June 1976 (Table 9.4). Corresponding samples were not collected for December 1976 at Fundo Santa Laura, but the data for late October indicate that percent phosphorus was similar or lower in the Chilean species measured than in the Californian species at the phenologically corresponding date, assuming the normal drop in phosphorus concentration in new leaves with increasing leaf age. In general, the percentages of phosphorus measured at Fundo Santa Laura for 1975-1976 were similar to those of Mooney et al. (1977).

At Echo Valley, the percent phosphorus in mature leaves and stems increased not only during the winter but also through the spring and into the summer in all species except *A. fasciculatum*. Mature leaves of *A. fasciculatum* followed a pattern similar to that reported by Mooney and Rundel (1979), with a sharp increase in early winter followed by a decline in spring and summer (Table 9.4). In 1976, the spring-summer decline was less steep than that observed in 1973, and the concentration of phosphorus in mature leaves of *A. fasciculatum* was higher by a factor of about 0.5 throughout 1975-1976 (Mooney and Rundel, 1979; Table 9.4).

9.2.3 Potassium

At Echo Valley, potassium concentrations were consistently highest in leaves, roots, and especially stems of *R. ovata* (Tables 9.5 and 9.6). The other three Californian species were similar to each other with percent potassium ranging from about 0.3% to 0.8% in leaves and stems and from 0.1% to 0.2% in roots. At Fundo Santa Laura, potassium concentrations were much higher than in all of the Californian species except *R. ovata.* In mature tissues at Fundo Santa Laura, leaf potassium concentrations averaged about 0.6% to 0.8% in leaves, 0.7% to 0.9% in stems, and 0.5% in roots, Variation among species was also much greater at Fundo Santa Laura than at Echo Valley.

Nutrient analyses of leaves and stems collected on phenologically and developmentally comparable dates at Echo Valley and Fundo Santa Laura were compared (Tables 9.5-9.7) for samples collected the first fall after the leaves and stems were produced, early in the second spring, and late in the second fall. Root nutrient concentrations were compared on samples collected during late spring and midsummer, when a complete set of samples was available from Fundo Santa Laura. Percent potassium increased in *A. fasciculatum* and *R. ovata* leaves at Echo Valley and in *L. caustica* leaves at

Table 9.5. Potassium, calcium, and magnesium percentages in leaves and stems of California shrubs, November 1975 and March 1976

Species	Percent potassium			Percent calcium			Percent magnesium		
	Age 2		Age 3	Age 2		Age 3	Age 2		Age 3
	Late fall 14 Nov	Spring 19 Mar	Late fall 14 Nov	Late fall 14 Nov	Spring 19 Mar	Late fall 14 Nov	Late fall 14 Nov	Spring 19 Mar	Late fall 14 Nov
Leaves									
Adenostoma fasciculatum	0.47	0.57		0.78	1.00		0.15	0.20	
Arctostaphylos glauca	0.46	0.43	0.32	0.50	0.52	1.00	0.15	0.22	0.18
Ceanothus greggii	0.48	0.33		1.00	1.20		0.21	0.29	
Rhus ovata	0.66	0.76	0.52	0.70	0.71	1.50	0.18	0.17	0.19
Mean	0.52	0.52	0.42	0.75	0.86	1.25	0.17	0.22	0.19
Stems									
Adenostoma fasciculatum	0.41	0.24	0.31	0.32	0.29	0.37	0.09	0.10	0.08
Arctostaphylos glauca	0.73	0.43	0.61	0.51	0.48	0.48	0.10	0.14	0.07
Ceanothus greggii	0.69	0.33	0.61	0.67	0.81	0.67	0.15	0.19	0.07
Rhus ovata	2.10	2.00	1.60	0.62	1.00	0.75	0.12	0.12	0.09
Mean	0.98	0.94	0.78	0.53	0.65	0.57	0.12	0.14	0.08

Table 9.6. Potassium, calcium, and magnesium concentrations in roots more than 3 years old, or 1 to 3 cm in diameter, May 1976 and July 1976

Species	Percent potassium		Percent calcium		Percent magnesium	
	Spring 13-27 May	Summer 20 Jul	Spring 13-27 May	Summer 20 Jul	Spring 13-27 May	Summer 20 Jul
California						
Adenostoma fasciculatum	0.14	0.16	1.10	1.10	0.06	0.11
Arctostaphylos glauca	0.19	0.15	0.24	0.25	0.06	0.09
Ceanothus greggii	0.14	0.19	1.50	1.15	0.08	0.08
Rhus ovata	0.48	0.43	1.30	1.25	0.10	0.22
Mean	0.24	0.23	1.04	0.94	0.08	0.13
	Percent potassium		Percent calcium		Percent magnesium	
	Spring 10-28 Nov	Summer 7-25 Jan	Spring 10-28 Nov	Summer 7-25 Jan	Spring 10-28 Nov	Summer 7-25 Jan
Chile						
Colliguaya odorifera	0.58	0.45	0.70	0.35	0.13	0.09
Lithraea caustica	0.68	0.23	1.60	1.60	0.18	0.09
Quillaja saponaria	0.40	0.38	1.48	1.38	0.16	0.33
Trevoa trinervis	0.30	0.40	2.50	2.00	0.10	0.15
Cryptocarya alba	0.50	0.45	1.03	1.25	0.04	0.07
Kageneckia oblonga	0.40	0.45	1.13	1.18	0.21	0.25
Satureja gilliesii	0.85	0.93	0.90	0.70	0.23	0.32
Mean	0.53	0.47	1.33	1.21	0.15	0.19

Table 9.7. Potassium, calcium, and magnesium concentrations in leaves and stems of Chilean shrubs, October 1975, May 1976, and September 1976

Species	Percent potassium				Percent calcium				Percent magnesium			
	Age 1		Age 2		Age 1		Age 2		Age 1		Age 2	
	Late fall 5-26 May	Spring 21 Sep	Spring 24-29 Oct	Late fall 5-26 May	Late fall 5-26 May	Spring 21 Sep	Spring 24-29 Oct	Late fall 5-26 May	Late fall 5-26 May	Spring 21 Sep	Spring 24-29 Oct	Late fall 5-26 May
Leaves												
Colliguaya odorifera	1.17	0.53	0.70		0.80	1.12	1.15		0.33	0.33	0.24	
Lithraea caustica	0.43	0.64	0.55		1.32	1.98	1.63		0.40	0.51	0.37	
Quillaja saponaria	1.10	0.72	0.60		1.60	2.83	1.55		0.70	0.76	0.59	
Trevoa trinervis												
Cryptocarya alba			0.75				2.30				0.25	
Kageneckia oblonga			1.13				1.75				0.29	
Satureja gilliesii												
Mean	0.90	0.63	0.75		1.24	1.98	1.68		0.48	0.53	0.35	
Stems												
Colliguaya odorifera	1.55	1.40	1.38	1.29	0.86	1.18	0.80	0.88	0.20	0.29	0.16	0.21
Lithraea caustica	1.65	0.91	0.68	0.52	1.33	2.12	1.33	2.00	0.26	0.27	0.13	0.14
Quillaja saponaria	0.72	0.56	0.50	0.57	0.94	1.05	1.40	1.43	0.60	0.41	0.46	0.57
Trevoa trinervis	0.51	0.52	0.50	0.43	1.46	1.13	1.45	1.62	0.14	0.11	0.09	0.13
Cryptocarya alba			0.58				1.85				0.08	
Kageneckia oblonga			0.65				2.30				0.30	
Satureja gilliesii			1.00				0.40				0.15	
Mean	1.11	0.85	0.77	0.70	1.15	1.37	1.36	1.48	0.30	0.27	0.20	0.26

Fundo Santa Laura from fall to spring. No increases in the percent potassium in stems were observed in any species from fall to spring. In most species, percent potassium in roots declined from the late spring to midsummer.

9.2.4 Calcium

The percent calcium in species measured at Echo Valley ranged from 0.3% to 1.2% in leaves and stems and from 0.2% to 2.2% in roots (Tables 9.5 and 9.6). *Ceanothus greggii* and *R. ovata* were highest in calcium content, while *A. glauca* and *A. fasciculatum* consistently were lower. In contrast to the results for nitrogen, phosphorus, potassium, or magnesium, leaf and stem percent calcium increased with age of tissue; this result was expected because of the relative immobility of calcium in plants (Epstein, 1972; Gauch, 1972).

The average percent calcium in leaf and stem tissue was much higher and more variable among the Chilean species measured than among the Californian species (Tables 9.6 and 9.7). The variation about the mean was by a factor of 5 to 10 times for species at Echo Valley and 3 to 5 times for species at Fundo Santa Laura. At Fundo Santa Laura, percent calcium averaged 1.2% to 2.0% in mature leaves, 1.0% to 1.5% in mature stems, and 1.2% to 1.3% in mature roots. Concentrations of calcium in leaves rose rapidly through the first year after their production but less thereafter. However, calcium concentrations in 1- and 2-year-old leaves and stems collected in spring 1976 were, in most cases, lower than calcium concentrations in the current year's tissue. This lack of steady increase from age class to age class may have been the result of varying environmental conditions or calcium availability during the formation of leaves or stems.

Seasonal variation in percent calcium in leaves and stems at Echo Valley was small and inconsistent, but at Fundo Santa Laura percent calcium in stems and leaves usually dropped during October-November, the period of most rapid growth. There was no corresponding decrease in root calcium content, but there was a fairly consistent low point in percent calcium in roots in June to October, which is mid- to late winter at Fundo Santa Laura. The percent calcium in roots also was low in mid- to late winter at Echo Valley.

9.2.5 Magnesium

Magnesium concentrations in Californian species averaged about 0.17% to 0.22% in leaves, 0.08% to 0.14% in stems, and 0.8% to 0.13% in roots (Tables 9.5 and 9.6). There was little strong or consistent seasonal variation. Percent magnesium generally decreased in both leaves and stems from age class to age class.

In the Chilean species measured, magnesium concentrations were considerably higher and more variable in all three tissue types (Tables 9.6 and 9.7), ranging from 0.24% to 0.76% in leaves, 0.09% to 0.60% in stems, and 0.04% to 0.33% in roots. Strong or consistent seasonal trends were not apparent in any tissue, nor was there a consistent decline or increase in magnesium content of the different age classes.

9.2.6 Total Nonstructural Carbohydrate

Total nonstructural carbohydrate in leaf, stem, and root tissue of the four Californian species measured at Echo Valley followed a fairly uniform seasonal pattern (Tables 9.8 and 9.9). The seasonal pattern of the percent total nonstructural carbohydrate in roots was similar to the seasonal pattern reported by Mooney and Chu (1974) for *H. arbutifolia*, although the percentages recorded in the present study were usually higher. The increase in total nonstructural carbohydrate through the late fall-winter months is consistent with the interpretation that accumulation of photosynthate took place at that time (Chapter 7) and that use of photosynthate exceeds its accumulation during the period of rapid growth in spring (Chapter 4). The fact that roots showed the greatest seasonal variation in percent total nonstructural carbohydrate indicates that they play a major storage role in all four Californian species.

Although among the Californian species *R. ovata* consistently had the highest percent total nonstructural carbohydrate in its roots (Table 9.9), it was lowest in percent total nonstructural carbohydrate in leaves and had an intermediate percentage in stems. *Arctostaphylos glauca* and *A. fasciculatum* had very high percent total nonstructural carbohydrate in leaves and stems but an intermediate percentage in roots. Each species apparently allocated total nonstructural carbohydrate among its tissues differently within the same seasonal pattern.

Except for *L. caustica*, the percent total nonstructural carbohydrate in roots of the species measured at Fundo Santa Laura followed a seasonal pattern similar to the Californian species, i.e., having highest total nonstructural carbohydrate levels in fall or winter (Table 9.9). None of the Chilean species measured were consistently high or low in percent total nonstructural carbohydrate present in leaves, stems, or roots (Tables 9.9 and 9.10).

The percent total nonstructural carbohydrate in roots was similar in species measured at both Echo Valley and Fundo Santa Laura, ranging from 5% to 16% of dry weight. In leaves and stems, however, there was little correspondence between the sites. Yearly average percentages of total nonstructural carbohydrate in 1-year-old leaves and stems at Echo Valley were approximately 15% and 9%, respectively, while at Fundo Santa Laura, average percentages were about 5% in both tissue types.

9.2.7 Crude Fat

Crude fat concentrations did not vary greatly through the year in leaves and roots of the Californian species measured (Tables 9.9 and 9.11). However, in the Age 2 stem class, crude fat concentrations were high in fall 1975 and low in spring and fall 1976. The cause of seasonal changes in percent crude fat in stem tissue of the Californian shrubs was unclear. As with total nonstructural carbohydrate concentrations, crude fat content was highest in leaves of *A. fasciculatum* and *A. glauca*. *Rhus ovata* was highest in fat content in roots but was lowest in leaves and intermediate in stems.

In the Chilean species measured, there was little seasonal variation in percent crude fat in leaves or stems. Unlike the Californian species, percent crude fat in roots of the Chilean species did not vary greatly or consistently. There also was less interspecific

Table 9.8. Total nonstructural carbohydrate percentages in leaves and stems of California shrubs from November 1975 to September 1976

Species	Age 1		Age 2				Age 3		
	Summer 27 Jun	Fall 30 Sep	Late fall 14 Nov	Spring 19 Mar	Summer 27 Jun	Fall 30 Sep	Late fall 14 Nov	Summer 27 Jun	Fall 30 Sep
Leaves									
Adenostoma fasciculatum	14.8	21.2	16.1	13.8	14.8	21.0			
Arctostaphylos glauca	14.7	22.1	19.2	20.4	15.3	21.1	9.6	14.2	
Ceanothus greggii	7.8	17.3	17.1	11.3	12.6	17.0			
Rhus ovata	4.2	12.1	6.1	9.9	12.5	21.6	6.7	11.7	
Mean	10.4	18.2	14.6	13.9	13.8	17.9	8.2	13.0	
Stems									
Adenostoma fasciculatum	5.5	9.6	4.8	6.2	3.8	9.0	4.3	2.4	8.0
Arctostaphylos glauca	10.0	14.3	12.5	15.6	8.8	12.4	9.6	6.7	12.1
Ceanothus greggii	7.4	11.6	9.2	7.1	6.3	11.0	6.9	5.5	9.7
Rhus ovata	5.2	11.9	10.7	8.6	7.3	11.5	8.6	6.5	12.0
Mean	7.0	11.9	9.3	9.4	6.6	11.0	7.4	5.3	10.5

Table 9.9. Total nonstructural carbohydrate and crude fat percentages in roots more than 3 years old, or 1 to 3 cm in diameter from November 1975 to December 1976

California — Percent total nonstructural carbohydrate

Species	Late fall 17 Nov	Winter 17 Jan	Spring 19 Mar	Spring 13-27 May	Summer 20 Jul	Fall 30 Sep-5 Oct	Winter 21 Dec
Adenostoma fasciculatum	6.3	10.1	6.5	5.1	3.4	8.1	2.2
Arctostaphylos glauca	6.0	10.0	10.2	7.3	5.4	12.8	8.6
Ceanothus greggii	5.0	9.8	9.3	5.5	3.4	7.3	5.0
Rhus ovata	9.2	20.4	12.5	11.1	9.3	13.6	10.5
Mean	6.6	15.7	13.6	7.3	5.4	10.5	6.6

California — Percent fat

Species	Winter 17 Jan	Spring 19 Mar	Spring 13-27 May	Summer 20 Jul	Fall 30 Sep-5 Oct	Winter 21 Dec
Adenostoma fasciculatum	0.5		0.5	0.2	0.9	
Arctostaphylos glauca	0.5	0.4	0.2		0.1	0.3
Ceanothus greggii	0.4		2.1	1.2	1.0	0.4
Rhus ovata	2.6	1.3	1.9	1.9	1.7	1.9
Mean	1.0	0.9	1.2	1.1	0.9	1.2

Chile — Percent total nonstructural carbohydrate

Species	Spring 10-28 Nov	Summer 7-25 Jan	Fall 4-19 Mar	Fall 5-26 May	Winter 6-10 Jul	Spring 21 Sep
Colliguaya odorifera	4.8	5.9	12.6	14.0	10.1	
Lithraea caustica	6.2	7.5	2.6	5.3	9.1	5.9
Quillaja saponaria	6.3	8.1	12.7	8.0	9.5	7.0
Trevoa trinervis		11.9	11.4	9.3	6.8	5.5
Cryptocarya alba	6.2	6.6				
Kageneckia oblonga	4.8	5.2				
Satureja gilliesii	3.6	3.3				
Mean	5.3	6.9	9.8	9.2	8.9	6.1

Chile — Percent fat

Species	Spring 10-28 Nov	Summer 7-25 Jan	Fall 4-19 Mar	Fall 5-26 May	Winter 6-10 Jul	Spring 21 Sep
Colliguaya odorifera	1.7	0.7	1.3	1.4	1.2	1.0
Lithraea caustica	0.8	0.6	1.0	1.4	1.2	2.3
Quillaja saponaria	0.3	0.8	1.0	1.3	0.9	1.0
Trevoa trinervis		0.7	0.2	0.5	0.5	1.0
Cryptocarya alba	1.6	1.6				
Kageneckia oblonga	0.7	0.4				
Satureja gilliesii	0.6					
Mean	1.0	0.8	0.9	1.2	1.0	1.3

Table 9.10. Total nonstructural carbohydrate and crude fat percentages in leaves and stems of Chilean shrubs from October 1975 to September 1976

Species	Percent total nonstructural carbohydrate						Percent fat					
	Age 1			Age 2		Age 3	Age 1			Age 2		Age 3
	Spring 24-29 Oct	Late fall 5-26 May	Spring 21 Sep	Spring 24-29 Oct	Late fall 5-26 May	Spring 24-29 Oct	Spring 24-29 Oct	Late fall 5-26 May	Spring 21 Sep	Spring 24-29 Oct	Late fall 5-26 May	Spring 24-29 Oct
Leaves												
Colliguaya odorifera	2.6	5.3	4.9	3.0			0.53	2.20	1.70	1.10	2.80	
Lithraea caustica	3.8	2.7	2.2	4.0	7.2	4.7	2.02	1.70	2.10	1.96	1.10	1.44
Quillaja saponaria	3.2	6.9	11.1	5.1	5.2	8.5	0.32	0.80	1.80	1.02	1.00	1.30
Trevoa trinervis	5.0				4.9		1.00					
Cryptocarya alba	8.6			6.5	8.4	7.2	2.94			6.30		5.58
Kageneckia oblonga	7.0			6.9		5.9	5.16			5.44		0.66
Satureja gilliesii	7.4						2.24					
Mean	5.4	5.0	6.1	5.1	6.4	6.6	2.03	1.57	1.87	3.10	1.63	2.25
Stems												
Colliguaya odorifera	2.6	6.6	2.8	2.0		2.9	1.86	3.70	2.60	1.64		1.04
Lithraea caustica	5.1	4.2	6.1	5.6		6.4	2.02	1.70	2.10	1.96		0.60
Quillaja saponaria	1.7	5.4	3.8	3.8		4.0	0.60	0.80	1.40	0.42		0.46
Trevoa trinervis	4.2	8.3	5.7	3.5		4.3	0.76	0.90	0.90	0.58		0.62
Cryptocarya alba	5.4			5.9		6.3	2.26			1.06		0.86
Kageneckia oblonga	5.3			5.1		4.9	1.10			2.38		0.56
Satureja gilliesii	4.6			3.9		3.8	1.44			1.04		
Mean	3.4	6.1	4.6	4.3		4.7	1.43	1.78	1.75	1.30		0.59

Table 9.11. Crude fat percentages in leaves and stems of Californian shrubs from September 1975 to September 1976

Species	Age 1		Age 2					Age 3			
	Summer 27 Jun	Fall 30 Sep	Fall 20 Sep	Late fall 14 Nov	Spring 19 Mar	Summer 27 Jun	Fall 30 Sep	Fall 20 Sep	Late fall 14 Nov	Summer 27 Jun	Fall 30 Sep
Leaves											
Adenostoma fasciculatum	7.0	9.0	8.4	9.4	7.5	7.8	9.5	9.4			
Arctostaphylos glauca	5.8	5.6	7.7	6.9	7.0	7.6	7.4	10.5	9.1	8.0	
Ceanothus greggii	4.6	5.1	5.2	5.1	5.4	6.1	5.7				
Rhus ovata	2.8	2.9	3.0	3.6	2.6	3.7	3.1	4.0	4.2	3.0	
Mean	5.1	5.6	6.1	6.3	6.1	6.3	6.4	8.0	6.7	5.5	
Stems											
Adenostoma fasciculatum	5.5	6.2	6.2	5.9	4.3	6.2	5.4	8.9	5.1	4.4	2.6
Arctostaphylos glauca	2.4	2.7	3.7	3.3	1.9	2.0	1.9	1.9	1.7	0.5	0.5
Ceanothus greggii	3.8	3.7	4.0	3.6	3.1	4.1	2.7	2.2	2.1	1.3	1.7
Rhus ovata	3.4	3.9	5.9	3.5	3.2	3.3	3.1	5.3	2.8	2.2	2.7
Mean	3.8	4.1	5.0	4.1	3.2	3.9	3.3	4.6	2.9	2.1	1.9

variation in the fat concentration of all three tissue types in the species from Fundo Santa Laura.

The percent crude fat of species at Echo Valley averaged above 5% to 8% in leaves and 3% to 5% in stems, which was consistently higher than the 3% measured in both leaves and stems of species from Fundo Santa Laura (Tables 9.10 and 9.11). In roots, the average fat percentages were about the same in all species from both sites, but if *R. ovata* was excluded, fat concentration were higher in the roots of the Chilean species (Table 9.9).

9.3. Influences of Nutrients on Plant Growth

9.3.1 Mineral Nutrient Storage and Growth

The seasonal patterns of nitrogen and phosphorus content at Echo Valley and Fundo Santa Laura closely resembled the results of Mooney and Rundel (1979) for percent nitrogen and phosphorus in leaves of *A. fasciculatum.* These results also supported the conclusions of Mooney et al. (1977) that tissue phosphorus concentrations were generally lower at Echo Valley than at Fundo Santa Laura and that nitrogen concentrations were similar in both countries. In 1975-1976, mean percent nitrogen was slightly higher in the species measured at Fundo Santa Laura, while in 1972-1974, they were higher at Echo Valley (Mooney et al., 1977). This close similarity to previously published and independently analyzed results from the same sites strengthens the reliability of the newly presented data on other species, on stem and root nitrogen and phosphorus contents, and on percent potassium, calcium, and magnesium in leaves, stems, and roots. Year-to-year variation within the sites apparently is less than consistent similarities or differences between species at Echo Valley and Fundo Santa Laura.

The argument for nitrogen storage during nongrowth periods in chaparral species may or may not be supported by the percent nitrogen concentration in plant tissue collected during the winter at Echo Valley. In *A. fasciculatum*, percent nitrogen in stems actually decreased during nongrowth periods. Using the 1973 biomass data of Mooney and Rundel (1979) and assuming that all stems behaved as 1-year-old stems, there would be a loss of 1.7 g m^{-2} nitrogen in stems from September to March and an increase of 0.7 g m^{-2} nitrogen in leaves, resulting in a net loss of about 1.0 g m^{-2} in the fall and winter in aboveground tissues alone. However, it may be unreasonable to use biomass data from 1973 for these calculations or to assume that all stems are as variable in percent nitrogen as young stems. Only one Californian species, *C. greggii*, showed a seasonal pattern in leaf nitrogen content similar to that of *A. fasciculatum.* In *R. ovata* and *A. glauca*, both leaf and stem percent nitrogen were low in winter-spring and apparently there was no uptake during the nongrowing season.

Analysis of stem and root tissue for percent phosphorus strongly supported the hypothesis that, in the species measured at Echo Valley, phosphorus was taken up and stored during the winter not only in leaves but also in all mature tissues. Because stems and roots together constituted a greater biomass than leaves (Mooney et al., 1977; Chapter 4), phosphorus storage in leaves alone is, in fact, an underestimate of whole-plant phosphorus storage. Because phosphorus concentrations in stems increased even during the growing season and root phosphorus concentrations increased in early summer, there may be a net gain in phosphorus throughout the period from late fall to early summer. Losses in phosphorus were concentrated in late summer-early fall, when soils are very dry and uptake presumably is severely restricted (Chapter 2). The decreases in percent phosphorus came at a time when leaf and stem loss was greatest and presumably back translocation of phosphorus from dying tissue was also high. Apparently, back translocation did not compensate for phosphorus losses.

The species measured at Fundo Santa Laura did not show clear evidence of uptake and storage of any nutrient examined during the winter, although data were available only for the period from June to October for youngest leaves and stems and for roots. In most species, concentrations of nitrogen, phosphorus, and potassium in leaves and stems declined or did not change during the winter. Percent magnesium in aboveground parts behaved erratically. Calcium increases may have been due mostly to the normal accumulation of calcium with tissue age.

9.3.2 Mobile Carbon and Nutrient Interactions

In general, the most striking difference between the plants from Echo Valley and those from Fundo Santa Laura was the consistently higher nutrient concentrations at Fundo Santa Laura, particularly of potassium, calcium, and magnesium. This was in contrast to the much higher total nonstructural carbohydrate and fat concentrations in plant tissues at Echo Valley. The higher nutrient levels at Fundo Santa Laura can be, in part, explained by the more fertile soils (Miller et al., 1977), but the differences in mobile carbon pools were not reflected by dramatic differences in photosynthetic rates or the climate for photosynthesis (Chapters 5 and 7).

The essentially opposite behavior of the mineral versus carbon pools suggests an interaction between them. Similar interactions have been demonstrated in fertilization experiments on numerous agronomic species (Fleming, 1973; Evans, 1975a) and in the Alaskan tundra (Shaver and Chapin, 1980). In general, fertilization tends to decrease total nonstructural carbohydrate levels. Low nutrient levels lead to accumulation of total nonstructural carbohydrate. The interpretation of this behavior usually is based on the observation that growth is more sensitive than photosynthesis to variation in mineral nutritional status. In other words, with high levels of available mineral nutrient, growth is source-limited or photosynthesis-limited, while under mineral nutrient stress, growth is sink-limited for carbon (Evans, 1975b). The same interaction appears to be taking place in the species studied at both Echo Valley and Fundo Santa Laura.

9.3.3 Convergence and Nutritional Limits to Growth

Convergence of plant or ecosystem function between southern California and central Chile was not strongly apparent in this comparison of seasonal cycles in mineral nutrient and mobile carbon pools. A more reasonable interpretation is that mineral nutrient limitation to shrub growth was stronger at Echo Valley than at Fundo Santa Laura.

The much higher total nonstructural carbohydrate and fat content of the species at Echo Valley probably resulted from their inability to acquire the minerals needed to metabolize photosynthate as rapidly as it was produced. Strong mineral nutrient limitation at Echo Valley also was apparent in the greater and more uniform seasonal cycling of phosphorus in particular. Apparently, the species measured at Fundo Santa Laura were capable of supplying most of their mineral needs through uptake during the growing season, while at Echo Valley, storage of at least phosphorus and probably also magnesium was necessary to meet annual growth demands.

The longer growing season at Fundo Santa Laura may contribute to the higher soil nutrient levels by providing a longer period when the soils are moist and conditions are favorable for mineralization and for uptake as well as growth (Miller, 1979; Cody and Mooney, 1978; Chapter 2). However, the lower shrub productivity at Fundo Santa Laura, less than half that of shrubs at Echo Valley, is anomalous in view of the more favorable soil conditions at Fundo Santa Laura. Yearly production per shrub at Fundo Santa Laura was only two-thirds that of individual Californian shrubs (Mooney et al., 1977; Chapter 4). Much of the difference in community production between the two sites was made up for by the extensive annual herb cover at Fundo Santa Laura, but vegetation at Fundo Santa Laura was still less productive on an area basis than was vegetation at Echo Valley. Clearly, the effects and importance of mineral nutrient limitation to plant production differs between these mediterranean type ecosystems, but it is not the sole factor controlling their productivity and composition.

9.4. Conclusions

Shrub nutrient concentrations were lower overall at Echo Valley than at Fundo Santa Laura. Total nonstructural carbohydrate and fat concentrations were higher at Echo Valley. At Echo Valley, the seasonal variation in plant nutrient concentration was greater than at Fundo Santa Laura, particularly for phosphorus. Phosphorus at Echo Valley may be taken up from the soil and stored during the nongrowing season; at Fundo Santa Laura, there was no evidence for uptake during periods of no growth. Seasonal cycles of nonstructural carbohydrates were similar, however. The seasonal cycles of mineral nutrients do not indicate convergence of these functions in the mediterranean regions of southern California and central Chile.

10. Nutrient Cycling in Mediterranean Type Ecosystems

JOHN T. GRAY and WILLIAM H. SCHLESINGER

This chapter reviews studies of nutrient pools and cycling processes in mediterranean type ecosystems of the world. Inter- and intrasystem cycles are compared to similar cycles in forest ecosystems.

10.1. Introduction

Major research efforts during the past decade have provided evidence for structural and functional similarity in mediterranean type ecosystems, particularly between California and Chile (Mooney and Dunn, 1970b; di Castri and Mooney, 1973; Mooney, 1977a; Cody and Mooney, 1978; Specht, 1979). Attention has focused on aspects of carbon balance in these ecosystems; similarities between the chaparral of southern California and the matorral of central Chile have been shown to exist in vegetative structure, biomass distribution, and plant seasonality (Mooney et al., 1977; Thrower and Bradbury, 1977; Chapters 3 and 4). To a lesser extent, ecologists have compared other mediterranean type ecosystems including the garrigue of southern France (Specht, 1969a, b; Lossaint, 1973), the maquis of Israel (Naveh, 1967), and the mallee and heathlands of southern Australia (Specht, 1969a, b, 1973). Nutrient cycling in mediterranean type ecosystems has not been studied in the same detail. In these ecosystems, it would be both instructive and interesting to see if the degree of similarity found in ecosystem structure and resource utilization is also found in nutrient cycles.

The similar climate and vegetation in mediterranean regions of the world suggest that aspects of the nutrient cycles should also be similar. However, because of the lack of long-term, detailed studies on nutrient cycles in mediterranean type ecosystems, this question cannot yet be fully answered. Major factors influencing nutrient cycles, such as the hydrology, geology, and soil conditions, have only recently been studied in these regions on a comparative basis (e.g., Paskoff, 1973; Zinke, 1973; Miller et al., 1977; Thrower and Bradbury, 1977).

Although there have been few detailed studies of complete nutrient cycles in mediterranean type ecosystems, many individual studies have been made of specific aspects of these cycles. Early ecologists and agricultural scientists in California were concerned with soil fertility and showed through experimental work that chaparral soils are generally nitrogen poor (Jenny et al., 1950; Vlamis et al., 1954; Hellmers et al., 1955a; Schultz et al., 1958). Further research has examined nutrient input by symbiotic nitrogen fixation (Hellmers and Kelleher, 1959; Vlamis et al., 1964; Delwiche et al., 1965; Kummerow et al., 1978b) and by deposition of nutrient-rich ash from fire (Vlamis and Gowans, 1961; Biswell, 1974; Christensen and Muller, 1975; DeBano and Conrad, 1978). In Australian plant communities, the generally low phosphorus content of soils and the predominance of sclerophyllous species suggested that xeromorphic characters were adaptations to low levels of available soil nutrients (Beadle, 1954, 1962, 1966, 1968; Ashton, 1976). Several dominant heath and woodland species are morphologically and physiologically adapted to low phosphorus levels, e.g., by storing phosphorus as polyphosphate in roots (Purnell, 1960; Jeffery, 1964, 1967, 1968; Specht and Groves, 1966; La Mont, 1972, 1973; Barrow, 1977). In the French garrigue, attention has focused on soil metabolic and nutrient processes (Lossaint, 1973). Microbial activity and decomposition with special regard to nitrogen mineralization have been studied in considerable detail (Billès et al., 1971a, b, 1975; Cortez et al., 1972; Bottner and Peyronel, 1977). The first complete nutrient budget for a mediterranean type ecosystem was calculated in the garrigue (Lossaint and Rapp, 1971; Lossaint, 1973).

The patterns of nutrient utilization and cycling in mediterranean type ecosystems provide an interesting contrast to forest ecosystems, where the greatest knowledge of nutrient cycles exists. Many of the steady state assumptions commonly used in the study and modeling of nutrient cycling in forests cannot be freely applied to mediterranean type ecosystems. Mediterranean type ecosystems, such as chaparral, are not mature, homeostatic communities (in the sense of Odum, 1969). They are relatively short-lived, continually aggrading ecosystems, periodically leveled by fire. Fire plays a dominant role as both a catalytic agent initiating succession and as a nutrient-recycling mechanism. The low biomass and production of mediterranean type ecosystems are intermediate between temperate deserts and forests (Whittaker and Likens, 1975). Nutrient pools and fluxes in mediterranean type ecosystems are generally small in magnitude compared to those found in most forests, but there appear to be exceptions.

This chapter reviews the current knowledge of nutrient-cycling processes in mediterranean type ecosystems, particularly the southern California chaparral. A comparison is made of nutrient pools and fluxes in mediterranean type ecosystems and forest ecosystems of the world. Finally, the unique characteristics of nutrient cycles in mediterranean type ecosystems are discussed.

In the analysis of nutrient cycles, inter- and intrasystem cycles are distinguished (Bormann and Likens, 1967; Duvigneaud and Denaeyer-DeSmet, 1970). Intersystem cycles include nutrient inputs by meteorological events, biological nitrogen fixation, and geological weathering and include nutrient outputs by various hydrological, geological, and pyric processes. Intrasystem cycles of mineral elements are plant- and soil-dominated and include uptake by vegetation and return by litterfall, foliar and stem leaching, decomposition, and ashfall from fire.

10.2. Intersystem Cycles

10.2.1 Nutrient Input by Atmospheric Deposition

Atmospheric precipitation contains many dissolved ions, including most of those that are essential for plant growth. In many areas, considerable quantities of plant nutrients are also deposited from the atmosphere by dry fallout, but ecologists have only recently appreciated the importance of separating the nutrient load received by wet and dry processes (e.g., Galloway and Likens, 1976). In the study of natural ecosystems, the deposition of nutrients received in dissolved and particulate form from the atmosphere is usually measured by capture in open funnels and is often called bulk precipitation (Whitehead and Feth, 1964). The composition of bulk precipitation is strongly influenced by regional factors, including climate, geology, and the proximity of oceans or industrial pollution sources.

Only a limited amount of data has been gathered for the nutrient inputs in bulk precipitation in mediterranean type ecosystems (Table 10.1). Generally, the deposition of nutrients from the atmosphere is small compared to the annual uptake by the plant community, as discussed later in this chapter. These data vary considerably; although rainfall quantities are similar for the chaparral and garrigue communities, annual inputs are considerably greater in France, particularly for nitrogen and calcium. Additional data for three coastal and one inland forest ecosystems are also presented for comparative purposes.

The position of mediterranean type ecosystems near the coast of continents means that these systems are subject to relatively high deposition of ions such as sodium, which are important in sea water (Junge, 1963; Art et al., 1974). There is considerable evidence that magnesium is derived from oceanic sources as well (e.g., Schlesinger and Hasey, 1980). While calcium and potassium may also be derived from maritime sources, the major portion of the calcium and potassium content of precipitation is generally derived from clay minerals in soil dust and increases with increasing continental influence (Gorham, 1961). In a northern California eucalypt forest, McColl and Bush (1978) estimated that the oceanic contribution to the annual inputs of calcium and potassium were only 12% to 13% of the total values (cf., Schlesinger and Hasey, 1980). The long periods of drought in mediterranean type ecosystems allow for drying of the upper soil layers and the potential for soil dust to be suspended in the atmosphere by wind. To the extent that the calcium and potassium concentrations measured in bulk precipitation are derived from dust suspended from the local area, estimates of the atmospheric deposition of these ions are greater than the actual input of new available quantities of these elements for plant growth. There is some evidence that much of the nitrate ion in bulk precipitation in southern California chaparral may also be derived from soil dust (Christensen, 1973; Schlesinger and Hasey, 1980).

Fog water collected by vegetation can be an important moisture input to terrestrial ecosystems along the western coast of the U.S. (Byers, 1953; Oberlander, 1956; Fritschen and Doraiswamy, 1973; Azevedo and Morgan, 1974). There is evidence for the same phenomenon along the western coast of Chile (Kummerow, 1962; Rundel and Mahu, 1976). Only a few of these studies have specifically examined fog precipita-

Table 10.1. Measurements of atmospheric nutrient deposition in bulk precipitation in several ecosystems

Locale and reference	Annual precipitation (mm)	Distance from ocean (km)	Deposition (kg ha⁻¹ yr⁻¹)					
			N	P	K	Ca	Mg	Na
Mediterranean type ecosystems								
California chaparral (Schlesinger and Hasey, 1980)	450-770	5-10	1.0	-	0.4	1.4	0.8	6.1
California chaparral (Schlesinger, *unpubl. data*)	756	10						
Wet deposition			-	-	0.05	0.2	0.4	3.4
Dry deposition			-	-	0.46	1.2	0.5	1.9
Total			-	-	0.5	1.5	0.9	5.2
California coastal scrub (Clayton, 1972)	635	2	-	-	14.9	7.6	3.8	68.6
France garrigue (Lossaint and Rapp, 1971; Lossaint, 1973)	770	20	14.6	1.0	2.0	10.5	1.5	22.6
Coastal forests								
California eucalyptus forest (McColl and Bush, 1978)	568	20	2.0	0.8	1.7	2.2	0.3	6.6
Oregon coniferous forest (Tarrant et al., 1968)	2286	10	1.5	-	-	-	-	-
New York barrier forest (Art et al., 1974)	1163	0.3	-	-	7.3	9.8	19.1	141.5
Inland forest								
New Hampshire temperate forest (Likens et al., 1977)	1250	116	20.7	0.04	0.9	2.2	0.6	1.6

tion in the mediterranean type ecosystems of these continents (e.g., Schlesinger and Hasey, 1980). Although fog drip is often impressive at low elevations along the coast (e.g., del Moral and Muller, 1969), in the chaparral of the Santa Ynez Mountains, California, significant interception of fog precipitation is important only above 1050 m elevation (Schlesinger and Hasey, 1980). Schlesinger and Hasey also presented strong evidence for the interception of dry aerosols by vegetation during periods of no precipitation. Both wet and dry interception processes may yield higher depositions of nutrients from the atmosphere than those measured by bulk precipitation (Table 10.1). The importance of these phenomena as a source of nutrients in mediterranean type ecosystems needs further investigation.

10.2.2 Nitrogen Input by Symbiotic Fixation

Symbiotic nitrogen fixation can represent a significant source of nitrogen for terrestrial ecosystems, particularly in nutrient-poor, pioneer habitats (Crocker and Major, 1955; Youngberg and Wollum, 1976; Lepper and Fleschner, 1977). Nitrogen fixation would be of an adaptive advantage where low quantities of soil nitrogen limit productivity in mediterranean type ecosystems. Thus, it is not surprising that it is from the relatively nitrogen-poor soils of the southern California chaparral that the greatest number of nodulated species have been reported. Early experimental work by Hellmers and Kelleher (1959) and Vlamis et al. (1964) indicated that two chaparral shrubs, *Ceanothus leucodermis* and *Cercocarpus betuloides*, could increase the nitrogen content of soils in greenhouse pots. More than a dozen California species of *Ceanothus* can possess root nodules of the endophyte *Frankia* (Vlamis et al., 1958; Furman, 1959; Hellmers and Kelleher, 1959; Delwiche et al., 1965; Russell and Evans, 1966; Youngberg and Wollum, 1976; Kummerow et al., 1978b), although most of these species are found in more mesic habitats than chaparral.

Nitrogen fixation appears to vary along a latitudinal sequence. In forests of northern California and Oregon, nitrogen fixation by *Ceanothus velutinus* has been estimated to be from 71 to 108 kg ha^{-1} yr^{-1} (Youngberg and Wollum, 1976). Delwiche et al. (1965) estimated that various nodulated *Ceanothus* species in northern California can fix up to 60 kg ha^{-1} yr^{-1}. *Cercocarpus ledifolius*, a species related to the chaparral shrub *C. betuloides*, was estimated to fix 6.9 kg ha^{-1} yr^{-1} in montane forests (Lepper and Fleschner, 1977). Kummerow et al. (1978b) recently measured root nodules from the chaparral shrub *Ceanothus greggii* at Echo Valley for rates of acetylene reduction. The rates per nodule were comparable to those found by Delwiche et al. (1965); however, because root nodule density per unit area was significantly lower than those observed in northern communities, Kummerow et al. (1978b) estimated that only 0.1 kg ha^{-1} yr^{-1} of nitrogen is fixed in the chaparral. They suggested that the low values are due to the generally more arid conditions in the southern California chaparral ecosystems. Preliminary investigations indicate that nonsymbiotic soil microorganisms may also fix up to 1 to 3 kg ha^{-1} yr^{-1} in chaparral ecosystems (J. Kummerow, *pers. comm.*).

Nitrogen fixation by leguminous species in the chaparral has also been noted for several subshrubs, including *Lotus scoparius*, *Lupinus exoubitus*, and *Pickeringia montana* (Vlamis et al., 1964; Dunn and DeBano, 1977). These plants are common post-fire invaders (Horton and Kraebel, 1955; Hanes, 1971) and may play an important role in replacing soil nitrogen lost by volatilization during a fire (DeBano and Conrad, 1978).

At Fundo Santa Laura, Rundel and Neel (1978) observed nitrogen fixation by the nodulated shrub, *Trevoa trinervis*, a common species in disturbed areas of the matorral; however, they did not find any positive correlation between soil nitrogen content and the distribution of *Trevoa*. Root nodules in two species of the legume, *Phyllota*, have been observed by Specht et al. (1958) in young stands of Australian heath. *Casuarina*, a dominant species in the heath, has been reputed to have mycorrhizae capable of fixing atmospheric nitrogen (Specht and Rayson, 1957).

These studies indicate that several nonleguminous species in mediterranean type ecosystems possess the ability to fix nitrogen. The fact that several mediterranean type ecosystems have analogous early successional species capable of nitrogen fixation indicates a functional similarity in nitrogen cycling. However, the aridity of mediterranean climates may limit the annual nitrogen inputs to the ecosystem by fixation (Kummerow et al., 1978b). More research and estimates of input based on both acetylene reduction rates and measured root nodule density are needed to evaluate fully the importance of nitrogen fixation in these ecosystems. In particular, it will be important to quantify the net addition of nitrogen by fixation in post-fire or disturbed areas in relation to ecosystem losses.

10.2.3 Nutrient Outputs by Hydrologic and Geologic Processes

Significant nutrient losses can occur when soil, organic debris, and dissolved substances are removed from ecosystems by hydrologic and geologic processes, including stream drainage, erosion, deep percolation, and soil slippage. Studies in experimental watersheds in several forests have quantified these nutrient losses on an annual and seasonal basis (Likens et al., 1977; Henderson et al., 1978). In most ecosystems, the gross export of nutrients is a difficult flux to measure but is a value necessary to formulate a total ecosystem budget. Experimental watersheds have not been used in studies of mediterranean type ecosystems with the exception of the San Dimas Experimental Forest in the San Gabriel Mountains of southern California (Mooney and Parsons, 1973). Although nutrient cycling was not the goal of the San Dimas research, a considerable amount of data has been gathered over the past several decades on hydrological and geological processes that characterize much of the southern California chaparral ecosystem. Similar studies are lacking in other mediterranean climate regions, thus, only the chaparral will be discussed in this section.

Streamflow outputs of nutrients represent gross ecosystem losses only when deep soil percolation is minimal (Bormann and Likens, 1967). Work by Rowe and Colman (1951) indicated that 41% to 47% of the annual rainfall in San Dimas is normally lost to underground waters by deep percolation. Losses by deep percolation represent an unknown, but significant, loss of nutrients through underground leaching of decomposed and weathered material. Because the amount of deep percolation in chaparral watersheds is not known, chaparral watersheds have not been used for the study of nutrient cycling on an ecosystem basis (Mooney and Parsons, 1973).

Intact, terrestrial ecosystems develop tight nutrient cycles as they mature and lose only small amounts of available nutrient capital by streamflow export (Likens and

Bormann, 1972; Likens et al., 1977, 1978). The regulation of ecosystem nutrient losses is largely a function of the intrasystem cycles of nutrient uptake and return by the
biotic portion of the ecosystem (Bormann et al., 1974). However, abiotic factors such
as climate, geology, and topography interact to limit the extent of this biotic regulation. The marked seasonality and great annual variation in precipitation, combined
with open vegetation and low biomass, promote high runoff and erosion from chaparral communities (Miller et al., 1977).

In the southern California chaparral, potentially large amounts of nutrients are lost
annually because of high erosion rates, even in undisturbed mature chaparral. Various
estimates of annual soil erosion losses for undisturbed chaparral range from 0 to 8000
kg ha^{-1} yr^{-1} (Rowe et al., 1951; Sinclair, 1954; Anderson et al., 1959; Krammes, 1965;
DeBano and Conrad, 1978). In contrast, typical sediment yields for mature hardwood
forests in the northeastern United States range from 25 to 53 kg ha^{-1} yr^{-1} (Wolman
and Schick, 1967; Lull and Sopper, 1969; Likens and Bormann, 1972). A tremendous
range in values for the chaparral ecosystems reflects the local differences in rainfall,
slope, vegetative cover, and geology.

High rates of erosion in southern California chaparral can be attributed to several
factors. Recently uplifted slopes in the San Gabriel Mountains (Anderson et al., 1959;
Scott and Williams, 1978) and the sandstone nature of much of the Santa Ynez Mountains (Dibblee, 1966) promote soil erosion. Chaparral occurs on extremely steep
slopes; the average slope in the San Dimas Forest is 68% (Biswell, 1974). Shallow-
rooted, deciduous shrubs (Horton and Kraebel, 1955; Kummerow et al., 1977) are
commonly found on steep slopes in the coastal mountains of southern California
(Mooney, 1977b). In the San Gabriel Mountains, equator-facing, sage-covered slopes
have a greater number of soil slippages than the more mesic slopes with deeper-rooted,
broad-leaved shrubs (Rice et al., 1969). Slope is the major factor contributing to erosional losses, followed in importance by vegetation type and the occurrence of unusually intense winter rains (Rice et al., 1969; Rice and Foggin, 1971).

Anderson et al. (1959) studied soil erosion in a number of sites in the San Gabriel
Mountains and found that erosion was usually greater in the dry season when the soil
had low cohesion. Noncohesive soils on steep slopes creep and slide with disturbances
created by wind or animals. On typical slopes of 55% to 60%, annual erosion losses
amounted to 440 to 1584 kg ha^{-1} yr^{-1}, of which 17% to 36% was organic matter
(Anderson et al., 1959; Mooney and Parsons, 1973). The high amount of organic matter in the fractions of soil that are lost by erosion, soil slippage, and winter runoff may
result in significant nutrient losses from these ecosystems.

DeBano and Conrad (1976, 1978) measured nutrient losses by erosion and surface
runoff on unburnt chaparral during the rainy season (Table 10.2). On slopes of 20%,
no measureable erosion occurred, and surface runoff volumes and dissolved nutrient
losses were insignificant. On steep slopes of 50%, a larger amount of debris was lost,
though only a small fraction (3.4%) was organic material. The annual amount of nutrients lost by combined erosion and surface runoff was less than 0.5% of the total nutrient pool in the surface litter and upper 10 cm of soil. Although the mass of eroded debris appears high, the gross loss of nutrients from unburned chaparral is similar to annual
losses from undisturbed, northeastern hardwood forests (Likens and Bormann, 1974).

The erosion rates observed by DeBano and Conrad (1976) were significantly less
than those measured in the San Gabriel Mountains (e.g., Sinclair, 1954, Krammes,

Table 10.2. Losses of nutrients by soil erosion and surface runoff from unburned and burned chaparral on different slopes[a,b]

Percent Slope	Debris (kg ha^{-1})	Percent organic matter in debris	Runoff (liters × 10^4 ha^{-1})	Nutrients (kg^{-1} ha^{-1})							
				In debris					In runoff		
				N	P	K	Ca	Mg	K	Ca	Mg
Unburned											
20	0	0.5	00	0	0	0	0.01	0.04	0.01		
50	210	3.4	2.4	0.3	0.1	0.5	0.5	0.5	0.09	0.41	0.07
Burned											
20	2848	10.9	58.4	7.5	1.0	7.6	18.5	6.2	3.26	9.14	1.91
50	7340	5.6	78.6	15.1	3.4	19.3	47.4	28.0	7.67	20.0	3.63

[a] Measured the first winter after a prescribed burn.
[b] Data from DeBano and Conrad (1976).

1965), and the percent organic matter in erodible debris was an order of magnitude smaller (Anderson et al., 1959). Local conditions of rainfall, topography, and geology can exert an overriding influence on nutrient losses in cháparral ecosystems. It appears that in favorable site conditions, biotic regulation by mature chaparral minimizes nutrient losses at levels similar to those measured in forest ecosystems. In other, extreme conditions (e.g., steep slopes), the chaparral ecosystem can be leaky, and nutrient losses may be a significant fraction of the ecosystem nutrient capital. Extreme habitats may not permit the development of high vegetative biomass that could reduce nutrient losses.

10.2.4 Nutrient Outputs by Erosion, Runoff, and Volatilization Following Fire

Fire has long been recognized as a dominant ecological factor in most mediterranean type ecosystems (Biswell, 1974; Naveh, 1974, 1977; Mooney and Conrad, 1977). It is a major environmental catalyst that abruptly alters the distribution and flux of nutrients within an ecosystem. It also promotes the movement of nutrients to other ecosystems by atmospheric and hydrologic vectors (Lewis, 1974, 1975; Clayton, 1976, McColl and Grigal, 1977). Although naturally caused fires occur in all mediterranean type ecosystems, the frequency varies. There has been a long history of controversy concerning the "natural" frequency of fire and the effect of man's activities. Climatic conditions in different regions impose broad limits on the occurrence of fire; for example, the absence of summer thunderstorms in central Chile reduces the number of lightning fires (Aschmann and Bahre, 1977). Recently, Byrne et al. (1977) described fossil charcoal sequences in oceanic sediments from the coast of southern California that indicate large fires occurred every 20 to 40 years. Aschmann (1977) suggested that the archaeological and historical records indicate that prehistoric man increased the fire frequency in all regions of mediterranean type climate.

Fire destroys vegetation and organic matter in the upper soil and deposits soluble nutrients in the ash layer (Smith, 1970; White et al., 1973; Lewis, 1974; Viro, 1974). This makes the ecosystem highly vulnerable to nutrient losses by hydrologic export or leaching (Likens and Bormann, 1974). Nutrient losses can be reduced by rapidly growing vegetation (Likens et al., 1978), either fire-resistant stump-sprouting shrubs or early successional species (e.g., Marks and Bormann, 1972).

Scant information exists on the extent of nutrient losses after fires in mediterranean type ecosystems. Naveh (1974) noted that local conditions, i.e., grazing and fire frequency, make it difficult to generalize about nutrient losses after a fire. In undisturbed maquis in Israel, soil erosion after a fire was not observed even on slopes of 30% to 40% (Naveh, 1974). However, post-fire erosion can be accelerated by other effects including grazing and road construction (Biswell and Schultz, 1957; Naveh, 1974). In southern California, enormous post-fire erosion has been reported by various authors (Rowe et al., 1951; Sinclair, 1954; Krammes, 1965; Biswell, 1974; DeBano and Conrad, 1976). In a representative chaparral area of the San Gabriel Mountains, pre-burn erosion rates were 7980 kg ha^{-1} yr^{-1} (Rowe et al., 1951; Mooney and Parsons, (1973). These rates increased to 230,000 kg ha^{-1} in the first year after the fire and decreased to 16,600 kg ha^{-1} by the third year.

The greatest amount of soil, debris, and dissolved material is lost during the first winter rains after a fire (Biswell, 1974). DeBano and Conrad (1976, 1978) have shown that on both gentle and steep chaparral slopes, erosional and runoff losses increased dramatically during the first winter (Table 10.2). Most nutrients were lost in the form of debris. Nutrient concentrations in debris were relatively constant as the rainy period progressed, but runoff concentrations dropped quickly after the first rains leached soluble nutrients from the ash and upper soil (DeBano and Conrad, 1976). In the chaparral studied by DeBano and Conrad (1976, 1978), nitrogen, phosphorus, and potassium lost by erosion in the first winter represented 3.4%, 2.4%, and 9.7%, respectively, of the nutrients in the ash and upper 2 cm of soil. If erosion is reduced by lush post-fire herbaceous growth and regeneration of chaparral shrubs, these losses may diminish greatly in subsequent years. Fire on steep, highly erodible slopes in the San Gabriel Mountains, where post-fire erosion rates are very high, can cause much larger nutrient losses from the ecosystem. Repeated fire would certainly deplete the pool of available soil nutrients and potentially limit ecosystem development.

Nitrogen losses occur not only by erosion but also by volatilization of organic matter. Volatilization of nitrogen occurs when temperatures exceed 200°C (White et al., 1973; DeBano et al., 1979). Evidence also indicates that potassium may be volatilized (White et al., 1973; DeBano and Conrad, 1978). DeBell and Ralston (1970) observed losses of 58% to 85% of the total nitrogen from pine needles burned in the lab (cf., Lewis, 1974; Christensen, 1977). However, these percentage losses cannot be used to estimate ecosystem losses because particulate ash may return to the ecosystem in rain or dry fallout (Lewis, 1975; Clayton, 1976).

DeBano and Conrad (1978) found that net ecosystem losses of nitrogen and potassium occurred by volatilization during fire in the chaparral and were significantly greater than erosional or runoff losses. Nitrogen loss by volatilization was 146 kg ha^{-1}, over 10% the total ecosystem nitrogen before fire. Potassium losses to the atmosphere were 48 kg ha^{-1}, 12% of the total potassium pool before fire.

Nutrient losses by volatilization, particularly of nitrogen, are significant in light of the relatively low inputs of nutrients by bulk precipitation and nitrogen fixation found in the southern California chaparral (Kummerow et al., 1978b; Schlesinger and Hasey, 1980). Chaparral ecosystems must have efficient nutrient acquisition and conservation mechanisms that enable them to recover from repeated nutrient losses by fire. Certainly, the potential importance of nitrogen fixation by post-fire herbs, woody shrubs, and soil microorganisms cannot be overestimated. Quantitative studies of these inputs are generally lacking and need further attention. Nitrogen input by rain and dust may be of a greater magnitude than nitrogen fixation; however, current estimates of net ecosystem inputs by bulk precipitation (Schlesinger and Hasey, 1980) indicate that 100 to 150 years may be needed to replace the nitrogen loss in fire such as was observed by DeBano and Conrad (1978).

10.3. Nutrient Pools

10.3.1 Nutrient Pools in the Vegetation

Nutrient pools in vegetation are a function of the standing biomass, the nutrient concentration of the vegetation, and the relative allocation of biomass to different vegetative components. In shrub-dominated, mediterranean type ecosystems, the amount of biomass does not equal that found in most forest ecosystems (Whittaker and Likens, 1975). The age of a mediterranean type ecosystem is an important factor in predicting biomass. Most of these ecosystems are considered mature at the age of 20 to 25 years, although individual shrub longevity may be much longer. Standing biomass is also a function of vegetative density and cover and of the weight of individual shrub species. The standing biomass in the chamise (*Adenostoma fasciculatum*) chaparral at Echo Valley was 23,079 kg ha^{-1}, while standing biomass in the matorral at Fundo Santa Laura was only 7380 kg ha^{-1} (Mooney et al., 1977). The smaller value in Chile can be explained by the low vegetative cover in the matorral and by the fact that evergreen shrubs in the matorral typically had less than half the dry weight of their chaparral counterparts (Mooney et al., 1977). In southern California, evergreen shrubs from the chaparral are considerably heavier than drought deciduous species. Equal-aged, adjacent pure stands of evergreen, *Ceanothus* chaparral, and semideciduous coastal sage scrub in southern California differed in biomass by a factor of five (Gray, *unpubl. data*; Table 10.3).

The standing biomass and nutrient pools vary in mediterranean type ecosystems. The average age of the vegetation is approximately 25 years and the mean standing biomass is 37,297 kg ha^{-1} (Table 10.3). The lowest observed biomass in an evergreen, chaparral stand (14,000 kg ha^{-1}) was measured in an old, senescing stand of chamise in the interior mountains of California (Rundel and Parsons, 1979). Chamise chaparral shows great variability in standing biomass, ranging from 14,000 up to 49,000 kg ha^{-1}. The biomass of the latter ecosystem was 40% dead wood, primarily *Ceanothus crassifolius* (Specht, 1969b). Some species of *Ceanothus* have been observed to die in chaparral stands older than 30 to 40 years (Hanes, 1977), although the longevity of other species may be up to 90 years (Keeley, 1975). The chamise chaparral stand in northern California had a large standing crop, perhaps the result of more favorable rainfall conditions. High biomass values for broad-leaved shrubs were observed for *Ceanothus* chaparral, *Quercus coccifera* garrigue, and *Banksia* scrub. The former two communities were characterized by high-density monocultures; the Australian ecosystem experiences a realtively high annual precipitation (Maggs and Pearson, 1977a). Large standing nutrient pools also were found in these three ecosystems. The largest calcium pool was in the French garrigue, situated on a highly calcareous soil (Lossaint and Rapp, 1971). In the *Banksia* scrub, the relatively low phosphorus pool in relation to biomass probably reflects plant adaptations to the low availability of phophorus in Australian soils (Beadle, 1966, 1968; Barrow, 1977).

The relative allocation of biomass to foliage influences nutrient standing pools in a community, as a result of the high nutrient concentration of leaves compared to wood. Leaves of dominant, evergreen species at Echo Valley and Fundo Santa Laura com-

Table 10.3. Aboveground standing biomass and nutrient pool size[a,b]

Ecosystem, locale, and reference	Age (yr)	Biomass	N	P	K	Ca	Mg
Mediterranean type ecosystems							
Coastal sage. So. Calif. (Gray and Schlesinger 1981, Gray, *unpubl. data*)	22	14,170	77	12	79	62	19
Chamise chaparral. So. Calif. (Rundel and Parsons, 1979, 1980)	37	14,000	83	17	60	56	6
Chamise chaparral, So. Calif. (Mooney and Rundel, 1979)	27	21,270	135	19	-	-	-
Mixed chamise, So. Calif.[c] (Specht, 1969b)	37	49,091	145	10	105	90	35
Chamise chaparral, No. Calif. (Sampson, 1944)	>25	31,460	-	14	141	156	-
Redshank chaparral, So. Calif. (DeBano and Conrad, 1978)	25	30,394	134	11	113	234	19
Mixed chaparral, No. Calif. (Sampson, 1944)	>25	31,001	-	16	158	222	-
Ceanothus chaparral, So. Calif. (Schlesinger, *unpubl. data*)	12	41,380	264	20	116	225	29
Ceanothus chaparral, So. Calif. (Gray, *unpubl. data*)	22	75,060	472	33	196	389	47
Ceanothus chaparral, So. Calif. (Schlesinger, *unpubl. data*)	21	63,370	411	29	168	348	45
Quercus garrigue, So. France[c] (Specht, 1969b)	13	49,693	240	18	250	663	80
Quercus garrigue, So. France (Lossaint and Rapp, 1971)	17	23,500	160	9	85	485	22
Banksia scrub, Australia (Maggs and Pearson, 1977a)	28	50,400	305	16	206	279	68
Heath scrub, Australia[c] (Specht et al., 1958)	25	27,170	116	6	67	112	52
Mean	24	37,297	212	16	134	255	38

prised 17% and 28% of standing biomass, respectively (Mooney et al., 1977). Other mediterranean type ecosystems showed lower allocation to foliage (Table 10.4). Nutrient-poor, dead wood in shrub ecosystems can make up a significant percentage of standing biomass. In southern California, chaparral dead wood tends to accumulate in older, senescing stands. Parsons (1976a) found that the percent cover by dead stems in chamise chaparral increases to 63% as stands reach 55 years of age. The dead wood reported in the mixed chamise was primarily due to dead individuals (Specht, 1969b), while in the coastal sage scrub and the *Ceanothus* chaparral, dead wood was mostly composed of attached, lower stems (Schlesinger and Gill, 1978, 1980; Gray, *unpubl. data*).

Table 10.3. (Continued)

Ecosystem, locale, and reference	Age (yr)	Biomass	N	P	K	Ca	Mg
Temperate forest ecosystems							
Beach-maple, New Hampshire (Likens et al., 1977; Gosz et al., 1976)	55	133,074	351	34	155	383	36
Oak-hickory, No. Carolina (Henderson et al., 1978)	-	139,000	995	-	400	830	-
Oak-hickory, Tennessee (Henderson et al., 1978)	-	122,000	470	-	340	980	-
Oak, Missouri (Rochow, 1975)	40-90	100,070	204	20	115	601	35
Oak, Oklahoma (Johnson and Risser, 1974)	80	194,379	902	75	1093	3894	230
Jack pine, Ontario (Foster and Morrison, 1976)	30	90,700	165	14	82	112	18
Loblolly pine, Mississippi (Switzer and Nelson, 1972)	20	90,000	174	19	99	91	24
Subalpine fir, Washington (Turner and Singer, 1976)	175	468,512	372	67	980	1046	160
Douglas-fir, Oregon (Henderson et al., 1978)	430	718,000	560	-	360	750	-
Mean		228,471	466	38	403	965	84

[a] Measured in kg ha^{-1}, rounded to the nearest kilogram.
[b] Standing biomass in shrub ecosystems includes dead wood.
[c] Estimated from curves in Specht (1969b) and Specht et al. (1958).

The weighted average nutrient concentration of the vegetation was calculated to permit a comparison of the nutrient pools in different-aged mediterranean type ecosystems. The percent nitrogen, phosphorus, and potassium is surprisingly similar in all scrub ecosystems and suggests a similarity in nutrient utilization.

Aboveground standing biomass and nutrient pool sizes for selected temperate forests can be compared to mediterranean type ecosystems (Table 10.3). The mean biomass of mediterranean type ecosystems is 84% lower than the mean biomass for temperate forests; however, the standing nutrient pools are only 55% to 75% lower. Nitrogen and phosphorus pools in *Ceanothus* chaparral, garrigue, and *Banksia* scrub were greater than in some pine ecosystems with twice the biomass. Potassium pools of several shrub ecosystems were also similar in magnitude to several of the forests. Shrub ecosystems have large nutrient pools in relation to biomass, due in part to a greater allocation to foliage. Typically, the dry matter allocation to leaves in hardwood and coniferous forests is only 1% to 7% (Rodin and Bazilevich, 1967). Weighted mean nutrient concentrations of nitrogen, phosphorus, and potassium for the forests listed are 0.21%, 0.02%, and 0.18%, respectively. These values are approximately half of the concentrations measured in shrub ecosystems studied (Table 10.4).

Table 10.4. Percent of standing biomass in leaves and dead wood, and nutrient concentration in standing biomass in mediterranean type ecosystems

Ecosystem, locale, and reference	Age (yr)	Percent leaf	Percent dead wood	Nutrient concentration[a]		
				N	P	K
Coastal sage, So. Calif. (Gray, *unpubl. data*)	22	13	18	0.55	0.08	0.56
Chamise chaparral, So. Calif. (Rundel and Parsons, 1979, 1980)	37	11	-	0.59	0.12	0.42
Chamise chaparral, So. Calif. (Mooney and Rundel, 1979)	27	17	-	0.63	0.09	-
Mixed chamise, So. Calif. (Specht, 1969b)	37	-	44	0.29	0.02	0.21
Redshank chaparral, So. Calif. (DeBano and Conrad, 1978)	25	-	34	0.44	0.03	0.37
Ceanothus chaparral, So. Calif. (Schlesinger, *unpubl. data*)	21	7	23	0.65	0.05	0.27
Ceanothus chaparral, So. Calif. (Gray, *unpubl. data*)	22	6	15	0.63	0.04	0.27
Quercus garrigue, So. France (Specht, 1969b)	13	-	-	0.48	0.04	0.50
Quercus garrigue, So. France (Lossaint and Rapp, 1971)	17	17	-	0.67	0.04	0.36
Banksia scrub, Australia (Maggs and Pearson, 1977a)	28	14	10	0.61	0.03	0.41
Heath scrub, Australia (Specht et al., 1958)	25	-	-	0.42	0.02	0.24
Mean		12	24	0.54	0.05	0.36

[a]Nutrient concentration = [total nutrient pool in vegetation (Table 10.3)/total standing biomass (Table 10.3)] × 100%.

10.3.2 Nutrient Pools in the Litter Layer

The nutrient pool in litter is found within the layer of accumulated dead organic matter that is in various stages of decomposition on the surface of the mineral soil. The elemental content of this detrital organic matter is influenced by many factors, including annual litter deposition, maturity of the community, and the rates of decomposition and of leaching losses. Litter mass typically increases with age in developing scrub communities. In montane chamise chaparral in California, Parsons (1976a) found a litter mass of 2180 kg ha^{-1} in a 4-year-old stand and 23,270 kg ha^{-1} in a stand over 60 years of age. In mixed chaparral at the San Dimas Experimental Forest, mean litter mass for stands 11, 31, and 50 years old was 9139, 12,844, and 37,394 kg ha^{-1}, respectively (Kittredge, 1955). The mean litter mass calculated for seven mediterranean type ecosystems is 11,900 kg ha^{-1} (Table 10.5). The highest observed values were for

Ceanothus chaparral and *Banksia* scrub, both with large, standing, vegetative biomass (Table 10.3).

The nutrient pools in the litter layer of mediterranean type ecosystems are more variable than are the nutrient pools in vegetation, primarily because of great differences in the concentration of nutrients in the litter in these ecosystems (Table 10.5). Although *Ceanothus* chaparral litter had a greater litter mass and nitrogen pool, the redshank chaparral had larger pools of most other elements due to the high nutrient concentration in the litter. The low litter pool of phosphorus in the two Australian scrub ecosystems may be a result of the generally low phosphorus availability throughout the system. Efficient foliar resorption of phosphorus by native shrubs, including *Banksia*, (Specht and Groves, 1966) may also contribute to the low phosphorus pool in the litter.

The percent of the aboveground nutrient pool that is contained in litter varies considerably in the shrub ecosystems studied (Table 10.6). The aboveground pools of nitrogen and phosphorus, limiting elements in most ecosystems, are held primarily in the vegetation. Nonlimiting elements like calcium and magnesium are shared more equally in both the vegetation and litter.

Nitrogen and phosphorus are immobilized quickly and preferentially by soil microorganisms in the litter layer of terrestrial ecosystems (Alexander, 1976). The relatively large nitrogen and phosphorus pools in the forest floors of temperate forests [60% to 70% of the total ecosystem pool for most forests listed in Lang and Forman (1978)] reflect the importance of microbes in regulating nutrient cycling and storage in forest ecosystems. Most shrub ecosystems do not usually have the thick, well-defined humus layers found in forests. Although only a few shrub ecosystems have been studied, litter pools of nitrogen and phosphorus are considerably lower than in most temperate forests (Table 10.5), while potassium, calcium, and magnesium pools may approach or exceed those in hardwood forest floors. Mediterranean type ecosystems have smaller nitrogen and phosphorus pools in litter because of their lower productivity and low accumulation of detrital mass. Microbial regulation of nutrient storage and release in mediterranean type ecosystems may be of less importance than in forests.

Nutrient pools of potassium, calcium, and magnesium in litter are strongly influenced by hydrologic factors. The comparatively high potassium pool in the litter of shrub ecosystems (Table 10.5) may be favored by low leaching rates associated with the semiarid climate. Elements like calcium and magnesium are structurally bound in cell walls and may tend to accumulate in the litter of shrub ecosystems.

10.3.3 Soil Nutrient Pools

Nutrient pools in soil are a function of the local parent material, climate, vegetation, and time. Zinke (1973) has shown that analogies can be made with soil and vegetation types in Italy, Greece, and California (cf., Paskoff, 1973). Low plant productivity, steep or dissected topography, and high erosion rates in southern California, central Chile, and in the area around the Mediterranean Sea limit soil development and fertility. In general, soil nutrient deficiencies have been noted in most mediterranean climate areas and may have played an important role in the evolution and migration of sclerophyllous species to these regions of the world (Small, 1973).

Table 10.5. Litter mass and nutrient pools in mediterranean type ecosystems and in a range of values for nine hardwood forests in the eastern United States[a]

Ecosystem, locale, and reference	Age (yr)	Litter mass	N	P	K	Ca	Mg
Coastal sage, So. Calif. Gray and Schlesinger, 1981; Gray, *unpubl. data*)	22	6,167	47	4	18	89	31
Chamise chaparral, So. Calif. (Mooney and Rundel, 1979)	27	4,170	25	4	-	-	-
Redshank chaparral, So. Calif. (DeBano and Conrad, 1978)	25	9,550	147	22	174	465	172
Ceanothus chaparral, So. Calif. (Schlesinger, *unpubl. data*)	12	16,650	230	12	82	641	45
Ceanothus chaparral, So. Calif. (Gray, *unpubl. data*)	22	20,270	205	6	47	261	67
Banksia scrub, Australia (Maggs and Pearson, 1977a)	28	19,000	178	7	20	119	31
Heath scrub, Australia[b] (Specht et al., 1958)	25	7,410	42	1	3	47	14
Mean		11,888	124	8	57	270	60
Hardwood forests, U.S.A. (Lang and Forman, 1978)		7,100–323,730	113–1654	16–49	20–42	117–573	16–38[c]

[a] Values measured in kg ha^{-1}, rounded to the nearest kilogram.
[b] Estimated from curves in Specht et al. (1958).
[c] Minimum value from oak forest (Rochow, 1975), maximum value from maple-beech forest (Likens et al., 1977).

Mediterranean type ecosystems have similar climates and vegetation, but differences in underlying rock limit the degree of similarity in soil nutrient status. Soils in Australia are typically low in phosphorus and plant productivity is nutrient limited (Beadle, 1966, 1968; Specht, 1966, 1973). Miller et al. (1977) described the soils at Echo Valley and Fundo Santa Laura and found similarities in soil development, depth, and clay content (see also Thrower and Bradbury, 1977). Southern California and central Chile have similar geologic histories and share at least one common parent material, quartz diorite, at Echo Valley and Fundo Santa Laura. However, the soils at Fundo Santa Laura are mostly derived from another material, andesite, which gives rise to a soil richer in nutrients and clays. Overall, soils at Fundo Santa Laura have greater concentrations of phosphorus and cations and less nitrate-nitrogen than soils at Echo Valley (Miller et al., 1977).

Although California chaparral soils have greater nitrogen content than Chilean soils, the marked response of native California shrubs to nitrogen fertilization indicates that nitrogen is still limiting (Vlamis et al., 1954; Hellmers et al., 1955a; Schultz et al., 1958; Christensen and Muller, 1975; Ball, 1977; J. Kummerow, *pers. comm.*). California chaparral soils have intermediate soil concentrations of total nitrogen and phosphorus in comparison to other mediterranean type ecosystems (Table 10.7). The Australian heath and mallee have extremely low levels of nitrogen and phosphorus. The

Table 10.6. Percent of aboveground mass (litter + vegetation) held within the litter in mediterranean type ecosystems

Ecosystem, locale, and reference	Age (yr)	Organic matter	N	P	K	Ca	Mg
Coastal sage, So. Calif. (Gray, *unpubl. data*)	22	30	38	25	19	59	62
Chamise chaparral, So. Calif. (Mooney and Rundel, 1979)	27	16	16	18	-	-	-
Redshank chaparral, So. Calif. (DeBano and Conrad, 1978)	25	24	52	68	61	66	90
Ceanothus chaparral, So. Calif. (Schlesinger, *unpubl. data*)	12	29	47	38	41	74	61
Ceanothus chaparral, So. Calif. (Gray, *unpubl. data*)	22	21	30	15	20	40	59
Banksia scrub, Australia (Maggs and Pearson, 1977a)	28	28	37	29	9	30	31
Heath scrub, Australia (Specht, 1969b)	25	21	27	16	4	30	21
Mean		24	35	29	25	50	54

calcareous soils of the French garrigue were found to be very high in nitrogen, phosphorus, and calcium (Specht, 1969b; Lossaint and Rapp, 1971). The total soil nitrogen content under chamise chaparral ranged from 900 to 3230 kg ha^{-1} in different localities in southern California; the lowest value was from Echo Valley. More data are needed to estimate available pools of nitrogen and phosphorus in mediterranean type ·ecosystems in order to make a more rigorous evaluation of the soil nutrient regimes in these regions of the world.

10.4. Intrasystem Cycles

10.4.1 Nutrient Return by Litterfall

Litterfall is the most important process for returning nitrogen, phosphorus, calcium, and magnesium to the soil in most ecosystems. The magnitude of this return is a function of the amount of annual litterfall and its elemental concentration. The annual litterfall for several mediterranean type ecosystems is a mean of 3835 kg ha^{-1} yr^{-1} (Table 10.8), which is larger than the 2726 kg ha^{-1} yr^{-1} observed at Echo Valley or the 1629 kg ha^{-1} yr^{-1} at Fundo Santa Laura (Mooney et al., 1977). Litterfall in the Californian chaparral, the Chilean matorral, and the French garrigue is strongly seasonal, peaking after the initiation of new canopy growth in the spring (Lossaint and Rapp, 1971; Mooney et al., 1977). Considerable year-to-year variation in litterfall is found in the chaparral (Kittredge, 1955; Mooney et al., 1977). For several years after a fire, annual litterfall increases as the vegetation regrows. Kittredge (1955) found that in mixed chaparral stands, litterfall reaches a plateau after the first 14 years of growth.

Table 10.7. Soil nutrient pools or concentrations in mediterranean type ecosystems

Ecosystem, locale and reference	Age (yr)	Sample depth (cm)	Percent total "		Total contents (kg ha⁻¹)	
			N	P	N	P
Chamise chaparral, So. Calif. (Mooney and Rundel, 1979)	27	0-10	-	-	900	750
Chamise chaparral, So. Caif. (Rundel and Parsons, 1980)	37	0-10	0.17	0.05	2580	780
Chamise chaparral, So. Calif. (Christensen and Muller, 1975)	40	0-2	0.30	0.05	-	-
Mixed chamise, So. Calif. (Specht, 1969b; Mooney and Parsons, 1973)	37	0-15	0.10	0.07	3230	1680
Redshank chaparral, So. Calif. (DeBano and Conrad, 1978)	25	0-10	-	-	1143	536
Matorral, Chile (Hoffmann and Kummerow, 1978)	-	0-20	0.15	0.09	-	-
Quercus garrigue, So. France (Lossaint and Rapp, 1971)	17	0-30	-	-	6000	550
Quercus garrigue, So. France (Specht, 1969b)	13	0-8	0.48	0.04	-	-
Banksia scrub, Australia (Maggs and Pearson, 1977a)	28	0-70	0.07	-	4280	-
Heath scrub, Australia (Specht, 1969b)	-	0-25	0.02	0.001	-	-
Mallee scrub, Australia (Specht, 1969b)	-	0-8	0.04	0.006	-	-

Leaves are the predominate plant material in litterfall. Mooney et al. (1977) found that 78% to 85% of the litter dry weight at Echo Valley and Fundo Santa Laura was leaves. Although nutrient concentrations in leaves are relatively high, reabsorption before abscission partially reduces these concentrations in litterfall. The percent phosphorus in the litterfall of the Australian coastal scrub (Maggs and Pearson, 1977b) was considerably lower than for California chaparral where soil phosphorus is not as limiting (Table 10.8). The highest return of nutrients to the soil by litterfall reported in mediterranean type ecosystems was in the *Ceanothus* chaparral and the *Banksia* scrub. The litterfall in the chamise chaparral showed the lowest overall nutrient return and elemental concentration. However, annual litterfall of up to 2130 kg ha⁻¹ yr⁻¹ in the same chamise stand has been reported in other years (Mooney and Rundel, 1979), indicating great year-to-year variation and potential for a higher nutrient flux.

The percent of aboveground biomass that becomes annual litterfall in mediterranean type ecosystems ranges from 6% to 14% (Table 10.8); most elements in the litterfall represent 10% to 15% of the aboveground nutrient pools. However in the *Ceanothus*

Table 10.8. Annual litterfall and nutrient return in several ecosystems[a]

Ecosystem, locale, and reference	Age (yr)	Litter fall	N	P	K	Ca	Mg
Mediterranean type ecosystems							
Coastal sage, So. Calif. (Gray and Schlesinger, 1981; Gray, *unpubl. data*)	22	1940	15.0	1.4	11.2	25.4	7.8
Chamise chaparral, So. Calif. (Mooney and Rundel, 1979)[b]	27	830	3.8	0.5	-	-	-
Ceanothus chaparral, So. Calif. (Schlesinger, *unpubl. data*)[c]	12	8126	89.3	10.5	34.5	106.3	16.1
Banksia scrub, Australia (Maggs and Pearson, 1977b)	28	5980	63.0	2.5	20.7	38.7	9.3
Quercus garrigue, So. France (Lossaint and Rapp, 1971)	17	2300	22.2	0.8	9.7	36.5	2.7
Mean		3835	38.7	3.1	19.0	51.7	8.9
Forested ecosystems							
Maple-beech, New Hampshire (Likens et al., 1977)	55	5702	54.2	4.0	18.3	40.7	5.9
Oak-hickory, No. Carolina (Cromack and Monk, 1975)	50	4369	33.9	5.0	18.1	44.5	6.6
Oak-hickory, Missouri (Rochow, 1975)	35-92	3490	31.6	3.2	6.4	70.3	4.9
Oak, Oklahoma (Johnson and Risser, 1974)	80	5539	66.7	7.4	45.9	84.4	12.8
White pine, No. Carolina (Cromack and Monk, 1975)	14	3253	26.5	3.7	5.5	19.2	2.7
Jack pine, Ontario (Foster and Morrison, 1976)	30	4000	25.2	1.6	7.1	15.0	2.1
Subalpine fire, Washington (Turner and Singer, 1976)	175	3017	16.3	2.0	7.3	39.7	2.3
Mean		4196	36.3	3.8	15.5	44.8	5.3

[a] Values measured in kg ha^{-1} yr^{-1}.
[b] Low value, see text.
[c] High values from a moist year.

chaparral, the values for nitrogen, phosphorus, and potassium were 33%, 52%, and 29%, respectively. These data indicated that a significant portion of the nutrient pool in this community was moving through the vegetation and back to the soil. This high nutrient return by litterfall and relatively rapid decomposition (see next section) may play an important role in maintaining the high productivity of *Ceanothus* chaparral (Schlesinger and Gill, 1980).

The mean annual litterfall for mediterranean type ecosystems (Table 10.8) is lower than the range given by Bray and Gorham (1964) and Rodin and Bazilevich (1967) for temperate deciduous and coniferous forests, 4000 to 7000 kg ha^{-1} yr^{-1}. Litterfall and nutrient return for seven temperate forests are also given in Table 10.8 for comparison. The nutrient return in litterfall for *Ceanothus* chaparral and *Banksia* scrub exceeded many of the forest values, but these are probably maximum values for any shrub ecosystem. In typical mature, deciduous forests, 8% to 10% of aboveground nutrient pools are returned to the soil by annual litterfall (Rodin and Bazilevich, 1967). In young, developing forests like the Hubbard Brook Forest in New Hampshire, these values increase to 10% to 15% for nitrogen, phosphorus, potassium, calcium, and magnesium (Likens et al., 1977). In older forests and in coniferous forests, the percentages decrease to 3% to 5% (Rodin and Bazilevich, 1967). Although data are limited, the relatively high return of nutrients to the soil by litterfall in several mediterranean type ecosystems characterize them as young, developing ecosystems.

10.4.2 Decomposition

Because the organic matter on the surface of the soil is a major site of nutrient remineralization in terrestrial plant communities, studies of decomposition processes are crucial to our understanding of nutrient cycling in mediterranean type ecosystems. Unfortunately, such studies are scarce, although a considerable amount of work is now in progress.

One might expect relatively low rates of decomposition in mediterranean type ecosystems, inasmuch as conditions for microbiological activity are often poor (Schaefer, 1973). During the winter wet season, microbial activity is limited by low temperature; during the summer drought, soil moisture is usually inadequate for decomposers. Notwithstanding, several studies suggest that the annual decomposition of litterfall is surprisingly rapid. Kittredge (1955) found that the annual disappearance of litter was 13% to 32% of the detrital mass in chaparral stands below 1000 m in elevation that had achieved steady state conditions of development, 8 to 12 years after fire. More recently, Yeilding (1977) reported some preliminary results of a large project in San Diego County, California, utilizing litterbags and analyzing for releases of nitrogen, phosphorus, and calcium over a 2-year period. In her study, foliage of *C. greggii*, *Quercus dumosa*, and *A. fasciculatum* placed in litterbags on the soil surface lost 11% to 24% of its original weight in 1 year; decomposition of buried litter was even greater. With a steady state model of the detrital mass (Olson, 1963; Schlesinger, 1977), the residence or turnover time for organic matter and nutrients in the litter layer of the chaparral near Santa Barbara, California was calculated (Table 10.9). Turnover times measured in southern California and other mediterranean type ecosystems are not greatly different from those measured in temperate forests at similar latitudes. In fact, turnover of nitrogen and phosphorus in mediterranean type ecosystems is rapid compared to a latitudinal range of values for temperate forests compiled by Lang and Forman (1978). As in other biomes, turnover of detrital mass in the chaparral is slower at higher elevations, particularly above 1000 m (Kittredge, 1955; Table 10.9).

Winn and Dunn (*unpubl. data*) found that low soil moisture often limits microbial activity in the chaparral of southern California. In their study and in the work of Billès et al. (1971a, b), a marked microbial response was observed after wetting dry soils. Relatively high rates of nutrient turnover must be due to the ability of microorganisms

Table 10.9. Mean residence or turnover time[a] in years for surface organic matter on the soil and its mineral elements in several ecosystems

Ecosystem, locale, and reference	Organic matter	N	P	K	Ca	Mg
Mediterranean type ecosystems						
Chamise chaparral, So. Calif. (Mooney and Rundel, 1979)	5.0	6.5	8.4	-	-	-
Mixed chaparral, So. Calif. (Kittredge, 1955)						
760-1060 m elevation	2.4 - 7.9	-	-	-	-	-
1370-1650 m elevation	11.4 - 23.0	-	-	-	-	-
Ceanothus chaparral, So. Calif. (Schlesinger, unpubl. data) 350-1050 m elevation	2.1	2.6	1.2	2.4	6.0	2.8
Quercus ilex woodland, So. France (Lossaint and Rapp, 1971)	3.0	3.8	1.3	0.6	5.6	1.8
Banksia scrub, Australia (Maggs and Pearson, 1977a, b)	3.8	4.0	3.3	1.3	3.4	3.8
Desert ecosystem						
Cold-winter desert, Utah (West and Skujins, 1977)	-	6.8	-	-	-	-
Temperate forests						
Hardwood forests, eastern U.S. (Lang and Forman, 1978) Range of 8 forests	1.7 - 14.1	2.6 - 37.8	2.1 - 27.6	0.8 - 5.0	2.3 - 9.2	-
Tropical forest						
Moist tropical, Panama (Golley et al., 1975)	1.8	-	1.6	0.3	1.3	0.9

[a] For each study, turnover time = detrital mass/annual litterfall.

to respond strongly and rapidly within short periods when conditions are favorable for decomposition. Repeated drying and rewetting of chaparral soils may also increase the overall decomposition (Schaefer, 1973), as has been observed in laboratory (Stevenson, 1956; Birch, 1958; Sorensen, 1974) and other field studies (Reiners, 1968; de Jong et al., 1974).

The foliage of sclerophyllous shrubs often contains relatively high lignin concentration. Phenolic compounds including tannins are abundant in the foliage of both deciduous and evergreen chaparral species (Muller and Muller, 1964; McPherson et al., 1971; Dement and Mooney, 1974; Christensen and Muller, 1975). These organic compounds are resistant to decomposition and may control the rates of loss of other, normally labile constituents by forming relatively resistant complexes (Phillips et al., 1930; Peevy and Norman, 1948; Williams and Gray, 1974). The pattern of litter decomposition in the chaparral may be more closely controlled by the decomposition of these refractory organic constituents than by the carbon:nitrogen ratio. Similar controls on decomposition have been reported from other terrestrial ecosystems (Cromack and Monk, 1975; Fogel and Cromack, 1977; Meentemeyer, 1978). However, while resistant organic constituents may control the pattern of decomposition, their effects are exerted in the context of the relatively rapid overall annual turnover of organic matter in mediterranean type ecosystems (Table 10.9).

Ecologists traditionally have regarded fire as the primary means of remineralization of detritus in the chaparral; the importance of decomposition processes has been minimized or ignored. In addition to producing a substantial nutrient turnover in well-developed stands, microbial processes are important, especially in nitrification, in the ash-bed after fire (Dunn et al., 1979). Margaris (1977) showed rapid microbial recolonization of the soils in burned areas; the work of Christensen (1973) suggested that a substantial portion of the nitrogen in ash was not available until it was processed by ammonifying and nitrifying bacteria (Christensen and Muller, 1975; Dunn et al., 1979). There is some evidence that nitrification is inhibited in undisturbed, mature chaparral, possibly by allelopathic compounds released from the shrubs (Christensen, 1973). In France, however, the mineralization of nitrogen in undisturbed garrigue proceeds year-round; nitrification is particularly strong in the summer months (Billès et al., 1971b, 1975). Microbial processes deserve further study in mediterranean type ecosystems. The importance of mycorrhizae, associated with many species of chaparral shrubs, is only currently being appreciated.

10.4.3 Net Primary Production and Nutrient Uptake by Vegetation

Nutrient uptake by vegetation is usually determined by the amount of a particular element in the annual primary production. Net primary production in mediterranean type ecosystems varies considerably. Annual aboveground net primary production in chamise chaparral at Echo Valley ranged from 3620 to 6709 kg ha^{-1} yr^{-1} (Mooney and Rundel, 1979; Mooney et al., 1977, respectively). Smaller values were measured at Fundo Santa Laura, 2525 kg ha^{-1} yr^{-1} (Mooney et al., 1977), and in the coastal sage scrub of southern California, 2551 kg ha^{-1} yr^{-1} (Gray and Schlesinger, 1980). These two communities also had very low standing biomass in comparison with other mediterranean type ecosystems (see Table 10.3). Aboveground production in the

Ceanothus chaparral, which is characterized by high density and biomass, was measured at 8500 kg ha^{-1} yr^{-1}, with considerable year-to-year variation (Schlesinger and Gill, 1980). The *Quercus* garrigue of southern France had an intermediate production of 3400 kg ha^{-1} yr^{-1} (Lossaint and Rapp, 1971).

Net primary production may change as a shrub ecosystem develops from rapidly growing initial stages to older stands where productivity may be very low. The biomass accumulation curves from several mediterranean type ecosystems indicate that net annual increment slows or stops after 15 to 20 years in chamise chaparral but continues at high rates in *Ceanothus* chaparral, *Quercus* garrigue, and heathlands (Figure 10.1). Net annual increment in mature *Ceanothus* chaparral was 2650 kg ha^{-1} yr^{-1} (Schlesinger and Gill, 1980); in the garrigue, it has been reported at 1100 (Lossaint and Rapp, 1971) and 4000 kg ha^{-1} yr^{-1} (Specht, 1969b). Rundel and Parsons (1979) found that after 16 years there was a slight decrease in standing biomass in montane chamise chaparral (Figure 10.1). Other estimates of net annual increment in mature chamise chaparral, ranging from 1500 (Specht, 1969b) to 2790 kg ha^{-1} yr^{-1} (Mooney and Rundel, 1979), indicate that high accumulation rates may continue beyond a stand age of 16 years. Biomass accumulation ratios in mediterranean type ecosystems range from 3 to 10, considerably lower than the values of 25 to 35 found in forest ecosystems (Whittaker, 1975).

Only a few studies have measured directly the gross nutrient uptake in a shrub ecosystem. Nitrogen can be used as an example to compare patterns of nutrient uptake.

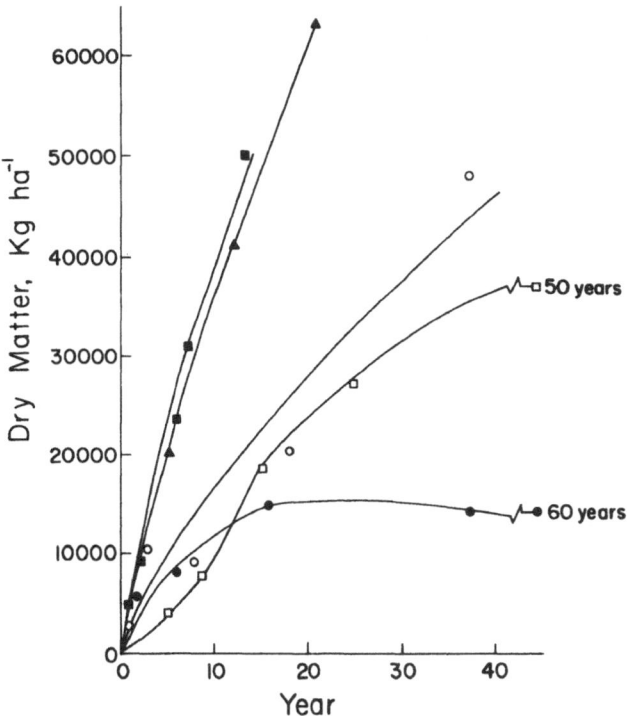

Figure 10.1. Standing biomass in various mediterranean type ecosystems as a function of age, including *Ceanothus* chaparral (▲) (Schlesinger and Gill, 1980), *Quercus* garrigue (■) (Specht, 1969b), heath scrub (□) (Specht et al., 1958), mixed chamise chaparral (o) (Specht, 1969b), and montane chamise chaparral (●) (Rundel and Parsons, 1979).

Gross nitrogen uptake in aboveground biomass in a chamise chaparral community was reported to be 30 kg ha^{-1} yr^{-1} (Mooney and Rundel, 1979). A value of 25 kg ha^{-1} yr^{-1} has been measured in the coastal sage scrub (Gray, *unpubl. data*). Gross nitrogen uptake in *Ceanothus* chaparral was estimated at 85 kg ha^{-1} yr^{-1} (Gray, *unpubl. data*), a relatively high value due primarily to the high net primary production of this ecosystem. Lossaint and Rapp (1971) reported that nitrogen uptake in *Quercus coccifera* garrigue was 29 kg ha^{-1} yr^{-1}.

Nitrogen accumulation in several shrub ecosystems follows a similar pattern as biomass (Figure 10.2). The garrigue and *Ceanothus* chaparral have steep curves in contrast to the chamise and heath ecosystems that tend to reach constant values at 20 to 25 years.

Despite a potentially greater annual production in coniferous forests, the low nitrogen concentrations in conifers result in relatively low nitrogen uptake. Thus, nitrogen uptake rates in coniferous forests, particularly in older communities, are comparable to rates measured in mediterranean type ecosystems. A mature Douglas fir forest had a nitrogen uptake of 23 kg ha^{-1} yr^{-1} (Henderson et al., 1978), while a 30-year-old jack pine forest had a value of 36 kg ha^{-1} yr^{-1} (Foster and Morrison, 1976). Hardwood forests have greater annual nitrogen uptake (Henderson et al., 1978). The maple-beech forest at Hubbard Brook had an annual uptake of 68 kg ha^{-1} yr^{-1} (Likens et al., 1977). An even greater value of 103 kg ha^{-1} yr^{-1} of nitrogen was reported for a midwestern oak forest (Johnson and Risser, 1974). The greater nutrient uptake is largely due to the higher productivity of these hardwood forests. Mean net primary production in temperate hardwood and coniferous forests was estimated at 12,500 kg ha^{-1} yr^{-1} (Whittaker and Likens, 1975), more than twice the production values of most scrub ecosystems with the exception of *Ceanothus* chaparral. The *Ceanothus* chaparral also has an annual nitrogen uptake comparable to several hardwood forest ecosystems.

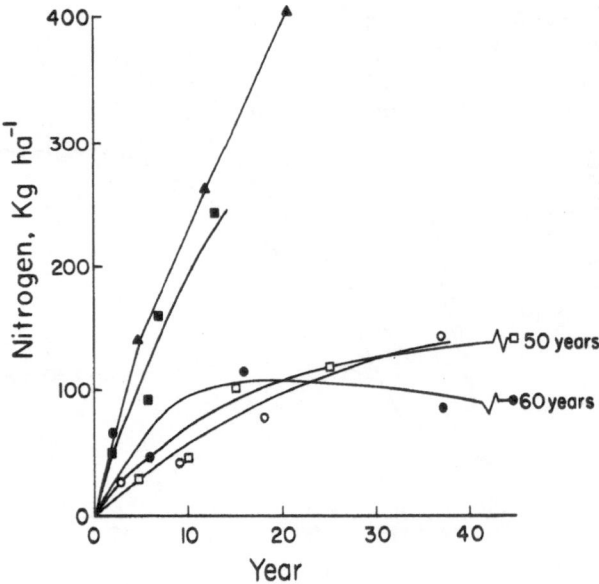

Figure 10.2. Standing nitrogen pool in biomass in various mediterranean type ecosystems, including *Ceanothus* chaparral (▲) (Schlesinger and Gill, 1980) *Quercus* garrigue (■) (Specht, 1969b), heath scrub (□) (Specht et al., 1958), mixed chamise chaparral (○) (Specht, 1969b), and montane chamise chaparral (●) (Rundel and Parsons, 1980).

10.4.4 Nutrient Return by Ashfall

Fire is important not only in causing potentially large losses of nutrients from the ecosystem by volatilization and erosion, but also in causing a drastic alteration of the intrasystem nutrient cycle. Nutrient pools in vegetation are reduced, while there is an enrichment of mineral elements and organic matter in the soil due to ashfall. In older communities where soil nutrient availability is limited, the effect of fire may be to promote productivity by making soil nutrients more available (Christensen and Muller, 1975; Groves, 1977; Vogl, 1977; Zinke, 1977).

There is scant quantitative information on the effect of fire on soil nutrients in mediterranean type ecosystems, with the exception of several studies in the southern California chaparral. Vlamis and Gowans (1961) noticed that the greater nutrient availability in burnt chaparral soils made fertilization ineffective. Christensen and Muller (1975) experimentally burned chamise foliage and branches and estimated ash deposition of 3000 kg ha^{-1}, which included 1140 kg ha^{-1} of organic matter, 21 kg ha^{-1} of nitrogen, 1 kg ha^{-1} of phosphorus, and 51 kg ha^{-1} of potassium. The ash layer adds considerable quantities of elements in both mineral form and in a readily available organic reservoir on the soil surface.

Christensen (1973) found negligible nitrate was added to the soil after fire, but large amounts of ammonium were deposited with the ash (Christensen and Muller, 1975; DeBano et al., 1979). With favorable moisture conditions, rapid nitrification by soil microorganisms increased the nitrate levels dramatically (cf., Dunn et al., 1979). Although the total soil nitrogen pool in the burned site was reduced by fire, the pool of available nitrogen was increased. In contrast, the nitrate and ammonium pools in unburnt chamise chaparral soils were very small (Christensen, 1973).

The intensity of a fire greatly affects the amount of organic matter in the litter and upper soil layers that is consumed (e.g., White et al., 1973; Viro, 1974). DeBano and Conrad (1978) studied nutrient changes in chaparral in southern California after a prescribed burn and quantified the changes in the nutrient pools in the litter and upper soil layers (Table 10.10). There was a net loss in nitrogen and phosphorus from these pools when the intensity of the fire was such that more than 4000 kg ha^{-1} of standing biomass was destroyed, but other elements showed variable increases (Table 10.10).

Table 10.10. Post-fire nutrient changes[a] in litter and upper soil pools in redshank chaparral after a prescribed burn in southern California[b]

	Organic matter	N	P	K	Ca	Mg
Litter	−4350	− 8	+9	+44	+136	+62
0-1 cm soil	− 120	−22	−6	0	− 23	0
1-2 cm soil	− 53	−15	−5	− 3	− 3	+ 5
Total	−4523	−45	−2	+41	+112	+66
Percent of pre-fire pool size	− 42%	−10%	−2%	+17%	+ 6%	+30%

[a]Measured in kg ha^{-1}.
[b]From DeBano and Conrad (1978).

Despite large nutrient losses from the ecosystem by volatilization and erosion, there is an increase in several soil nutrient pools after a fire. Mineral elements like potassium, calcium, phosphorus, and magnesium are deposited directly by ashfall. Although the soil pool of total nitrogen may be decreased, there is an immediate increase in available forms of nitrogen. The soil environment may also become more favorable for mineralization processes by microorganisms as the pH increases, as nutrients and organic matter become more plentiful, and as possible toxins in the soil are eliminated (Christensen and Muller, 1975; Dunn and DeBano, 1977; Margaris, 1977; DeBano and Conrad, 1978). The effect of this increased nutrient availability decreases with time as new growth occurs. The net loss of soil nitrogen may be replenished by atmospheric inputs and nitrogen fixation before soil nitrogen becomes limiting. Thus, fire acts as an effective, but not a conservative, recycling agent. It converts old, low-production ecosystems with great amounts of nutrients tied up in the vegetation, into young, developing ecosystems with large soil nutrient pools and potentially greater rates of nutrient movement through the system.

10.5. Conclusions Regarding Nutrient Dynamics During Ecosystem Development

Several concepts of nutrient cycling have been formulated that predict that nutrient losses from ecosystems are minimized as succession proceeds. Odum (1969) suggested that, as a community progresses to a climax condition, nutrient cycles tend to be closed. In contrast, Vitousek and Reiners (1975) proposed that intermediate successional stages show the greatest conservation of nutrients. Although these ideas are largely from conceptual models, data from experimental watersheds such as the Hubbard Brook Experimental Forest indicated that the biotic regulation of nutrient losses was extremely effective during the recovery from disturbance (Bormann et al., 1974; Likens et al., 1978).

The successional development of community structure and nutrient cycles in mediterranean type ecosystems is dominated by the role of fire. Evidence from geological, archeological, and historical records show that fire has long been an important ecological and evolutionary factor in these ecosystems (Naveh, 1974, 1977; Aschmann, 1977; Byrne et al., 1977). Fire produces a massive perturbation in which potentially large stores of nutrients, especially nitrogen, may be lost. However, fire also releases those nutrients tied up in the standing vegetation and in resistant, undecomposed litter.

Post-fire nitrogen fixation and atmospheric inputs of nutrients may be critical sources that replenish the nutrient capital of the ecosystem and contribute to the annual nutrient uptake. Recent evidence also suggests that decomposition and nutrient release in soils of mediterranean type ecosystems may be faster than originally thought, thus promoting a rapid movement of nutrients in the young, developing stands (Yeilding, 1977; Schlesinger, *unpubl. data*).

Fire acts not only to produce a radical recycling of mineral elements, but also to initiate the successional process. When fires are relatively frequent in a region, the shub communities will be in various intermediate stages of development, with concomitant high rates of nutrient accumulation (e.g., Vitousek and Reiners, 1975; Woodmansee, 1978). In most mediterranean type ecosystems, this period of high biomass and nutrient accumulation continues for the first 15 to 20 years (Figures 10.1 and 10.2). In some shrub ecosystems, like *Ceanothus* chaparral and *Quercus coccifera* garrigue, favorable climatic and soil conditions appear to promote long-term high rates of production and nutrient accumulation.

When fire is excluded for long periods of time, shrub communities like chamise chaparral become decadent (Hanes, 1971). Production decreases and dead wood, undecomposed litter, and soil toxins may accumulate (Specht, 1969b; Hanes, 1971, 1977; Parsons, 1976a; Rundel and Parsons, 1979). Nutrients may be limiting in old chamise chaparral communities (Rundel and Parsons, 1980), while in *Ceanothus* chaparral, light may be the limiting factor (McPherson and Muller, 1967; Schlesinger and Gill, 1980). Although biomass will eventually level off, it does not appear that self-perpetuating, homeostatic communities arise in these or other mediterranean type ecosystems.

11. Models of Plant and Soil Processes

MARTHA B. JACOBSON, WAYNE A. STONER, and SUSAN P. RICHARDS

This chapter describes the structure of the simulation models CAPS and MEDECS, explains how data from earlier chapters and from the literature were used in the models, and shows the validity of using the simulation models to calculate resource utilization at Echo Valley in southern California and Fundo Santa Laura in central Chile.

11.1. Introduction

Data from literature and from the field research reported in earlier chapters were synthesized into two models of primary production processes, the Canopy Process Simulator (CAPS) and the Mediterranean Ecosystem Simulator (MEDECS), in order to test the hypothesis that vegetations in regions with similar environments display similar patterns of resource utilization and similar patterns of nitrogen and carbon allocation. Both models were based on mathematical equations that reflect hypotheses concerning the way environmental processes affect primary production. Individual processes were compartmentalized into several submodels, where each submodel expressed mathematically defined hypotheses about a process that eventually affected the behavior of the simulated ecosystem. The submodel structure allowed easy modification of any hypothesis and exploration of the consequences of the modification.

Both models were developed as FORTRAN computer programs. Rather than presenting the detailed program statements, the discussion centers instead on the mathematical formulas used to express the biological concepts.

The CAPS model simulated diurnal patterns of physical and physiological processes in the vegetation canopy under steady state conditions. The canopy consisted of leaves and both live and dead stems. The unit for interactions at the individual leaf level was a square centimeter of leaf area, and irradiance was calculated per square meter of ground surface (Figure 11.1). CAPS had several submodels that simulated different processes, including: radiation transfer; air, soil surface, and soil temperatures; humidity;

transpiration; respiration; and net photosynthesis. The model was run in 5-min time steps for a 24-hour period and produced simulated diurnal patterns of each process. A shorter time step of 2 min was used to simulate a day of severe water stress during the summer, because the shorter time interval was needed to keep the simulated plant water content from oscillating and becoming unstable.

Parameters describing the physical environment during a 24-hour period were entered as input at the beginning of a CAPS simulation. Because steady state conditions were assumed for all physical and environmental processes except plant water content, the conditions of the system at any one time step did not depend on conditions of the previous time step but did depend on the time of day.

CAPS was useful in predicting steady state interactions between the environment and the plant canopy at the individual leaf level. In those areas where more research was needed, CAPS failed to predict accurately field conditions. CAPS was used to generate some of the table functions used in MEDECS and to simplify the MEDECS submodels. Validation of the model as a whole was not done because the shrub form and land surface topography hindered measurement of carbon dioxide fluxes throughout the canopy. Validation of photosynthesis and plant submodels in CAPS by field experiments was easier because fluxes of carbon dioxide and water vapor through leaves are measurable over short periods of time.

MEDECS integrated hypotheses from the organism and community levels. It simulated annual patterns of resource utilization by vegetation in response to the environment and ran in 1-day time steps for 1 year. MEDECS integrated simplified versions of some CAPS submodels with additional submodels of plant growth and death of above- and belowground plant components, decomposition, soil water, and nitrogen uptake

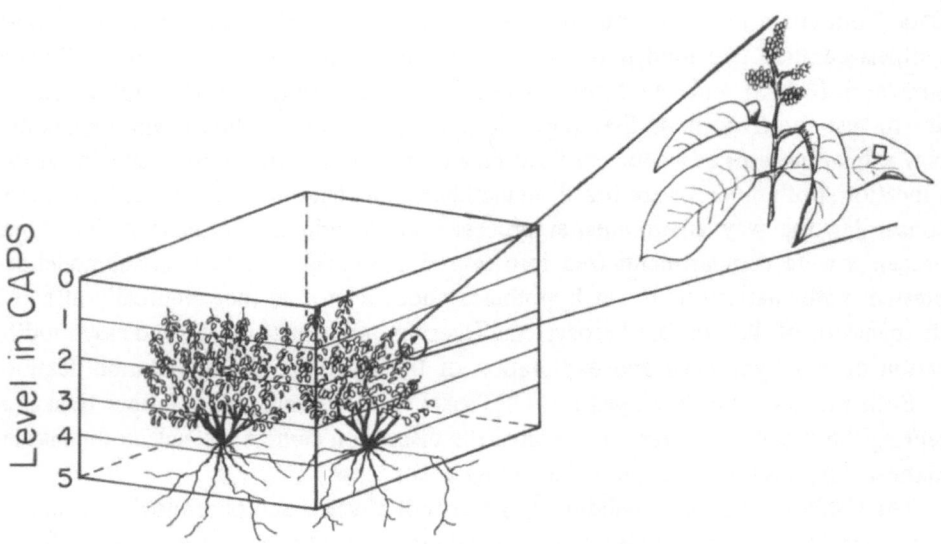

Figure 11.1. A plant and its environment in the Canopy Process Simulator (CAPS). The canopy is divided into strata 0.25 cm thick. Calculations for plant processes are done per square centimeter leaf, those for the environment are per square meter ground surface.

and utilization. The plant components used in MEDECS were old and new leaves, old and new stems, and conducting and absorbing roots. The unit of calculation was a square meter of ground surface (Figure 11.2). MEDECS was more difficult to validate than CAPS because it was a more complex model, contained more submodels, covered a longer time span, and included as many as four different plant species living together and competing for similar resources. MEDECS and CAPS contained both state and parameter variables. A state variable is one whose value in one time step affects its value in the next time step and one which defines the condition or state of the system at a given point in time. CAPS had only one state variable, plant water content, but MEDECS contained many, including leaf biomass and the amount of nitrogen in the soil. Some of the MEDECS state variables had initial values, while others were initially zero (Tables 11.1 and 11.2). Environmental variables, such as precipitation, were not state variables because the value for one day did not determine the value for the next day.

The rate of change of a state variable within a time step depended not only on the current value of the state variable but also on its associated parameter variables (Tables 11.3 to 11.5). The amount of change for a time step was added to the current state value to determine the new value of the state variable. Some parameter values for the environment, such as the soil water content at saturation and the albedo of the canopy, varied among Camp Pendleton and Echo Valley in southern California and Fundo Santa Laura in central Chile, while some were the same for all three sites (Tables 11.3 and 11.4). Some of the parameter values were regression coefficients obtained by linear or polynomial regression analysis of field data. Because the field data were incomplete for all three research sites, table functions were used to determine the remaining parameter values. For instance, from a table of the relationship of soil water

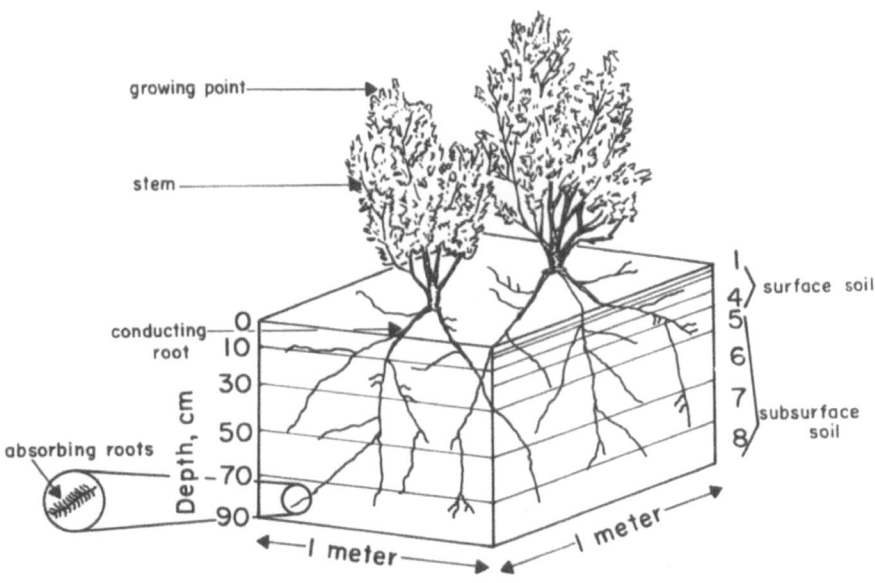

Figure 11.2. A plant and its environment in the Mediterranean Ecosystem Simulator (MEDECS). The soil is divided into eight layers of varying thickness. Calculations are done per square meter ground surface.

Table 11.1. Initial state values for the environment in a MEDECS simulation[a]

	Units	Soil layer	Camp Pendleton	Echo Valley	Fundo Santa Laura
Soil water content (θ_i)	cm³ H$_2$O cm^{-3} soil	1	0.0467	0.0467	0.0450
		2	0.0489	0.0489	0.0500
		3	0.0526	0.0526	0.0600
		4	0.0565	0.0565	0.0600
		5	0.0660	0.0660	0.0700
		6	0.1190	0.1190	0.0800
		7	0.1310	0.1310	0.1300
		8	0.1180	0.1180	0.1700
Soil organic matter (SOM$_i$)	g O.M. cm^{-3} soil	1	0.0482	0.0482	0.120
		2	0.0502	0.0502	0.090
		3	0.0474	0.0474	0.075
		4	0.0528	0.0528	0.045
		5	0.0149	0.0149	0.015
		6	0.0106	0.0106	0.010
		7	0.0103	0.0103	0.006
		8	0.0103	0.0103	0.003
Percent of soil organic matter that is nitrogen (SOMN$_i$)	percent	1	25	12.9	12.9
		2	25	12.9	12.9
		3	25	12.9	12.9
		4	25	12.9	12.9
		5	12.9	8.4	12.9
		6	0.9	4.0	4.0
		7	0.9	2.5	0.4
		8	0.9	0.618	0.4

Soil soluble (available) nitrogen (SSN_i)	g N cm^{-3} soil		
1	0.0000279	0.0000279	0.0000375
2	0.0000240	0.0000240	0.0000330
3	0.0000252	0.0000252	0.0000270
4	0.0000276	0.0000276	0.0000180
5	0.00000577	0.00000577	0.00001200
6	0.00000701	0.00000701	0.00000900
7	0.00000729	0.00000729	0.00000600
8	0.00000766	0.00000766	0.00000450

[a] Terms in parentheses are used in equations in other tables.

Table 11.2. Initial state values for species in a MEDECS simulation[a]

Variable	Units	Soil layer	Adenostoma fasciculatum	Arctostaphylos glauca	Ceanothus greggii	Rhus ovata	Colliguaya odorifera	Satureja gilliesii	Trevoa trinervis	Lithraea caustica
Conducting root biomass, total	g root stem^{-1}		39.9	52.0	86.9	49.8	16.9	8.35	24.2	64.8
Carbon in roots	g C stem^{-1}		35.8	46.8	78.2	44.8	15.2	7.51	21.8	58.3
Nitrogen in roots	g N stem^{-1}		0.0200	0.0260	0.0435	0.0249	0.0840	0.00417	0.0121	0.0324
Total nonstructural carbohydrate in roots	g TNC stem^{-1}		3.99	5.20	8.69	4.98	1.69	0.835	2.42	6.48
Live absorbing roots	g root layer^{-1}	5	0	0	0	0	0	10	0	0
		6	4.9	6.0	5.0	5.0	10.0	0	0	10.0
		7	1.0	6.0	5.0	5.0	0	0	0	5.0
		8	1.0	0	0	0	0	0	0	5.0
Dead roots	g dead root layer^{-1}	1	1.50	4.24	3.50	0.30	4.40	3.00	3.0	1.00
		2	5.35	8.00	4.90	0.90	6.60	10.50	10.5	2.00
		3	9.85	20.90	10.50	2.00	11.00	18.80	18.8	5.00
		4	26.60	33.00	18.90	3.60	22.00	34.00	33.0	15.00
		5	17.70	33.10	17.90	9.40	28.60	9.00	9.75	16.00
		6	9.70	0.95	10.50	6.00	28.60	0	0	21.00
		7	2.50	0.05	2.10	6.00	0	0	0	15.00
		8	0	0	0	0	0	0	0	14.90
Stem biomass	g stem stem^{-1}		71.8	118.0	140.0	82.9	31.8	32.7	33.5	28.6
Carbon in stems	g C stem^{-1}		64.6	106.0	126.0	74.6	28.6	29.4	30.1	25.7
Nitrogen reserves in stems	g N stem^{-1}		0.0359	0.0640	0.0701	0.0415	0.0159	0.0163	0.0167	0.0143

Species

Total nonstructural carbohydrate reserves	g TNC stem^{-1}	7.18	11.8	14.0	8.29	3.18	3.27	3.35	2.86
Old leaf biomass	g leaf stem^{-1}	14.5	38.4	28.9	23.0	7.0	5.7	0	14.7
Stem area index (SAI)	m^2 stem m^{-2} ground stem^{-1}	0.032	0.090	0.0929	0.0844	0.0350	0	0.066	0.05
Labile nitrogen in old leaves (NLABO)	g N stem^{-1}	0.03620	0.07680	0.11600	0.05180	0.00854	0	0	0.02820
Stable nitrogen in old leaves (NSTABO)	g N stem^{-1}	0.10900	0.19200	0.23100	0.15500	0.06850	0	0	0.08940

[a]Terms in parentheses are used in equations in other tables.

Table 11.3. Environmental parameter values used in MEDECS simulations that differ among study sites[a]

Parameter	Units	Soil level	Camp Pendleton	Echo Valley	Fundo Santa Laura
Extinction coefficient for effect of	FAI^{-1}				
Extinction coefficient for effect of foliage on solar irradiance (K)	FAI^{-1}		0.371	0.371	0.400
Extinction coefficient for effect of soil depth on infiltration of precipitation (C1)	cm^{-1}		−0.05	−0.0101	−0.01
Extinction coefficient for effect of soil depth on evaporation from soil (C2)	cm^{-1}		−0.0969	−0.0969	−0.1000
Coefficient of cloudiness (COEFC)	fraction		0.2	0.923	0.9
Emissivity of ground (EGRD)	fraction		1.02	1.02	0.97
Emmisivity of leaf	fraction		1.03	1.03	0.97
Average canopy albedo (ALBEDO)	fraction		0.136	0.136	0.150
Albedo of soil surface (ALBEDOG)	fraction		0.208	0.208	0.200
Number of times the water storage capacity of foliage is filled per rainy day (NFILL)	day^{-1}		0.956	0.956	1.0
Diffusion coefficient of nitrogen through soil	cm^2 day^{-1}		0.00101	0.00101	0.00060
Reduction of decay rate due to soil lignin [f(PLIGNIN)]	fraction		0.00182	0.00182	0.00400

Maximum water-holding capacity of soil (θ_{max}) — cm^3 H$_2$O cm^{-3} soil

1	0.450	0.400	0.400
2	0.450	0.435	0.400
3	0.450	0.370	0.400
4	0.450	0.388	0.400
5	0.450	0.439	0.400
6	0.450	0.421	0.400
7	0.450	0.409	0.400
8	0.450	0.431	0.400

Field capacity of soil — cm^3 H$_2$O cm^{-3} soil

1	0.400	0.262	0.300
2	0.400	0.250	0.300
3	0.400	0.239	0.300
4	0.400	0.239	0.300
5	0.400	0.230	0.300
6	0.400	0.229	0.300
7	0.400	0.255	0.300
8	0.400	0.266	0.300

[a] Terms in parentheses are used in equations in other tables.

Table 11.4. Environmental parameter values used in MEDECS simulations that are the same for all study sites[a]

Parameter		Units	Value
Altitude		m	1000
Thickness of soil	1	m	0.02
level (DZ_i)	2		0.03
	3		0.05
	4		0.10
	5		0.10
	6		0.20
	7		0.20
	8		0.20
Water storage on foliage due to rain (SC)		mm H_2O cm^{-2} foliage	0.2
Latent heat of evaporation (L)		J cm^{-3} H_2O	2448
Volumetric heat capacity of air (ρCP)		J cm^{-3} air °C^{-1}	0.00178
Stefan-Boltzmann constant (σ)		J cm^{-2} s^{-1} °K^{-4}	5.67×10^{-8}
Psychrometer constant (γ)		MPa	0.0067
Factor to calculate slope of vapor pressure vs temperature curve		MPa °C^{-1}	17.4
Factor to calculate vapor pressure at saturation		MPa	0.000611
Days needed to remove all available nitrogen from soil (TAU)		day	1
Maximum decay rate of soil organic matter (RDKMAX)		g decay g^{-1} organic matter day^{-1}	0.045
Ratio of CO_2 taken up to TNC produced in photosynthesis (CO2TNC)		g CO_2 g^{-1} TNC	0.64
Soil water content at wilting point		cm^3 H_2O cm^{-3} soil	0.055
Minutes in a day		min day^{-1}	1440
Seconds in a day		s day^{-1}	86400

[a]Terms in parentheses are used in equations in other tables.

potential to selected soil water contents, the soil water potential for any given soil water content could be found by interpolation (Tables 11.6 to 11.8). For some table functions only a few points were known or could be reliably estimated, so the other points were projected from a graph of the known points (Figures 11.3 to 11.6). Some table functions were constructed from literature values for agricultural species if data on chaparral or matorral species did not exist. Unless specifically stated, CAPS and MEDECS used the same table functions for all species at the three research sites.

Table 11.5. Species parameters for MEDECS and CAPS simulators. Note: Terms in parentheses are used in equations in other tables.

Parameter	Units	Adenostoma fasciculatum	Arctostaphylos glauca	Ceanothus greggii	Rhus ovata	Colliguaya odorifera	Satureja gilliesii	Trevoa trinervis	Lithraea caustica
MEDECS									
Maximum potential leaf growth rate per growing shoot (GMAX)	g dry wt. leaf growing shoot^{-1} day^{-1}	0.0110	0.02	0.0169	0.022	0.02	0.02	0.02	0.10
Maximum potential stem growth rate per growing stem (STMAX)	g dry wt. stem growing shoot^{-1} day^{-1}	0.0011	0.002	0.0017	0.0022	0.01	0.005	0.02	0.08
Number of growing shoots per stem (SHTPERST)	growing shoot stem^{-1}	190 (PFS)[a] 1036 (EFS)[b]	100	118	90	100	100	100	80
Number of stems per plant (STPERPL)	stem plant^{-1}	3.4 (PFS)[a] 10 (EFS)[b]	2	6	16	16	16	16	16
Leaf width (WIDTH)	cm leaf	0.1	2	1	2	1	0.3	1	4
Grams leaf per m^2 leaf surface (GTA)	g leaf m^{-2} leaf	417	270	285	250	179	122	91	260
Fraction of new leaf growth going to labile N (C1N)	fraction	0.60	0.50	0.60	0.60	0.30	0.5	0.5	0.5
Fraction of new leaf growth going to stable N (C2N)	fraction	0.40	0.50	0.40	0.40	0.70	0.5	0.5	0.5
Fraction of available N reserves resynthesized into labile N (C3N)	fraction	0.07	0.10	0.05	0.05	0.12	0.05	0.07	0.07
Fraction of new stem growth going to labile N	fraction	0.50	0.50	0.50	0.50	0.50	0.50	0.50	0.60

Table 11.5. (Continued)

Parameter	Units	Species							
		Adenostoma fasciculatum	Arctostaphylos glauca	Ceanothus greggii	Rhus ovata	Colliguaya odorifera	Satureja gilliesii	Trevoa trinervis	Lithraea caustica
Fraction of new stem growth going to stable N	fraction	0.50	0.50	0.50	0.50	0.50	0.50	0.50	0.40
Fraction of available N reserves resynthesized into labile N	fraction	0.05	0.05	0.05	0.07	0.05	0.05	0.05	0.10
Grams N required for 1 gram dry wt. new leaf growth (GRAMN)	g N g^{-1} dry wt. new leaf	0.200	0.020	0.033	0.030	0.017	0.025	0.068	0.018
Grams N required for 1 gram dry wt. stem growth	g N g^{-1} dry wt. stem growth	0.010	0.010	0.010	0.010	0.021	0.015	0.030	0.012
Grams N required for 1 gram dry wt. root growth	g N g^{-1} dry wt. stem	0.020	0.020	0.030	0.020	0.002	0.002	0.002	0.002
Distribution of conducting	fraction 1	0.086	0.065	0.070	0.086	0.050	0.040	0.012	0.0115
	2	0.129	0.098	0.110	0.129	0.075	0.140	0.017	0.0172
	3	0.215	0.163	0.180	0.215	0.125	0.250	0.029	0.0286
	4	0.430	0.325	0.360	0.430	0.250	0.410	0.057	0.0572
	5	0.060	0.155	0.140	0.060	0.245	0.130	0.179	0.179
	6	0.070	0.175	0.090	0.070	0.255	0	0.443	0.443
	7	0.010	0.020	0.050	0.010	0	0	0.264	0.264
Number of absorbing roots per length of conducting root (GROPT)	cm^{-1} conducting root	1	2	0.9	1	2	2	2	1.6

Description	Units								
Length of conducting root per gram (CMSGR)	cm g^{-1} conducting root	281	239	239	563	100	100	100	100
Correction factor for relative growth of absorbing roots	fraction	2.0	1.5	1.2	2.0	15.0	1.0	6.0	5.0
Maximum possible fraction of leaves dying due to low light	day^{-1}	0.1	0.1	0.1	0.1	0.1	0.0	0.0	0.1
Initial estimate of fraction of leaves dying due to low light	day^{-1}	0.10	0.10	0.13	0.10	0.13	0.13	0.13	0.13
Factor to multiply by solar irradiance at bottom of canopy; result is added to initial estimate of leaf death	m^2 ground MJ^{-1}	-0.001	-0.0048	-0.0033	-0.0027	-0.0033	-0.0015	-0.0015	-0.0033
Minimum light required to prevent leaf death	MJ m^{-2} ground day^{-1}	4.10	0.88	1.67	1.55	1.67	2.09	3.51	1.67
Degree hours above 15° (10° for Chile) needed to initiate growth	°h	860	1050	860	800	500	600	700	1880
Number of days between stem initiation and leaf growth initiation	day	0	0	7	6	0	0	0	0
Number of days of water stress endured before leaf and stem growth stop	day	1000	1000	1000	1000	1000	1000	1000	1000
Fraction of nitrogen removed from roots before death	fraction	0.5	0.5	0.5	0.5	0.5	0.5	0.5	0.5
Growth respiration for leaves	g TNC g^{-1} tissue	1.7	1.7	1.7	1.7	1.7	1.7	1.7	1.7

Table 11.5. (Continued)

Parameter	Units		Species							
			Adenostoma fasciculatum	Arctostaphylos glauca	Ceanothus greggii	Rhus ovata	Colliguaya odorifera	Satureja gilliesii	Trevoa trinervis	Lithraea caustica
Last possible date for stem growth initiation	day		9 Apr	19 Apr	9 Apr	18 Jul	10 Jan	10 Jan	1 Jan	15 Dec
Maximum root elongation of a single fine root (RELONG)	cm root day^{-1}		1.0	0.27	0.05	0.35	0.34	0.34	0.34	0.14
Grams fine roots per cm fine root growth (GR\$CM)	g dry wt. root cm^{-1} root		0.000200	0.000224	0.000167	0.000200	0.000180	0.000130	0.000100	0.000200
Number of potential fine roots in level i (RUT$_i$)	root stem^{-1}									
Level #		Depth (cm)								
1		0.02	73	281	335	169	315	80	480	53
2		0.05	73	365	503	253	615	280	1680	79
3		0.10	219	844	1074	422	915	500	3000	288
4		0.20	364	1618	1933	845	2070	880	5280	480
5		0.30	515	1869	2148	845	1800	60	360	509
6		0.50	5	358	1074	135	150	0	0	365
7		0.70	5	125	215	135	0	0	0	281
8		0.90	0	0	0	0	0	0	0	0
Coefficients relating soil water potential in rooting zone to plant water potential										
CΨ1	MPa		-0.917	-0.712	-0.753	-0.313	-0.942	0.219	-0.192	-1.474
CΨ2	fraction		0.567	-2.180	-1.080	1.540	5.53	6.52	4.281	3.040

Parameter	Units								
CΨ3	MPa^{-1}	1.551	1.855	2.810	4.936	0.050	-6.400	-6.800	-2.990
CΨ4	MPa^{-2}	0.402	0.300	0.498	1.850	0	-0.700	1.400	0.010
Maximum leaf conductance of water (RC1)	cm s^{-1}	0.440	0.76	0.81	0.65	0.280	0.240	0.480	0.31
Slope of leaf conductance vs. ΨPLANT curve (RC2)	cm s^{-1} MPa^{-1}	0.110	0.250	0.193	0.163	0.122	0.037	0.080	0.056
Minimum possible leaf stomatal conductance of water during daylight period (HLMIN)	cm s^{-1}	0.010	0.02	0.02	0.02	0.0083	0.0083	0.0083	0.0083
CAPS			$\dfrac{\text{leaf}}{75}\ \dfrac{\text{stem}}{200}$						
Light compensation point for photosynthesis (LC)	μE cm^{-2} ground day^{-1}	50	50	50	50	33	50	19	88
Root resistance to water uptake (RR)	s cm^{-1}	7	5	5	7	15	2	2	2
Leaf width (differs from MEDECS) (WIDTH)	cm leaf	3	$\dfrac{\text{leaf}}{1}\ \dfrac{\text{stem}}{0.1}$	1	1	4	1	2	1
Water content per leaf area at turgidity (WT)	g H$_2$O m^{-2} leaf	229	142	277	264	292	412	241	251

[a] PFS is pole-facing slope.
[b] EFS is equator-facing slope.

Table 11.6. Table functions used in MEDECS for soil moisture content versus soil water potential

Camp Pendleton and Echo Valley		Fundo Santa Laura	
Water content (cm^3 H$_2$O cm^{-3} soil)	Ψsoil (MPa)	Water content (cm^3 H$_2$O cm^{-3} soil)	Ψsoil (MPa)
0.0092	−9.68	0.010	−9.60
0.0184	−7.62	0.018	−8.30
0.0277	−7.38	0.027	−6.80
0.0369	−5.10	0.036	−5.50
0.0461	−4.49	0.046	−4.10
0.0553	−2.81	0.055	−2.80
0.0646	−1.69	0.062	−1.56
0.0740	−0.948	0.065	−1.00
0.110	−0.316	0.078	−0.30
0.152	−0.0979	0.092	−0.10
0.235	−0.0504	0.164	−0.040
0.305	−0.0312	0.200	−0.034
		0.220	−0.031
		0.240	−0.027
		0.260	−0.024
		0.280	−0.021
		0.300	−0.018
		0.400	−0.001

11.2. Canopy Process Simulator

The Canopy Process Simulator modeled the energy, water, and carbon dioxide balance of a continuous canopy. The climate above the canopy was defined from input data consisting of measurements of a 24-hour course of incident direct and diffuse solar and infrared irradiances, temperature and humidity of the air, and velocity of the wind. The canopy itself was divided into a series of strata each 0.25 m thick, which were numbered from just above the canopy to immediately above the ground surface. Levels were planes that separated the strata; they were numbered so that a level was the lower boundary of the associated stratum (Figure 11.1). The distance between levels was small enough so that the microclimate calculated using discrete strata was close to the microclimate that would be calculated using differential calculus methods in a continuously varying canopy. The plant properties of each stratum were determined by the area indices of leaves, stems, and dead material within the stratum. Area indices were calculated as the square meter or leaf area per square meter of ground surface. Leaves and stems within each stratum were also defined by their inclinations, absorptances, widths, and table functions for water and photosynthetic relationships. The species modeled were *Adenostoma fasciculatum*, *Arctostaphylos glauca*, *Ceanothus greggii*, and *Rhus ovata* at Echo Valley and *Colliguaya odorifera*, *Satureja gilliesii*, *Trevoa trinervis*, and *Lithraea caustica* at Fundo Santa Laura.

Solar and infrared irradiances interacted with the canopy structure and microclimate to produce profiles within the canopy of direct and diffuse solar and infrared irradiances. The air temperature, humidity, and wind velocity above the canopy and the temperature and humidity at the soil surface then interacted with the canopy structure and irradiance profiles to produce canopy profiles of air and leaf temperature, humidity, and wind velocity. For both sunlit and shaded leaves or photosynthetic stems, the model calculated profiles of solar and infrared radiation, wind, air and leaf temperature, vapor density, transpiration, and photosynthesis at each level in the canopy.

The fluxes of energy at any level in a vegetation canopy were summarized in the energy budget equation for leaves and stems (Table 11.9:Eq. 1). The soil surface was defined as the level below which conduction was the main mechanism of heat transport and above which irradiance and turbulent exchange were the main mechanisms of heat transport. Absorbed solar and infrared irradiance was either dissipated by convection, transpiration, and infrared irradiance or was retained, thus increasing plant temperature. Leaf and stem temperatures were calculated from the energy budget equation.

The rates of canopy photosynthesis were based on rates of individual photosynthetic organs, which were influenced by a series of feedback reactions. Solar and infrared irradiances, air temperature, humidity, and wind velocity affected leaf temperature and transpiration which, in turn, affected the plant water status. Because of convectional heat exchange, leaf temperatures affected air temperatures. Soil moisture and soil tempeature controlled water uptake, which then affected the plant water status. Because the movement of carbon dioxide through the leaf is affected by the same factors that control water movement, the plant water status influenced not only respiration and leaf temperature but also photosynthesis.

11.2.1 Development

CAPS incorporated new ideas into a synthesis of already existing models, which became submodels within the larger CAPS model. The first version of CAPS was developed for the mangrove vegetation of southern Florida (Miller, 197a, 1975), and later versions were then applied to arctic and alpine tundra (Miller and Tieszen, 1972; Tieszen et al., *in press*; Miller et al., 1976a; Stoner et al., 1978a) and to mediterranean shrub vegetation (Lawrence, 1975; Miller and Mooney, 1976; Miller and Stoner, 1979). Earlier versions of the CAPS submodels were used to explore ecosystem function and environment-plant interactions in the arctic tundra (Miller et al., 1976a; Stoner et al., 1978b, c) and the effects of thermal addition via power plant effluents on a mangrove ecosystem (Miller et al., 1976b).

Models of canopy photosynthesis are often based on the photosynthetic relations of a single leaf within the canopy. Monsi and Saeki (1953) and Davidson and Philip (1958) used a simplified model to calculate stand photosynthesis. De Wit's (1965) model used leaf inclination and direct and diffuse solar irradiances at various canopy heights to calculate the annual course of potential production for crops in different agricultural regions. The photosynthesis submodel of CAPS is based on refinements made to earlier models by Monteith (1965), Anderson (1966), Duncan et al. (1967), and Lemon (1967).

Table 11.7. Table function used in MEDECS for alpha factor for transpiration

Leaf area index (m² leaf m⁻² ground)	Solar irradiance (MJ m⁻² day⁻¹)	Alpha							
		Adenostoma fasciculatum	*Arctostaphylos glauca*	*Ceanothus greggii*	*Rhus ovata*	*Colliguaya odorifera*	*Satureja gilliesii*	*Trevoa trinervis*	*Lithraea caustica*
0.2	3.1	0.880	0.840	0.900	0.890	0.840	0.840	0.840	0.840
	4.2	0.900	0.870	0.920	0.910	0.870	0.870	0.870	0.870
	8.4	0.940	0.920	0.950	0.940	0.920	0.930	0.920	0.920
	12.6	0.950	0.940	0.960	0.950	0.940	0.940	0.940	0.940
	16.7	0.970	0.950	0.970	0.970	0.950	0.950	0.950	0.950
	20.9	0.975	0.960	0.975	0.975	0.960	0.960	0.960	0.960
	25.1	0.980	0.970	0.980	0.980	0.980	0.970	0.970	0.970
	28.5	0.980	0.980	0.990	0.980	0.980	0.980	0.980	0.980
0.5	3.1	0.860	0.840	0.900	0.890	0.840	0.840	0.840	0.840
	4.2	0.890	0.860	0.910	0.900	0.860	0.860	0.860	0.860
	8.4	0.930	0.910	0.940	0.940	0.910	0.910	0.910	0.910
	12.6	0.950	0.930	0.950	0.950	0.930	0.930	0.930	0.930
	16.7	0.960	0.950	0.970	0.960	0.950	0.950	0.950	0.950
	20.9	0.965	0.955	0.975	0.965	0.955	0.955	0.955	0.955
	25.1	0.970	0.960	0.980	0.970	0.960	0.960	0.960	0.960
	28.5	0.980	0.970	0.980	0.980	0.970	0.970	0.970	0.970
2.0	3.1	0.790	0.730	0.860	0.840	0.730	0.730	0.760	0.730
	4.2	0.830	0.780	0.880	0.860	0.780	0.780	0.780	0.780
	8.4	0.890	0.850	0.900	0.890	0.850	0.850	0.850	0.850
	12.6	0.910	0.890	0.920	0.920	0.890	0.890	0.890	0.890
	16.7	0.930	0.910	0.940	0.930	0.910	0.910	0.910	0.910
	20.9	0.935	0.915	0.945	0.940	0.915	0.915	0.915	0.915

3.8	25.1	0.940	0.920	0.950	0.950	0.920	0.920	0.970	0.970
	28.5	0.960	0.940	0.970	0.960	0.940	0.940	0.940	0.940
	3.1	0.690	0.630	0.780	0.760	0.630	0.630	0.630	0.630
	4.2	0.750	0.690	0.800	0.780	0.690	0.690	0.690	0.690
	8.4	0.830	0.790	0.850	0.840	0.790	0.790	0.790	0.790
	12.6	0.870	0.830	0.880	0.870	0.830	0.830	0.830	0.830
	16.7	0.890	0.860	0.910	0.900	0.860	0.860	0.860	0.860
	20.9	0.900	0.870	0.915	0.910	0.870	0.870	0.870	0.870
	25.1	0.910	0.880	0.920	0.920	0.880	0.880	0.880	0.880
	28.5	0.930	0.910	0.940	0.940	0.910	0.910	0.910	0.910
7.7	3.1	0.550	0.500	0.590	0.540	0.500	0.500	0.500	0.500
	4.2	0.620	0.540	0.660	0.640	0.570	0.570	0.570	0.570
	8.4	0.730	0.690	0.760	0.750	0.690	0.690	0.690	0.690
	12.6	0.790	0.750	0.810	0.800	0.750	0.750	0.750	0.750
	16.7	0.830	0.780	0.850	0.830	0.780	0.780	0.780	0.780
	20.9	0.840	0.800	0.860	0.845	0.800	0.800	0.800	0.800
	25.1	0.850	0.810	0.870	0.860	0.810	0.810	0.810	0.810
	28.5	0.870	0.840	0.890	0.880	0.840	0.840	0.870	0.840

Table 11.8. Table function used in MEDECS for photosynthesis[a]

			Photosynthesis			
Solar	Temp	LAI	Adenostoma fasciculatum	Arctostaphylos glauca	Ceanothus greggii	Rhus ovata
1.15	9	2	8.80	5.25	4.45	2.65
		4	5.70	4.17	2.88	2.51
		8	3.25	3.03	1.85	2.37
		10	1.80	1.97	1.08	2.02
	14	2	8.90	5.90	4.43	1.89
		4	5.70	4.38	2.86	1.82
		8	3.23	3.12	1.85	1.72
		10	1.80	2.01	1.08	1.51
	19	2	8.60	6.04	4.33	1.30
		4	5.60	2.19	2.82	1.25
		8	3.16	3.12	1.83	1.19
		10	1.70	2.01	1.06	1.07
	29	2	7.80	5.84	4.17	0.78
		4	5.20	4.25	2.74	0.75
		8	2.97	3.05	1.80	0.72
		10	1.60	1.97	1.04	0.65
1.94	9	2	11.00	6.28	6.16	3.29
		4	7.70	5.15	4.01	3.10
		8	4.99	3.99	2.73	2.86
		10	2.80	2.76	1.65	2.44
	14	2	11.70	7.09	6.00	2.44
		4	7.80	5.51	3.98	2.32
		8	5.00	4.14	2.72	2.17

			Photosynthesis			
Solar	Temp	LAI	Colliguaya odorifera	Satureja gilliesii	Trevoa trinervis	Lithraea caustica
0.42	9	2	6.90	4.10	6.64	4.80
		4	4.30	2.62	4.18	6.40
		8	2.60	1.64	2.01	3.60
		10	1.40	0.90	1.60	3.60
	14	2	7.00	4.34	6.61	2.50
		4	4.30	2.66	4.13	2.60
		8	2.60	1.65	2.59	2.60
		10	1.40	0.91	4.58	2.50
	19	2	6.90	4.28	6.38	1.70
		4	4.30	2.63	4.03	1.20
		8	2.50	1.64	2.53	1.70
		10	1.40	0.90	1.54	1.60
	29	2	6.40	4.08	5.77	0.90
		4	4.10	2.54	3.73	0.90
		8	2.40	1.60	2.37	0.90
		10	1.30	0.88	1.45	0.20
1.56	9	2	9.70	4.92	8.55	5.60
		4	5.90	3.45	6.02	7.20
		8	3.90	2.39	4.52	4.80
		10	2.30	1.46	2.96	4.40
	14	2	9.90	5.46	8.48	3.20
		4	6.00	3.55	5.97	3.40
		8	3.90	2.43	4.44	3.30

Left panel:

		n				
2.36	19	10	2.80	2.82	1.65	1.87
		2	11.30	7.39	5.62	1.78
		4	7.50	5.56	3.85	1.71
		8	4.88	4.15	2.65	1.62
		10	2.70	2.82	1.62	1.42
	29	2	10.20	7.09	5.32	1.13
		4	6.90	5.37	3.70	1.09
		8	4.53	4.04	2.57	1.03
		10	2.60	2.75	1.57	0.92
	9	2	11.70	6.88	6.83	3.57
		4	8.70	5.78	4.59	3.30
		8	6.03	4.59	3.21	3.05
		10	3.50	3.26	2.01	2.58
	14	2	12.80	7.77	6.41	2.86
		4	8.80	6.18	4.47	2.53
		8	6.00	4.76	3.16	2.37
		10	3.50	3.33	1.98	2.02
	19	2	12.30	8.14	6.08	2.02
		4	8.60	6.23	4.33	1.92
		8	5.86	4.77	3.09	1.81
		10	3.40	3.33	1.95	1.57
	29	2	11.00	7.78	5.73	1.33
		4	7.80	5.99	4.13	1.27
		8	5.39	4.62	2.97	1.20
		10	3.10	3.24	1.88	1.06
2.85	9	2	13.60	7.88	8.21	4.17
		4	11.00	7.09	5.78	3.86

Right panel:

		n				
	19	10	2.30	1.48	2.90	3.10
		2	9.60	5.32	8.04	2.40
		4	5.80	3.50	5.74	2.40
		8	3.90	2.40	4.29	2.40
		10	2.30	1.46	2.81	2.30
	29	2	8.70	5.06	7.15	1.50
		4	5.40	3.35	5.21	1.50
		8	3.70	2.32	3.95	1.50
		10	2.20	1.42	2.59	0.90
2.34	9	2	13.60	6.49	11.29	6.80
		4	9.10	5.02	8.78	8.40
		8	6.50	3.70	6.97	6.40
		10	4.00	2.36	4.66	5.60
	14	2	14.30	7.22	11.13	4.70
		4	9.30	5.20	8.60	3.60
		8	6.60	3.76	6.75	4.80
		10	4.10	2.37	4.52	3.40
	19	2	13.60	7.10	10.53	3.60
		4	9.00	5.12	8.21	3.70
		8	6.50	3.71	6.47	3.60
		10	4.00	2.34	4.33	3.50
	29	2	12.10	6.72	9.35	2.40
		4	8.30	4.88	7.37	2.40
		8	6.10	3.56	5.85	2.30
		10	3.80	2.26	3.83	1.90
3.13	9	2	15.80	7.31	12.85	8.00
		4	11.40	6.01	10.59	8.80

Table 11.8. (Continued)

			Photosynthesis							Photosynthesis			
Solar	Temp	LAI	Adenostoma fasciculatum	Arctostaphylos glauca	Ceanothus greggii	Rhus ovata	Solar	Temp	LAI	Colliguaya odorifera	Satureja gilliesii	Trevoa trinervis	Lithraea caustica
		8	7.91	5.75	4.13	3.57			8	8.10	4.62	8.63	7.20
		10	4.70	4.10	2.60	3.03			10	5.10	3.06	5.98	6.80
	14	2	15.60	9.09	7.65	3.29		14	2	17.10	8.25	12.82	6.50
		4	11.40	7.66	5.60	3.11			4	11.60	6.29	10.39	5.90
		8	8.00	6.01	4.05	2.91			8	8.40	4.71	8.42	5.80
		10	4.70	4.22	2.57	2.49			10	5.20	3.08	5.78	5.40
	19	2	15.20	9.54	7.29	2.53		19	2	16.20	8.34	12.24	4.50
		4	11.20	7.76	5.45	2.41			4	11.20	6.22	9.95	4.60
		8	7.85	6.04	3.97	2.27			8	8.20	4.65	8.06	4.50
		10	4.60	4.22	2.52	1.98			10	5.10	3.04	5.53	4.30
	29	2	13.70	9.19	6.82	1.68		29	2	14.40	7.86	10.83	3.00
		4	10.20	7.46	5.18	1.61			4	10.20	5.92	8.90	3.00
		8	7.22	5.84	3.81	1.52			8	7.60	4.45	7.16	2.90
		10	4.30	4.10	2.44	1.34			10	4.90	2.88	4.27	2.60

[a]Solar irradiance is in MJ m^{-2} day^{-1}, air temperature in °C, leaf area index (LAI) in m^2 leaf per m^2 ground surface, and photosynthesis in g CO_2 per m^2 leaf.

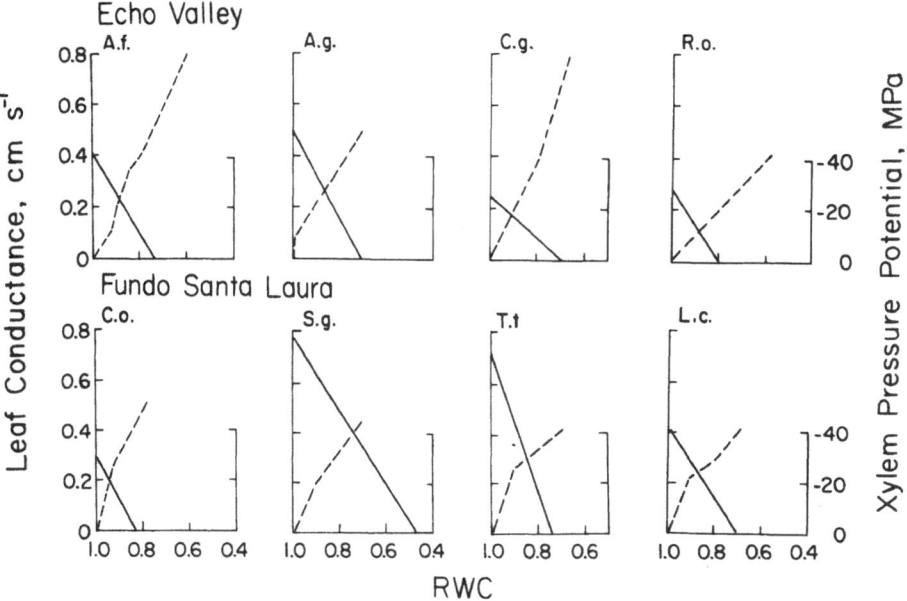

Figure 11.3. Leaf conductance (–) and xylem pressure potential (- - -) versus relative water content (RWC) for *Adenostoma fasciculatum* (A.f.), *Arctostaphylos glauca* (A.g.), *Ceanothus greggii* (C.g.), and *Rhus ovata* (R.o.) at Echo Valley; and *Colliguaya odorifera* (C.o.), *Satureja gilliesii* (S.g.), *Trevoa trinervis* (T.t.), and *Lithraea caustica* (L.c.) at Fundo Santa Laura.

Irradiance models developed from concepts of light penetration in continuous vegetation canopies were used for temperate zone crops (Monsi and Saeki, 1953; Davidson and Philip, 1958; de Wit, 1965; Anderson and Denmead, 1969; Yim et al., 1969; Acock et al., 1970; Duncan et al., 1967; Idso and de Wit, 1970). Models of irradiance penetration in noncontinuous vegetation were developed in agricultural research to investigate the effects of row spacing on the light environment of the crops (Allen, 1974; Fukai and Loomis, 1976). These models were synthesized into the CAPS irradiance submodel. CAPS included infrared irradiance exchange within the canopy, a component not found in previous models. The model also included the effects of stems on irradiance distribution, a factor only rarely considered earlier.

Hypotheses concerning the transfer of heat, water vapor, and carbon dioxide within crop canopies and natural vegetation have been expressed mathematically (Denmead, 1964; Waggoner and Reifsnyder, 1968; Waggoner et al., 1969; Murphy and Knoerr, 1970, 1972; Stewart and Lemon, 1972). The algorithms in CAPS that calculated the steady state profiles of air temperature and humidity from profiles of net irradiance an turbulent transfer throughout the canopy were basically those proposed by Waggoner and Reifsnyder (1968) and Waggoner et al. (1969). In CAPS, this canopy exchange submodel was linked to the soil surface temperature and energy budget submodel. Wind and energy exchange due to air turbulence were not linked, following Monteith (1973).

Models of plant water relations were proposed by Honert (1948) and Rawlins (1963) and were described in nonmathematical terms by Jarvis and Jarvis (1963). Re-

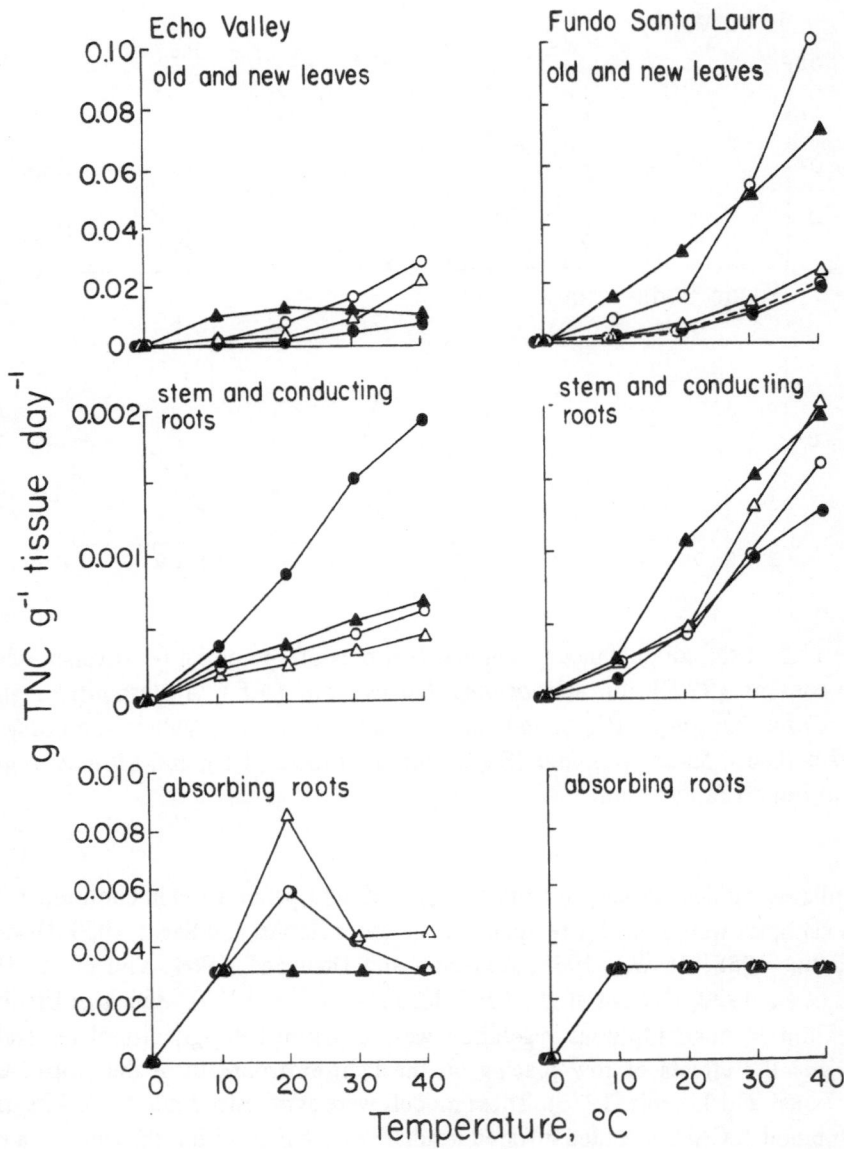

Figure 11.4. Maintenance respiration as a function of temperature for old and new leaves, stems and conducting roots, and absorbing roots for *Arctostaphylos glauca* (●), *Adenostoma fasciculatum* (▲), *Ceanothus greggii* (○), and *Rhus ovata* (△) at Echo Valley and *Colliguaya odorifera* (●), *Satureja gilliesii* (▲), *Trevoa trinervis* (−○−), *Trevoa trinervis* photosynthetic stems (---○---), and *Lithraea caustica* (△) at Fundo Santa Laura.

finements in concepts led to the formulation of more complex models (Cowan, 1965; Phillip, 1966). These models, drawn as electrical circuit analogues, attempted to explain the regulation by plant organs and the environment of water movement from the soil through the plant into the atmosphere. The CAPS plant water relations submodel differed from these previous models because it dealt with nonsteady state plant water conditions. It was used in studies of mangrove (Miller, 1975), alpine tundra (Ehleringer and Miller, 1975a, b), arctic tundra (Stoner and Miller, 1975), and chaparral (Miller an Mooney, 1976; Miller and Poole, 1979).

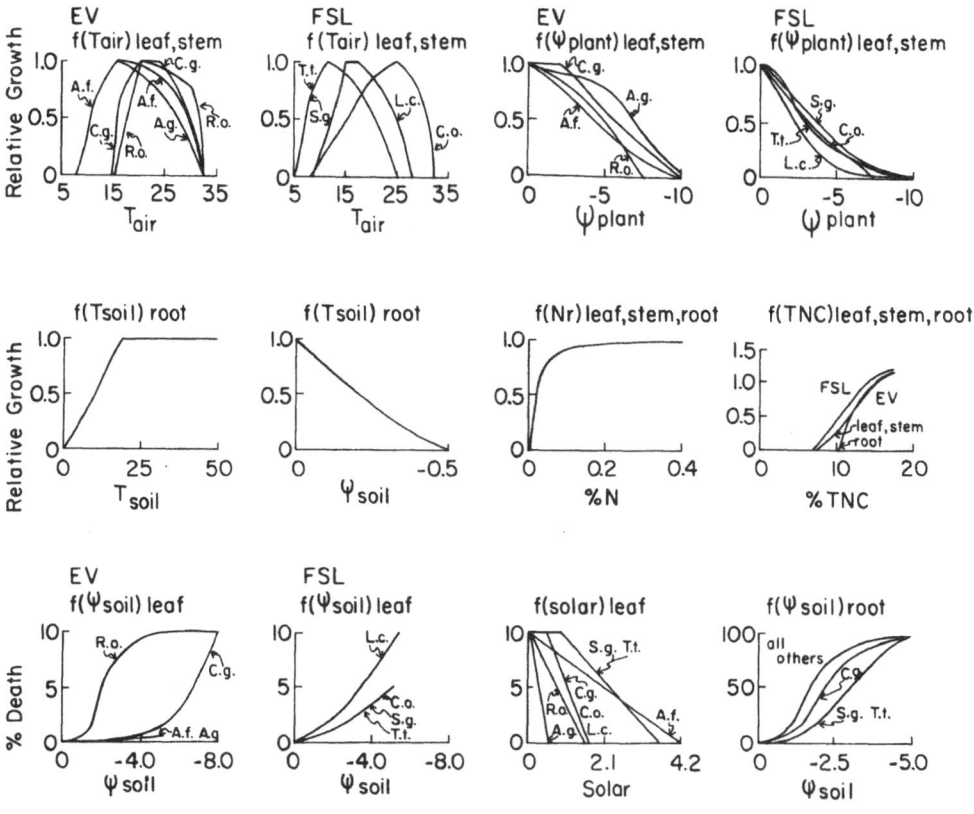

Figure 11.5. Table functions used in MEDECS. Relative growth rates of leaves and stems as functions of air temperature in °C [f(Tair)] and plant water potential in MPa [f(Ψplant)] by species at Echo Valley (EV) and Fundo Santa Laura (FSL). Relative growth rates of roots as functions of soil temperature in °C [f(Tsoil)] and soil water potential in MPa [f(Ψsoil)]. Relative growth rate of leaves, stems, and roots as a function of nitrogen reserves [f(N r)] and total nonstructural carbohydrate [f(TNC)]. Relative death rate of leaves as a function of soil water potential in MPa [f(Ψsoil)] and solar (visible) radiation at the bottom of the canopy [f(solar)] in MJ m⁻² day⁻¹. Relative death rate of absorbing roots as a function of soil water potential in MPa [f(Ψsoil)]. Species are *Adenostoma fasciculatum* (A.f.), *Arctostaphylos glauca* (A.g.), *Ceanothus greggii* (C.g.), *Rhus ovata* (R.o.), *Lithraea caustica* (L.c.), *Trevoa trinervis* (T.t.), *Colliguaya odorifera* (C.o.), and *Satureja gilliesii* (S.g.). Where species are not indicated, the table function applied to all species and where a site is not indicated, the table function applied to both sites (Miller et al., 1978).

Energy exchange processes at the soil surface were modeled by Goudriaan and Waggoner (1972) and Denmead (1973), by means of simplified assumptions about soil moisture status. A model that simulated soil temperature profiles in the arctic (Nakano and Brown, 1972) was later refined (Ng and Miller, 1975; Miller and Ng, 1977) and used as a submodel in CAPS to calculate surface and soil temperature profiles. The soil moisture submodel in CAPS was similar to earlier models (Rubin et al., 1964; Rubin and Steinhardt, 1964; Bresler et al., 1969; Hanks et al., 1969a, b) and was coupled with the soil temperature submodel to make a surface temperature submodel that calculated the evaporation rate from the surface.

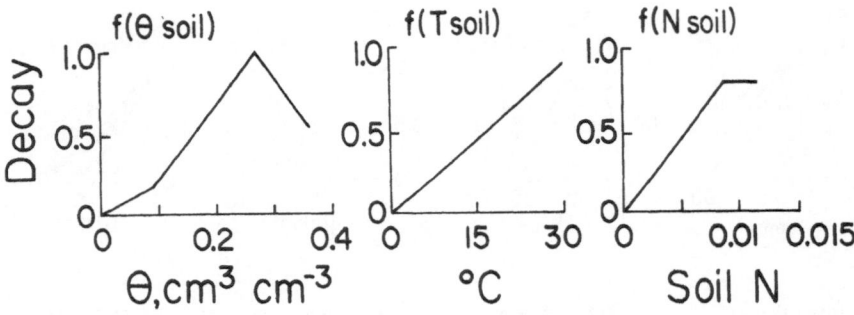

Figure 11.6. Table functions used in MEDECS. Relative decay rate of soil organic matter as a function of soil water content in cubic centimeters water per cubic centimeters soil [f(θsoil)] and temperature in °C [f(Tsoil)]. Relative rate of nitrogen release as a function of soil organic nitrogen in percent [f(Nsoil)].

11.2.2 Canopy Irradiance

Radiation exchange within a vegetation canopy involves direct solar irradiance, downward diffuse solar irradiance, upward diffuse solar irradiance, downward infrared irradiance, and upward infrared irradiance (Figure 11.7). The word *solar* was used to distinguish visible and shortwave infrared radiation emitted by the sun in wavelengths 300 to 3000 nm from longwave radiation emitted by the environment in wavelengths 9000 to 11,000 nm. Direct irradiance comes directly from the sun and diffuse irradiance is direct irradiance reflected by the atmosphere, environment, or canopy. These fluxes were affected by the vertical distribution of live and dead foliage area, the average leaf and stem inclinations, the reflectances of live and dead material to solar and infrared irradiances, the incoming direct and diffuse solar and infrared irradiances at the top of the canopy, and the reflectances of the soil surface to solar and infrared irradiances.

The direct solar irradiance at level z in the canopy was a function of the direct solar flux at level z − 1, the foliage area index, the inclination of the foliage within stratum z, and the sun inclination (Table 11.9:Eq. 2). The inclinations of the foliage and of the sun determined the extinction coefficient for direct solar irradiance. The equation (Duncan et al., 1967; Table 11.9:Eq. 3) assumes that the light-intercepting canopy components are randomly dispersed. To adjust for nonrandom distribution of leaves and stems in a vegetation canopy, a coefficient or dispersion that weighted the relative clustering within the canopy was included in the exponent of Equation 2 (Acock et al., 1970). The amount of direct solar irradiance intercepted by foliage in the zth stratum was the difference between the direct irradiance at level z − 1 and level z. Intercepted direct irradiance was either absorbed or became diffuse irradiance when it was reflected by or transmitted through foliage.

The diffuse solar irradiance downward at level z was composed of five fluxes: (1) the direct solar irradiance at level z − 1 intercepted by stratum z and reflected down; (2) the direct solar irradiance at level z − 1 intercepted by stratum z and transmitted down; (3) the downward diffuse solar irradiation passing through stratum z unattenuated; (4) the diffuse solar irradiance entering stratum z from above, which was reflected down after interception; and (5) the diffuse solar irradiance entering stratum z from below, which was reflected down (Table 11.9:Eq. 4). The upward diffuse irradiance at

Table 11.9. Summary of equations used in the Canopy Production Simulator (CAPS).

Equations

Energy budget of the whole plant canopy

$$RNET = IRC + HEATCONV + LATENT\,(\lambda) + HEATCOND \tag{1}$$

Direct solar radiation in the canopy

$$SB_Z = SB_{Z-1} \exp\,[(-KBL)(CL)(AL_Z) - (KBST)(CST)(AST_Z) - (KBD)(CD)(AD_Z)] \tag{2}$$

Extinction coefficient for direct solar radiation

$$\text{if } \beta \geqslant \alpha, KBL = \cos \alpha \tag{3}$$
$$\text{if } \beta < \alpha, KBL = [(2/\pi)\sin \alpha \cos \beta \sin \theta + (1 - 2\theta/\pi)\cos \alpha \sin \beta\phi$$
$$\text{if } \beta < \alpha, KBL = [(2/\pi)\sin \alpha \cos \beta \sin \theta + (1 - 2\theta/\pi)\cos \alpha \sin \beta]\,/\sin \beta$$
$$\text{where } \theta = \cos^{-1}\,[(\cos \alpha)(\tan \beta)]$$

KBST and KBD are similarly calculated.

Downward flux of diffuse solar radiation

$$SDDN_Z = \rho BDN_Z\,[\rho_Z\,(FB_{Z-1} - FB_Z)\,SB_0] + (FD_Z)\,(SDDN_{Z-1}) + \rho DDN_Z\,[\rho_Z\,(1 - FD_Z)\,(SDDN_{Z-1} + SDUP_Z)] \tag{4}$$

Upward flux of diffuse solar radiation

$$SDUP_Z = (\rho BUP_{Z+1})\,[\rho_Z\,(FB_Z - FB_{Z+1})\,SB_0] + (FD_{Z+1})\,(SDUP_{Z+1}) + \rho DUP_{Z+1}\,[\rho_{Z+1}\,(1 - FD_{Z+1})\,SDDN_Z + SDUP_{Z+1}] \tag{5}$$

Unintercepted fraction of diffuse solar or infrared radiation passing through a stratum

$$FD_Z = \sum_{j=1}^{q} \phi_j \exp\left\{[-KL_j\,(AL_Z) - KST_j\,(AST_Z) - KD_j\,(AD_Z)]\,/\sin j\right\} \tag{6}$$

Reflectance of leaves, stems, and dead material in a stratum to solar radiation

$$\rho_Z = [AL_Z(\rho L) + AST_Z(\rho ST) + AD_Z(\rho D)]\,/\,[AL_Z + AST_Z + AD_Z] \tag{7}$$

Downward flux of infrared radiation

$$IRDN_Z = (FD_Z)\,(IRDN_{Z-1}) + (1 - FD_Z)\,IRC_Z + \rho DDN_Z\,(1 - \epsilon)\,(1 - FD_Z)^2 \tag{8}$$
$$(IRDN_{Z-1} - IRUP_Z)$$

Upward flux of infrared radiation

$$IRUP_Z = (FD_{Z+1})\,IRUP_{Z+1} + (1 - FD_{Z+1})\,IRC_{Z+1} + \rho DUP_{Z+1}(1 - \epsilon) \tag{9}$$
$$(1 - FD_{Z+1})^2\,(IRDN_Z + IRUP_{Z+1})$$

Net radiation

$$RNET_Z = FB_Z(SB_0) + SDDN_Z - SDUP_Z + IRDN_Z - IRUP_Z \tag{10}$$

Table 11.9. (Continued)

Equations

Energy budget for each stratum

$$\triangle RNET_Z = HEATCONV_Z + LATENT_Z(\lambda) \tag{11}$$

Convectional exchange between leaf and air

$$HEATCONVL_Z = 0.006(WIND \{ exp [(-KT) \sum_{k=1}^{z} AF_k] \})^{\frac{1}{2}} (WIDTH)^{-\frac{1}{2}} \tag{12}$$
$$(TL_Z - TA_Z)$$

Convectional exchange of energy between strata z and z-1

$$HEATCONV_Z = 60(CP)(\rho A)(KH_Z)(\triangle T_Z)/\triangle Z \tag{13}$$

Leaf transpiration

$$TRANLEAF_Z = 60(\chi L_Z - \chi A_Z)/(RA_Z + RLX_Z) \tag{14}$$

Water vapor flux between strata z and z-1

$$LATENT_Z = 60(KW_Z)(\triangle CHIA_Z)/\triangle Z \tag{15}$$

Convectional (16) and water vapor (17) fluxes

$$HEATCONVL_Z + HEATCONV_{Z+1} - HEATCONV_Z = 0 \tag{16}$$
and
$$AL_Z(TRANLEAF_Z) + LATENT_{Z+1} - LATENT_Z = 0 \tag{17}$$

Turbulent exchange in the canopy

$$K_Z = (KZERO) exp [(-KT) \sum_{k=1}^{z} AF_k] \tag{18}$$

Energy budget equation for leaves

$$SABSX_Z + (\epsilon)(IRC_Z) = IRCL_Z + HEATCONVL_Z + (\lambda)TRANLEAF_Z \tag{19}$$

Absorbed solar radiation for sunlit leaves

$$SABSUN_Z = [(1 - REFL)/2] [(KBL)(SB_0) + SDDN_Z + SDUP_Z] \tag{20a}$$

Absorbed solar radiation for shaded leaves

$$SABSHADE_Z = [(1 - REFL)/2] [SDDN_Z + SDUP_Z] \tag{20b}$$

Table 11.9. (Continued)

Equations

Infrared radiation emitted by leaves

$$IRCL_Z = \epsilon(\sigma)(TL_Z + 273)^4 \qquad (21)$$

Leaf-air boundary layer resistance to water vapor

$$RA = 0.05(WIDTH)^{1/2}(WIND \{ \exp [(-KT) \sum_{k=1}^{Z} AF_k] \})^{1/2} \qquad (22)$$

Water deficit of the leaves

$$WD_{t+1} = WD_t + (TRANLEAF_Z - WATUP)(\triangle t)/WT \qquad (23)$$

Movement of water from soil to leaf

$$WATUP = (\Psi SOIL - \Psi LEAF)/(RR + RS) \qquad (24)$$

Leaf resistance to water vapor

For sunlit and shaded leaves, calculate RLSUN and RLSHADE using SABSUN and SABSHADE for SABSX, respectively.

$$RLX_Z = RLMIN + b_1/[b_2 + b_3(SABSX_Z)] + f(WD) \qquad (25)$$

Net photosynthesis

For sunlit and shaded leaves, calculate PSUN and PSHADE using either SABSUN, RLSUN and RISUN or SABSHADE, RLSHADE and RISHADE, respectively, for SABSX, RLX, and RIX. For each calculation, use equation (26) if SABSX > LC or equation (27) if SABSX $\leqslant$ LC.

$$PSX_Z = \triangle C/[1.56(RA_Z + RLX_Z) + RIX_Z] \qquad (26)$$

$$PSX_Z = [(LC - SABSX_Z)/LC](RZERO)2^{0.1(TL - TZERO)} \qquad (27)$$

Internal leaf resistance to CO_2 incorporation

For sunlit and shaded leaves, calculate RISUN and RISHADE using SABSUN and SABSHADE for SABX, respectively.

$$RIX_Z = RMIN + b_4/SABSX_Z \qquad (28)$$

Photosynthesis including sunlit and shaded leaves

$$PS_Z = (PSUN_Z)(FB_Z)(AL_Z) + (PSHADE_Z)(1 - FB_Z)(AL_Z) \qquad (29)$$

Table 11.9. (Continued)

Equations

Soil temperature flow equation

$$CS(\partial T/\partial t) = (\partial/\partial Z)\,[KS(\partial T/\partial Z)] \tag{30}$$

Definition of Symbols

AL_z	Leaf area index in zth stratum of canopy (if unsubscripted, for all strata)	m^2 leaf m^{-2} ground
AST_z	Stem area index in zth stratum of canopy (if unsubscripted, for all strata)	m^2 stem m^{-2} ground
AD_z	Dead area index in zth stratum of canopy (if unsubscripted, for all strata)	m^2 dead m^{-2} ground
AF_z	Foliage area index in zth stratum of canopy (if unsubscripted, for all strata)	m^2 foliage m^{-2} ground
b_1, b_2	b_1/b_2 approximates the cuticular resistance to passage of water vapor which occurs at zero light	s cm^{-1} MPa^{-1}
b_3	Sensitivity of stomatal opening to radiation absorbed by the leaf	cm min MJ^{-1}
b_4	Constant related to internal leaf resistance to CO_2 versus radiation absorbed by the leaf	MJ cm^{-3}
CL	Coefficient of dispersion of radiation by leaves	fraction
CD	Coefficient of dispersion of radiation by dead material	fraction
CP	Specific heat of air	MJ $°C^{-1}$ g^{-1}
CS	Volumetric heat capacity of the soil	J m^{-3} soil $°C^{-1}$
CST	Coefficient of dispersion of radiation by stems	fraction

Table 11.9. (Continued)

Definition of Symbols		
FB_z	Fraction of direct solar (beam) radiation which passes through canopy stratum z unintercepted	fraction
FD_z	Fraction of diffuse solar or infrared radiation which passes through canopy stratum z unintercepted	fraction
HEATCOND	Heat exchanged by conduction between canopy and air	$MJ\ cm^{-2}$ ground min^{-1}
$HEATCONV_z$	Heat exchanged by convection between strata z and z-1 (if unsubscripted, for whole canopy)	$MJ\ cm^{-2}$ ground min^{-1}
$HEATCONVL_z$	Heat exchanged by convection between leaves and air in canopy stratum z (if unsubscripted, for whole canopy)	$MJ\ cm^{-2}$ ground min^{-1}
IRC_z	Infrared radiation emitted by leaves, stems, and dead material in canopy stratum z (if unsubscripted, for whole canopy)	$MJ\ cm^{-2}$ ground min^{-1}
$IRCL_z$	Infrared radiation emitted by leaves in stratum z	$MJ\ cm^{-2}$ ground min^{-1}
$IRDN_z$	Downward flux of infrared radiation at canopy level z	$MJ\ cm^{-2}$ ground min^{-1}
$IRUP_z$	Upward flux of infrared radiation at canopy level z	$MJ\ cm^{-2}$ ground min^{-1}
K_z	Turbulent transfer coefficient for air packets at canopy level z	$cm^2\ s^{-1}$
KBD	Extinction coefficient for direct solar radiation depending on the altitude of the sun and inclination of dead material	m^2 ground m^{-2} dead

Table 11.9. (Continued)

	Definition of Symbols	
KBL	Extinction coefficient for direct solar radiation depending on the altitude of the sun and inclination of leaves	m^2 ground m^{-2} leaf
KBST	Extinction coefficient for direct solar radiation depending on the altitude of the sun and inclination of stems	m^2 ground m^{-2} stem
KD_j	Extinction coefficient of dead material for solar radiation coming from band j of the sky	m^2 ground m^{-2} dead
KH_z	Turbulent transfer coefficient for heat at canopy stratum z; assumed equivalent to K_z	cm^2 s^{-1}
KL_j	Extinction coefficient of leaves for solar radiation coming from band j of the sky	m^2 ground m^{-2} leaf
KS	Thermal conductivity of soil layer	J m^{-1} s^{-1} $°C^{-1}$
KST_j	Extinction coefficient of stems for solar radiation coming from band j of the sky	m^2 ground m^{-2} stem
KT	Extinction coefficient for turbulent transfer or wind in the canopy	m^2 ground m^{-2} foliage
KW_z	Turbulent transfer coefficient for water vapor; assumed equal to K_z	cm^2 s^{-1}
KZERO	Turbulent exchange coefficient above the canopy	cm^2 s^{-1}
$LATENT_z$	Water vapor exchanged due to transpiration from canopy in canopy stratum z (if unsubscripted, from whole canopy)	g H_2O cm^{-2} ground min^{-1}
$LATENTL_z$	Water vapor exchanged due to turbulent transfer between strata z and z + 1	g H_2O cm^{-2} ground min^{-1}

Table 11.9. (Continued)

	Definition of Symbols	
LC	Light compensation point, where photosynthesis (uptake of CO_2) equals respiration (production of CO_2) in the leaf	MJ cm^{-2} ground min^{-1}
n	Number of strata in the canopy where 1 is highest stratum, n is the lowest stratum	number
PS_z	Photosynthesis for a stratum, including sunlit and shaded leaves	g CO_2 cm^{-2} ground min^{-1}
PSHADE	For shaded leaves, equals PSX calculated by either Eq. (26) or (27)	g CO_2 cm^{-2} ground min^{-1}
PSUN	For sunlit leaves, equals PSX calculated by either Eq. (26) or (27)	g CO_2 cm^{-2} ground min^{-1}
PSX	Net photosynthesis	g CO_2 cm^{-2} ground min^{-1}
q	Number of sky bands, each $10°$ width (9)	number
RA_z	Resistance of the leaf-air boundary layer to passage of water vapor	s cm^{-1}
REFL	Reflectance of leaf to solar radiation	fraction
RISUN, RISHADE	Actual internal resistance of leaf to incorporation of CO_2, for sunlit and shaded leaves, respectively	s cm^{-1}
RIX_z	Equals either RISUN or RISHADE for stratum z	s cm^{-1}
RLSUN, RLSHADE	Resistance of stomata and cuticle of leaf to passage of water vapor for sunlit and shaded leaves, respectively	s cm^{-1}
RLX_z	Equals either RLSUN or RLSHADE for stratum z	s cm^{-1}

Table 11.9. (Continued)

	Definition of Symbols	
RLMIN	Minimum possible resistance of stomata and cuticle of leaf to passage of water vapor, a function of temperature and light	s cm^{-1}
RMIN	Minimum internal resistance to CO_2 incorporation	s cm^{-1}
RNET	Net solar and infrared radiation (if subscripted, for stratum z)	MJ cm^{-2} ground min^{-1}
RR	Resistance of roots to uptake of water	s cm^{-1} MPa^{-1}
RS	Boundary layer resistance of soil to passage of water vapor from soil surface to air	s cm^{-1} MPa^{-1}
RZERO	Leaf respiration rate at base temperature TZERO	g CO_2 cm^{-2} ground min^{-1}
SABSHADE$_z$	Solar radiation absorbed by shaded leaves	MJ cm^{-2} ground min^{-1}
SABSUN$_z$	Solar radiation absorbed by sunlit leaves	MJ cm^{-2} ground min^{-1}
SABSX$_z$	Absorbed solar radiation, equals either SABSUN$_z$ or SABSHADE$_z$	MJ cm^{-2} ground min^{-1}
SB$_o$	Incoming direct solar radiation at the top of the canopy	MJ cm^{-2} ground min^{-1}
SB$_z$	Direct solar radiation at canopy stratum z	MJ cm^{-2} ground min^{-1}
SDDN$_z$	Downward flux of diffuse solar radiation at canopy level z	MJ cm^{-2} ground min^{-1}
SDUP$_z$	Upward flux of diffuse solar radiation at level z	MJ cm^{-2} ground min^{-1}
t	Time	min

Table 11.9. (Continued)

	Definition of Symbols	
T	Soil temperature	°C
TA	Temperature of air	°C
TL	Temperature of leaf	°C
TRANLEAF_z	Transpiration of water from leaves in canopy stratum z	g H_2O cm^{-2} leaf min^{-1}
TZERO	Base temperature from which a leaf respiration rate is calculated	°C
WATUP	Water removed from soil to leaf	g H_2O min^{-1} cm^{-2} leaf
WD	Water deficit of the leaves; subscript t refers to time	fraction
WIDTH	Width of leaf surface	cm
WIND	Wind velocity	cm s^{-1}
WT	Weight of water per area in a turgid leaf	g H_2O cm^{-2} leaf
Z_z	Height interval of stratum z or soil layer z	cm
α	Angle of leaf (or stem or dead material) to horizontal	degrees
β	Altitude of sun	degrees
$\triangle C$	Difference between CO_2 concentrations of air and the site of carboxylatin	g CO_2 cm^{-2} ground min^{-1}
$\triangle\text{CHIA}_z$	Difference between humidities of canopy stratum z and canopy stratum z$-$1	10^{-6} g H_2O cm^{-3} air
$\triangle\text{RNET}_z$	Net radiation absorbed by canopy stratum z	MJ cm^{-2} ground min^{-1}

Table 11.9. (Continued)

	Definition of Symbols	
ΔT_z	Difference between temperatures of canopy stratum z and canopy stratum z-1	°C
Δt	Time interval between calculations	min
ΔZ	Distance between midheights of canopy stratum z and canopy stratum z-1	cm
ϵ	Emissivity for (infrared) radiation of the leaves, stems, and dead material (= absorptance of infrared for the leaves, stems, and dead material)	fraction
λ	Latent heat of evaporation	MJ g^{-1} H_2O
ρ_z	Fraction of the intercepted direct or diffuse beam (solar radiation) which is reflected by canopy stratum z	fraction
ρA	Density of air	g cm^{-3} air
ρBDN_z	Fraction of the reflected direct beam (solar radiation) in canopy stratum z which is reflected down	fraction
ρBUP_z	Fraction of the reflected direct beam in canopy stratum z which is reflected up	fraction
ρD	Reflectance of dead material to solar radiation	fraction
ρDDN_z	Fraction of the reflected diffuse beam (solar radiation) in canopy stratum z which is reflected down	fraction
ρDUP_z	Fraction of the reflected diffuse beam in canopy stratum z which is reflected up	fraction
ρL	Reflectance of leaves to solar radiation	fraction

Table 11.9. (Continued)

Definition of Symbols		
ρST	Reflectance of stems to solar radiation	fraction
σ	Stefan-Bolzmann constant (3.4×10^{-16})	MJ cm^{-2} min^{-1} °K^{-4}
ϕ_j	Fraction of all radiation received on a horizontal surface from a 10° band of the hemisphere with a mean altitude j	fraction
χA	Saturation vapor density of the bulk air	10^{-6} g H$_2$O cm^{-3} air
χL	Saturation vapor density at leaf temperature	10^{-6} g H$_2$O cm^{-3} air
ΨSOIL	Soil water potential	MPa
ΨLEAF	Leaf water potential	MPa

any canopy level was similar to the downward flux but had only four fluxes because direct solar irradiance cannot come from below (Table 11.9:Eq. 5). The diffuse solar or infrared irradiance passing through a stratum without interception was related to the fraction of the hemisphere visible within the stratum. The fraction of the hemisphere visible was expressed as the irradiance received on a horizontal surface from any 10° band of the hemisphere (de Wit, 1965; Anderson, 1966; Duncan et al., 1967; Table 11.9:Eq. 6). For both direct and diffuse irradiance, the reflectance of a stratum was the weighted average reflectance of all leaves, live stems, and dead stems (Table 11.9:Eq. 7). Reflectances were estimated from Billings and Morris (1951), Birkebak and Birkebak (1964), and Gates et al. (1965a, b). It was assumed that 60% of the reflected irradiance passed downward and 40% upward; the amounts are similar to those determined by Duncan et al. (1967) and Idso and de Wit (1970).

Infrared irradiance at a canopy level was the sum of the infrared fluxes from the sky, foliage, and ground. The downward flux at level z consisted of unintercepted infrared irradiance from level z − 1, unintercepted infrared irradiance emitted by the foliage in stratum z, the infrared irradiance entering stratum z from above, which was intercepted and reflected down, and the infrared irradiance entering stratum z from below, which was intercepted and reflected down (Table 11.9:Eq. 8). The upward flux at level z consisted of the analogous components moving in the opposite direction, substituting ground for the sky (Table 11.9:Eq. 9). In both equations, the infrared irradiance emitted by leaves, stems, and dead material was calculated from the sky or

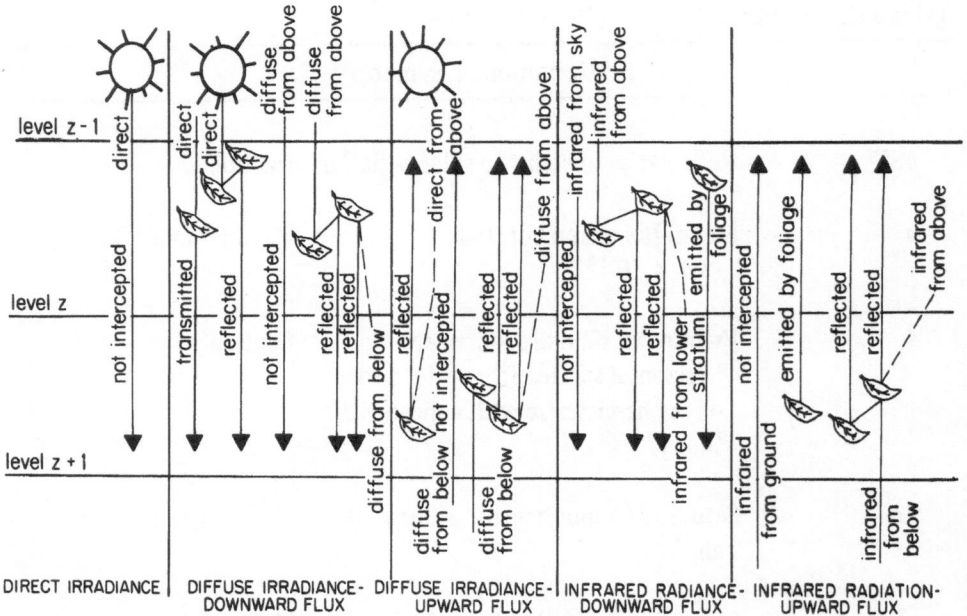

Figure 11.7. Radiation fluxes in CAPS.

foliage temperature, respectively, by using the Stefan-Boltzmann law. The foliage temperature was assumed uniform within a stratum because the few seconds required to equilibrate any temperature gradients were negligible compared to the time step of 2 or 5 min used in the model.

11.2.3 Air Temperature and Air Humidity

The canopy affects air temperature and air humidity inside the canopy by convectional and evaporational heat exchange with the foliage and by modification of turbulent air exchange. In the model, net irradiance at a level was calculated by summing downward fluxes and subtracting upward fluxes of direct solar, diffuse solar, and infrared radiation (Figure 11.7; Table 11.9:Eq. 10). Irradiance absorbed in a stratum was the difference between net irradiance at the top (level z − 1) and that at the bottom (level z) of the stratum and equaled the heat exchange between the plant and the air within the stratum and between adjacent strate (Table 11.9:Eq. 11). The two methods of heat exchange were convection and transpiration. Convectional heat exchange occurred between the leaf and air (Table 11.9:Eq. 12) and also between strata (Table 11.9:Eq. 13). Water vapor moved from leaf to air by transpiration within a stratum (Table 11.9: Eq. 14) and between strata by turbulent exchange (Table 11.9:Eq. 15). The sums of the convectional fluxes and water vapor fluxes between leaf and air and between strata were assumed zero (Table 11.9:Eqs. 16-17). For each stratum, Equations 11, 13, and 15 to 17 (Table 11.9) were solved simultaneously for air temperature and water vapor content or humidity (Waggoner et al., 1969).

Turbulent air exchange affected convection and water vapor movement between strata. The turbulent transfer coefficient was assumed to decrease exponentially with foliage area index (Table 11.9:Eq. 18). The turbulent transfer coefficients for heat and water vapor exchange were assumed equal (K of Table 11.9:Eq. 18) and were converted to resistances, so that the equations could be solved for temperature as in an electrical resistance network (Waggoner et al., 1969). Wind affected convection and water vapor movement between leaf and air within a stratum. Wind also decreased exponentially with foliage area index but was not linked to turbulent air exchange because of the complexity of these interactions in the canopy (Monteith, 1973).

11.2.4 Plant Temperature

Leaf temperatures were calculated for sunlit and shaded leaves at each canopy level. The proportion of sunlit leaves within a canopy was calculated from the ratio of direct to total solar irradiance within the canopy. Shaded and sunlit leaves received similar irradiances except that shaded leaves did not receive direct solar irradiance. The energy budget equation (Gates, 1962, 1965; Table 11.9:Eq. 19) balanced absorbed solar and infrared irradiance with emitted infrared radiation, convection, and latent heat fluxes. The temperature of each plant component was calculated to balance the energy budget equation for each component. The irradiance absorbed by leaves in a stratum depended on the flux densities within the stratum and the leaf absorptances for solar or infrared (Table 11.9:Eq. 20). The infrared loss from a leaf was calculated by using the Stefan-Boltzmann relation (Table 11.9:Eq. 21). Loss due to convection was determined by the velocity of the wind, the leaf width, and the difference between air and leaf temperature (Table 11.9:Eq. 12). Loss of latent heat by water vaporization during transpiration was calculated as a diffusion process across the resistances of the leaf stomata, the leaf cuticle, and the air boundary layer (Table 11.9:Eqs. 14 and 22). Heat conduction between the leaf surface and the leaf mass was negligible and was assumed zero.

11.2.5 Plant Water

Plant water content in a given time step was the only state variable in CAPS. The water deficit (Barrs, 1968) of the leaves was a function of the previous water deficit; of the turgid weight, fresh weight, and dry weight; and of the water transpired and taken up (Table 11.9:Eq. 23). Water moved from the soil to the leaf because of a difference in soil and leaf water potentials and was impeded by resistances of the roots and the soil (Table 11.9:Eq. 24). Table functions were used to relate soil water content to soil water potential (Table 11.6) and leaf water content to leaf water potential. The resistance of the root system varied with temperature and the rate of water uptake (Kuiper, 1964; Kramer, 1969; Stoner and Miller, 1975). Absorbed solar irradiance, leaf water deficit, leaf temperature, and a minimum possible resistance determined leaf resistance (Stoner and Miller, 1975; Table 11.9:Eq. 25) which, in turn, affected the water loss by transpiration (Table 11.9:Eq. 14).

11.2.6 Photosynthesis

Photosynthesis was calculated for sunlit and shaded leaves for each canopy stratum. Except for absorbed solar irradiance (Table 11.9:Eqs. 20a, b), calculations for sunlit and shaded leaves were identical. When the absorbed solar irradiance was above the light compensation point, net photosynthesis was calculated as a diffusion of carbon dioxide into the leaf (Table 11.9:Eq. 26), where the concentration of carbon dioxide at the carboxylation site was assumed to be zero. The flux of carbon dioxide was impeded by the leaf area boundary layer resistance, the leaf resistance, and an internal resistance which in CAPS was assumed to depend on solar irradiance and temperature. The minimum possible internal resistance under the given leaf temperature and light saturation conditions was calculated from a table function. Then, the actual internal resistance was determined from the absorbed solar irradiance (Table 11.9:Eq. 28), but was not allowed to be less than the calculated minimum. When solar irradiance was below the compensation point, net photosynthesis equaled a respiration rate calculated from a table function according to leaf temperature. Respiration was assumed to increase linearly from zero at the light compensation point to a maximum at zero solar irradiance (Table 11.9:Eq. 27). After calculating the photosynthetic rate for sunlit and shaded leaves, photosynthesis for the entire stratum was calculated as a weighted average of net photosynthesis for sunlit and shaded leaves (Table 11.9:Eq. 29).

11.2.7 Soil Temperature and Moisture

In CAPS, heat flow between two points was related to the distance, temperature difference, and thermal conductivity of the medium, which in this case was the soil (Table 11.9:Eq. 30). The greater the temperature gradient or amplitude of the temperature fluctuations, the shorter the distance required between points or nodes for stability of the numerical solution. The thermal conductivity varied because of the nature of the soil, either mineral or organic, and its moisture content. Heat entered or left the soil through the surface. Surface temperature was related to conduction from below and irradiance and convection from above.

11.3. The Mediterranean Ecosystem Simulator

The Mediterranean Ecosystem Simulator (MEDECS) describes resource utilization by mediterranean type vegetation in southern California and central Chile. The overall modeling goal was to integrate processes determining the utilization of light energy, water, and nitrogen, starting with physiological and morphological details and proceeding to the community level and to predict annual patterns and annual totals of the utilization of light energy, water, and nitrogen. Within MEDECS, submodels of plant growth, decomposition of soil organic matter, and inorganic nutrient release in the soil provided a means to relate growth to temperature, plant water status, and carbohydrate and nitrogen storage in the plant and, thus, indirectly to relate growth to soil water availability, photosynthesis, respiration, nutrient release in the soil, and nutrient uptake by the plant. MEDECS could simulate the effects of changes in annual precipitation on vegetation production by direct and indirect feedback reactions among the above processes.

MEDECS was run with up to four species: *A. fasciculatum* for the equator-facing slope and *A. fasciculatum, A. glauca, C. greggii*, and *R. ovata* for the pole-facing slope at Echo Valley; and *C. odorifera, S. gillessii, T. trinervis*, and *L. caustica* for both slopes at Fundo Santa Laura. Modifications were made to simulate *Artemisia californica, Saliva mellifera, Eriogonum fasciculatum*, and *Vulpia (Festuca) octoflora* in the coastal sage scrub vegetation at Camp Pendleton.

The unit was 1 m² of ground surface. The soil was divided into eight layers (Figure 11.2), each of which had a temperature, soil moisture, soil organic matter, and exchangeable inorganic nitrogen. The sequence of calculations in each time step was water absorption, nitrogen uptake, root growth, photosynthesis, and leaf growth, which allowed new leaves to utilize the total nonstructural carbohydrate reserves first and absorbing roots to utilize nitrogen reserves first (Table 11.10).

In MEDECS, the simulated year began in early fall, 1 October for California and 1 April for Chile. It was assumed that, because of the summer drought, soil water potentials were at the minimum, no growth was occurring, there were no leaves on deciduous shrubs, conducting roots were present, but there were no absorbing roots. In general, the four species at Echo Valley and the four species at Fundo Santa Laura were modeled the same way, except that the species' growth form was taken into account. Each species had six compartments: main roots, which included conducting roots through which water and nutrients moved and supportive roots plus a burl, if present, in the species being considered; absorbing roots; main stems; new stems; old leaves; and new leaves (Figure 11.8). Initial state values and some parameter values differed between species while other values were the same for all species (Figures 11.3 to 11.6; Tables 11.2 to 11.5).

11.3.1 Development

Community-level integrative models based on simplifications of physiological and morphological details were developed in agronomy and forestry (Loomis et al., 1979). Brouwer and de Wit (1969) proposed a model of plant growth in which allocation to shoots or roots was controlled by the availability of carbohydrate reserves and water. De Wit et al. (1970) elaborated upon this model, which was then further developed for other agricultural crops and compared against data on corn (de Wit et al., 1978). Thornley (1972a, b) proposed general models describing the growth of shoots and roots by logistic equations. Botkin (1975) developed a model of growth of several forest trees that projected realistically the process of succession in a northeastern hardwood forest with as simple formulations as possible to describe responses to light and water. Fick et al. (1975) described a model for the growth and carbohydrate allocation in sugar beets that was used to explore effects of respiration on system production (Hunt and Loomis, 1979). Penning de Vries et al. (1975) presented a model of nitrogen dynamics and growth of loblolly pine over a multiyear time interval. Van Keulen (1975) described models of the growth of annual grasses in an arid region, which were extended to include nitrogen (van Keulen et al., 1975). Harpaz (1975) included the nitrogen dynamics for an arid region in a model similar to van Keulen's. However, these previous models included only a superficial treatment of nitrogen dynamics.

Table 11.10. Sequence of calculations for one day in MEDECS model

Sequence of calculations

Foliage area indices
Climate
Evaporation from soil surface
Soil water content after rain, percolation and evaporation from within soil levels; use
  this water content to determine if growth could start
Soil nitrogen after leaching
Nitrogen uptake into plant root reserves
Decay and mineralization, soil nitrogen
Transpiration, soil water content
Root growth, growth respiration, death, decay, reserves of nitrogen and total non-
  structural carbohydrate
Soil organic matter, organic nitrogen
Soil water potential
Photosynthesis
Maintenance respiration
Net photosynthesis
Phenology
Leaf growth, growth respiration, reserves of nitrogen and total nonstructural carbo-
  hydrate
Stem growth, growth respiration, reserves of nitrogen and total nonstructural carbo-
  hydrate
Leaf death
Litterfall
Nitrogen reserves in stem due to reabsorption before dead leaves fall from plant
Soil nitrogen after litterfall
Translocation of nitrogen and total nonstructural carbohydrate reserves from roots to
  stems
Sums for computing resource use and efficiencies to be printed at end of simulation
  period

Three models of growth and allocation for the single-shooted graminoid, with different levels of sophistication, were developed during the US/IBP Tundra Biome research (Miller, 1972b; Lawrence et al., 1978; Miller et al., 1976a; Stoner et al., 1978c).
In these models, total biomass was controlled by the seasonal patterns of growth in
various structures and was affected by temperature and plant sugar, nitrogen, phosphorus, and calcium status. The carbon and inorganic nutrients associated with the
growth of new tissue were calculated from the composition of young mature tissue
and the growth respiration costs of creating different biochemical constituents (Penning
de Vries, 1972a, b, 1973, 1974; Penning de Vries et al., 1974). A similar model was developed for *Ledum palustre*, an evergreen shrub; *Salix pulchra*, a deciduous shrub; and
Eriophorum vaginatum, a tundra tussock grass (Stoner et al., 1978d).

Models of translocation of carbon and nutrients in relation to plant growth and
development have been either detailed and mechanistic (Horowitz, 1958; Dainty, 1965)

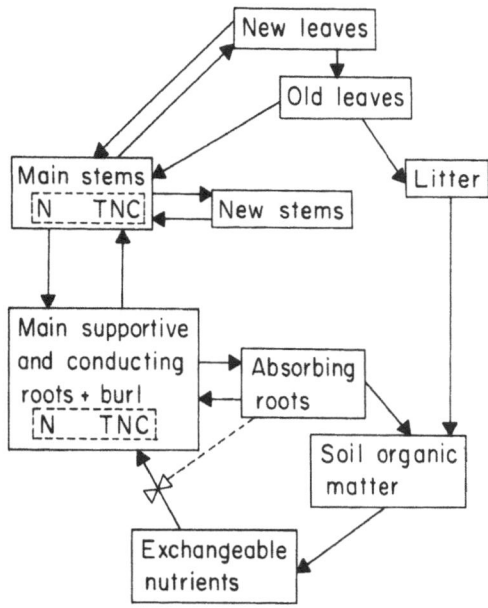

Figure 11.8. Diagram of plant compartments of biomass, nitrogen (N), and total non-structural carbohydrate (TNC) and transfers in MEDECS for evergreen shrubs. Solid line indicates a flow of material. Interrupted line indicates control on a flow.

or descriptive (Warren Wilson, 1964, 1966; Blackman, 1968; Kvet et al., 1971). Thornley (1972a, b), Miller et al. (1976a), and Stoner et al. (1978d) modeled translocation as being driven by a concentration gradient impeded by resistances. Brouwer and de Wit (1969), de Wit et al. (1970), and Fick et al. (1975) bypassed the problem by pooling all reserves into one plant pool, which was drawn upon for growth by each plant part. All formulations ignored the possibility of sugars being transported against a concentration gradient. In MEDECS, the plant reserves available for growth in any organ were assumed to be one plant pool of carbohydrate and nitrogen.

The methods developed by Penman (1948) were used in MEDECS to calculate evaporation from the soil surface, while a modified version of the Penman equation was used for estimating potential transpiration.

Net photosynthesis is difficult to calculate in the context of other plant processes. Photosynthesis exerts a strong control on growth (Mooney, 1972) and should be described as completely as possible in order to incorporate feedback relations that make the system respond correctly to factors affecting the process. However, simplification of the photosynthetic response is often necessary for expedience. De Wit et al. (1970) argued that there is a constant transpiration-photosynthetic rate, at least within a species, and modeled photosynthesis by modeling transpiration. Another method used by van Keulen (1975) and Loomis et al. (1979) was to simulate the photosynthetic rates for a variety of foliage area distributions and climates by using a complex radiation interception and photosynthesis canopy model and construct a response surface to predict photosynthesis for the integrative model. This technique is particularly useful when the vegetation canopy is viewed as a monolayer because the relation of the canopy photosynthesis to total canopy foliage area and the effect of the mutual shading of leaves is described by the radiation penetration model. The method works well in monocultures such as agricultural systems (Loomis et al., 1979). Problems arise when there is the possibility of many foliage distributions, as in natural communities. MEDECS used a monolayer canopy and employed a similar technique to calculate photosynthesis (Miller et al., 1978). Regression equations from CAPS relating photosynthesis to absorbed light and air temperature were used to calculate a photosynthetic rate for the canopy (Chapter 7). This rate was not corrected for water stress.

The submodels in MEDECS for soil nutrients, organic matter, decay, and mineralization were based on the work of de Wit and van Keulen (1972) and Frissel and Reinger (1974), which modeled the release of exchangeable nutrients and their movement through the soil. The effect of low soil moistures on nutrient movement has been described empirically and theoretically (Olsen and Kemper, 1968). Generalized simulations of the dynamics of nitrogen in the soil have been presented in the context of the overall nitrogen cycle (Beek and Frissel, 1973; Harpaz, 1975). The microbial activity involved in the decomposition of organic matter was simulated in several ecological systems, including tundra and grasslands (Bunnell and Tait, 1974; Hunt, 1977). Nitrogen mineralization was modeled for the grasslands (Reuss and Innis, 1977; Woodmansee, 1978), and the control of decomposition and mineralization by the microflora was simulated for the wet meadow tundra (Barkley et al., 1978). Models for irrigation and fertilizer application (Jensen, 1972) have been extended to simulate chemical and biological processes in agricultural soils (Dutt et al., 1972).

11.3.2 Climate

In MEDECS, variables describing the climate above the canopy were calculated for each day by interpolation from monthly averages of field data which were taken in 1972-1976 at Echo Valley and Fundo Santa Laura and which were entered as input to the simulations (Table 11.11). The variables were day length, the day's maximum and minimum temperatures, total solar irradiance, relative humidity, and cloud cover. Temperatures and solar irradiance were then adjusted for site inclination by using regression coefficients, which related values from a horizontal site to those for a slope. The average temperature for the day was calculated as 0.5 of the sum of the minimum and maximum temperatures for the day. The average daytime temperature was calculated as 0.75 times the sum of the minimum and maximum temperatures. The daytime vapor pressure equaled the saturation vapor pressure at the nighttime temperature (Table 11.12:Eq. 1) on the assumptions that the dewpoint is reached at night and that water vapor is not lost during the day. Infrared irradiance from the sky was calculated from the average temperature and cloud cover for the day (Table 11.12:Eq. 2), following Monteith (1973). The amount of precipitation received each day was entered as part of the input information.

11.3.3 Water

The processes in MEDECS involving water included precipitation, interception and evaporation of precipitation by the canopy, throughfall and stemflow, infiltration of water into and through soil layers, subsurface drainage, and removal of water from each layer by soil evaporation and transpiration (Chapters 2 and 6).

Potential evaporation from the soil surface was calculated from net radiation, vapor pressure, and temperature at the surface by the Penman equation. The drying power of the air depended on the vapor pressure difference between air and soil and on the wind speed (Table 11.12:Eqs. 3 to 9). The potential evaporation was reduced by the fraction of available water in the surface layer to find the estimated evaporation from the soil surface (Penman, 1948; Miller and Mooney, 1976; Table 11.12:Eq. 10).

Loss of precipitation through interception by the canopy and subsequent evaporation depended on the storage capacity of the canopy and the number of times the storage capacity was filled in each day of rainfall (Chapter 6). The storage capacity was the amount of water retained on a square centimeter of foliage (Leyton et al., 1967) times the foliage area index. To allow for intermittent sunshine and drying during a storm, the storage capacity was assumed to be filled, then emptied, by evaporation twice during each day of rain. Precipitation reached the ground by direct passage through the foliage and by drip from leaves. Effective precipitation was equal to the incoming precipitation minus the stored and evaporated water (Table 11.12:Eq. 11).

Water movement downward through the soil layers depended on the effective precipitation and was calculated by two different methods. The first method (van Keulen, 1975) was used in MEDECS for storms with precipitation less than 12.5 mm. During a 1-day time step, water entered a layer until the soil water content of that layer reached saturation; if more water was added, it moved into the layer below or out the bottom of the profile as deep drainage. Then, any water in excess of the field capacity in a soil layer was drained to the next lower layer. Because the coarse soil at Echo Valley allowed water from large storms to move downward more rapidly than the above method predicted, precipitation from storms exceeding 12.5 mm was distributed through the soil profile by using an exponential function of depth (Table 11.12:Eqs. 12 and 13). The soil was then allowed to drain to field capacity as in the first method. The upward movement of water through the soil levels due to evaporation was simulated by distributing the soil evaporation from the surface through several layers. Surface evaporation, the amount of available water in each layer, and the depth of the layer determined the amount of water removed from each layer (Table 11.12:Eqs. 14 and 15). The proportion removed from each layer decreased exponentially with depth, with the result that almost all evaporated water came from the upper 0.2 m of soil (Ng, 1974).

Extraction of soil water by transpiration was determined by several factors (van Keulen, 1975; Table 11.12:Eqs. 16 to 24). Potential transpiration of each species, which was expressed as transpiration per centimeter of absorbing root of the species, was calculated for the daylight period by using a modified Penman equation (Table 11.12:Eq. 16). The maximum leaf conductance to water vapor diffusion (Table 11.12: Eq. 21) varied according to species; the parameters defining the relation between leaf conductance and plant water potential were determined by linear regression analysis of field data (Chapter 6). Transpiration was affected by the drying power of the air and by the factor alpha, which took into account the closure of the stomata in response to low light within the canopy (van Keulen, 1975; Table 11.12:Eq. 23). Alpha was reduced by increasing foliage area or by decreasing solar irradiance and was calculated by interpolating a three-way table function (Table 11.7), which was generated by the CAPS model using field data (Poole and Miller, 1975, 1978; Miller and Poole, 1979; Oechel, *unpubl. data*; Chapter 7). For each soil layer, potential transpiration was multiplied by a fraction calculated from a table function of root activity versus soil

Table 11.11. Environmental data used in MEDECS for simulations of the Camp Pendleton, Echo Valley, and Fundo Santa Laura study sites. Except for precipitation, for a given day, each environmental parameter was found by interpolation. Precipitation was determined by a pre-processing program which randomly assigned precipitation to specific days so that the appropriate amount fell in the given time intervals.[a]

Julian day number[b]	Camp Pendleton	Echo Valley	Julian day number[b]	Fundo Santa Laura
Daylength (DAYL) (hours)				
31	10.3	10.3	1	14.07
76	12.0	12.0	15	14.07
107	13.1	13.1	76	12.4
137	14.0	14.0	106	11.7
199	14.07	14.07	183	10.3
260	12.4	12.4	259	12.0
290	11.7	11.7	290	13.1
321	10.3	10.3	320	14.0
351	10.3	10.3	365	14.08
Maximum temperature (°C)				
1	22.1	25.8	1	21.1
16	21.0	23.8	16	20.1
46	17.6	19.4	46	16.3
77	16.1	16.4	77	12.2
107	14.9	15.9	107	11.3
138	16.8	16.6	138	12.7
166	16.6	16.3	166	14.7
197	19.3	18.5	197	16.6
227	20.1	21.6	227	19.9
258	21.3	27.8	258	21.8
289	23.3	31.8	289	23.6
320	24.4	32.2	320	23.6
350	23.3	27.9	350	22.2
365	22.2	25.9	365	21.2

Minimum temperature (°C)

Day			
1	7.2	6.9	8.1
16	6.0	5.3	7.4
46	4.6	2.4	6.3
77	4.1	1.8	4.3
107	3.1	1.2	3.3
138	3.0	1.0	3.3
166	3.8	2.6	3.8
197	2.3	1.6	4.8
227	4.9	5.0	6.8
258	7.7	6.9	8.7
289	8.5	10.1	10.9
320	9.0	9.3	10.6
350	8.5	8.7	8.9
365	7.3	7.1	8.2

Solar irradiation (SOLAR) (MJ m^{-2} ground day^{-1})

Day			
1	13.9	16.4	15.8
16	12.6	14.8	12.9
46	10.5	11.7	8.1
77	8.1	9.7	6.3
107	8.9	10.9	7.9
138	11.7	12.6	11.2
166	13.3	15.3	14.9
197	19.1	11.9	20.2
227	16.9	20.9	24.3
258	18.5	23.2	26.4
289	19.2	22.3	26.0
320	17.9	21.4	23.7
350	14.1	18.0	19.0
365	14.0	16.5	16.0

Table 11.11. (Continued)

Julian day number[b]	Camp Pendleton	Echo Valley	Julian day number	Fundo Santa Laura
Relative humidity (RELHUM) (%)				
1	62	30	1	44
16	60	30	16	46
46	65	32	46	54
77	67	39	77	60
107	67	36	107	61
138	68	35	138	53
166	66	40	166	49
197	59	32	197	48
227	61	31	227	44
258	66	28	258	44
289	63	27	289	45
320	65	26	320	44
350	64	30	350	43
365	62	30	365	44
Fraction of sky overcast (OVF) (fraction)				
1	0.72	0.75	1	0.76
16	0.74	0.75	16	0.69
46	0.79	0.74	46	0.64
77	0.72	0.67	77	0.66
107	0.63	0.72	107	0.68
138	0.79	0.69	138	0.66
166	0.68	0.70	166	0.70
197	0.79	0.76	197	0.77
227	0.67	0.80	227	0.87
258	0.77	0.84	258	0.90
289	0.75	0.85	289	0.90
320	0.78	0.88	320	0.88

Julian day	PPT	PPT	Julian day	PPT
350	0.69	0.76	350	0.83
365	0.71	0.76	365	0.76

Precipitation (PPT) (mm)

Julian day	PPT	PPT	Julian day	PPT
274	14.0	43.1	91	13.8
305	37.0	48.5	121	80.6
335	38.5	65.2	152	142.8
1	22.8	57.0	182	181.4
32	22.6	114.2	213	83.7
60	49.2	59.8	244	30.0
91	3.9	38.1	274	36.7
121	3.3	16.2	305	23.1
152	6.4	3.6	335	0
182	0.2	9.0	1	0
213	0	8.3	32	0
244	3.0	11.1	60	0.4

[a] The julian day number is the day number within a year, starting from 1 January.

Table 11.12. Equations for water used in the MEDECS simulator[a]

Equations

Evaporation from soil surface

$$\text{VAPOR} = (6.11) \exp [(17.4) \text{ TA}/(\text{TA} + 239)] \tag{1}$$
$$(\text{RELHUM})/100$$

$$\text{IRSKY} = 1440 \, (\sigma) \, (\text{TMEAN} + 273)^4 \, [0.44 + (0.08) \sqrt{\text{VAPOR}}] \tag{2}$$
$$[1 + \text{COEFC} \, (1 - \text{OVF})]$$

$$\text{PE} = (1/\text{L}) \, [\text{SLOPEG} \, (\text{RNETGRD})/\gamma + \text{AIRTERMG}] / \tag{3}$$
$$(1 + \text{SLOPEG}/\gamma)$$

$$\text{SLOPEG} = 17.4 \, (\text{ESTG}) \, [(1) - \text{TG}/(\text{TG} + 239)] / (\text{TG} + 239) \tag{4}$$

$$\text{ESTG} = (6.11) \exp [(17.4) \, (\text{TG})/(\text{TG} + 239)] \tag{5}$$

$$\text{AIRTERMG} = 0.35 \, [0.95 \, (\text{ESTG}) - \text{ESTA}] \, (0.5 + \text{WINDG}/160) \tag{6}$$

$$\text{ESTA} = (6.11) \exp [(17.4) \, (\text{TMEAN})/(\text{TMEAN} + 239)] \tag{7}$$

$$\text{RNETGRD} = \text{SOLAR} \, (1 - \text{COVER}) \, (1 - \text{ALBEDOG}) \tag{8}$$
$$+ \text{IRSKY} \, (1 - \text{COVER}) \, (\text{EGRD})$$
$$+ \text{COVER} \, (\text{ELEAF}) \, (1440) \, (\sigma) \, (\text{TMEAN} + 273)^4 \, (\text{EGRD})$$
$$- 1440 \, (\sigma) \, (\text{TMEAN} + 273)^4 \, (\text{EGRD})$$

$$\text{COVER} = 1 - \exp [-\text{K(AF)}] \tag{9}$$

$$\text{ES} = \text{PE}(\theta\text{SURF} - \theta\text{MIN})/(\theta\text{MAX} - \theta\text{MIN}) \tag{10}$$

Effective precipitation

$$\text{EP} = \text{PPT} - \text{NFILL} \, (\text{SC}) \, (\text{TAF}) \tag{11}$$

Addition of water to soil level for rainy day with more than 12.5 mm precipitation

$$\text{PA}_i = (\theta\text{MAX}_i - \theta_i) \exp [-\text{C1}(\text{ZD}_i)] \tag{12}$$

$$\text{ACTUALADD}_i = 0.1 \, (\text{EP}) \, (\text{PA}_i / \sum_{i=1}^{m} \text{PA}_i) \tag{13}$$

Removal of water from soil levels because of soil evaporation

$$\text{PEL}_i = (\theta_i - \theta\text{MIN}_i) \exp [-\text{C2}(\text{ZD}_i)] \tag{14}$$

$$\text{AEL}_i = \text{ES}(\text{PEL}_i / \sum_{i=1}^{m} \text{PEL}_i) \tag{15}$$

Transpiration for each species

$$\text{PET} = (1/\text{L}) \, [\text{SLOPE} \, (\text{COVER}) \, (\text{RNET}) \, (\text{AL}/\text{AF}) + \text{AIRTERM}] / \tag{16}$$
$$[\text{SLOPE} + \gamma(\text{HA} + \text{HL})/\text{HL}]$$

$$\text{SLOPE} = 17.4 \, (\text{ESTAIR}) \, [(1) - \text{TA}/(\text{TA} + 239)] / (\text{TA} + 239) \tag{17}$$

$$\text{ESTAIR} = (6.11) \exp [(17.4) \, (\text{TA}/(\text{TA} + 239)] \tag{18}$$

$$\text{RNET} = (1 - \text{REFF}) \, (\text{SOLAR}) + (\text{DAYL}/24) \, [\text{IRSKY} - 1440(\sigma) \tag{19}$$
$$(\text{TA} + 273)^4]$$

Table 11.12. (Continued)

Equations

$$HA = 0.003045 \sqrt{WIDTH/WIND} + 63/WIND \tag{20}$$

$$HL = [RC1 + RC2(\Psi PLANT)] 86400 \tag{21}$$
Note: if HL < HMIN, then HL = HMIN

$$\psi PLANT = C\psi 1 + C\psi 2(\psi SOIL) + C\psi 3(\psi SOIL)^2 + C\psi 4(\psi SOIL)^3 \tag{22}$$

$$AIRTERM = a(AL)\,(\gamma)\,(\rho CP)\,(HA)\,(ESTAIR - VAPOR)\,(DAYL/24) \tag{23}$$

$$AT = \sum_{i=1}^{m} (AL/TAL)\,(PET)\,f\,(\theta_i)\,(ROOTLENGTH_i/ \tag{24}$$
TOTROOTLENGTH)

Definition of Symbols		
AF	Foliage area index for one species	m^2 foliage m^{-2} ground
AL	Leaf area index for one species	m^2 leaf m^{-2} ground
ACTUALADD$_i$	Actual amount of water added to soil level i	cm^3 H_2O day^{-1}
AEL$_i$	Actual evaporation of water from soil level i	cm^3 H_2O day^{-1}
AIRTERM	Contribution of drying power of atmosphere to transpirational demand	MJ m^{-2} ground day^{-1}
AIRTERMG	Contribution of drying power of atmosphere to evaporative demand from the ground	MJ m^{-2} ground day^{-1}
ALBEDOG	Albedo of ground	fraction
AT	Actual transpiration for one species	cm^3 H_2O cm^{-2} ground day^{-1}
C1	Extinction coefficient of soil water addition with depth (0.1)	cm^{-1}
C2	Extinction coefficient for soil evaporation with depth (0.1)	cm^{-1}
COEFC	Coefficient of cloudiness	fraction

Table 11.12. (Continued)

Definition of Symbols		
COVER	Fraction of ground surface covered by foliage	fraction
$C\psi 1$ to $C\psi 4$	Coefficients relating soil water potential in rooting zone to plant water potential	MPa, fraction, MPa^{-1}, MPa^{-2}
DAYL	Daylight period	h
EA	Actual vapor pressure of air	MPa
EGRD	Emissivity of ground	fraction
ELEAF	Emissivity of leaf	fraction
EP	Precipitation striking surface of soil	mm H_2O day^{-1}
ES	Estimated soil evaporation	cm^3 H_2O day^{-1} cm^{-2} ground
ESTA	Saturation vapor pressure at air temperature (TMEAN)	MPa
ESTAIR	Saturation vapor pressure at air temperature (TA)	MPa
ESTG	Saturation vapor pressure at ground temperature (TG)	MPa
$f(\theta_i)$	Reduction of water absorbed due to soil dryness. (Interpolate from table below.)	fraction
HA	Average air boundary layer conductance of water during daylight period	cm day^{-1}
HL	Average leaf stomatal conductance of water during daylight period	cm day^{-1}
HMIN	Minimum possible leaf stomatal conductance of water during daylight period	cm day^{-1}

Table 11.12. (Continued)

Definition of Symbols		
IRSKY	Infrared radiation from sky received by top of canopy	MJ m^{-2} ground day^{-1}
K	Extinction coefficient of total sky radiation with leaf area (0.4)	m^2 ground m^{-2} foliage
L	Latent heat of evaporation	J cm^{-3} H$_2$O
m	Number of soil levels (8)	number
NFILL	Number of times SC is filled per rainy day	day^{-1}
OVF	Cloud cover	fraction
PA$_i$	Potential infiltration and addition of water to soil level$_i$	cm^3 H$_2$O day^{-1}
PE	Potential evaporation from soil surface	cm^3 H$_2$O cm^{-2} ground day^{-1}
PEL$_i$	Potential evaporation from soil level i	cm^3 H$_2$O day^{-1}
PET	Potential transpiration by one species	cm^3 H$_2$O cm^{-2} ground day^{-1}
PPT	Precipitation	mm H$_2$O day^{-1}
RC1	Maximum leaf conductance of water	cm s^{-1}
RC2	Slope of leaf conductance vs. ψPLANT curve	cm s^{-1} MPa^{-1}
REFF	Reflectance of foliage to solar radiation (0.97)	fraction
RELHUM	Relative humidity	number
RNET	Net radiation received by whole canopy	MJ m^{-2} ground day^{-1}

Table 11.12. (Continued)

Definition of Symbols		
RNETGRD	Net radiation (solar and infrared) absorbed by ground surface in a 24-hour period)	MJ m^{-2} ground day^{-1}
ROOTLENGTH$_i$	Total length of absorbing roots in soil level i for one species	cm root
SLOPE, SLOPEG	Slope of vapor pressure vs. temperature curve for air in canopy and at ground surface, respectively	MPa $°C^{-1}$
SOLAR	Incoming solar radiation measured just above canopy	MJ m^{-2} ground day^{-1}
SC	Water storage per foliage area (0.2)	mm H_2O cm^{-2} foliage
TA	Air temperature for daylight period	°C
TAF	Foliage area index for all species	m^2 foliage m^{-2} ground
TAL	Leaf area index for all species	m^2 leaf m^{-2} ground
TG	Temperature of ground surface	°C
TMEAN	Mean temperature for the day	°C
TOTROOTLENGTH	Total length of absorbing roots in all soil levels for one species	cm root m^{-2} ground
VAPOR	Vapor pressure of the air	MPa
WIDTH	Leaf width	cm leaf
WIND	Wind speed	km day^{-1}
WINDG	Wind speed at the ground	km day^{-1}
ZD$_i$	Depth from ground surface to middle of soil level i	cm soil

Table 11.12. (Continued)

Definition of Symbols		
a	Correction of transpiration due to shading obtained from a table function	fraction
γ	Psychrometric constant	MPa $^{\circ}$C^{-1}
θ_i	Actual average water in soil level i	cm^3 H$_2$O cm^{-3} soil
θMAX	Maximum water holding capacity of soil (0.4)	cm^3 H$_2$O cm^{-3} soil
θMIN	Minimum water holding capacity of soil)0.04)	cm^3 H$_2$O cm^{-3} soil
θSURF	Water content of surface soil (soil level 1) (equals θ1)	cm^3 H$_2$O cm^{-3} soil
ρCP	Volumetric heat capacity of air	J m^{-3} air $^{\circ}$C^{-1}
σ	Stefan-Boltzmann constant	MJ m^{-2} s^{-1} $^{\circ}$K^{-4}
ψPLANT	Plant water potential	MPa
ψSOIL	Soil water potential in root zone	MPa

Relation of water content to the water factor of the soil							
$(\theta_i - \theta\text{MIN})/(\theta\text{MAX} - \theta\text{MIN})$	0	0.055	0.083	0.10	0.15	0.30	0.50
$f(\theta_i)$	0	0.001	0.010	0.15	0.25	0.70	0.95

aTable functions are indicated by "f(x)" and are presented in Tables 11.6 through 11.8 and Figures 11.3 through 11.6.

water content (Figure 11.6) and then multiplied by root length in the layer. If the total potentially transpired water in a layer exceeded the amount of water available, the available water was apportioned to each species according to its proportion of the total transpiration demand. Total actual transpiration for a species was the sum of its transpiration for each soil level (Table 11.12:Eq. 24).

11.3.4 Photosynthesis and Respiration

Because the maximum photosynthetic rate cannot be determined in the field, data are only available for rates with a given solar irradiance, temperature, and leaf area index. In the model, the rate was determined from a three-way table function relating these factors (Table 11.8). To correct for water stress effects on carbon dioxide availability via its effect on stomatal opening, the photosynthetic rate was then multiplied by the ratio of actual to potential transpiration (Table 11.13:Eq. 1).

Energy from respiration was used both to maintain existing live biomass and to synthesize new biomass. If there was not enough carbohydrate available for both functions, maintenance was assumed to take precedence over growth. Therefore, total maintenance costs for each day were subtracted from the total nonstructural carbohydrate reserves before growth was calculated. Maintenance respiration for any plant part was calculated by a table function involving temperature (Chapter 7; Figure 11.4; Table 11.13:Eqs. 2 to 6). Maintenance respiration for leaves was calculated for the dark period only, because respiration during the daylight hours was included in photosynthesis. For all other plant parts, the day and night respiration rates were based on day and night temperature, respectively, and a weighted average based on the length of the daylight period constituted the respiration rate for the 24-hour period. Growth respiration was based on the relative amounts of various constituents of new tissue (Miller, 1979) assuming standard carbohydrate costs (Penning de Vries, 1972b, 1974) for each constituent. In general, the synthesis of 1 g of new, evergreen leaf tissue required 1.7 g of carbohydrate, and the synthesis of 1 g of new stem, root, or deciduous leaf tissue required 1.3 g of carbohydrate (Table 11.13:Eqs. 7 to 9).

11.3.5 Plant Growth

The parameters determining growth varied for each species (Table 11.5). Once the simulation began, bud dormancy was broken and growth began only after two requirements were met. The first requirement was that the soil moisture content be above a certain minimum. The second requirement was that either the plant experienced a certain number of degree hours above a critical temperature (Chapter 4) or a certain day in the season be reached. The critical temperature was $15°C$ for Echo Valley (Chapter 4) and $10°C$ for Fundo Santa Laura. Degree hours were counted only on those days when the soil water content at a depth of 0.2 to 0.3 m was above the wilting point of 0.055 cm^3 H_2O cm^{-3} soil.

Stem growth commenced as soon as bud dormancy was broken, and leaf growth began a certain number of days later. Water stress determined the length of the growing season and occurred whenever the soil water potential was low enough to stop growth for the day (Figure 11.5). Once the number of consecutive water stress days equaled a critical number, growth of stem and leaf stopped for the season. The rate of transfer of nitrogen from the reserves to new leaves depended on the growth rate. Nitrogen reserves were considered as one nitrogen pool located throughout the plant. The growth rate of leaves and stems was calculated by multiplying several factors, all of which were calculated from table functions, by the maximum possible growth rate

Table 11.13. Equations of nutrient production and utilization used in the shrub sub-model of the MEDECS simulator

Equations

Photosynthesis

$$PS = PSEFF \ (AT/PET) \ (BIOL + BIOLO)/GTA \tag{1}$$

Respiration

Maintenance

$$RML = CRML \ (BIOL) \ (1 - DAYL/24) \tag{2}$$

$$RMLO = CRMLO \ (BIOL) \ (1 - DAYL/24) \tag{3}$$

$$RMS = CRMS \ (BIOS) \tag{4}$$

$$RMR = CMR \ (BIOR) \tag{5}$$

$$RMRF = CRMRF \ (BIORF) \tag{6}$$

Growth

$$GRESPL = 0.7(GLEAF) \tag{7}$$

$$GRESPS = 0.7 \ (GSTEM) \tag{8}$$

$$GRESPR = 0.3 \ (SUMTROOTG) \tag{9}$$

Leaf nitrogen dynamics (similar equations used for stem)

Update of nitrogen reserve

$$RESERVN = RESERVNY + BRDOWN + TLOCN - GRON + DEATHL + \\ DEATHS - NSYN \tag{10}$$

$$GRON = GLEAF \ (GRAMN) \tag{11}$$

$$NSYN = C3N \ (RESERVNY + BRDOWN + TLOCN) \tag{12}$$

$$BRDOWN = C4N \ (NLABY) \tag{13}$$

Labile nitrogen pool in new leaves

$$NLAB = C1N \ (GRON) - DEATHL - BRDOWN + NSYN + NLABY \tag{14}$$

Stable nitrogen pool in new leaves

$$NSTAB = C2N \ (GRON) - DEATHS + NSTABY \tag{15}$$

Translocation of reserves

Average available nitrogen content of whole plant

$$AA = (ROOTN + RESERVNS)/(BIOR + BIOS) \tag{16}$$

Translocation of N to or from root

$$TLOCN = ROOTN - BIOR \ (AA) \tag{17}$$

Table 11.13. (Continued)

	Definition of Symbols	
AA	Average percent available N in whole	g N g^{-1} dry wt plant
AT	Actual transpiration	cm^3 H$_2$O cm^{-2} ground day^{-1}
BIOL	New leaf biomass	g dry wt m^{-2} ground
BIOLO	Old leaf biomass	g dry wt m^{-2} ground
BIOR	Total biomass of root	g dry wt m^{-2} ground
BIORF	Fine root biomass	g dry wt fine root m^{-2} ground
BIOS	Total biomass of stem	g dry wt m^{-2} ground
BRDOWN	Labile N which breaks down into available N	g N stem^{-1} day^{-1}
CRML	Maintenance respiration per gram of leaf; depends on temperature	g TNC[a] g^{-1} tissue
CRMLO	Maintenance respiration per gram of old leaf; depends on temperature	g TNC g^{-1} tissue
CRMR	Maintenance respiration per gram of root; depends on temperature	g TNC g^{-1} tissue
CRMRF	Maintenance respiration per gram of fine root; depends on temperature	g TNC g^{-1} tissue
CRMS	Maintenance respiration per gram of stem; depends on temperature	g TNC g^{-1} tissue
C1N	Fraction of total N added by growth which is labile N	fraction
C2N	Fraction of total N added by growth which is stable N	fraction
C3N	Fraction of available N reserves re-synthesized into labile N per day	day^{-1}
C4N	Fraction of labile N breaking down per day (0.1)	day^{-1}

Table 11.13. (Continued)

Definition of Symbols		
DAYL	Daylight period	h
DEATHL	Labile N lost with dead leaves	g N stem^{-1} day^{-1}
DEATHS	Stable N lost with dead leaves	g N stem^{-1}
GLEAF	New leaf growth	g leaf dry wt stem^{-1} day^{-1}
GRAMN	Grams of N required for 1 gram of new leaf growth	g N g dry wt^{-1}
GRESPL ⎫ GRESPR ⎬ GRESPS ⎭	Growth respiration of leaves (L), stems (S), and fine roots (R)	g TNC m^{-2} day^{-1}
GRON	Total N added with new growth	g N stem^{-1} day^{-1}
GSTEM	New stem growth	g dry wt m^{-2} day^{-1}
GTA	Weight to area ratio of leaves	g leaf m^{-2} leaf
NLAB	Labile N in new leaves today	g N stem^{-1}
NLABY	Labile N in new leaves from yesterday	g N stem^{-1}
NSTAB	Stable N in new leaves today	g N stem^{-1}
NSTABY	Stable N in new leaves from yesterday	g N stem^{-1}
NSYN	Labile N resynthesized from available N	g N stem^{-1}
PET	Potential transpiration	cm^3 H$_2$O cm^{-2} ground day^{-1}
PS	Photosynthesis	g CO$_2$ day^{-1}
PSEFF	Photosynthesis rate, depends on foliage area index, temperature, and solar radiation; obtained from a table function	g CO$_2$ FAI^{-1}
RESERVN, RESERVNS	Reserve of N available for growth in leaves, stems	g N stem^{-1}

Table 11.13. (Continued)

Definition of Symbols		
RESERVNY	Reserve of N in leaves from yesterday	g N stem^{-1}
RML RMLO RMR RMRF RMS	Maintenance respiration of new (L) and old leaves (LO), stems (S), main roots (R), and fine roots (RF)	g TNC m^{-2} ground
ROOTN	Available N reserves in roots	g N m^{-2} ground
SUMTROOTG	Sum of growth of absorbing roots in all soil levels	g dry wt m^{-2} day^{-1}
TLOCN	N translocated from root reserves to stem reserves	g N stem^{-1}

^{a}TNC = total nonstructural carbohydrate.

(Table 11.14:Eqs. 1 to 3). Labile nitrogen was defined as the nitrogen turning over in the course of normal metabolism. The protein broken down was calculated from a leaf nitrogen turnover rate of 10% a day (Penning de Vries, 1973). Twenty percent of the labile nitrogen in the leaf was transferred to reserves each day because of protein breakdown. Nitrogen reserves were assumed to limit growth only when the reserve was less than 0.1% of leaf total dry weight (table function; Figure 11.5). Total nonstructural carbohydrate reserves affected growth similarly, with maximum growth at reserves above 10% and reductions of growth below this level (Chapter 4; Figure 11.5). The effect of temperature on growth differed by species (Chapter 4; table function; Figure 11.5). Growth was limited by available water according to the soil water potential experienced by the plant and depended on the water potential in each soil level and the absorbing root biomass of the soil level. Growth was maximal when the soil water potential was less than −0.1 MPa and decreased linearly to zero when the water potential reached a minimum value, which varied by species but was about −1.0 MPa (Chapter 4; Figure 11.5). All of the factors affecting growth were fractions which, since they acted simultaneously, were multiplied together to find the total effect on the maximum possible growth rate of leaves and stems (Table 11.14:Eqs. 1 and 3). The amount of nitrogen incorporated in growth depended on the nitrogen content of new mature tissue, which differed for each species (Table 11.4). Of the nitrogen incorporated in growth, 60% became incorporated into labile protein and 40% into stable protein. The changes in nitrogen contents of labile and stable protein and in reserve nitrogen depended on the balance of losses by tissue death and breakdown and the gains by growth (Table 11.13:Eqs. 13 and 14).

Analogous to leaf and stem nitrogen processes, the movement of nitrogen from plant reserves into the absorbing roots depended on the growth of absorbing roots and

Table 11.14. Equations related to growth and death used in the shrub submodels of the MEDECS simulator[a]

Equations

Growth rate of leaves on a growing shoot

$$\text{GLEAF} = (\text{GMAX}) \, f\,(\text{RESERVN}) \, f\,(\text{RESERVTNC}) \, f\,(\psi\text{PLANT}) \quad (1)$$
$$f\,(\text{MEAN})$$

Growth of all leaves on a plant

$$\text{TLEAFG} = \text{NSTEM} \, (\text{SHTPERST}) \, (\text{GLEAF}) \quad (2)$$

Growth rate of stem on a growing shoot

$$\text{GSTEM} = (\text{STMAX}) \, f\,(\text{RESERVN}) \, f\,(\text{RESERVTNC}) \, f\,(\psi\text{PLANT}) \quad (3)$$
$$f\,(\text{TMEAN})$$

Growth of fine roots on a root growing point in soil level i

$$\text{GROOT}_i = (\text{RELONG}) \, (\text{GR\$CM}) \, f\,(\text{ROOTN}) \, f\,(\text{ROOTNC}) \, f\,(\psi\text{SOIL}_i) \quad (4)$$
$$f\,(\text{TG}) \, (\text{ELNFC})$$

Growth of all roots in soil level i

$$\text{TROOTG}_i = (\text{GROOT}_i) \, (\text{RUT}_i) \quad (5)$$

Live root density in soil level i

$$\text{RUTDEN}_i = \text{RTBIO}_i / [\text{DZ}_i (\text{GR\$CM}) \, 10000] \quad (6)$$

Death of old leaves on a branch

$$\text{RDEATH} = \text{BIOLO} \, [f(\text{SOLAR}_1) + f(\psi\text{PLANT})] \quad (7)$$

Death of absorbing roots in soil level i

$$\text{DROOT}_i = (\text{RTBIO}_i) \, f\,(\psi\text{SOIL}_i) \quad (8)$$
$$\text{DEADRT}_i = \text{DEADRTY}_i + \text{DROOT}_i \quad (9)$$

Dead root density in soil level i

$$\text{DRTDEN}_i = \text{DEADRT}_i / [\text{DZ}_i (\text{GR\$CM}) \, 10000] \quad (10)$$

Definition of Symbols		
BIOLO	New leaf biomass	g dry wt m^{-2} ground
DEADRT$_i$	Total dead roots in soil level i today	g dry wt root m^{-2} soil layer^{-1}
DEADRTY$_i$	Total dead roots in soil level i yesterday	g dry wt root m^{-2} soil layer^{-1}

Table 11.14. (Continued)

Definition of Symbols		
$DROOT_i$	Root death in soil level i	g dry wt root m^{-2} soil layer^{-1} day^{-1}
$DRTDEN_i$	Average dead root density in soil level i	cm root cm^{-3}
DZ_i	Depth interval of soil level i	cm soil
GLEAF	Leaf growth rate	g dry wt leaf growing shoot^{-1} day^{-1}
GMAX	Maximum growth of leaves on a growing shoot	g dry wt leaf growing shoot^{-1} day^{-1}
$GROOT_i$	Root growth	g dry wt root fine root^{-1} day^{-1}
GR$CM	Grams per centimeter for a single fine root	g dry wt root cm^{-1} root
GSTEM	Stem growth rate	g dry wt stem growing shoot^{-1} day^{-1}
NSTEM	Number of stems in square meter	stem m^{-2} ground
RDEATH	Death rate of old leaves	g dry wt m^{-2} day^{-1}
RELONG	Maximum elongation rate of a single fine root	cm root day^{-1}
RESERVN	Reserve nitrogen available for growth in leaves (or stems)	g N stem^{-1}
RESERVTNC	Reserve total nonstructural carbohydrate available for growth in leaves (or stems)	g TNC stem^{-1}
ROOTN	Available N reserves in roots	g N m^{-2} ground
ROOTNC	Available total nonstructural carbohydrate in roots	g TNC m^{-2} ground
$RTBIO_i$	Total absorbing root biomass in level i	g dry wt root m^{-2} ground

Table 11.14. (Continued)

	Definition of Symbols	
RUT_i	Number of potential absorbing roots in soil level i	# root
$RUTDEN_i$	Average root density in soil	cm root cm^{-3} soil
SHTPERST	Number of growing shoots per stem	growing shoot stem^{-1}
SOLAR	Solar irradiance at the bottom of the canopy	MJ m^{-2} ground day^{-1}
STMAX	Maximum growth of stem on a growing shoot	g dry wt stem growing shoot^{-1} day^{-1}
STPERPL	Number of stems per plant	stem plant^{-1}
TG	Temperature of the ground	°C
TLEAFG	Total leaf growth	g dry wt leaf m^{-2} day^{-1}
TMEAN	Daily mean temperature of air	°C
$TROOTG_i$	Growth of absorbing roots in soil level i	g dry wt root m^{-2} day^{-1}
ψPLANT	Plant water potential	MPa
ψSOIL	Soil water potential in rooting zone at soil level i	MPa

[a]Table functions are indicated by "f(x)" and are presented in Tables 11.6 through 11.8 and Figures 11.3 through 11.6.

the nitrogen content of the new root tissue. The nitrogen content of new mature absorbing roots was 0.2% (Chapter 9). The maximum rate of root elongation varied by species (Chapter 4; Table 11.5) and was based on field data. Absorbing root growth rates were, in turn, limited by several factors. The limitation caused by the availability of nitrogen was assumed to be the same as for leaves and stems (Figure 11.5), although root growth actually required less nitrogen than did leaf and stem growth, an assumption based on the nitrogen content measured in leaves, stems, and roots (Chapter 9). The limiting levels of total nonstructural carbohydrates were higher for roots than for leaves (Figure 11.5). Root growth was slowed at soil water potentials above −0.1 MPa, was maximum at −0.1 MPa, and decreased linearly to zero at −0.7 MPa (table function, Figure 11.5). Soil water potential was calculated from soil water content by a table function (Ng, 1974; Chapter 6; Table 11.6). The temperature effect was such that root growth was not inhibited at soil temperatures above 20°C, but root growth was zero at 0°C (Figure 11.5). The calculated effects of available nitrogen, total nonstructural carbohydrate, soil water potential, and soil temperature were fractions that acted simultaneously and were multiplied together to find the total effect on the maximum possible growth rate of roots. Total root growth in a soil level was calculated from the product of growth rate of a single root in that level, the amount of conducting root in that level, and the number of growing points, i.e., absorbing roots, per centimeter of conducting root (Kummerow and Krause, 1976; Table 11.14:Eqs. 4 and 5).

11.3.6 Mortality and Litterfall

The death of shrub leaves was caused by low soil water potential and low solar irradiance. Second-year leaves died first. Death of old leaves due to water stress was zero when the soil water potential was above -0.1 MPa. At a water potential of -1.5 MPa, 0.3% of the old leaves died each day, and at -5.0 MPa, 1.0% died per day (Figure 11.5). The table function of leaf death versus soil water potential was obtained from phenological records (Kummerow and Fishbeck, 1977) and midday leaf water potentials (Poole and Miller, 1975; Figure 11.5). Leaf mortality caused by water stress was assumed the same for all species because specific data were lacking for individual species. If, because of shading within the canopy, a leaf was maintained below the light compensation point necessary for photosynthesis, the leaf was shed. When solar irradiance beneath the canopy was less than the light compensation point, old leaves died at a rate that increased to a maximum of 1% per day at zero solar irradiance (Figure 11.5). If air temperatures were below the minimum temperature for growth, no death was caused by low solar irradiance. The fraction of leaves dying from the lack of nitrogen was assumed to be zero. Even though lack of nitrogen contributes to leaf death, in the model it was assumed to have no effect because water stress was a much stronger influence and occurred earlier in the season. The factors affecting leaf mortality were assumed to act independently and simultaneously and were summed to calculate the amount of old leaf biomass dying each day (Table 11.14:Eq. 7). At death, all nitrogen from stable protein in shrub leaves was assigned to litter, and all nitrogen from labile protein was reabsorbed and returned to the general plant reserves.

Death of absorbing roots was caused by low soil water potentials in each soil layer. Root mortality was zero at soil water potentials above -0.5 MPa and increased with lower soil water potentials. When the soil dried to -2.5 MPa, all the absorbing roots died. At death, the absorbing root biomass, but not the conducting root biomass, became soil organic matter.

11.3.7 Soil Nitrogen Processes and Uptake

In MEDECS, the rate of nitrogen release into the exchangeable pool depended on the rate of decay of soil organic matter, on the nutrient content of the soil organic matter, and on the incorporation and immobilization of the released nitrogen in soil microorganisms. The rate of decay was affected by moisture, temperature, and lignin content

(Figure 11.5; Table 11.15:Eq. 1). Under optimum conditions, the maximum rate of decay was assumed to be 0.045 g g^{-1} organic matter day^{-1} (Waksman, 1929; Harpaz, 1975). The decay rate was reduced when soil water content deviated from 0.27 cm^3 cm^{-3} (Figure 11.6). With saturated soils, the decay rate was 55% of maximum. With soils dried to 0.09 cm^3 cm^{-3}, which was common during the summer, the decay rate was 10% of maximum. The decay rate decreased as soil temperature decreased; it was 90% of the maximum at 30°C and zero at 0°C (Figure 11.6). The effect of the lignin content on the decay rate was not known, so the lignin factor was held constant at 0.25 in the model. The nitrogen release factor indicates how much nitrogen was available for root uptake after accounting for the nitrogen incorporated and immobilized by bacteria. At a soil organic nitrogen content greater than 1.4%, i.e., a carbon:nitrogen ratio less than 40, a total of 50% of the nitrogen from decaying organic matter was released in an organic form available to plants, while the other 50% was incorporated into bacteria and was considered to remain in the soil organic nitrogen pool. At a soil organic nitrogen content lower than 1.5%, the released inorganic nitrogen available for root uptake was lower than 50%.

The uptake of nitrogen by the roots depended on the amount already in the soil, the amount released from soil organic matter, the nitrogen diffusion coefficient, and the root density in each soil stratum (Table 11.15:Eqs. 4 and 5). The absorbing root density was highest in the upper 0.3 m of the soil (Kummerow et al., 1977; Chapter 4). Since the plant was assumed to take up all of the nitrogen available to it, the rate of uptake was assumed to be limited by the ability of the soil to supply nitrogen to the root surface rather than by the ability of the plant to absorb nitrogen.

11.3.8 Annual Grass Submodel

MEDECS simulations of the Camp Pendleton site used three shrub species, *A. californica*, *S. mellifera*, and *E. fasciculatum*, and an annual grass, *V. (Festuca) octoflora*. To accommodate the grass growth form, an annual grass submodel was developed that was different from the shrub submodels (Table 11.16).

The annual grass submodel had five compartments: absorbing roots, stems, leaves, inflorescences, and seeds (Figure 11.9). The annual grass passed through seed, vegetative shoot, reproductive shoot, and litter stages.

Grass phenology within the submodel was quite different from that of shrubs. *Vulpia* was dormant as seed from 1 October, the beginning of the simulation period, to at least 1 January. After germination, each plant compartment except the seed compartment went through stages of growth, maturity, senescence, and death. *Vulpia* started from a seed with some nitrogen and nonstructural carbohydrate reserves. The seeds, located in the top two soil layers, were allowed by the model to germinate after 1 January as soon as the soil temperature reached 10°C and the soil moisture rose above 0.06 cm^3 water cm^{-3} soil. After a 5-day germination period, stems and roots in the upper three soil layers were assigned initial biomasses. Dates for phenological phases were counted from the last day of the germination period. The leaf started to grow on day 1, stem growth began on day 2, and inflorescence growth on day 10. Leaves, stems, and inflorescences grew until day 30, 40, and 30, respectively. After leaf and stem growth stopped at maturity and while the inflorescence was maturing, nitrogen taken

Table 11.15. Equations related to nitrogen release and uptake in the soil used in MEDECS simulator[a]

Equations

Rate of decay of soil organic matter in soil level i

$$RDK_i = SOM_i(RDKMAX) \, f(\theta_i) \, f(TG_i) \, f(PLIGNIN_i) \, f(N) \qquad (1)$$

Rate of release of nitrogen from soil level i

$$RNREL_i = (RDK_i)(SOMN_i/SOM_i)(RELNF) \qquad (2)$$

$$x = SOMN_i/SOM_i$$

$$RELNF = \exp[69(x^2) - 1] \qquad (3)$$

Note: RELNF is 1 if $x \geqslant 0.014$
RELNF is 0 if $x \leqslant 0.004$

Rate of uptake by the plant of nitrogen from soil level i

$$RNUP_i = [(SSN_i)/TAU](DIFFN) \, 1.28 \, (RUTDEN_i) \qquad (4)$$

$$SSN_i = SSNY_i + RNREL_i \, (1 \text{ day}) \qquad (5)$$

Definition of Symbols		
DIFFN	N diffusion coefficient	$cm^2 \, day^{-1}$
PLIGNIN$_i$	Percent of lignin in soil level i	fraction
RDK$_i$	Rate of decay of soil organic matter in soil level i	g dry wt cm^{-3} soil day^{-1}
RDKMAX	Maximum rate of decay of soil organic matter	g dry wt g dry wt^{-1} day^{-1}
RELNF	g N released per g soil organic matter N decayed	fraction
RNREL$_i$	Rate of N release from soil level i	g N cm^{-3} soil day^{-1}
RNUP$_i$	Rate of uptake of N from soil level i	g N cm^{-3} soil day^{-1}
RUTDEN$_i$	Root density per soil level i	cm root cm^{-3} soil
SOM$_i$	Soil organic matter in soil level i	g dry wt cm^{-3} soil
SOMN$_i$	Organic N in soil level i	g N cm^{-3} soil
SSN$_i$	Exchangeable inorganic soil N in soil level i today	g N cm^{-3} soil

Table 11.15. (Continued)

Definition of Symbols		
SSNY$_i$	Exchangeable inorganic soil N in soil level i yesterday	g N cm^{-3} soil
TAU	Theoretical number of days needed to remove all exchangeable nitrogen from soil	day
TG$_i$	Temperature of the soil at level i	°C
θ_i	Water content of soil level i	g H$_2$O cm^{-3} soil

^{a}Table functions are indicated by "f(x)" and are presented in Tables 11.6 through 11.8 and Figures 11.3 through 11.6.

up by the roots and photosynthate left over after maintenance respiration were transferred to the inflorescence for incorporation into seeds. The relative amounts of nitrogen and total nonstructural carbohydrates within the seed remained constant throughout this transfer (Harpaz, 1975).

Senescence for leaves, stems, and inflorescences began at day 50, 55, and 60, respectively. Nitrogen from broken down labile protein was reabsorbed by the inflorescence to produce seeds, leaving the nitrogen of stable protein behind. Leaves and stems were dead at day 70 and 80, respectively, and became standing dead material. Seed set occurred after all leaves, stems, and roots were dead and after all available nitrogen from labile protein and all available nonstructural carbohydrate had been incorporated into the seed. Nongerm seed material was considered part of the standing dead biomass. The number of seeds that started to grow the next spring was calculated by dividing the total seed biomass by the average seed weight.

Simultaneously with aboveground events, annual grass roots were growing, maturing, and dying. The model allowed grass roots to grow only if stems were also growing. Only root tips at the bottom of the root mass grew, and their growth rate depended on nitrogen and nonstructural carbohydrate reserves and on the soil water potential and soil temperature. On a given day, if the water content of any soil layer was less than 0.06 cm^3 water cm^{-3} soil, root growth occurring that day in that layer and in all lower layers was classified as dead roots, but roots already present in all layers at the beginning of the day remained alive. The water content of the second soil layer, 0.02 to 0.05 m, was critical; if it fell below 0.06 cm^3 H$_2$O cm^{-3} soil, all root growth stopped and death of all existing roots began. If the water content rose above 0.06 cm^3 H$_2$O cm^{-3} soil on a later day, root growth resumed. The whole root mass was considered mature, i.e., stopped growing, when it reached a depth of 0.2 m. Permanent root death began when stems senesced, after which a fraction of the roots died each day.

Because the grass was a subcanopy under the shrubs, the incoming solar irradiance available for grass photosynthesis was reduced by the shrub canopy. The high foliage area index within the grass subcanopy produced selfshading, further reducing the average solar irradiance within the annual grass canopy. The photosynthetic rate was

Table 11.16. Equations for annual grasses for photosynthesis, respiration, and growth used in MEDECS simulator[a]

Equations

Photosynthesis in annual grasses

$$PSL = AT\ (PSEFF/PET)\ (AL/TAI)\ (FACLF)\ DAYL/24 \tag{1}$$

$$PSST = AT\ (PSEFF/PET)\ (AS/TAI)\ (FACST)\ DAYL \tag{2}$$

$$PSIN = AT\ (PSEFF/PET)\ (AI/TAI)\ (FACIN)\ DAYL \tag{3}$$

$$PSTOT = PSL + PSST + PSIN \tag{4}$$

Respiration in annual grasses

Growth

$$GRESPL = 0.3\ (GLEAF) \tag{5}$$

$$GRESPS = 0.3\ (GSTEM) \tag{6}$$

$$GRESPI = 0.3\ (GINFLOR) \tag{7}$$

$$GRESPR = 0.3\ (GROOT) \tag{8}$$

Maintenance

$$RML = (BLEAF)\ (1 - DAYL/24)\ f\ (TEMP) \tag{9}$$

$$RMS = (BSTEM)\ (1 - DAYL/24)\ f\ (TEMP) \tag{10}$$

$$RMI = (BINFLOR)\ (1 - DAYL/24)\ f\ (TEMP) \tag{11}$$

$$RMF = (BROOTF)\ f\ (TEMP) \tag{12}$$

Growth rates in annual grasses

$$GLEAF = (GMAX)\ Min\ [f(n),\ f(TNC),\ f(TEMP),\ f(\psi SOIL)] \tag{13}$$

$$GSTEM = (STMAX)\ Min\ [f(N),\ f(TNC),\ f(TEMP),\ f(\psi SOIL)] \tag{14}$$

$$GINFLOR = (GINMAX)\ Min\ [f(N),\ f(TNC),\ f(TEMP),\ f(\psi SOIL)] \tag{15}$$

$$GROOT = (RMAX)\ Min\ [f(N),\ f(TNC),\ f(TEMP),\ f(\psi SOIL)] \tag{16}$$

Definition of Symbols		
AI	Inflorescence area index	m^2 inflorescence m^{-2} ground
AL	Leaf area index	m^2 leaf m^{-2} ground
AS	Stem area index	m^2 stem m^{-2} ground
AT	Actual transpiration	g H_2O day^{-1}
BINFLOR	Biomass of inflorescences	g indiv^{-1}

Table 11.16. (Continued)

Definition of Symbols		
BLEAF	Biomass of leaves	g indiv^{-1}
BROOTF	Biomass of fine roots, summed for all soil levels	g indiv^{-1}
BSTEM	Biomass of stems	g indiv^{-1}
DAYL	Daylength	hours
FACIN	Photosynthetic efficiency of inflorescence tissue	fraction
FACLF	Photosynthetic efficiency of leaf tissue	fraction
FACST	Photosynthetic efficiency of stem tissue	fraction
GINFLOR	Inflorescence growth rate	g inflor dry wt indiv^{-1} day^{-1}
GINMAX	Potential maximum growth rate of inflorescences	g inflor dry wt indiv^{-1} day^{-1}
GLEAF	Leaf growth rate	g leaf dry wt indiv^{-1} day^{-1}
GMAX	Potential maximum growth rate of leaves	g leaf dry wt indiv^{-1} day^{-1}
GRESPI	Growth respiration for inflorescence	g TNC indiv^{-1} day^{-1}
GRESPL	Growth respiration for leaves	g TNC indiv^{-1} day^{-1}
GRESPR	Growth respiration for fine roots	g TNC indiv^{-1} day^{-1}
GRESPS	Growth respiration for stems	g TNC indiv^{-1} day^{-1}
GROOT	Fine root growth rate	g root dry wt indiv^{-1} day^{-1}
GSTEM	Stem growth rate	g stem dry wt indiv^{-1} day^{-1}
N	Concentration of N in whole plant	g N g^{-1} indiv
PET	Potential transpiration	g H$_2$O FAI^{-1}
PSEFF	Photosynthesis rate; depends on foliage area index and solar radiation; obtained from a table function	g CO$_2$ FAI^{-1}

Table 11.16. (Continued)

	Definition of Symbols	
PSIN	Inflorescence photosynthesis	g CO_2 day^{-1}
PSL	Leaf photosynthesis	g CO_2 day^{-1}
PSST	Stem photosynthesis	g CO_2 day^{-1}
PSTOT	Total photosynthesis	g CO_2 day^{-1}
RMAX	Potential maximum growth rate of fine roots	g dry wt indiv^{-1} day^{-1}
RMI	Maintenance respiration for inflorescence	g TNC indiv^{-1} day^{-1}
RML	Maintenance respiration for leaves	g TNC indiv^{-1} day^{-1}
RMRF	Maintenance respiration for fine roots	g TNC indiv^{-1} day^{-1}
RMS	Maintenance respiration for stems	g TNC indiv^{-1} day^{-1}
STMAX	Potential maximum growth rate of stems	g dry wt indiv^{-1} day^{-1}
TAI	Total area index	m^2 foliage m^{-2} ground
TEMP	Temperature of air	°C
TNC	Concentration of total nonstructural carbohydrate in whole plant	g TNC g^{-1} indiv
ψSOIL	Water potential of soil; obtained from a table function	MPa

[a]Table functions are indicated by "f(x)" and are presented in Tables 11.6 through 11.8 and Figures 11.3 through 11.6.

calculated from a table function similar to van Keulen's (1975), which contained photosynthetic rates versus reduced solar irradiance values. Leaves, stems, and inflorescences were included in the photosynthesis calculation, since all of these components were green.

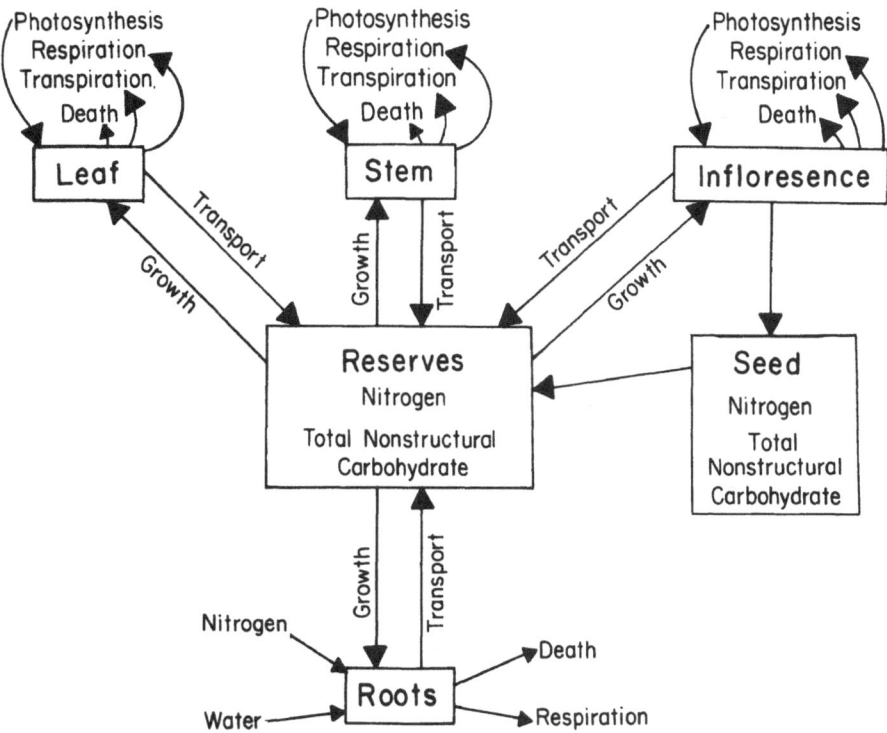

Figure 11.9. Diagram of compartments and transfers in MEDECS for the annual grass submodel.

11.4. Validation

11.4.1 Validation of the Canopy Process Simulator

A multiprocess model such as CAPS can be validated either by comparing the final outcome of the total model, in this case the flux of carbon dioxide into the canopy, with data as was done by Lemon (1960) or by validating each process or a combination of processes, such as radiation, water relations, or the light response of an individual leaf. The second method of validation was used. If each of the various submodels such as radiation extinction through the canopy and water relations are valid, results generated by the integration of the submodels should be valid.

Physical climatological processes were measured and simulated for profiles of radiation components, air temperature, and humidity within a canopy (Miller, 1973; Roberts, 1974; Lawrence, 1975; Stoner et al., 1978a; Chapter 5). The results indicated that profiles of total, downward diffuse, and upward reflected solar irradiance, air temperature; and humidity within the canopy could be reasonably simulated over a range of conditions. Infrared irradiance within the canopy was reasonably simulated but did not show a wide range of measured or simulated values (Figure 11.10).

Canopy leaf water potential involved processes such as energy exchange, water uptake, and transpiration (Chapters 5 and 6). The fact that CAPS accurately predicted

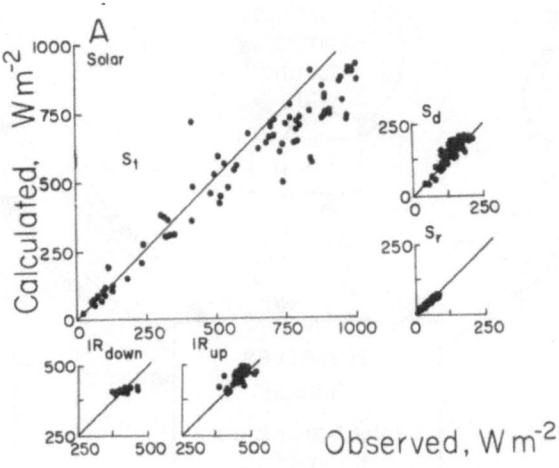

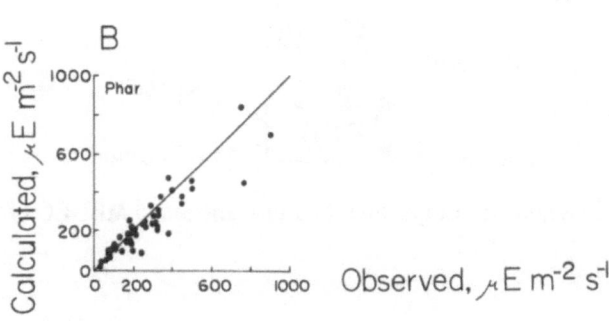

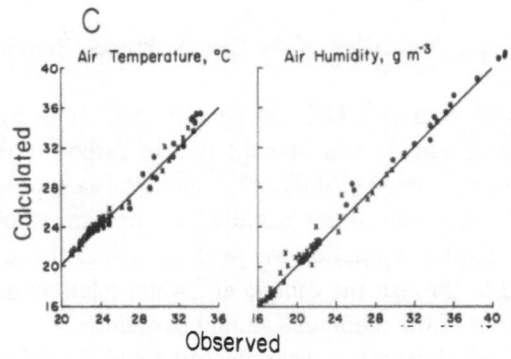

Figure 11.10. Comparison of observed values versus values calculated by a CAPS simulation using foliage area indices determined by Lawrence (1975). (A) Observed versus calculated radiation fluxes (W m^{-2}) for total solar radiation (S_t), diffuse solar irradiation (S_d), reflected solar radiation (S_r), and downward (IR_{down}) and upward (IR_{up}) infrared radiation. Observed values from Lawrence (1975) and Miller et al. (1977). (B) Observed versus calculated (μE m^{-2} s^{-1}) photosynthetically active radiation (Phar). (C) Observed versus calculated air temperature (°C) and air humidity (g m^{-3}) with a chaparral canopy for 4 July (X) and 5 July (•).

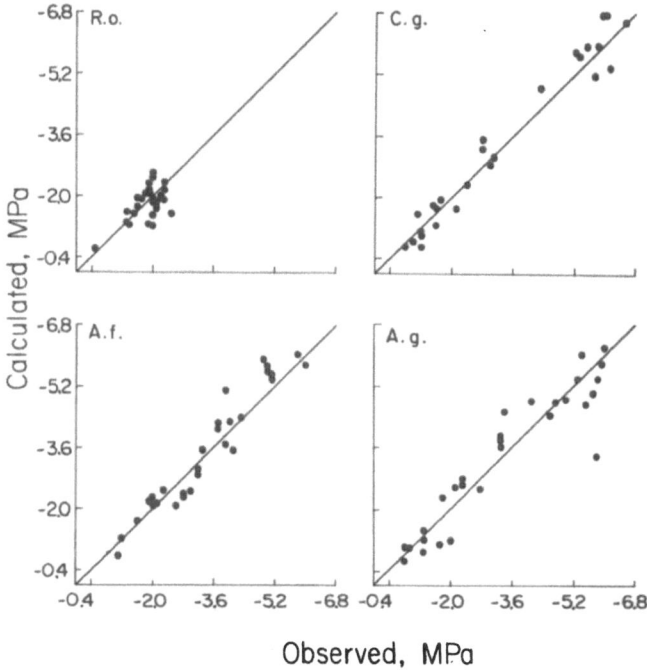

Figure 11.11. Observed (●) leaf water potential (MPa) versus values calculated (–) by a CAPS simulation for four Californian shrub species: *Rhus ovata* (R.o.), *Ceanothus greggii* (C.g.), *Adenostoma fasciculatum* (A.f.), *Arctostaphylos glauca* (A.g.). Observed data were physiological responses from Poole and Miller (1975, 1978) and Miller and Poole (1979). Foliage areas for CAPS were taken from average values on slopes (Thrower and Bradbury, 1977).

leaf water potentials (Figure 11.11) indicated that the plant water submodels were probably also correct.

The carbon dioxide uptake by a single leaf at the top of the canopy was calculated with CAPS and compared with measured data (Chapter 7). Overall, the predicted carbon dioxide values were close to those observed during the day (Figure 11.12). At the onset of the dark period, simulated values were below observed values. The disparity was due to higher observed than calculated leaf and air temperatures during the night. The disparity decreased as the night progressed.

Physical processes such as radiation exchange, air temperature, and humidity within a level in the canopy were simulated with reasonable accuracy. Water potentials based largely on these processes also were accurately simulated by using CAPS. The photosynthetic rates of a single leaf at the top of the canopy were predicted accurately by using data on light and temperature responses of individual leaves. In conjunction with accurate prediction of radiation, temperature, and humidity throughout the canopy levels, the carbon dioxide fluxes simulated through the canopy were reasonable.

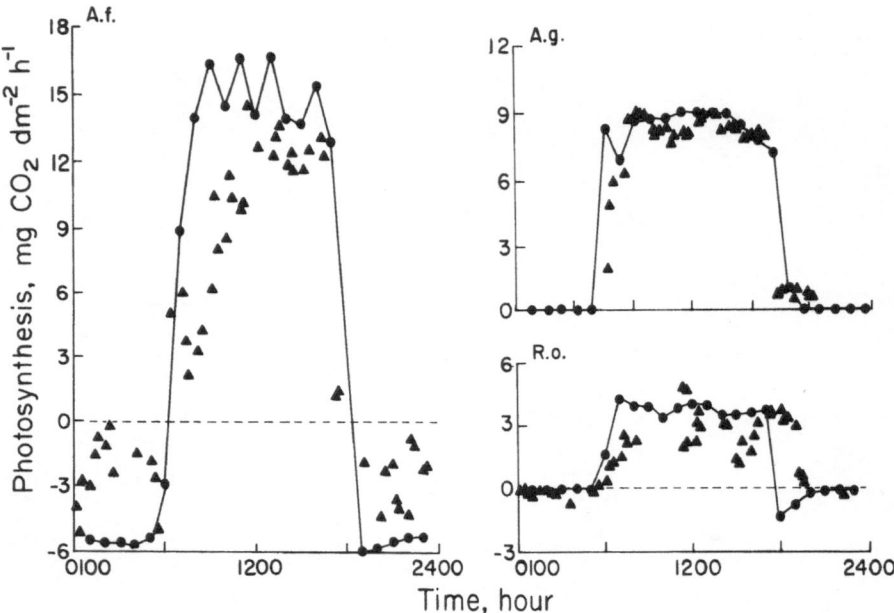

Figure 11.12. Diurnal progression of net photosynthesis (mg CO_2 dm^{-2} h^{-1}) for three Californian species: *Adenostoma fasciculatum* (A.f.), *Arctostaphylos glauca* (A.g.), and *Rhus ovata* (R.o.), observed (▲) and calculated by a CAPS simulation (●).

11.4.2 Validation of the Mediterranean Ecosystem Simulator

MEDECS contained several submodels, each of which could be validated. Simulations were made for 1 year at each site to indicate the accuracy of the submodels. Comparisons of the simulated results to field measurements were made on days when field data were available. When reporting the correlation coefficients below, the number of days for which field data were available is given in parentheses.

Close agreement between simulated and measured values for soil moisture at different study sites indicate the generality of the soil moisture submodel. The seasonal progression of soil water at different soil depths was simulated for the years 1972-1973 and 1975-1976 at Echo Valley and Fundo Santa Laura (Figure 11.13) and compared to field measurements (Miller and Poole, 1979; Ng and Miller, 1980; Chapter 2). Overall, the simulated soil water contents were reasonably close to observed values at both Echo Valley and Fundo Santa Laura. Correlation coefficients were: 0.86 (N = 131) for simulations using *A. glauca* and field data for solar irradiance and soil moisture for the pole-facing slope at Echo Valley, 0.84 (N = 141) for simulations using *A. fasciculatum* and field data for solar irradiance and soil moisture for the equator-facing slope at Echo Valley, and 0.70 (N = 100) for a simulated mixture of *L. caustica* and *C. odorifera* and field data for solar irradiance and soil moisture on the ridgetop at Fundo Santa Laura. Over an annual period, the simulated values varied less than the measured values. The slopes for linear regressions of simulated versus measured values were 0.91, 0.85, and 0.99 for the above three sites, respectively. Comparing values by soil depth indicated that errors in the simulations increased at deeper soil layers. The extraction of water from a soil layer was sensitive to the density of roots within that layer. An

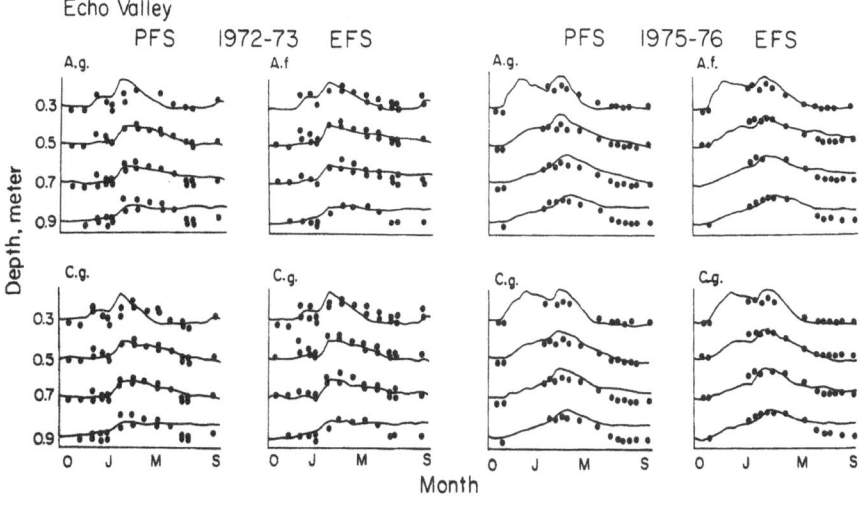

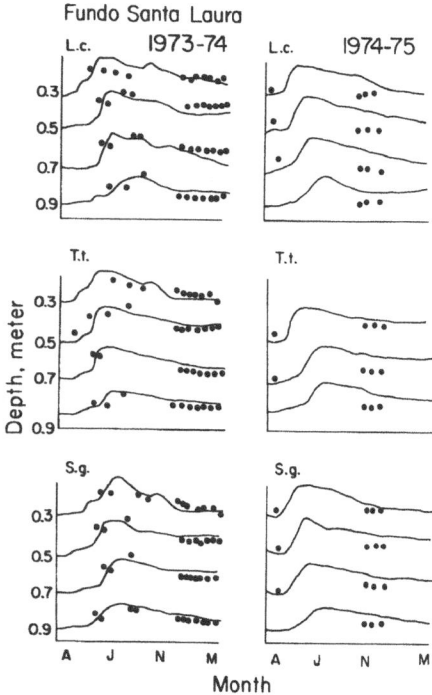

Figure 11.13. Comparison of observed (●) soil water content (cm³ cm⁻³) and values simulated by MEDECS (–) at four soil depths: 0.3, 0.5, 0.7, and 0.9 m through the water year for 1972-1973 and 1975-1976 under *Arctostaphylos glauca* (A.g.), *Adenostoma fasciculatum* (A.f.), and *Ceanothus greggii* (C.g.) on the pole-facing slope (PFS) and equator-facing slope (EFS) at Echo Valley, and through the water year 1973-1974 and 1974-1975 under *Lithraea caustica* (L.c.), *Trevoa trinervis* (T.t.), and *Satureja gilliesii* (S.g.) at Fundo Santa Laura.

underestimation of the absorbing root densities in the deeper soil layers or an underestimation of subsoil drainage would explain the underutilization of water at these depths. An underestimation of the absorbing root densities is plausible in view of problems of root extraction (Chapter 4) and variations in soil depths at Echo Valley and Fundo Santa Laura.

The relationship of annual transpiration to leaf area index calculated by CAPS differed from that calculated by MEDECS because of the lack of a soil water profile and of feedback between transpiration and soil moisture in CAPS (Figure 11.14). The CAPS water submodel uses the amount of water measured in the field to estimate the mean soil water in the root zone on a monthly basis. This soil water content is constant in

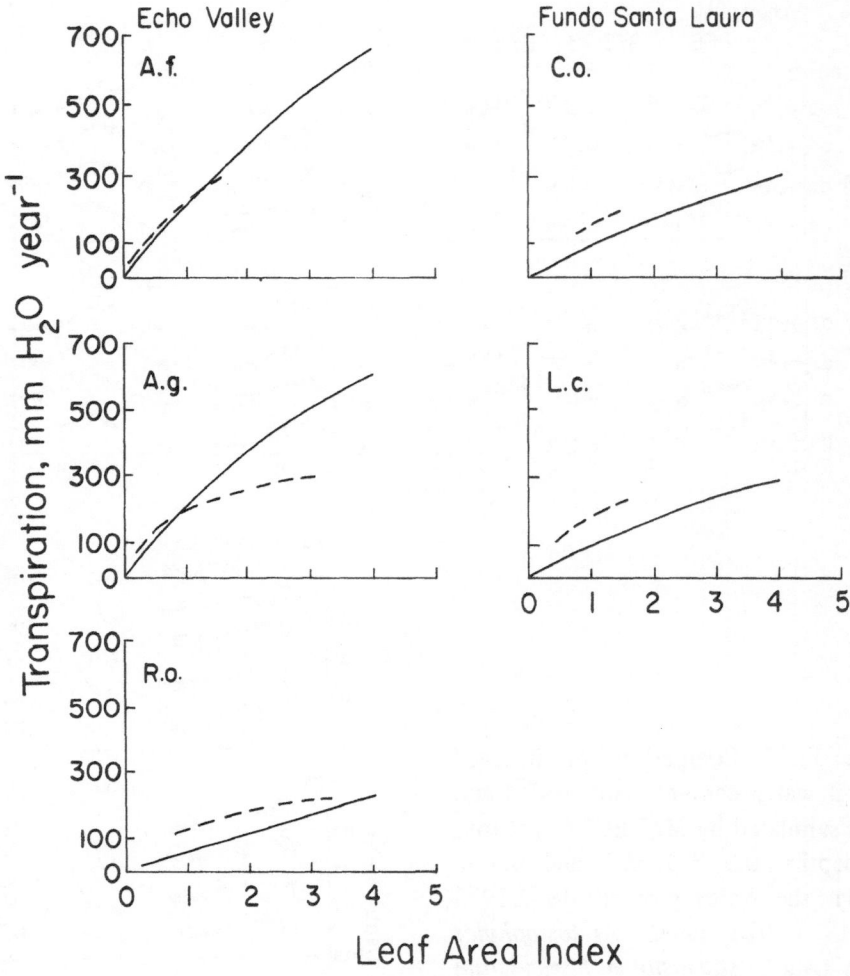

Figure 11.14. Comparison of annual transpiration (mm H_2O year^{-1}) at different leaf area indices as calculated by CAPS (–) and MEDECS (- - -) for *Adenostoma fasciculatum* (A.f.), *Arctostaphylos glauca* (A.g.), and *Rhus ovata* (R.o.) at Echo Valley and *Colliguaya odorifera* (C.o.) and *Lithraea caustica* (L.c.) at Fundo Santa Laura.

CAPS for any month regardless of previous transpiration rates. Because MEDECS has a soil profile and feedback controls between transpiration and soil water content, MEDECS probably provides a more accurate estimate of yearly transpiration. However, field data indicated that soil profiles dry out too fast in the MEDECS soil water submodel. The best estimate of yearly transpiration probably lies between the values calculated by the two models. Nevertheless, the two simulations were correlated (r = 0.87, N = 18).

The seasonal progression of new leaf and stem shrub growth and the seasonal progression of litter were measured and simulated for 1972-1973 at Echo Valley and 1974-1975 and 1975-1976 at Fundo Santa Laura (Mooney, 1977a; Thrower and Bradbury, 1977; Chapter 4). For these years, the correlation between measured and simulated values of new leaf biomass was 0.88 (N = 17) for the Californian species and 0.76 (N = 21) for the Chilean species. The correlation between measured and simulated values for litter was 0.94 (N = 27) for Californian species and 0.85 (N = 31) for Chilean species. MEDECS underestimated both canopy growth and litter production for species at Echo Valley and at Fundo Santa Laura, but especially at Echo Valley (Figure 11.15). The underestimation of growth was probably due to errors in the relationship of growth to total nonstructural carbohydrate (Chapter 8), while the underestimation of the litter production may be due to problems in relating leaf death to water potential as well as problems in simulating new leaf growth.

Root biomass data were collected throughout 1977-1979 by Kummerow et al. (*unpubl. data*; Chapter 4). In most cases, the values simulated were within one standard error of the observed values (Figure 11.16). The correlation between measured and simulated values was 0.95 (N = 6) for a mixture of *A. fasciculatum* and *A. glauca.*

The seasonal progression of total nonstructural carbohydrate and nitrogen contents simulated with MEDECS were compared with measured values for four species at Echo Valley and at Fundo Santa Laura (Chapters 8 and 9; Figures 11.17 and 11.18). In general, the simulated total nonstructural carbohydrate contents were lower than the observed and were not correlated. The simulated nitrogen contents of leaves and stems were close to the observed values. Correlation between the simulated and measured nitrogen contents in leaves and stems were 0.91 (N = 15, $P \leqslant 0.01$) and 0.75 (N = 14, $P \leqslant 0.01$) at Echo Valley and 0.91 (N = 9, $P \leqslant 0.01$) and 0.60 (N = 9, $P \leqslant 0.05$) at Fundo Santa Laura. For leaves and stems, the simulated nitrogen contents were less variable than the measured nitrogen contents.

For validation of the annual grass submodel, simulations of grass growth were compared with grass biomass data collected in 1973 at Echo Valley (Thrower and Bradbury, 1977). Simulated growth started earlier and was more rapid than observed growth (Figure 11.19). However, the peak season biomass was correctly calculated, and the pattern of the simulated growth was reasonable. The discrepancies between the observed and predicted curves may be due to an inadequate understanding of the role of temperature and precipitation in the process of annual grass seed germination. Simulated root biomasses were compared against data taken at Echo Valley in the spring of 1978. The simulated root biomasses were within the standard error limits of the mean measured values for the deeper soil layers but were low when compared to measured values for the upper three soil layers. The simulated water potential in the top three soil layers might be too low for root growth to take place.

11.5. Conclusions

The Canopy Process Simulator (CAPS) and Mediterranean Ecosystem Simulator (MEDECS) are complex models that simulate primary production. Each contains a synthesis of several submodels, which have been derived from earlier models of various plant processes. CAPS deals with the canopy of a single species in one diurnal period. MEDECS simulates a maximum of four species, with their canopies, roots, and soil, in a community over a period of 1 year. CAPS accurately predicted the physical environ-

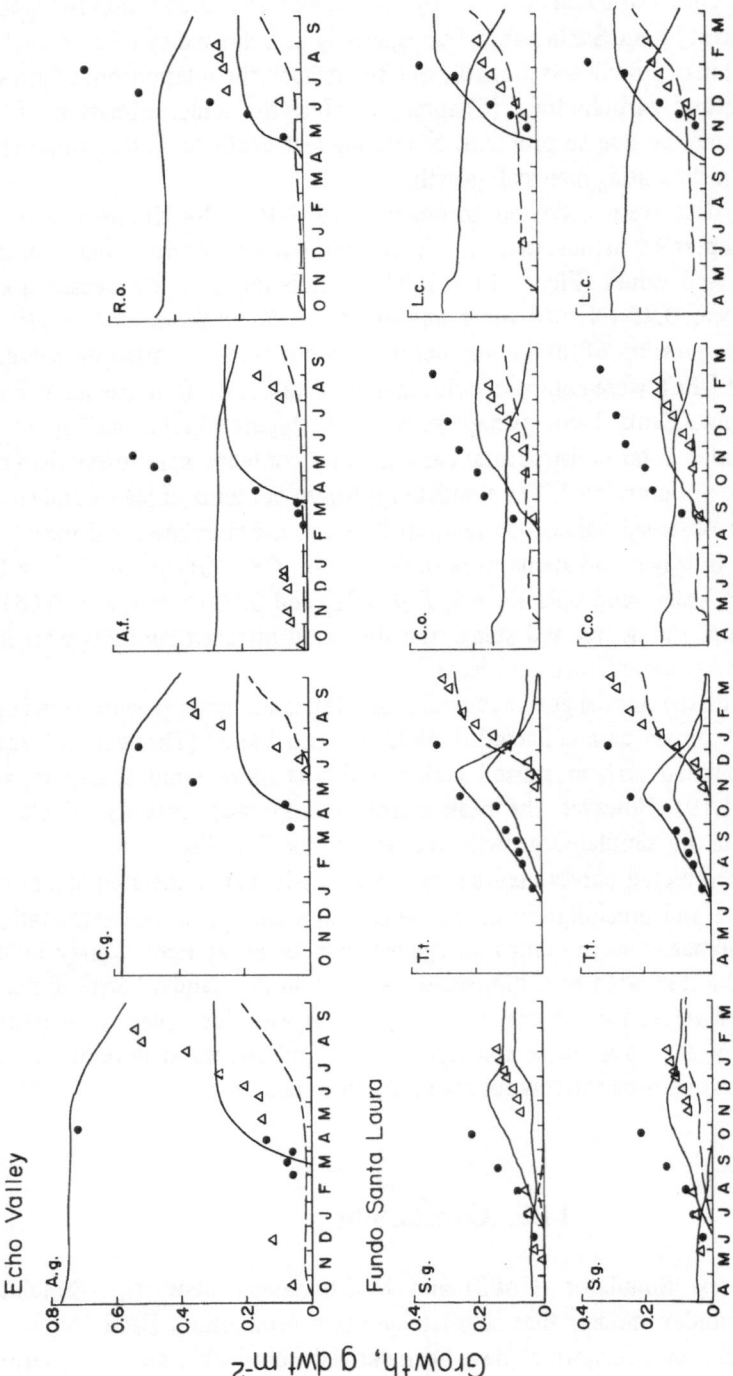

Figure 11.15. MEDECS simulations of seasonal courses of old leaf biomass (—), new leaf biomass (– –), and litter standing crop (- - -) with measured new biomass (●) and litter standing crop (△) in grams dry weight per square meter for *Arctostaphylos glauca* (A.g.), *Ceanothus greggii* (C.g.), *Adenostoma fasciculatum* (A.f.), and *Rhus ovata* (R.o.) at Echo Valley in 1975-1976 and for *Satureja gilliesii* (S.g.), *Trevoa trinervis* (T.t.), *Colliguaya odorifera* (C.o.), and *Lithraea caustica* (L.c.) at Fundo Santa Laura in 1973-1974 (middle row) and 1974-1975 (bottom row).

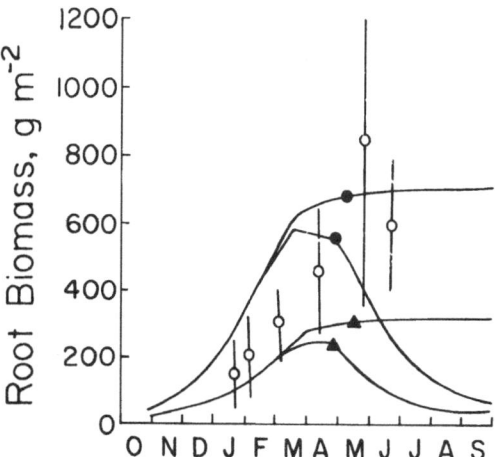

Figure 11.16. Simulated progression by MEDECS of absorbing root biomass (g m^{-2}) for *Adenostoma fasciculatum* (▲) and *Arctostphylos glauca* (●); top line is total live and dead biomass and bottom line is live biomass. Observed data (open symbols with standard error bars) is from Kummerow et al. (1978a).

ment within the canopy, water potentials, and photosynthetic rates and was used to generate simplified versions of some submodels for MEDECS. Because of the good comparisons of soil moistures and biomasses simulated by MEDECS with field data, MEDECS was assumed valid for comparing resource use between the southern Californian and central Chilean mediterranean type ecosystems.

(following pages: Figures 11.17-11.19)

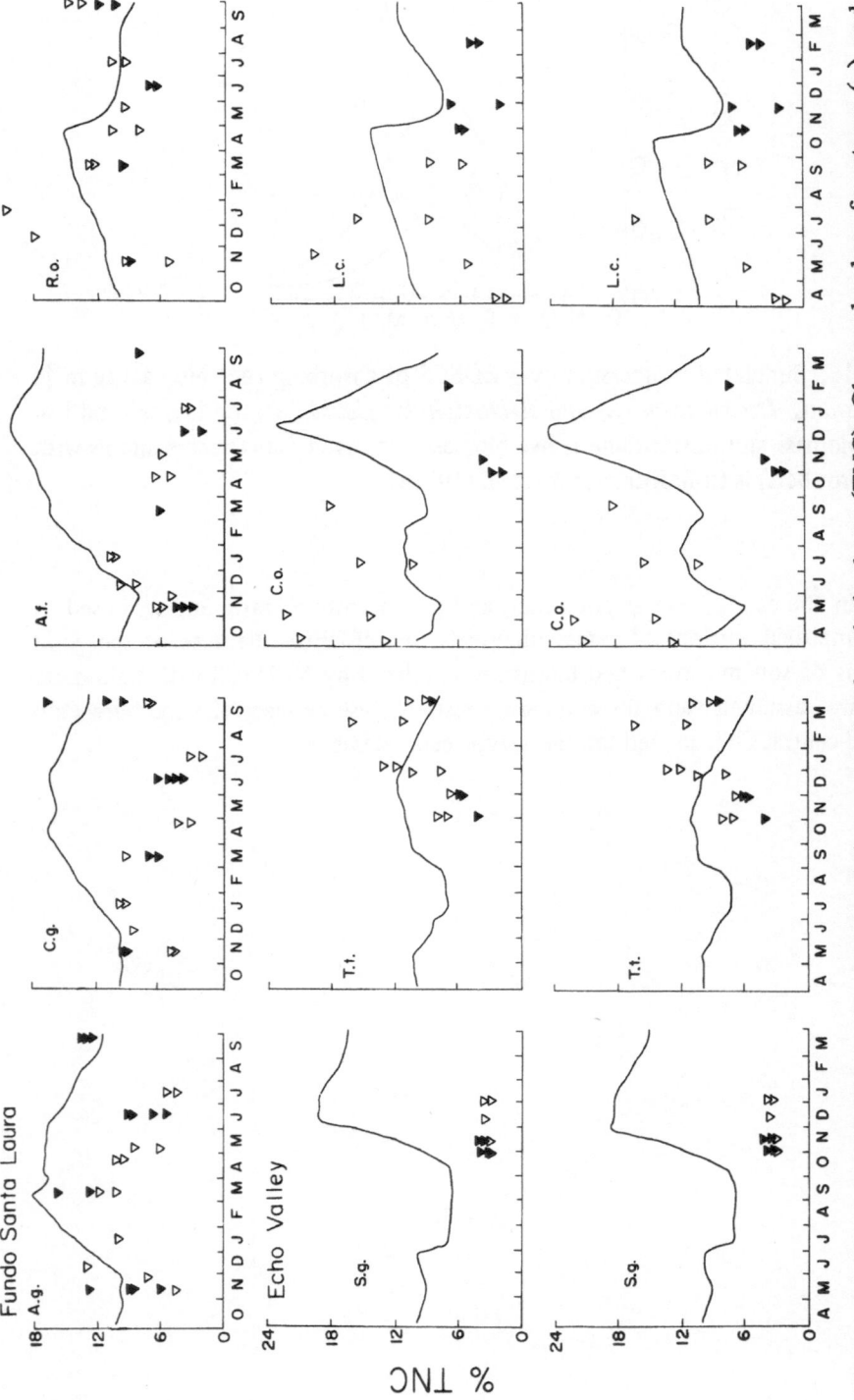

Figure 11.17. Annual course of percent total nonstructural. carbohydrate (TNC). Observed values for stems (▼) and roots (▽) and values calculated by MEDECS for stems and roots combined (–) for *Arctostaphylos glauca* (A.g.), *Ceanothus greggii* (C.g.), *Adenostoma fasciculatum* (A.f.), and *Rhus ovata* (R.o.) at Echo Valley in 1975-1976 and for *Saturejia gilliesii* (S.g.), *Trevoa trinervis* (T.t.), *Colliguaya odorifera* (C.o.), and *Lithraea caustica* (L.c.) at Fundo Santa Laura in 1973-1974 (middle row) and 1974-1975 (bottom row).

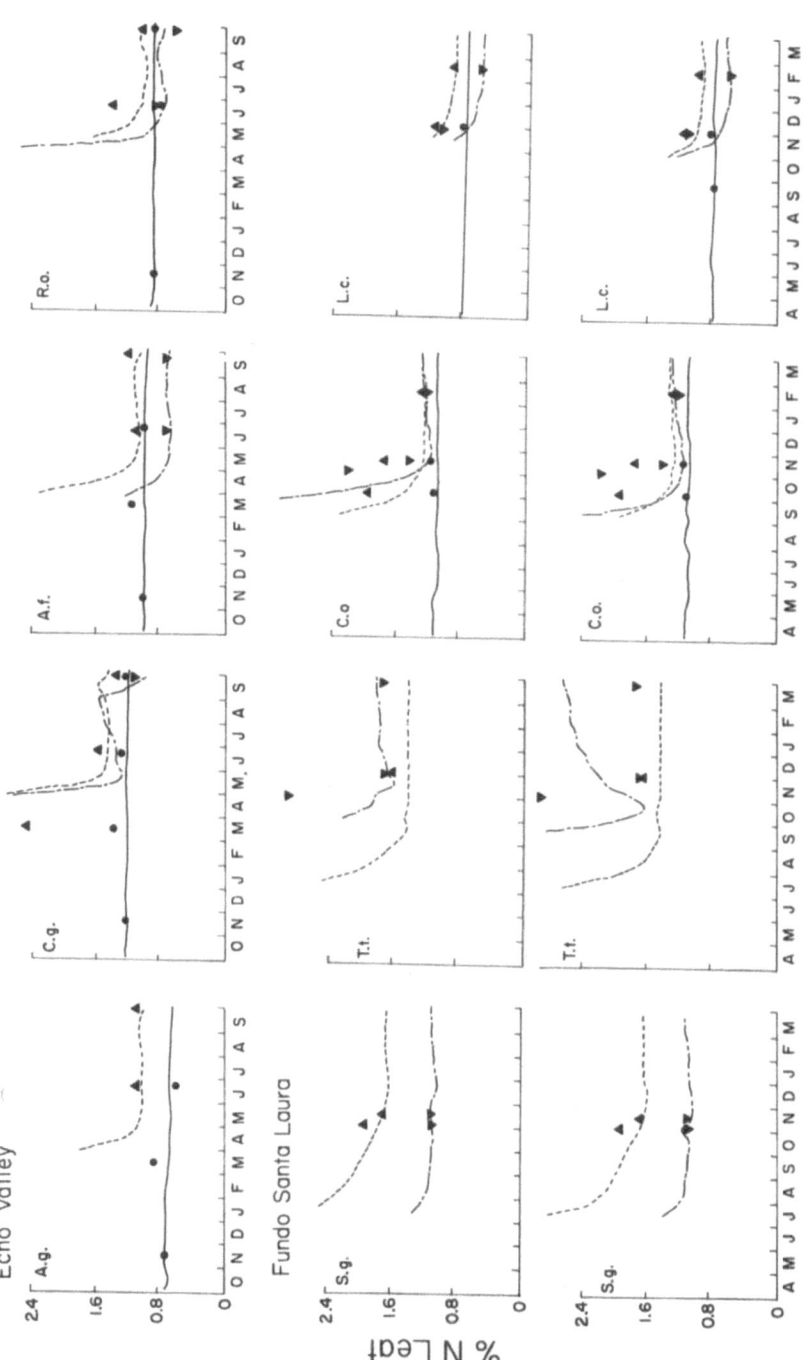

Figure 11.18. Annual course of observed percentage nitrogen and percentages calculated by MEDECS for old leaves simulated (—) and observed (●), new leaves simulated (- - -) and observed (▲), and new stems simulated (— - —) and observed (▼) for *Arctostaphylos glauca* (A.g.), *Ceanothus greggii* (C.g.), *Adenostoma fasciculatum* (A.f.), and *Rhus ovata* (R.o.) at Echo Valley in 1975-1976 and for *Satureja gilliesii* (S.g.), *Trevoa trinervis* (T.t.), *Colliguaya odorifera* (C.o.), and *Lithraea caustica* (L.c.) at Fundo Santa Laura in 1973-1974 (middle row) and 1974-1975 (bottom row). *Satureja gilliesii* (S.g.) and *Trevoa trinervis* (T.t.) do not have old leaves because they are drought semideciduous.

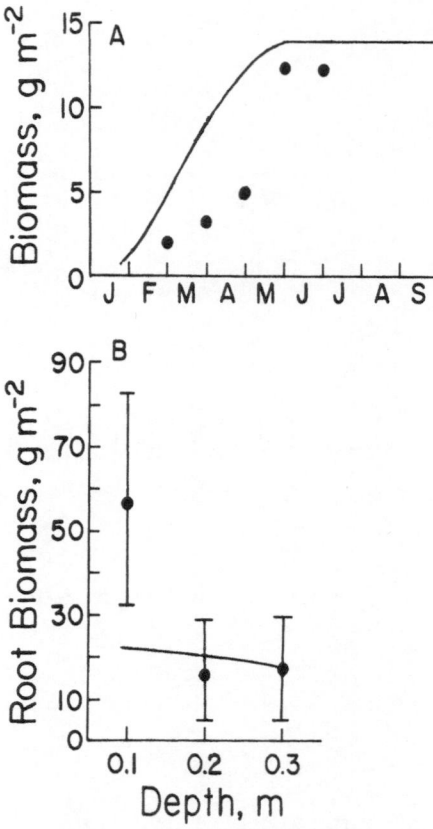

Figure 11.19. Annual course of observed annual grass biomass (g m^{-2} ground surface) versus values calculated by MEDECS. (A) Aboveground biomass observed (●) and calculated (−). Observed data were taken from Thrower and Bradbury (1977). (B) Root biomass, observed data with standard error bars and calculated (−) for depths of 0.1, 0.2, and 0.3 m below the ground surface.

12. Similarities and Limitations of Resource Utilization in Mediterranean Type Ecosystems

PHILIP C. MILLER

This chapter discusses the resource use and resource-use efficiencies of chaparral and matorral and evaluates some of the environmental and biotic influences on resource use as assessed by the simulation models.

12.1. Introduction

12.1.1 Background

The general hypothesis that the patterns of resource utilization are similar in the mediterranean type ecosystems of southern California and central Chile was evaluated directly from available data and indirectly from simulations. The simulation models were parameterized for the two systems with published data and data collected on this project. The data base has been described in detail in the previous chapters and in the discussion of the simulation models (Chapter 11). The following discussion summarizes the results of the simulations and develops the implications of the simulations with respect to the general hypothesis.

12.1.2 Methods of Simulation

The basic simulation experiments followed several schemes. In the first scheme, the seasonal progression of the plant and soil processes was simulated for 2 years at Echo Valley, 1972-1973 and 1975-1976; and for 3 years at Fundo Santa Laura, 1972-1973, 1973-1974, and 1974-1975; the years for which the best data to evaluate the model were available. These years included a large range of annual precipitations so that the seasonal progressions of the various processes were calculated within the range of natural variability. The seasonal changes of plant processes were simulated with the Canopy Process Simulator (CAPS) and the Mediterranean Ecosystem Simulator (MEDECS) by using the measured progression of climatic variables and the best estimates for species parameters. These simulations provide a framework within which the similarities and differences in resource use betweeen the research sites can be compared.

In the second scheme, the similarity of the resource use and resource-use efficiency in the different vegetation types was tested statistically, including the natural variation within the types and the variations due to errors of measurement. Resource uses and efficiencies were calculated on an annual basis. A vegetation type, e.g., coastal sage scrub, chamise chaparral, or mixed chaparral in southern California or matorral in central Chile, was selected. A stand within a vegetation type was defined by the mean relative composition of the predominant species for the type plus a random component chosen from the observed variation within the stand type (Chapter 3). The parameters that describe the structure and physiology of each species were defined from mean values determined from field and laboratory data plus random components appropriate to the precision of the measurement and to the natural variation that occurs in the field (Chapters 4 to 9). For example, the measurements of photosynthesis were expected to be accurate to within 25% of the mean measured value, so a value within the range of ± 25% of the mean was selected at random and added to the mean value before the simulation began. The measurements of root biomass were expected to be accurate to within 50% of the mean, so a larger random component was added to the mean initial root biomass at each soil level. The progression of climatic variables throughout the year was then selected from the average climate with random components based on the year-to-year variation in monthly values (Chapter 2). The annual progressions of the plant and soil processes were simulated, and the annual resource use and resource-use efficiency were calculated. Another stand within the vegetation type was selected, new random components added to the parameter values, and a new climate defined until a sample of five simulated stands was produced that included the variations inherent in the natural system and in the measurements. The mean resource uses and resource-use efficiencies were calculated, together with a standard error, for the sample of five simulated stands. Then a new vegetation type was selected and the procedure repeated until calculations were completed for the coastal sage scrub on pole- and equator-facing slopes, chamise chaparral, and mixed chaparral vegetation types in southern California and for the matorral on pole- and equator-facing slopes in central Chile. The significance of the difference among the vegetation types with respect to resource use and resource-use efficiency was tested with ANOVA and Duncan's Multiple Range Test (Bancroft, 1968).

12.2. Similarities of Resource Use and Resource-Use Efficiency

The resources considered were water, light energy, and nitrogen. Resource use is the water transpired, light energy converted into plant organic matter in photosynthesis, and nitrogen taken up from the soil by the plant. The test of the hypothesis involved the resources used by the vegetation type, not by the individual species. The resource use could vary because of climatic and soil characteristics, species structure and physiology, and the relative abundance within a vegetation type of species with differing structures and physiologies. Resource-use efficiency was defined as the ratio of the amount of resource taken up to the amount available; those calculated were: transpiration to precipitation, energy converted in photosynthesis to incident solar irradiance, and nitrogen taken up to nitrogen mineralized and made available to the plants. Additional ratios of aboveground growth to resources available, e.g., precipitation, solar irradiance, and nitrogen available, and aboveground growth to resources taken up, e.g., transpiration, photosynthesis, and nitrogen uptake, were also calculated to compare processes in mediterranean type ecosystems with those in other ecosystems. Annual uses summed over the year and patterns of use throughout the year were compared.

12.2.1 Similarities of Annual Resource Use and Resource-Use Efficiency

The simulation experiments indicated that the uptake and uptake efficiency of water, solar energy, and nitrogen differed by species, slope, and country. The uptake and uptake efficiency of mixed chaparral on the pole-facing slope and chamise chaparral on the equator-facing slope at Echo Valley differed more than did the uptake and uptake efficiency of mixed chaparral at Echo Valley and matorral at Fundo Santa Laura (Tables 12.1 and 12.2). Uptake and uptake efficiencies of matorral on both slopes at Fundo Santa Laura were similar. In general, mixed chaparral on the pole-facing slope and matorral on the pole- and equator-facing slopes were more similar and had higher uptakes and uptake efficiencies for water, solar irradiance, and nitrogen than did chamise chaparral on equator-facing slopes. Transpiration was similar, i.e., was not statistically different, in mixed chaparral and matorral, both of which had higher transpiration rates than did chamise chaparral. The water-capture efficiency, calculated as transpiration/precipitation, was highest in mixed chaparral and was similar in chamise chaparral and matorral. The higher efficiency of mixed chaparral than matorral, in spite of similar transpiration rates, was due to the lower annual precipitation at Echo Valley. Net carbon uptake, which is a measure of solar irradiance uptake, was similar in mixed chaparral and matorral and was higher than in chamise chaparral. The uptake efficiency, in joules production per joule incident solar irradiance, was similar in mixed chaparral and matorral on the pole-facing slopes. The uptake efficiency of matorral on the equator-facing slope was lower. Uptake efficiency for solar irradiance in chamise chaparral on the equator-facing slope was only half that of mixed chaparral on the pole-facing slope. The variation from slope to slope within a research site was greater than the variation between the two pole-facing slopes in each hemisphere. The nitrogen-uptake efficiency was similar in mixed chaparral and matorral on the two slopes and was higher than in chamise chaparral.

The resource-use efficiencies of the chamise chaparral at Echo Valley were generally more similar to those of the coastal sage scrub on the equator-facing slope at Camp Pendleton than to those of the mixed chaparral at Echo Valley or the matorral at Fundo Santa Laura.

In terms of using a resource after it has been taken up, the similarities between vegetation types changed. The total aboveground growth was similar for all sites, but chamise chaparral was more efficient than the others in the aboveground growth obtained per unit of absorbed resource. The growth ratios were usually similar between mixed chaparral and matorral on the two slopes but were one-half to two-thirds lower

Table 12.1. Summary of resources and resource use efficiencies for mature vegetation in the mediterranean scrub regions of California and Chile[a]

| | California | | | | Chile | |
| | Camp Pendleton | | Echo Valley | | Fundo Santa Laura | |
	Pole-facing slope	Equator-facing slope	Pole-facing slope	Equator-facing slope	Pole-facing slope	Equator-facing slope
Solar Irradiance						
Solar (MJ m^{-2} yr^{-1})	4585 ± 73	5490 ± 114	5364 ± 117	6359 ± 85	5357 ± 80	6651 ± 130
P$_{net}$ (g dry wt m^{-2} yr^{-1})	1160 ± 53	763 ± 43	1646 ± 118	1107 ± 55	1521 ± 41	1564 ± 19
Growth (g dry wt m^{-2} yr^{-1})	234 ± 11	189 ± 9	370 ± 34	366 ± 14	334 ± 29	337 ± 20
P$_{net}$/Solar (J J^{-1})	0.0042 ± 0.0002	0.0024 ± 0.0002	0.0052 ± 0.0004	0.0028 ± 0.0001	0.0048 ± 0.0001	0.0040 ± 0.0001
Growth/Solar (J J^{-1})	0.0008 ± 0.0000	0.0004 ± 0.0000	0.0012 ± 0.0000	0.0012 ± 0.0000	0.0012 ± 0.0001	0.0012 ± 0.0000
Growth/P$_{net}$ (g dry wt g dry wt^{-1})	0.20 ± 0.01	0.25 ± 0.01	0.22 ± 0.01	0.33 ± 0.01	0.22 ± 0.01	0.22 ± 0.01
Water						
Precip (mm yr^{-1})	208 ± 4	205 ± 7	444 ± 12	425 ± 3	599 ± 5	590 ± 13
Tran (mm yr^{-1})	163 ± 10	124 ± 9	350 ± 24	257 ± 7	369 ± 12	351 ± 15
Tran/ppt (mm mm^{-1})	0.79 ± 0.05	0.60 ± 0.04	0.79 ± 0.05	0.60 ± 0.02	0.61 ± 0.02	0.60 ± 0.02
Growth/ppt (g dry wt kg H$_2$O^{-1})	1.13 ± 0.05	0.92 ± 0.03	0.83 ± 0.07	0.86 ± 0.03	0.56 ± 0.05	0.58 ± 0.05
Growth/Tran (g dry wt kg H$_2$O^{-1})	1.47 ± 0.14	1.53 ± 0.05	1.05 ± 0.04	1.43 ± 0.07	0.91 ± 0.06	0.97 ± 0.09
Nitrogen						
N$_{rel}$ (g N m^{-2} yr^{-1})	5.64 ± 0.29	4.30 ± 0.29	3.53 ± 0.27	2.69 ± 0.12	4.80 ± 0.11	4.67 ± 0.11
N$_{up}$ (g N m^{-2} yr^{-1})	3.90 ± 0.38	3.46 ± 0.19	3.55 ± 0.54	1.94 ± 0.26	4.36 ± 0.13	4.74 ± 0.26
N$_{up}$/N$_{rel}$ (g N g N^{-1})	0.70 ± 0.06	0.82 ± 0.06	0.98 ± 0.08	0.72 ± 0.08	0.91 ± 0.01	1.02 ± 0.05
Growth/N$_{rel}$ (g dry wt g N^{-1})	41.9 ± 2.3	44.6 ± 3.1	104.4 ± 3.3	137.1 ± 6.5	69.8 ± 6.3	72.7 ± 5.8
Growth/N$_{up}$ (g dry wt g N^{-1})	61.2 ± 2.9	54.6 ± 1.4	108.4 ± 6.8	200.5 ± 24.3	76.8 ± 6.5	72.4 ± 7.1

[a] At Echo Valley, mixed chaparral occurs on the pole-facing slope and chamise chaparral occurs on the equator-facing slope. Variables are solar irradiance, net photosynthesis, aboveground growth, precipitation, transpiration, nitrogen released in mineralization, and nitrogen taken up by the vegetation. Means and standard errors are given for a sample size of 5.

Table 12.2. Summary of similarity of research sites[a]

	Ranking of sites regarding resource-capture efficiency and resource-use efficiency							
Variables	Low							High
	Resource-capture efficiency							
P_{net}/Solar	CP-E		EV-E		FSL-E	CP-P	FSL-P	EV-P
Tran/Ppt	FSL-E	EV-E	CP-E	FSL-P		CP-P	EV-P	
N_{up}/N_{rel}	CP-P	EV-E	CP-E	FSL-P	EV-P	FSL-E		
	Resource-use efficiency							
Growth/P_{net}	CP-P	FSL-E	FSL-P	EV-P	CP-E		EV-E	
Growth/Tran	FSL-P	FSL-E	EV-P		EV-E	CP-P	CP-E	
Growth/N_{up}	CP-E	CP-P	FSL-E	FSL-P	EV-P		EV-E	

[a] As indicated by Duncan's (Bancroft, 1968) multiple range test. CP-E: Camp Pendleton, California, equator-facing slope; CP-P: Camp Pendleton, California, pole-facing slope; EV-E: Echo Valley, California, equator-facing slope; EV-P: Echo Valley, California, pole-facing slope; FSL-E: Fundo Santa Laura, Chile, equator-facing slope; and FSL-P: Fundo Santa Laura, Chile, pole-facing slope. Underlined sites are not statistically different. Variables are net photosynthesis, solar irradiance, transpiration, precipitation, nitrogen taken up by the vegetation, nitrogen released in mineralization, and aboveground growth.

than the growth ratios of chamise chaparral. Chamise chaparral is less efficient in taking up resources but more efficient in using those than it does absorb. The simulated aboveground growth was slightly higher in the two chaparral types than in the matorral. The simulated gross production was 1646, 1107, 1521, and 1564 g dry weight m^{-2} yr^{-1} for mixed chaparral, chamise chaparral, matorral on the pole-facing slope, and matorral on the equator-facing slope. The simulated aboveground growth was 370, 366, 334, and 337 g dry weight m^{-2} yr^{-1} in the four vegetation types. Field data indicated that the annual aboveground growth is lower than the simulated values. Assuming the present aboveground biomasses have accumulated during the 22 years since the last major fire at Echo Valley and during the 15 years since the cessation of wood cutting at Fundo Santa Laura (Aschmann and Bahre, 1977), aboveground net accumulation rates were 132, 78, and 62 g dry weight m^{-2} yr^{-1} for mixed chaparral, chamise chaparral, and matorral vegetation types. Mooney et al. (1977) indicate that the aboveground production by shrubs at Echo Valley is twice that at Fundo Santa Laura.

Three patterns of resource-use efficiency occurred. Limitation by all three resources—water, solar irradiance, and nitrogen—was indicated by relatively high resource-use efficiencies for all three resources on the pole-facing slopes at Echo Valley and Fundo

Table 12.3. Summary of standing crops, annual fluxes and allocation patterns in mixed chaparral, chamise chaparral, and matorral

	Annual mean standing crops amount (m⁻²)			Annual fluxes amount (m⁻² yr⁻¹)				Fraction of uptake allocated to:				
								New biomass			Growth respiration	Secondary growth or change in storage
	Leaves	Stems	Roots	Resource available	Resource use	Resource-use efficiency	Maintenance respiration	Leaves	Stems	Roots		
Mixed chaparral, Echo Valley												
Biomass (g)	700	2200	1000	-	1200	-	-	0.15	0.04	0.08	0.14	0.05
Energy (MJ)	117	367	167	5527	20	0.0036	0.54	-	-	-	-	-
Water (kg)	1050	1474	670	450	320	0.71	0.	0.90	0.20	0.10	0	0
Nitrogen (g)	4.2	6.6	3.0	2.2	0.9	0.41	-	-	-	-	-	-
Chamise chaparral, Echo Valley												
Biomass (g)	290	1440	800	-	600	-	-	0.18	0.08	0.05	0.17	0.00
Energy (MJ)	48	240	134	6657	10	0.0015	0.52	-	-	-	-	-
Water (kg)	435	965	536	450	160	0.36	-	0.56	0.37	0.05	-	0.00
Nitrogen (g)	2.6	4.3	2.4	2.3	2.0	0.87	-	-	-	-	-	-
Matorral, Fundo Santa Laura												
Biomass (g)	220	700	800	-	800	-	-	0.19	0.06	0.08	0.18	0.05
Energy (MJ)	37	117	134	6113	13.4	0.0022	0.44	-	-	-	-	-
Water (kg)	330	470	536	590	150	0.25	-	1.05	0.25	0.10	-	0.00
Nitrogen (g)	2.0	3.9	4.8	3.3	2.0	0.61	-	-	-	-	-	-

Santa Laura. Limitation by only one resource, nitrogen, was indicated on the equator-facing slope at Fundo Santa Laura. Only modest limitation by these resources was indicated by the relatively low resource-use efficiencies for all three resources in the chamise chaparral on the equator-facing slope at Echo Valley. In spite of its apparently low resource-use efficiencies, chamise chaparral is the most abundant chaparral type in San Diego County (Chapter 3). *Adenostoma fasciculatum* may replace other chaparral species as the vegetation ages (Hanes, 1971).

The simulated patterns of allocation of photosynthate to different plant parts were similar in the two chaparral types and the matorral. About 18% of the new photosynthate was allocated to leaves, 6% to stems, and 7% to roots (Table 12.3). Maintenance respiration required over 50% of the photosynthate in the chaparral and 44% in the matorral. The patterns of allocation of nitrogen taken up from the soil and redistributed with the plant in mixed chaparral, chamise chaparral, and matorral were less similar. The percentage of nitrogen taken up that was incorporated in new tissue was 90%, 60%, and 105% to new leaves; 20%, 40%, and 25% to stems; and 10%, 5%, and 10% to roots in the three vegetation types, respectively. The percentages can total more than 100% because only part of the nitrogen incorporated into new growth came from the soil, the rest comes from old plant parts (Chapter 9).

Leaf dry weight turnover rates were lower in the chaparral, about 0.28 g g^{-1} yr^{-1} in chaparral versus 0.80 g g^{-1} yr^{-1} in matorral (Table 12.4). Stem and root turnover rates were similar for the three vegetation types, about 0.10 g g^{-1} yr^{-1}. Leaf nitrogen turnover rates were lower in the chaparral. Nitrogen rates, in general, were higher than were dry matter turnover rates. Average dry weight litterfall at both sites, based on new growth and annual accumulation rates, was similar, while nitrogen lost in litterfall was less at Echo Valley than at Fundo Santa Laura.

12.2.2 Similarities of Annual Patterns of Resource Use and Resource-Use Efficiency

The simulated seasonal progressions of the use of water, light energy, and nitrogen in mixed and chamise chaparral at Echo Valley and in matorral at Fundo Santa Laura indicated some of the plant and environmental constraints on resource use (Figure 12.1, pp. 378-381). During the winter, transpiration and net photosynthesis were higher at Echo Valley than at Fundo Santa Laura because of higher incoming solar irradiance at Echo Valley. The higher winter solar irradiance at Echo Valley compared to that at Fundo Santa Laura is caused by the greater distance of the Earth from the sun and by the greater cloud cover in the Southern Hemisphere (Chapter 2). Because these factors are typical of the Southern Hemisphere compared to the Northern Hemisphere, they may also cause lower water and production efficiencies in other mediterranean regions in the Southern Hemisphere (F. Kruger, *pers. comm.*; P. C. Miller and J. Kummerow, *pers. obs.*; Chapter 2).

At Echo Valley and Fundo Santa Laura, spring was the period of greatest activity for both soil and plant processes, including decomposition, mineralization, nutrient uptake, photosynthesis, transpiration, growth of shoots and roots, and reproduction. During spring, both evapotranspiration and photosynthesis increased as solar irradiance increased because of the higher solar intensities and longer day lengths. At both Echo Valley and Fundo Santa Laura, evaporation from the soil peaked earlier in the spring

Table 12.4. Carbon and nitrogen turnover rates, net aboveground accumulation rates assuming 22 years for Echo Valley and 15 years for Fundo Santa Laura, and litter production calculated as the difference between new biomass (from Table 12.3) and net accumulation for mixed chaparral, chamise chaparral, and matorral

	Turnover rates $(g\ g^{-1}\ yr^{-1})$			Net accumulation $(g\ m^{-2}\ yr^{-1})$			Litter production $(g\ m^{-2}\ yr^{-1})$		
	Leaves	Stems	Roots	Leaves	Stems	Total	Leaves	Stems	Total
Carbon (g dry weight)									
Mixed chaparral	0.29	0.12	0.10	32	100	132	168	0	168
Chamise chaparral	0.26	0.12	0.06	13	65	78	162	10	172
Matorral	0.80	0.07	0.09	15	47	62	160	3	163
Nitrogen (g)									
Mixed chaparral	0.43	0.06	0.09	0.19	0.30	0.49	1.51	0.10	1.61
Chamise chaparral	0.31	0.28	0.04	0.12	0.20	0.42	0.68	0.60	1.28
Matorral	1.05	0.13	0.04	0.13	0.26	0.39	1.97	0.24	2.21

than did transpiration. Soil evaporation was controlled by water near the surface; transpiration was controlled by water both near the surface and below. Soil evaporation decreased earlier than transpiration because the surface dried out earlier than did the deeper soil layers.

The simulations indicated differences in the patterns of water use and photosynthesis by the different species, which directly affected the overall patterns of water use and photosynthesis by the vegetation. The mixed chaparral at Echo Valley is composed predominantly of *Arctostaphylos glauca* and *Ceanothus greggii*. The chamise chaparral is almost wholly *A. fasciculatum*. The matorral at Fundo Santa Laura is predominantly *Lithraea caustica*. Of the Californian species. *A. glauca* had the highest rates of transpiration and photosynthesis. Its high rate of transpiration dried out the soil earliest, so that vegetation comprised of this species had the earliest decline of transpiration and photosynthesis. In contrast, *Rhus ovata*, which is rare in Californian chaparral, had the lowest transpiration rates, was the slowest to dry the soil, and had the highest photosynthetic rate in summer. *Adenostoma fasciculatum* had a low transpiration rate but still dried out the soil enough to depress transpiration and photosynthesis in summer. Growth ceased earlier in *A. glauca* than in *R. ovata* following the decline of moisture availability in early summer. In California, only *R. ovata* followed the pattern postulated by Specht (1957a, b, 1972b) for evergreen shrubs in mediterranean type ecosystems; a pattern of moderate water use to conserve water for the summer growth period. *Rhus ovata* also had the highest temperature requirements for growth. *Lithraea caustica*, the Chilean analogue of *R. ovata*, showed transpiration, photosynthesis, and growth later in the summer than did the other Chilean species. Consequently, summer soil moisture was higher under *L. caustica* than under the other Chilean species measured. This pattern is similar to the pattern of *R. ovata* in relation to the other Californian species studied. Of the Chilean species studied, *L. caustica* had some of the attributes of water conservation that Specht (1957a, b, 1972b) postulated for evergreen shrubs. Because *R. ovata* is a minor component of the Californian chaparral and *L. caustica* is a major component of the matorral, the indications are that Specht's hypothesis may apply to shrubs in the mediterranean type regions of the Southern Hemisphere but not to shrubs of the Northern Hemisphere.

The simulated nitrogen uptake declined in early summer as soil moisture declined because of the effect of low soil moisture on the diffusion of nitrogen to the roots, on the biomass of actively absorbing roots, and on the activity of the soil organisms involved in mineralization. The seasonal pattern of the mineralization of nitrogen was similar for all species except *L. caustica*. With *L. caustica*, mineralization was higher in the spring. The simulated patterns of nitrogen uptake varied more widely than did the patterns of mineralization. Absorbing root growth and nitrogen uptake preceded aboveground growth, similar to field measurements (Mooney and Rundel, 1979; Chapter 9). Nitrogen-uptake rates and root growth in *Colliguaya odorifera* started later but soon exceeded that of the other Chilean species. Nitrogen uptake by *L. caustica* was delayed because of its slower rate of root growth, but the slower initiation of roots was partly compensated for by the longer period of time that soil moisture was available under *L. caustica* during the summer. Root growth in *Trevoa trinervis* and *Satureja gilliesii*, drought semideciduous shrubs, stopped early in winter, which made the nitrogen uptake stabilize through the winter. Nitrogen fixation by *T. trinervis* was concentrated during the late spring season.

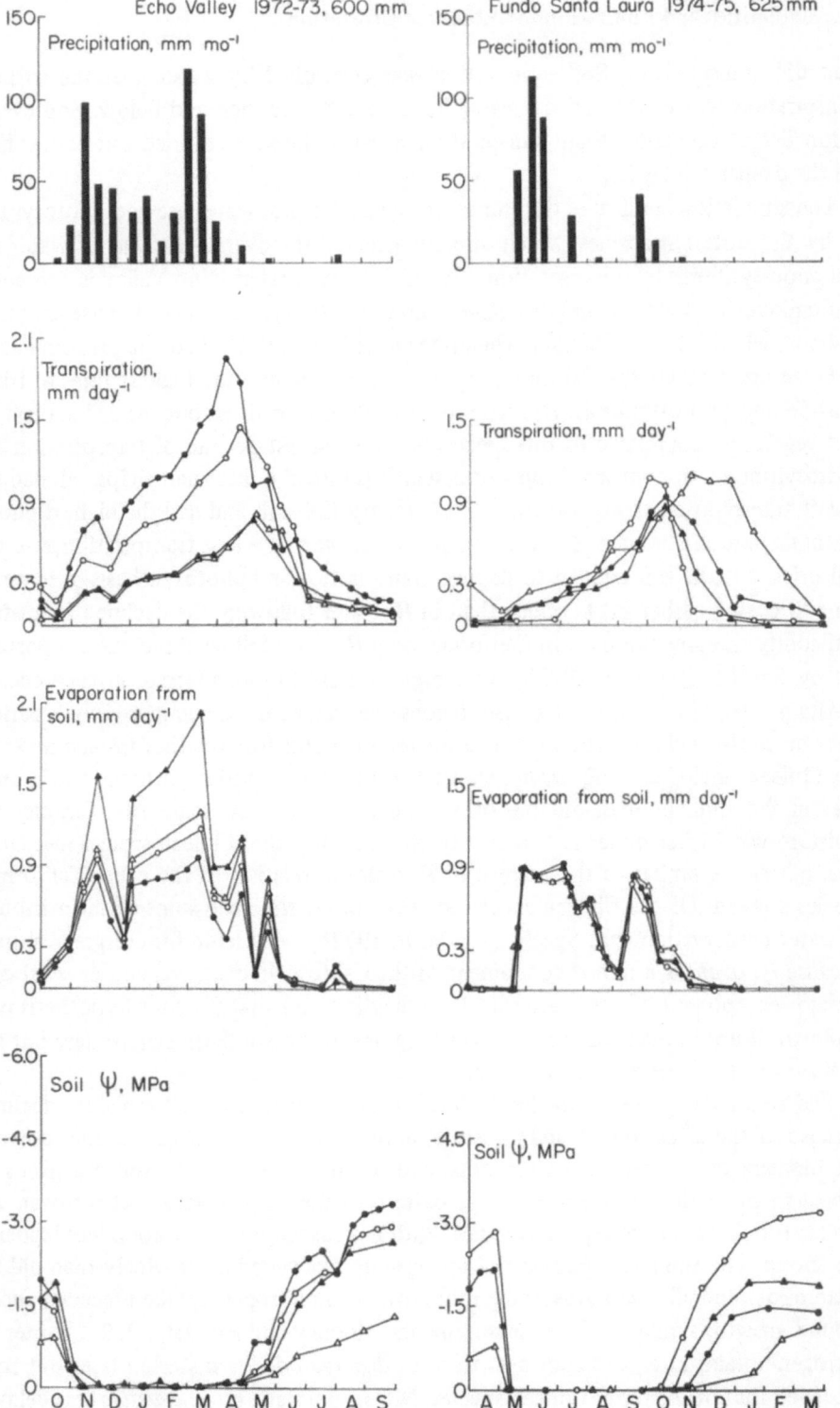

Figure 12.1. Seasonal progressions of the resources available to the vegetation (precipitation, solar irradiance, and nitrogen released by mineralization), capture of the resource by the vegetation [transpiration, net photosynthesis (P_n), and nitrogen uptake], and resource-use efficiencies [transpiration over precipitation (Tr/Ppt), net photosynthesis over solar irradiance (P_n/Solar), and nitrogen uptake over nitrogen release (N_{up}/N_{rel})] simulated by MEDECS for Echo Valley and Fundo Santa Laura.

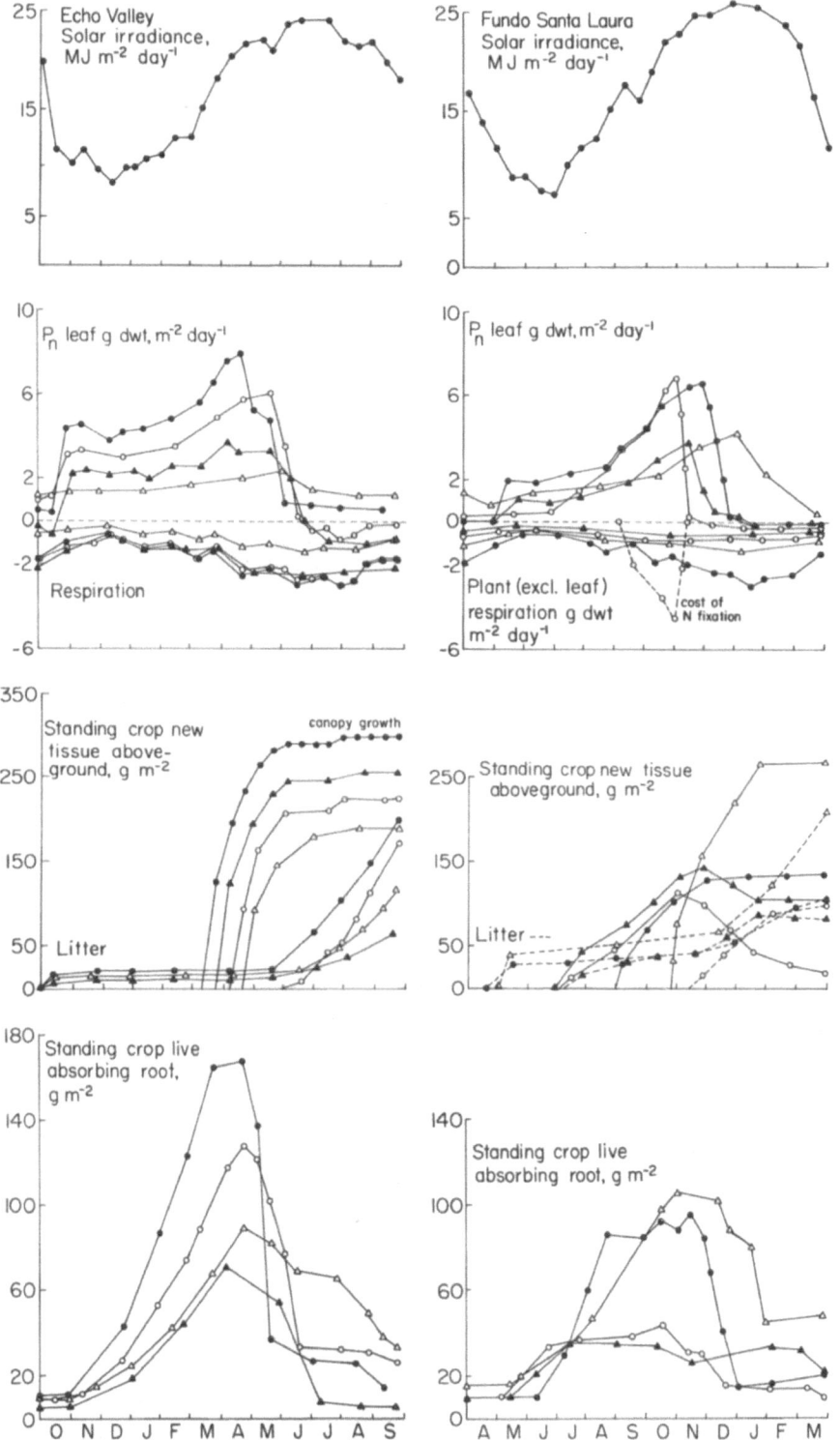

Figure 12.1. (Continued) In addition, the simulated seasonal progression of other indicators of plant and soil activity are given, including evaporation from the soil, soil water potential (Ψ), and standing crops of new biomass aboveground, litter mass, live absorbing root mass, nitrogen in new aboveground biomass, and nitrogen in live absorbing roots. Species at Echo Valley are *Adenostoma fasciculatum* (▲), *Arctostaphylos glauca* (●), *Ceanothus greggii* (○), and *Rhus ovata* (△); and at Fundo Santa Laura, *Colliguaya odorifera* (●), *Lithraea caustica* (△), *Satureja gilliesii* (▲), and *Trevoa trinervis* (○).

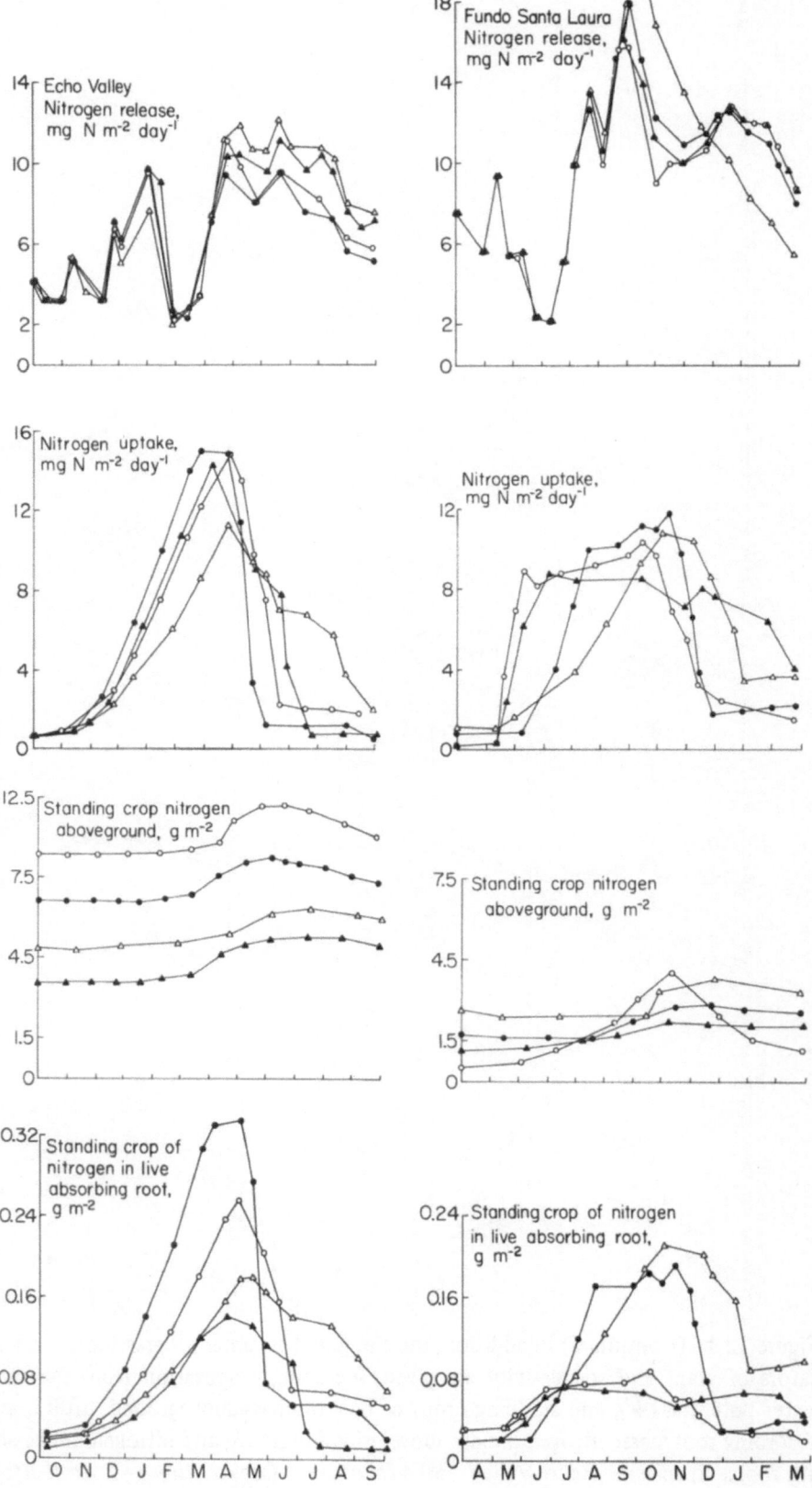

Figure 12.1. (Continued)

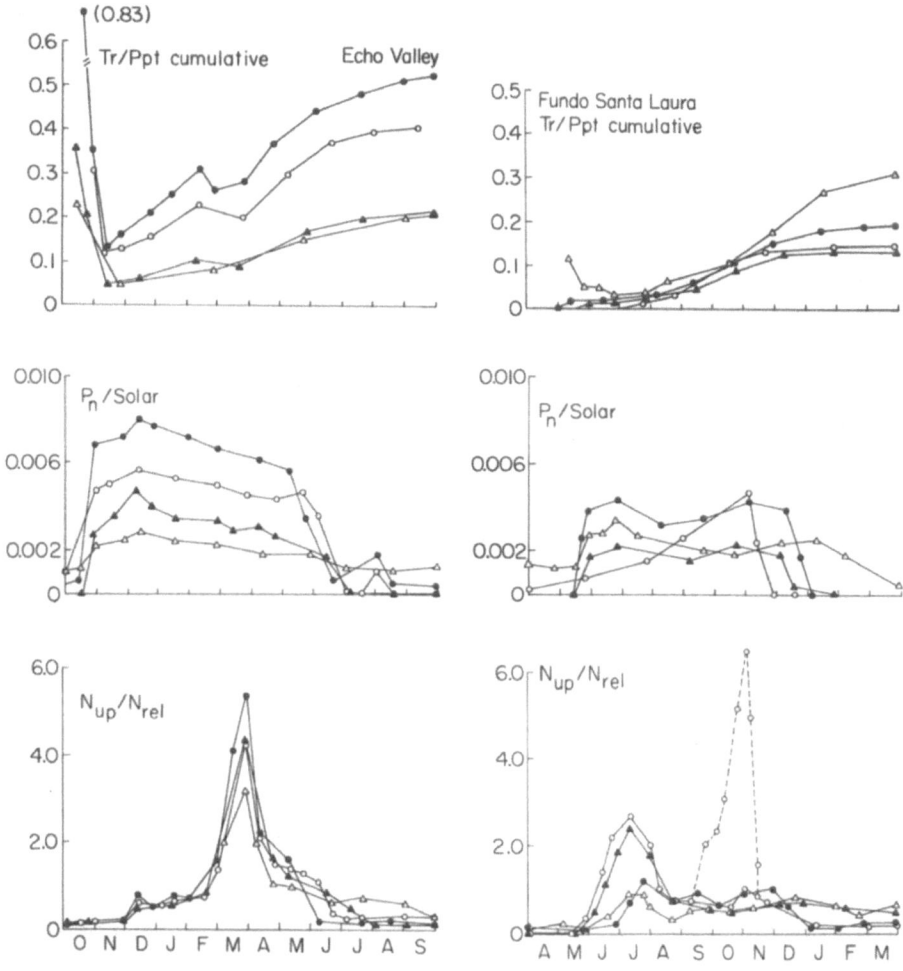

Figure 12.1. (Continued)

The season progression of resource-use efficiencies indicated peak production efficiencies, calculated as joules in production per joule incident solar irradiance, of 0.002 to 0.008 in the Californian species and 0.002 to 0.007 in the Chilean species. Production efficiencies were usually highest in midwinter, although in some species they were highest in late spring. Daily transpiration efficiencies, calculated as millimeters transpired per day per millimeter precipitation per day, were not instructive since the ratio varied widely depending on the amount of precipitation per day, which was usually zero. Transpiration efficiencies were relatively low in winter, extremely high in spring when transpiration used water stored in the soil, and varied widely in summer when transpiration was low and precipitation was usually zero. The cumulative transpiration per cumulated precipitation showed increasing efficiency through spring as the soil moisture was depleted. Nitrogen-uptake efficiencies, calculated as grams nitrogen taken up per grams nitrogen released through mineralization, showed peaks in the spring because uptake increased with root growth. The efficiency of *T. trinervis* was increased as a result of nitrogen fixation by its root nodules, which was a path of nitrogen uptake separate from and not in competition with the other species.

12.3. Environmental Limitations of Resource Use and Resource-Use Efficiency

Resource-use efficiencies are affected by environmental and plant properties. Inasmuch as the environments of southern California and central Chile are not identical, the different resource-use efficiencies might be due to environmental differences in precipitation, solar irradiance, and nitrogen availability. In addition, these environmental factors interact to constrain the vegetation properties that can exist in the mediterranean regions of each country and to produce the similar vegetative structures. An analysis of some of these interactions was conducted with the simulation models and with simplifications based on the simulation models.

12.3.1 Limitation by Water Availability

The annual precipitation constrains the steady state biomass and sets an upper limit to the resource-use efficiency attainable by the vegetation. Water is the currency by which dry matter is gained in photosynthesis. The rate of photosynthesis limits the maximum amount of biomass that can be maintained. In the steady state, photosynthate is completely used to maintain the biomass, where maintenance includes maintenance respiration and the replacement of shed parts. Thus, the steady state biomass (B) is approximately

$$B = (Ppt) \, (Tr/Ppt) \, (P_s/Tr)/(r_m + r_g \tau) \qquad (12.1)$$

where Ppt is annual precipitation, Tr is transpiration, P_s is photosynthesis, r_m is annual maintenance respiration rate, r_g is growth respiration, and τ is the turnover rate.

Because of the effect of shading, the amount of biomass affects the rate of soil evaporation and, therefore, the amount of water transpired annually and the transpiration efficiency (Tr/Ppt) (Chapter 6). With low precipitation, less than 350 mm yr^{-1}, only a low biomass can be maintained, relatively high soil evaporation occurs, and transpiration efficiency is low (Figure 12.2). Transpiration efficiencies increase as precipitation increases to moderate amounts of 350 to 550 mm yr^{-1} because biomass also increases, which reduces soil evaporation by shading the soil surface and increases transpiration. Because shrubs are limited in the maximum foliage area and transpiration they can support, transpiration increases only slightly, and transpiration efficiency decreases with high precipitation of greater than 550 mm yr^{-1}. A similar relation was noted for grasslands and other semiarid regions (Fischer and Turner, 1978; Sims and Singh, 1978). Fundo Santa Laura receives enough precipitation to place it above the range of relatively high transpiration efficiencies for shrubs. Thus, the matorral stands at Fundo Santa Laura have relatively low transpiration efficiencies because of relatively high precipitation. Chamise chaparral has a relatively low foliage area index, high soil evaporation, and relatively low transpiration efficiency. The openness of the chamise canopy relates to its shade intolerance, which is discussed later in this chapter.

Year-to-year variations in precipitation influence transpiration efficiencies because leaf growth occurs after the period of receipt of the precipitation and varies less than

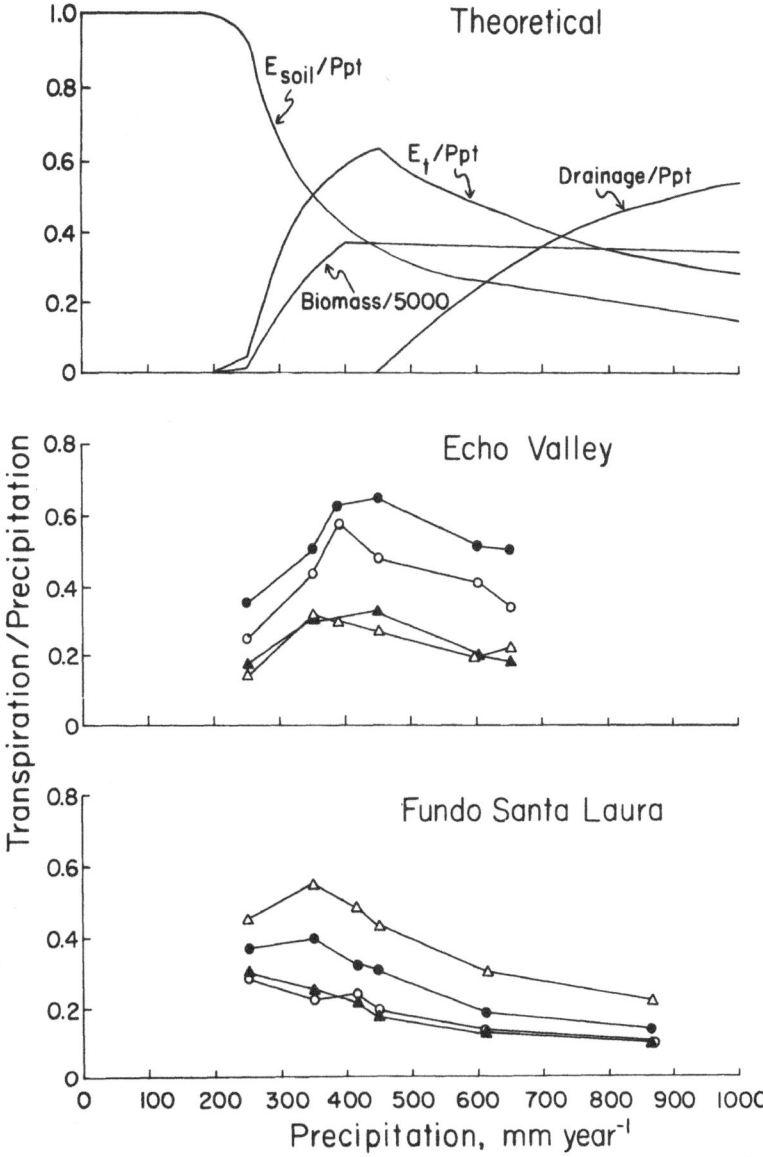

Figure 12.2. Relations between transpiration efficiency and annual precipitation for Californian and Chilean shrubs at their steady state foliage area indices for different precipitation amounts. Theoretical curves are calculated from the general relationships described in the text. E_{soil}/Ppt is the ratio of annual soil evaporation to annual precipitation, E_t/Ppt is transpiration to precipitation, Drainage/Ppt is the water drained from the soil in relation to precipitation. Biomass is given in g m^{-2} divided by 5000. Species at Echo Valley are *Adenostoma fasciculatum* (▲), *Arctostaphylos glauca* (●), *Ceanothus greggii* (○), and *Rhus ovata* (△); and at Fundo Santa Laura, *Colliguaya odorifera* (●), *Lithraea caustica* (△), *Satureja gilliesii* (▲), and *Trevoa trinervis* (○).

precipitation. The effect of increased precipitation on transpiration, net photosynthesis by leaves, aboveground growth, and carbon stored in carbohydrates was simulated by using data on four shrubs from Echo Valley and four from Fundo Santa Laura (Figure 12.3). The shrub parameters, especially those affecting root distribution, differed slightly from the parameters used in comparing chaparral and matorral. Although the trends in vegetation type and individual species efficiencies are similar, the exact numbers differ. The response of production to precipitation was flatter than the usual production-precipitation relations in arid regions (Chang, 1968; Whittaker, 1975) and was flatter at Fundo Santa Laura than at Echo Valley. The flatness of the response was due to the use of constant initial conditions for foliage area index and to the seasonal distribution of annual rainfall.

The maximum leaf area index that could be maintained with different annual precipitation amounts was estimated in simulations by increasing the initial leaf area index until the initial leaf area index was the same as the final leaf area index after a simulated year with a given annual precipitation (Table 12.5). The steady state leaf area indices increased as precipitation increased to about 400 mm yr^{-1} with the Californian shrubs and to 350 mm yr^{-1} with the Chilean shrubs and were similar to leaf area indices measured in the field. In the simulations, *A. fasciculatum* could support no more than about 1.0 m^2 leaf area m^{-2} ground, while *A. glauca* could support up to 3.5 m^2 m^{-2}. The Chilean species could support no more than 2.0 to 2.7 m^2 m^{-2}. Measured leaf area indices were about 1.0 m^2 m^{-2} for *A. fasciculatum*, 2 to 3.6 for *A. glauca*, and 0.9 to 2.3 for the Chilean species (Mooney et al., 1977; Murray, *unpubl. data*; Stuart, *unpubl. data*).

Transpiration tended to increase with annual precipitation, although the maximum biomass and maximum transpiration rates were limited by physiological characteristics of the different species. However, transpiration efficiency decreased rapidly as annual precipitation decreased below 350 mm yr^{-1}, except with the drought semideciduous shrubs. The transpiration efficiency of the drought semideciduous shrubs decreased with increasing precipitation. When the transpiration efficiency of a species was plotted against its elevational distribution from 0 to 1400 m (Chapter 3), the elevation at which maximum transpiration efficiency occurred was below the elevation of the maximum abundance of the species in every case simulated (Figure 12.4). The simulations supported the generality for mediterranean shrub vegetation that water is often the limiting factor determining the lower elevational limit of a species.

Several plant characteristics were correlated with annual precipitation. As precipitation decreased from that common in the center of the distribution of hard-leaved shrubs, roots became more shallow, hydrolability increased, maximum leaf conductance increased, and the proportion of seed-reproducing species increased (Chapter 6).

Net photosynthesis increased as precipitation increased up to an annual amount of 400 to 500 mm. Because of relatively high leaf conductances, the increase in net photosynthesis was greater in *A. glauca* and *C. greggii* than in the other Californian and Chilean species. With low precipitation, the Californian shrubs had lower production than did the Chilean shrubs. Growth increased with precipitation. In *A. glauca* and *C. greggii*, growth increased more than net photosynthesis. The net effect was that stored carbohydrates decreased with increased precipitation, which may be due to inaccurate formulation of growth-water content relations relative to the photosynthesis-water content relations because growth-water content relations are not well known for chaparral and matorral species.

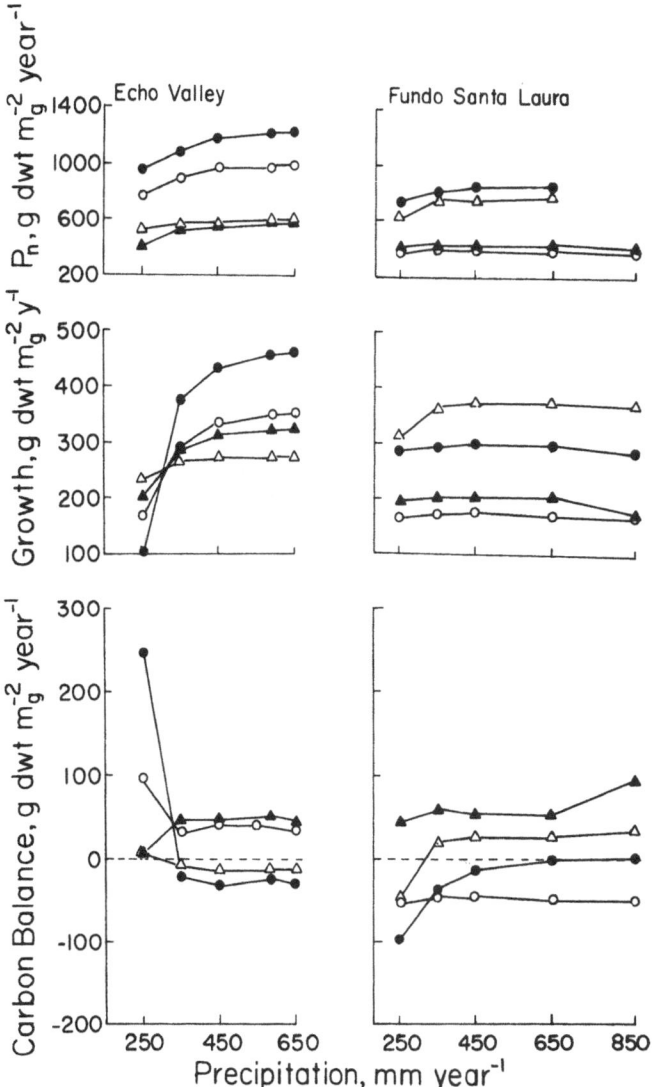

Figure 12.3. Simulated influence of precipitation on net photosynthesis, growth, and carbon balance of four species in each country. Units are grams dry weight per square meter of ground per year. Species at Echo Valley are *Adenostoma fasciculatum* (▲), *Arctostaphylos glauca* (●), *Ceanothus greggii* (○), and *Rhus ovata* (△); and at Fundo Santa Laura, *Colliguaya odorifera* (●), *Lithraea caustica* (△), *Satureja gilliesii* (▲), and *Trevoa trinervis* (○).

The seasonal pattern of rainfall in the mediterranean climate influences the sensitivity of transpiration, photosynthesis, and growth to increasing annual precipitation. The growing season and seasons of maximum transpiration, photosynthesis, and nitrogen uptake are controlled by temperatures at the beginning of the season and by the depleted soil moisture at the end of the season. Once temperatures become favorable the length of the growing season is related to the time necessary for evapotranspiration

Table 12.5. Steady state foliage area indices, net photosynthesis rates, transpiration rates, and transpiration efficiencies at different annual precipitation amounts[a]

Echo Valley

Ppt (mm yr⁻¹)	Foliage area index ($m_l^2\ m_g^{-2}$)				Net photosynthesis (kg dry wt m⁻² ground yr⁻¹)				Transpiration (mm yr⁻¹)				Water-uptake efficiency				Equivalent elevation (m)
	A.g.	C.g.	A.f.	R.o.	A.g.	C.g.	A.f.	R.o.	A.g.	C.g.	A.f.	R.o.	A.g.	C.g.	A.f.	R.o.	
650	7.0	3.8	1.7	4.0	1.30	1.00	0.64	0.62	332	221	143	117	0.51	0.34	0.22	0.18	1310
600									312	246	120	120	0.52	0.41	0.20	0.20	1150
550																	
500																	
450	7.0	3.6	1.7	4.0	1.25	0.96	0.60	0.62	293	216	144	122	0.65	0.48	0.32	0.27	671
390									246	226	121	105	0.63	0.58	0.31	0.30	479
350	4.0	2.5	1.2	3.7	0.98	0.83	0.52	0.59	179	151	105	109	0.51	0.43	0.30	0.31	351
300																	
250	1.9	1.1	0.5	1.2	0.63	0.52	0.30	0.39	88	63	43	35	0.35	0.25	0.17	0.14	32

Fundo Santa Laura

Ppt (mm yr⁻¹)	Foliage area index ($m_l^2\ m_g^{-2}$)				Net photosynthesis (kg dry wt m⁻² ground yr⁻¹)				Transpiration (mm yr⁻¹)				Water-uptake efficiency				Equivalent elevation (m)
	C.o.	T.t.	S.g.	L.c.	C.o.	T.t.	S.g.	L.c.	C.o.	T.t.	S.g.	L.c.	C.o.	T.t.	S.g.	L.c.	
867					-	0.37	0.43	-	121	78	78	191	0.14	0.09	0.09	0.22	2003
800																	
750																	
700																	
650	2.0	2.7	2.5	2.3	0.94	0.39	0.46	0.77									
625									119	88	81	188	0.19	0.14	0.13	0.30	1230
550																	
500																	

450	2.2	2.8	3.0	2.3	0.94	0.42	0.48	0.77	140	81	86	194	0.31	0.18	0.19	0.43	671
416									133	100	92	204	0.32	0.24	0.22	0.49	562
350	2.2	2.8	3.0	2.3	0.86	0.42	0.48	0.76	140	81	88	193	0.40	0.23	0.25	0.55	351
300																	
250	1.7	2.7	3.0	1.5	0.76	0.39	0.46	0.62	93	70	75	113	0.37	0.28	0.30	0.45	32

[a]Calculated for four species from Echo Valley, California: *Arctostaphylos glauca* (A.g.), *Ceanothus greggii* (C.g.), *Adenostoma fasciculatum* (A.f.), and *Rhus ovata* (R.o.) and for four species from Fundo Santa Laura, Chile: *Colliguaya odorifera* (C.o.), *Trevoa trinervis* (T.t.), *Satureja gilliesii* (S.g.), and *Lithraea caustica* (L.c.). The equivalent elevation on the coastal-facing slope of the coastal mountains is given.

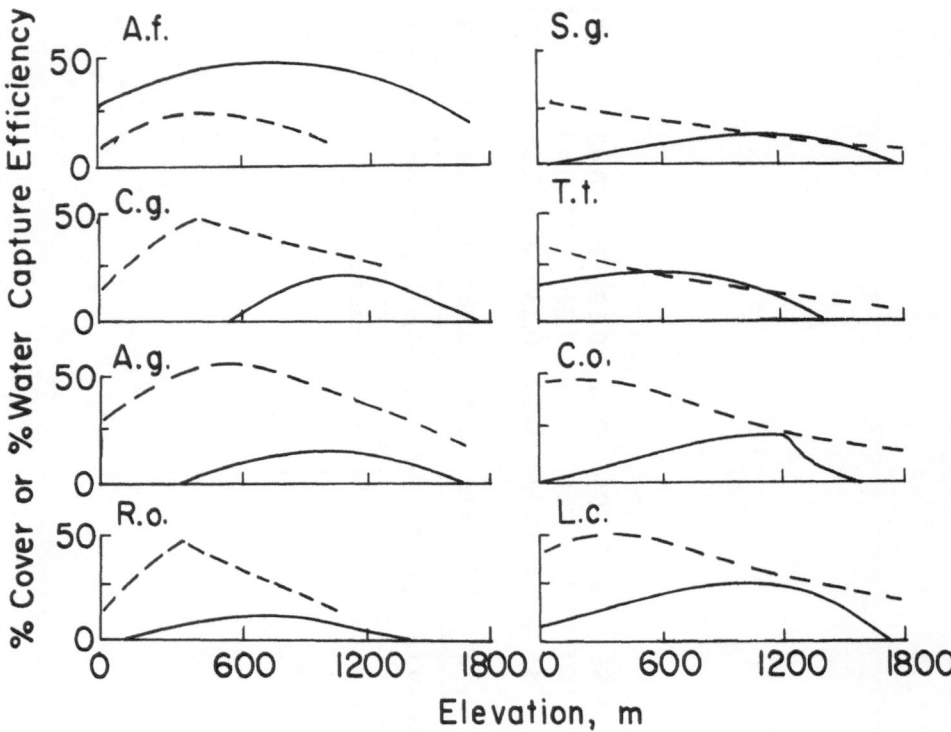

Figure 12.4. Elevational distributions (% cover, —) of eight shrub species in southern California and central Chile and the simulated water uptake efficiency (· · ·), transpiration over precipitation. *Adenostoma fasciculatum* (A.f.), *Ceanothus greggii* (C.g.), *Arctostaphylos glauca* (A.g.), *Rhus ovata* (R.o.), *Satureja gilliesii* (S.g.), *Trevoa trinervis* (T.t.), *Colliguaya odorifera* (C.o.), and *Lithraea caustica* (L.c.).

to remove all available water from the soil, which in turn depends on the amount of water held in the soil at the beginning of the growing season, the precipitation during the period, and the rate of evapotranspiration (Chapter 2). The period of active plant growth is more clearly defined at Echo Valley than at Fundo Santa Laura because of the greater diversity in the patterns of growth of the Chilean shrubs. Southern California receives, on the average, a greater fraction of the annual precipitation in the spring than does central Chile. In simulations with MEDECS, the growth rate of the evergreen shrubs was nearly constant during the growing season, about 4 g dry weight m^{-2} day^{-1} at Echo Valley and 1.5 to 2.5 g dry weight m^{-2} day^{-1} at Fundo Santa Laura. The length of the period of active growth greatly influences the annual production and depends on annual precipitation and the fraction of the precipitation that occurs in the spring. The small fraction of annual precipitation that occurs during the growing season desensitizes annual production from annual precipitation, less so at Echo Valley than at Fundo Santa Laura. According to the simulations and to the interrelations between precipitation and length of growing season, the growing season should lengthen about 8.5 days and growth should increase about 34 g dry weight m^{-2} yr^{-1} per 100-mm increase in annual precipitation at Echo Valley and about 4.5 days and 9 g dry weight m^{-2} yr^{-1} per 100-mm increase at Fundo Santa Laura.

The soil depth, by affecting the amount of water in the soil at the beginning of the growing season, affects the length of the growing season and the length of the dry season. The length of both seasons affects the distribution of species and of evergreen and drought deciduous leaf forms (see next section). Because precipitation increases with elevation, elevation compensates for shallow soils, and evergreen shrubs can be expected on shallower soils at higher elevations. Because precipitation increases less steeply with elevation in Chile, drought deciduous shrubs may be more widespread in central Chile than in southern California (Chapter 2).

Annual precipitation and nitrogen taken up by the vegetation should be interrelated, and a relationship can be derived from the steady state situation. The amount of nitrogen taken up by the vegetation limits the maximum biomass attainable, analogously to the limitations imposed by water. In the steady state, nitrogen taken up must equal that lost as plant parts are shed, which can be expressed as

$$B = (N_{up})/[(N_d)\tau], \tag{12.2}$$

where B is the biomass, N_{up} is the nitrogen taken up, N_d is the nitrogen content of plant parts at death, and τ is the turnover rate. The equation relating the maximum biomass attainable to precipitation (Eq. 12.1) and the equation relating the maximum biomass attainable to nitrogen uptake (Eq. 12.2) can be combined, given the plant turnover rates and the nitrogen contents of the shed parts, to calculate the nitrogen uptake expected with different annual precipitation amounts when the biomass is at a maximum (Figure 12.5). In the steady state, photosynthate is completely used to maintain the biomass and none is left over for net growth. Maintenance in the steady state includes maintenance respiration and growth costs to replace dying parts. The equations indicate that the steady state biomass should increase by 400 to 500 g m^{-2} per 100 mm yr^{-1} increase in annual precipitation and by about 400 g m^{-2} per g N m^{-2} yr^{-1} taken up. Turnover rates slower than 0.5 g g^{-1} yr^{-1}, which are equal to 2-year leaf longevities or more, have smaller effects on the steady state biomass than do more rapid turnover rates with leaf longevities shorter than 2 years. The nitrogen uptake demands follow the precipitation and water efficiency relationships. Within a precipitation regime, leaf longevity should increase in accordance with nitrogen limitations, because turnover rate is more closely related to nitrogen uptake than to precipitation.

12.3.2 Limitation by Solar Irradiance

The photosynthetic efficiency, calculated as photosynthesis/solar irradiance, is small because much of the incoming solar irradiance is lost as reflection, infrared irradiance, convection, and transpiration, processes over which the plant has little control. Transpiration occurs concomitantly with carbon dioxide diffusion into the plant. The water-use efficiency, calculated as either grams carbon dioxide incorporated per gram water lost or grams dry weight gain per gram water lost, is small because of the low concentration of carbon dioxide in the air relative to the amount of water vapor and because of the higher resistance in the path of carbon diffusion into the leaf than in the path of water vapor diffusion out of the leaf. Consequently, the highest rates of photosynthetic efficiency are about 5%, and in most natural communities are less than 2%.

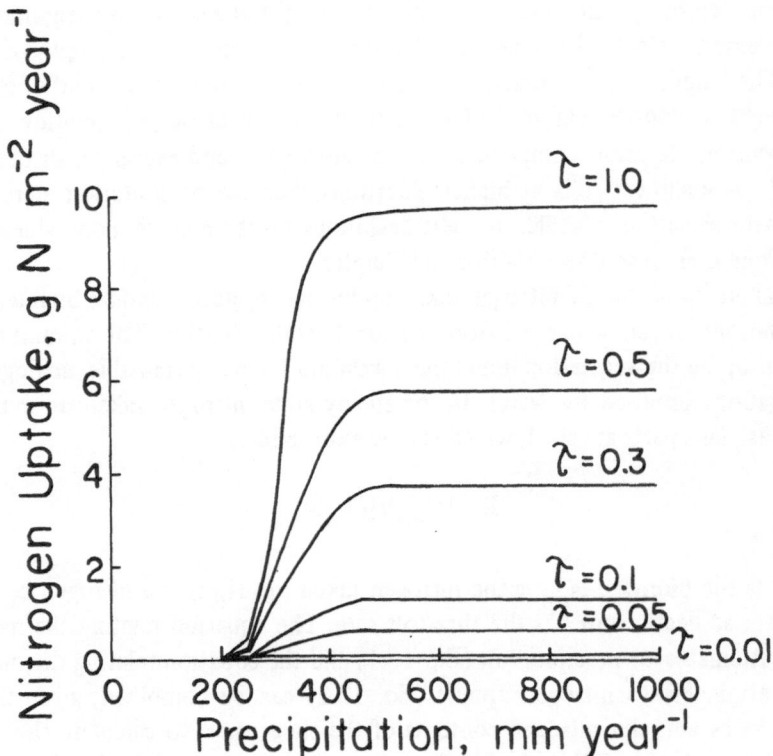

Figure 12.5. Relations between annual precipitation and nitrogen uptake for vegetation with different turnover rates of leaves and stems (τ, g g^{-1} year^{-1}). The lines indicate combinations giving full use of both nitrogen and precipitation.

Because the two countries differed in the seasonal progression of solar irradiance even though the annual totals were similar (Chapter 2), simulations were run to determine the effect of changing solar irradiance (Figure 12.6). Responses of the shrubs of the two countries were similar. Increasing irradiances increased net photosynthesis, growth, and carbon balance. The photosynthetic efficiency decreased with increasing solar irradiance because the foliage area index was constrained by the constant initial foliage area indices and by the availability of water. The low solar irradiance during the winter, when temperatures near the top of the canopy are adequate for photosynthesis but are too low for growth, should result in higher levels of storage carbohydrate near the top of the canopy than near the bottom. As a result, when growth is possible in spring, there should be higher growth rates at the top of the canopy. Moderate temperatures relative to the temperature response of photosynthesis, availability of water, and the low solar irradiances result in competition for light and create an advantage for taller shrubs during much of the year. Low solar irradiances in winter produce low evaporative demands, when water is available, and low transpiration rates. Any excess water during the winter is lost by drainage from the soil.

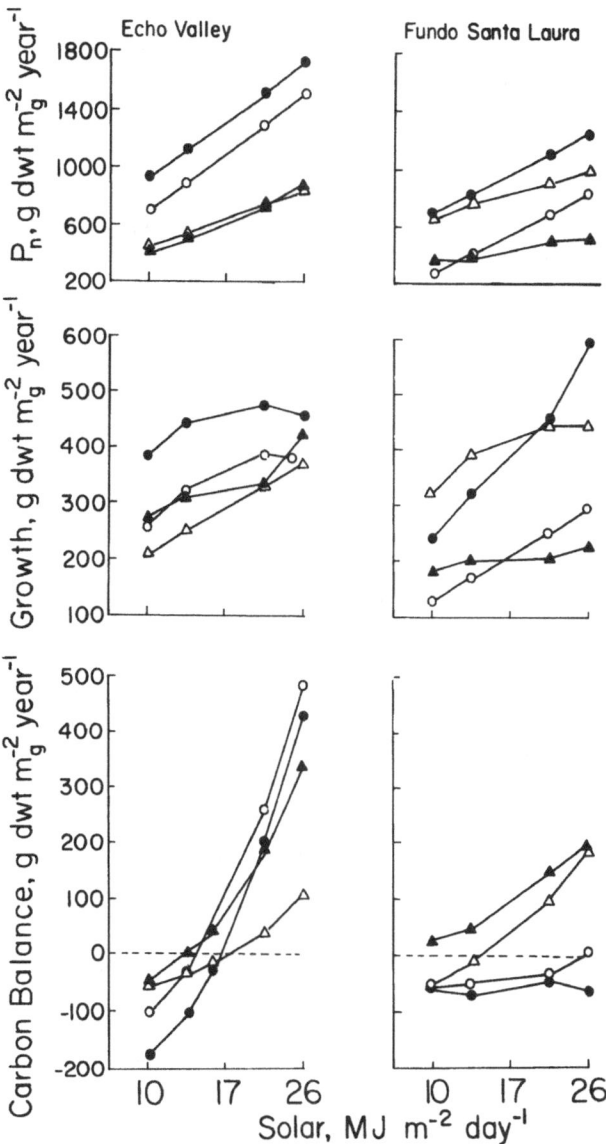

Figure 12.6. Simulated influence of mean annual solar irradiance on net photosynthesis of the leaves (P_n), growth, and carbon balance of four species in each country using MEDECS. Units are grams dry weight per square meter of ground per year. Species at Echo Valley are *Adenostoma fasciculatum* (▲), *Arctostaphylos glauca* (●), *Ceanothus greggii* (○), and *Rhus ovata* (△); and at Fundo Santa Laura, *Colliguaya odorifera* (●), *Lithraea caustica* (△), *Satureja gilliesii* (▲), and *Trevoa trinervis* (○).

12.3.3 Limitation by Nitrogen Availability

The simulated growth of shrubs increased with increased nitrogen availability. Nitrogen availability is expected to increase as community turnover rates increase (Stoner et al., 1978b; Miller and Stoner, 1979). Species with low leaf turnover rates produce low-quality litter for decomposition, leading to low rates of decomposition and minerali-zation. Because plant turnover rates and mineralization are interdependent, the resource-use efficiency for nutrients can stay constant over a range of nutrient availabilities. If the resource-use efficiency for nitrogen is not high, unabsorbed nitrogen can be lost from the soil by leaching. The vegetation-soil system can develop a realtively high nitrogen-use efficiency both by the vegetation developing efficient nitrogen-capturing mechanisms and by the loss of nitrogen through leaching. Because the nitrogen-use efficiency depends mainly on the behavior of the vegetation-soil system, it contrasts with the transpiration efficiency, which varies with the annual precipitation. Nitrogen additions should always increase growth regardless of the type of vegetation they are applied to, unless some other factor is strongly limiting. Partial evidence for this is the fact that fertilization increased growth in semiarid grasslands where water is generally limiting (Van Keulen, 1975; F. W. T. Penning de Vries, *pers. comm.*).

The availability of inorganic nutrients can affect water use. When Specht (1972b) added phosphorus to heath in Australia, the increased growth increased transpiration, thus increasing the severity of the drought, and eventually the *Banksia* in the heath was killed by drought. In simulations, increased precipitation increased the period when soil moisture was adequate for decomposition, which increased the amount of nitrogen mineralization. The precipitation and nitrogen interactions were quantified with relationships derived from simulations with MEDECS. A gram of nitrogen taken up and incorporated into aboveground tissue yields an increment of new biomass of about 166 g dry weight g^{-1} N. This increment of biomass increases leaf area by about 0.004 m^2 g^{-1} dry weight, although some of the new biomass is in stems. This incre-ment of leaf area increased transpiration by about 0.67 kg day^{-1} g^{-1} N. Allowing for the dry weight and water costs involved in the growth of the leaf area, the net increase in transpiration is about 0.33 kg day^{-1} g^{-1} N. The increased transpiration decreases the length of the season in which soil moisture is available. The length of the growing season changes with transpiration according to the amount of water available divided by the square of the daily evapotranspiration rate. Thus, the reduction in the growing season because of added nitrogen would be about 0.33 $\times$ 38 or 12 days per gram nitrogen taken up by the plants in southern California. In central Chile, the growing season would be reduced about 10 days per gram nitrogen taken up. This shortening of the growing season would reduce growth by about 48 g dry weight m^{-2} yr^{-1} in southern California and by 20 g dry weight m^{-2} yr^{-1} in central Chile, while the increased nitro-gen should increase growth by about 166 g dry weight. Net growth should increase about 118 g dry weight g^{-1} N taken up in California and 146 g dry weight g^{-1} N taken up in Chile, even though the growth period is limited by temperature and water. The mechanism of nitrogen influence is probably through increasing the rate of growth while water is available. Inasmuch as water is limiting and is being evaporated from the soil, increasing the leaf area growth rate increases the amount of water that is transpired, while reducing the amount evaporated. With the increased growth rate, more carbon is assimilated and is available for growth.

12.3.4 Limitation by Past Disturbances

Since the chaparral and matorral areas studied have been disturbed in the past (Chapters 1 and 3), the differences in resource-use efficiencies by species, slope, and country could be due to differences in the successional stage of the vegetation. Echo Valley was burned in 1952; Fundo Santa Laura was cut for firewood and charcoal until 1959. In order to determine the possible influence of the current successional stage of the vegetation on resource-use efficiency, the chaparral stands at Echo Valley and matorral stands at Fundo Santa Laura were simulated with different initial foliage area indices. The final foliage area index after one simulated year was noted. An increase in foliage area index at the end of the year indicated that the stand should still be in a growth phase. A decrease in foliage area index indicated that the initial foliage area of the simulated stand was too large to be supported (Figure 12.7). A steady state foliage area where the initial and final foliage areas were equal was determined. These simulated steady state foliage areas were similar to those measured in the field, indicating that the vegetation at Echo Valley and Fundo Santa Laura had recovered from past disturbances. In most cases, the simulated steady state foliage area index was greater than that expected on the basis of carbon balance calculations. Inasmuch as both foliage area index and carbon balance should be in the steady state, a range of foliage areas is considered to be near the steady state. The simulations used 606 mm yr^{-1} precipitation

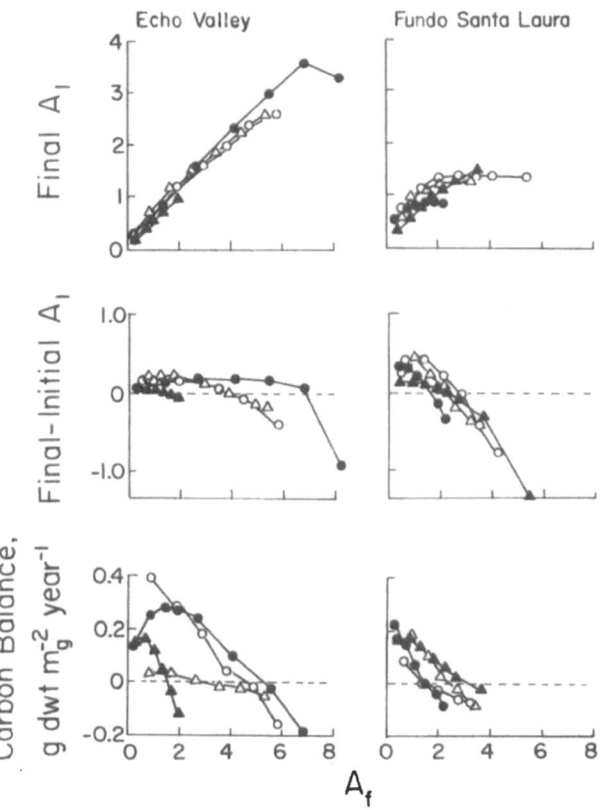

Figure 12.7. Simulated relations of initial foliage area index (A$_f$) on final leaf area indices (A$_l$), the difference between final and initial leaf area indices, and the carbon balance (grams dry weight per square meter of ground per year) of four shrub species. Species at Echo Valley are *Adenostoma fasciculatum* (▲), *Arctostaphylos glauca* (●), *Ceanothus greggii* (○), and *Rhus ovata* (△); and at Fundo Santa Laura, *Colliguaya odorifera* (●), *Lithraea caustica* (△), *Satureja gilliesii* (▲), and *Trevoa trinervis* (○).

in southern California and 804 mm yr^{-1} in central Chile; these amounts are higher than the expected normal precipitation in both countries.

In these simulations, transpirational water loss increased with foliage area (Figure 12.8). *Arctostaphylos glauca* transpired up to 350 mm yr^{-1} and *C. greggii* up to 300 mm yr^{-1}. *Adenostoma fasciculatum* and *R. ovata* transpired up to 150 mm yr^{-1}. The Chilean species transpired less, *L. caustica* transpired 200 mm yr^{-1}, *C. odorifera* 150 mm yr^{-1}, and *S. gilliesii* and *T. trinervis* less than 100 mm yr^{-1}. Soil evaporation decreased with foliage area and was 200 to 300 mm yr^{-1} in southern California and about 200 mm yr^{-1} in central Chile. Total evapotranspiration from the vegetation and soil increased when foliage area index was increased for species with high transpiration rates, but remained constant in species with moderate or low transpiration rates. In these latter species, increased transpirational losses were compensated for by decreased soil evaporation.

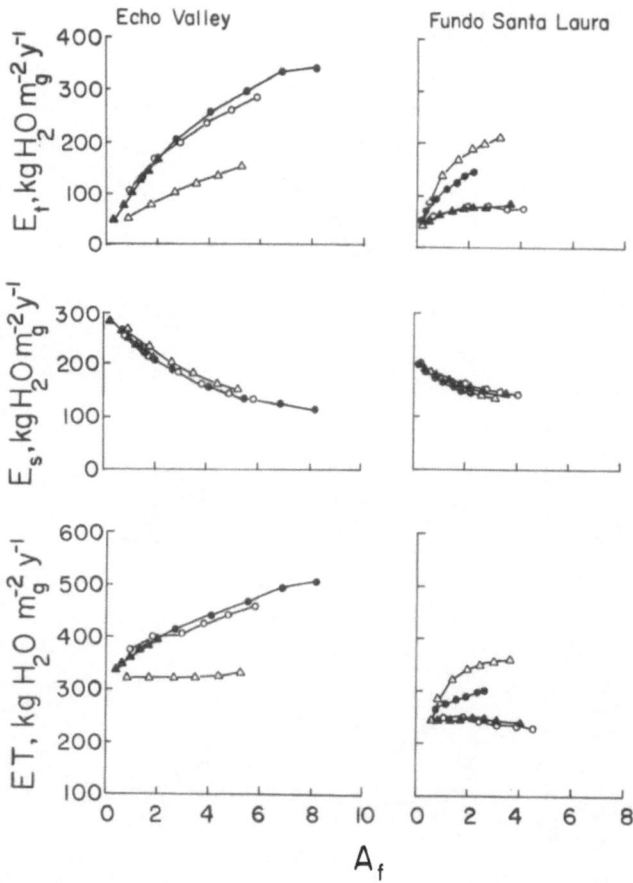

Figure 12.8. Effect of foliage area index on transpiration (E_t), soil evaporation (including interception losses, E_s) and total evapotranspiration (ET) for four shrub species at Echo Valley and Fundo Santa Laura. Units are kilograms water per square meter of ground per year. Species at Echo Valley are *Adenostoma fasciculatum* (▲), *Arctostaphylos glauca* (●), *Ceanothus greggii* (○), and *Rhus ovata* (△); and at Fundo Santa Laura, *Colliguaya odorifera* (●), *Lithraea caustica* (△), *Satureja gilliesii* (▲), and *Trevoa trinervis* (○).

Production increased with foliage area index and was near maximal at the steady state foliage areas (Figure 12.9). The steady state foliage area was usually slightly greater than that which would give maximum production.

Nitrogen release decreased as foliage area increased (Figure 12.10), because the increasing transpiration rates decreased the length of the decomposition season. Lower rates of nitrogen release were calculated for Echo Valley than for Fundo Santa Laura, corresponding with the lower herb cover at Echo Valley. For most species, nitrogen uptake increased to a plateau as foliage area index increased, except for *S. gilliesii* and *C. odorifera* in which nitrogen uptake peaked, then decreased. Nitrogen-uptake efficiency also increased to a plateau in most species similarly to nitrogen uptake.

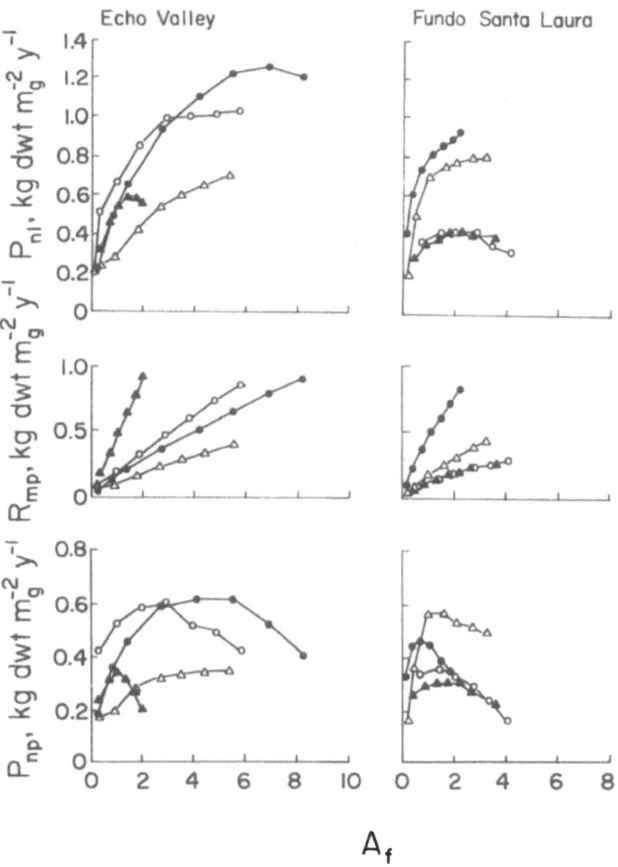

Figure 12.9. Effect of foliage area index (A_f) on net photosynthesis of leaves (P_{nl}), maintenance respiration of the plant except for leaves (R_{mp}), and net production (P_{np}) before growth of four species at Echo Valley and Fundo Santa Laura. Units are kilograms dry weight per square meter of ground per year. Species at Echo Valley are *Adenostoma fasciculatum* (▲), *Arctostaphylos glauca* (●), *Ceanothus greggii* (○), and *Rhus ovata* (△); and at Fundo Santa Laura, *Colliguaya odorifera* (●), *Lithraea caustica* (△), *Satureja gilliesii* (▲), and *Trevoa trinervis* (○).

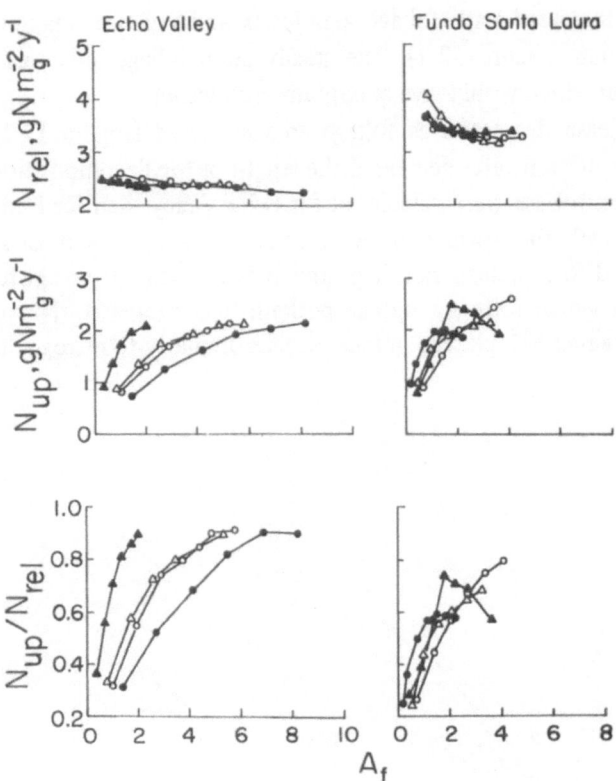

Figure 12.10. Effect of foliage area index on nitrogen released by decomposition (N_{rel}), nitrogen taken up by the plant (N_{up}), and nitrogen-uptake efficiency (N_{up}/N_{rel}) for four species at Echo Valley and Fundo Santa Laura. Units are grams nitrogen per square meter of ground per year. Species at Echo Valley are *Adenostoma fasciculatum* (▲), *Arctostaphylos glauca* (●), *Ceanothus greggii* (○), and *Rhus ovata* (△); and at Fundo Santa Laura, *Colliguaya odorifera* (●), *Lithraea caustica* (△), *Satureja gilliesii* (▲), and *Trevoa trinervis* (○).

12.4. Functional Interpretation of Morphological and Physiological Characteristics of Mediterranean Shrub Vegetation

Similar vegetation characteristics in mediterranean type ecosystems imply that these characteristics are important for using resources within the constraints of irradiance, temperature, annual precipitation, and soil moisture in mediterranean type climates, assuming that resource use is important for species survival. These similar vegetation characteristics include the shrub form, moderate leaf area indices, evegreen leaves, and small leaves and are related to physiological characteristics of the vegetation. An analysis of the interrelations of vegetational and physiological characteristics and resource use was carried out with simulation experiments.

12.4.1 Predominance of the Shrub Form

Shrub form affects light interception, especially during the winter, and affects convectional cooling during the summer. Miller and Mooney (1976) simulated canopy microclimates, plant temperatures, transpiration, and photosynthesis in mediterranean shrub canopies. In winter, photosynthesis was light limited; the highest rates of photosynthesis were at the top of the canopy. In early spring, photosynthesis was limited by low temperatures; the highest rates were at the bottom of the canopy because high soil temperatures produced near optimal temperatures there. In late spring, photosynthetic rates were uniform throughout the canopy. In summer, photosynthesis was water and temperature limited; photosynthetic rates were highest at the top of the canopy where temperatures were lower because of convectional cooling. These simulations used a relatively low leaf area index and data for *Heteromeles arbutifolia* (Harrison, 1971), which indicated a relatively large temperature dependence of photosynthesis. More recent data (Chapter 7) on other species from Echo Valley and Fundo Santa Laura indicated less temperature dependence. In simulations using these newer data, the rates of photosynthesis and transpiration and the photosynthesis:transpiration ratio were highest at the top of the canopy throughout the year. The shrub form has an advantage over shorter vegetation forms in the mediterranean scrub because, for much of the year, climatic conditions are favorable for higher photosynthetic rates at the top of the canopy, which favors taller growth forms. With higher photosynthetic rates at the top of the canopy, growth is greater at the top of the canopy because carbohydrate reserves are more available at the top.

The woody root systems of shrubs, in contrast to the fibrous root systems of grasses, undoubtedly provide an advantage for the shrubs on the rocky, coarsely textured soils prevalent in mediterranean type regions, whereas grasses tend to predominate on pockets of less rocky soil.

The relative advantage of the shrub form over the tree form is considered to relate to the short length of time required for shrubs to recover from periodic disturbance. The rapidity of regrowth of lucerne (alfalfa) after cutting was related to the number of shoots coming from the root crown, rather than to the storage carbohydrate content of the root crown (Leach, 1969). The regrowth of lucerne can be interpreted in a plant shoot-population context (White, 1979) and applied to the growth of shrubs. Biomass increase by an individual shrub shoot is limited by the length of the growing season and other limiting factors, yet biomass increase by the shrub is important for increasing photosynthesis and nitrogen uptake in competition with other individuals. Increasing the number of shoots per individual is an expedient way to rapidly increase biomass. The population of dormant buds that can potentially form active shoots is increased each year by the number of axillary buds produced each year by each active shoot. The population increase is more rapid with a higher initial population of axillary buds, which occurs in the shrub form in contrast to the tree form. Part of the advantage of the shrub form is in the rapidity of regrowth after disturbance.

12.4.2 Leaf Area Index

The leaf area indices of the vegetation in mediterranean regions are commonly 1 to 2 m^2 leaf m^{-2} ground and increase to 3 to 4 m^2 leaf m^{-2} ground as precipitation increases or the site water balance becomes more favorable (Lossaint, 1973; Cody and Mooney, 1978). In simulations with CAPS, canopy photosynthesis increased with increasing leaf area index and reached a plateau for all the mediterranean shrub species except *A. fasciculatum* (Figure 12.11). With *A. glauca*, the highest canopy photosynthetic rates occurred with a leaf area index of 3 to 4. With *C. greggii* and *R. ovata*, the highest rates occurred with leaf area indices of 2.0 to 2.5. Canopy photosynthesis rates were consistently high with leaf area indices from 2 to about 5. With *A. fasciculatum*, the highest rates occurred at a leaf area index of about 1.0. The leaf area indices measured for these species at Echo Valley were near the values associated with maximal photosynthetic rates. The overall leaf area index estimated for Fundo Santa Laura by Mooney (1977a) was near the leaf area index for high production in a mixed stand of the three Chilean species simulated, *L. caustica*, *C. odorifera*, and *S. gilliesii*, but the leaf area indices of the individual species were slightly below optimal.

The simulations indicated that *A. fasciculatum* can survive in open canopies where the leaf area index is less than 1 but cannot survive in closed canopies where the leaf area index is greater than 1, because it is intolerant of shading. *Arctostaphylos glauca* or *C. greggii* could occur in closed canopies. These conclusions are consistent with field observations of community structure. The three Chilean shrubs, *L. caustica*, *C. odorifera*, and *S. gilliesii*, should be able to coexist and do coexist at Fundo Santa Laura.

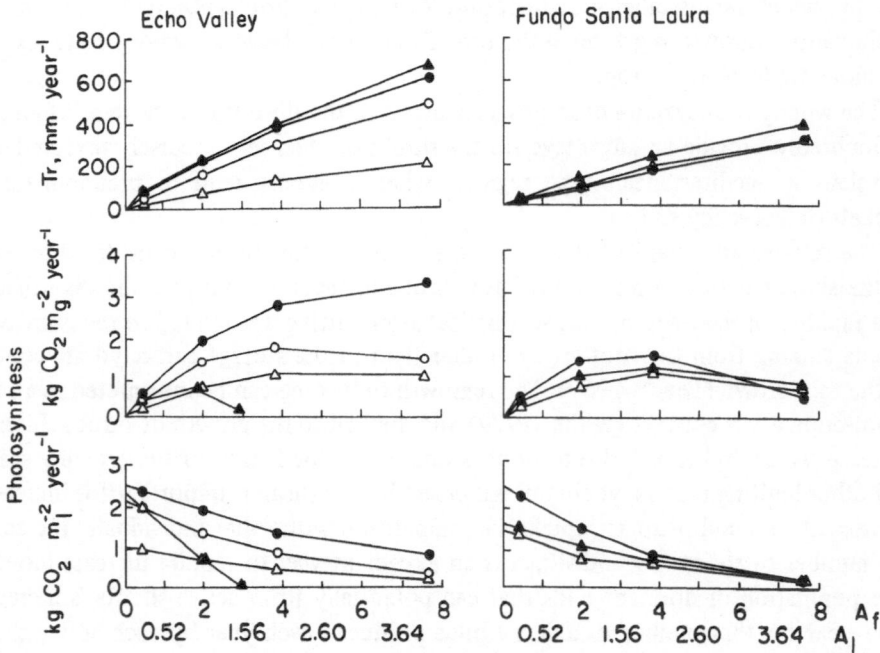

Figure 12.11. Transpiration (Tr) and photosynthesis on both ground area (kilograms carbon dioxide per square meter of ground per year) and leaf area (kilograms carbon dioxide per square meter of leaf per year) basis with increasing foliage area index of the four shrub species at Echo Valley and Fundo Santa Laura simulated with CAPS. Species at Echo Valley are *Adenostoma fasciculatum* (▲), *Arctostaphylos glauca* (●), *Ceanothus greggii* (o), and *Rhus ovata* (△); and at Fundo Santa Laura, *Colliguaya odorifera* (●), *Lithraea caustica* (△), *Satureja gilliesii* (▲), and *Trevoa trinervis* (o).

12.4.3 Evergreenness

Leaf longevity affects nutrient utilization by affecting the demand for nutrients and by affecting the quality of the litter available for decomposition and mineralization. Several physiological constraints affect leaf longevity in the vegetation in mediterranean climates in addition to nutrients.

Miller (1979) proposed that the distribution of evergreen and deciduous shrubs was controlled by the moisture and temperature requirements for growth and photosynthesis and by the seasonal progression of soil moisture and plant temperatures. Since growth requires higher temperatures than does photosynthesis, the moderate winter temperatures of the mediterranean climate suppress growth more than photosynthesis. At midelevations, in order for a leaf to be available for the winter photosynthetic season, it must be grown in the spring before the drought and must survive the drought. Leaf survival requires strengthening the leaf with structural tissue. The seasonal patterns of moisture and temperature limitations vary geographically. At lower elevations along the coast, temperatures during the winter are sufficiently high to allow some growth as well as photosynthesis. Summer deciduous plants are possible and are favored by the longer drought. With increased precipitation along the coast, the drought deciduous forms should be less favored and evergreen forms more favored because of the decreased length of drought. At higher elevations, the summer drought is shorter, low winter temperatures reduce photosynthesis, and winter deciduous forms are possible and advantageous. Miller (1979) calculated the annual dry weight gain of evergreen and deciduous leaves. The elevational patterns of the net dry weight gain of evergreen and deciduous leaf forms followed the elevational distribution of these leaf forms (Figure 12.12). Although many species can grow in temperatures below 10°C, species in the mediterranean type climates do not. Species such as *R. ovata* and Australian shrubs require temperatures above 17°C for growth. With lower temperature requirements, growth should be possible through the winter, and more summer deciduous species could occur. Evidently, the physiological adjustments required to increase growth at temperatures below 10°C are incompatible with the high summer temperatures and summer drought. Miller and Mooney (1976) proposed that the length of the drought affected the distribution of evergreen and deciduous shrubs; evergreen shrubs occurred where the carbon cost of leaf maintenance through the drought was less than the carbon cost of new leaf growth. The costs of maintenance and new leaf growth were equal with a drought of about 100 days. Evergreen forms should occur when the drought was less than 100 days and deciduous forms where the drought was more than 100 days, which agrees with the distribution of plant forms and precipitation at the coast and inland in southern California.

In addition to surviving the drought, the leaf must function until the carbon costs of its construction are recovered. The period required to recover the costs depends on the photosynthetic rates, the biochemical composition of the leaf, the amount of supportive stems and roots associated with the leaf, and the time of year when leaf growth takes place. Miller and Stoner (1979) indicated that *A. fasciculatum* required the shortest time to recover construction costs; *R. ovata* and *L. caustica*, the longest (Figure 12.13). *Adenostoma fasciculatum* should be favored under conditions of periodic catastrophic mortality, such as fire, because of its relatively low cost of leaf construction

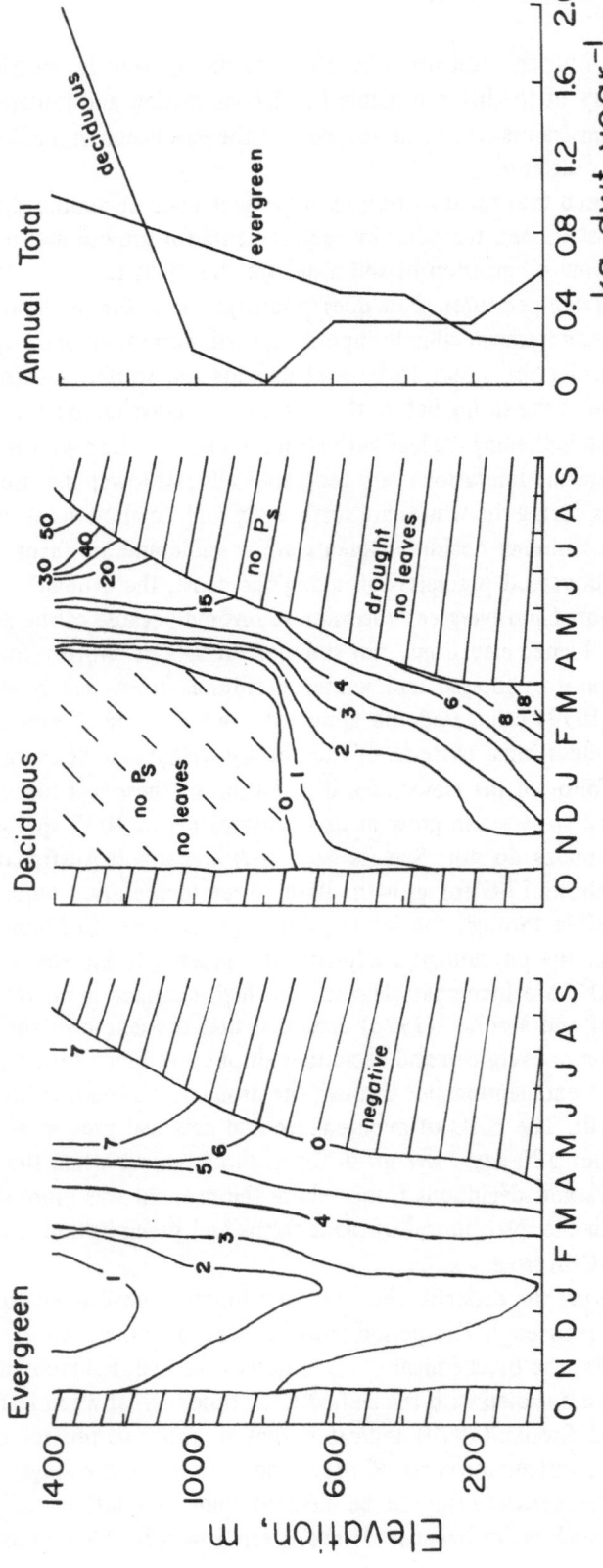

Figure 12.12. Simulated dry weight gain of evergreen and deciduous leaf species along an elevational gradient in southern California. Values given in the season progression are grams dry weight per square meter of leaf per day for evergreen species and grams dry weight per plant per day for deciduous species. Values for the annual totals are kilograms dry weight per year.

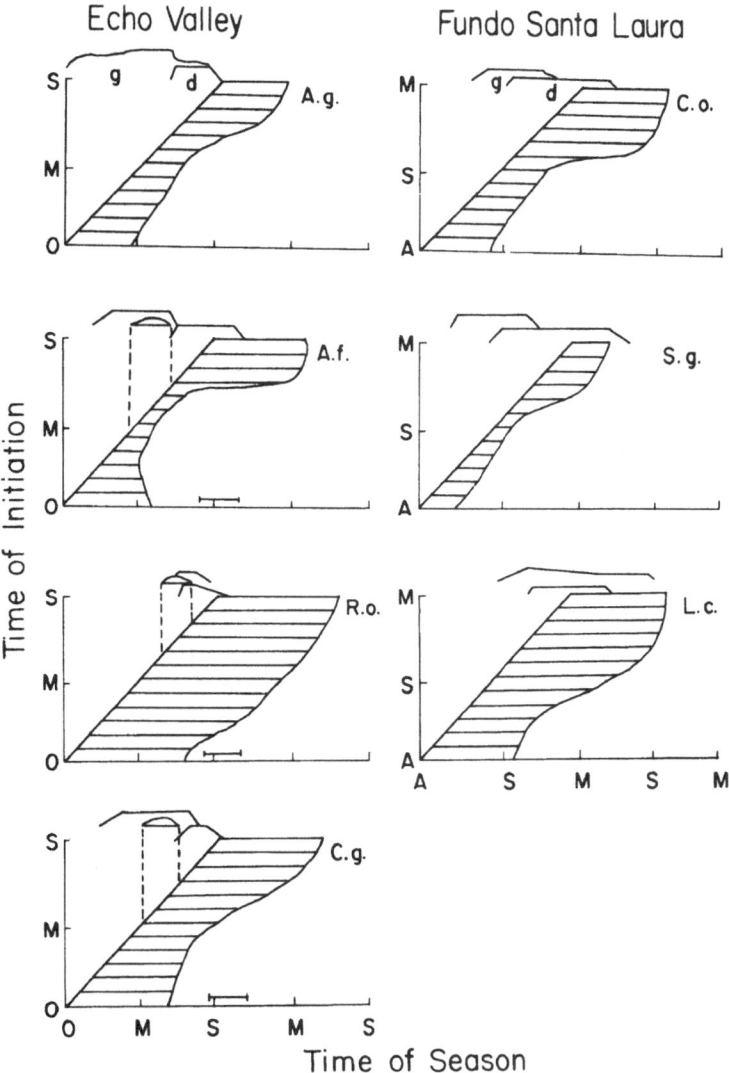

Figure 12.13. Length of time (horizontal lines) required to regain the cost of constructing a new leaf and associated stem and root biomass when initiated at different times of the year. Months are indicated for the time of the season. The observed periods of leaf growth (g) and leaf death (d) are given for each species. The period of most frequent fires for California is indicated by the heavy horizontal line (from Miller and Stoner, 1979). Species at Echo Valley are *Arctostaphylos glauca* (A.g.), *Adenostoma fasciculatum* (A.f.), *Rhus ovata* (R.o.), and *Ceanothus greggii* (C.g.); and at Fundo Santa Laura, *Colliguaya odorifera* (C.o.), *Satureja gilliesii* (S.g.), and *Lithraea caustica* (L.c.).

due to its low specific leaf density. Leaves of *A. fasciculatum*, initiated between October and early June, recovered their construction costs by September. Leaves initiated in May and June did not recover the leaf cost until the following spring. Other species at Echo Valley required longer periods to recover the cost of constructing new leaves. The longest required period was in *R. ovata*. *Rhus ovata* leaves initiated in spring or summer did not recover their construction costs until almost a year later. *Rhus ovata* branches cannot grow each year without developing a carbon deficit because of the long carbon recovery time of the species. Field observations confirm that a low number of *R. ovata* branches produce new growth each year (Kummerow and Fishbeck, 1977). Of the species at Fundo Santa Laura, *S. gilliesii* had the shortest recovery time, which is appropriate for a drought deciduous species. If leaves initiated in December remained active, they could not recover their growth costs until April. *Satureja gilliesii* leaves become brown and dry in summer (January), but some persist until the following fall (April). These persistent leaves are those initiated later in the spring (Montenegro et al., 1979).

Because photosynthetic rates decrease with shading, the required length of the recovery period increases with foliage area index. Leaf area index, stem area index, and leaf longevity are interrelated. In order for the species to coexist on the pole-facing slope at Echo Valley, leaves of *C. greggii*, *A. fasciculatum*, and *R. ovata* require at least 2.5 to 3.0 years to recover the costs of construction, but leaves of *A. glauca* require only 1 year. On the equator-facing slope, leaf longevities of *A. fasciculatum* should be at least 1.0 to 1.5 years. At Fundo Santa Laura, a leaf area index of 2.0 should be associated with leaf longevities of 1.0 in *C. odorifera*, 1.5 in *L. caustica*, and less than 1.0 in *S. gilliesii*. Leaf longevities calculated from Thrower and Bradbury (1977) were 1.8 years in *A. glauca*, 1.5 in *C. greggii*, 2.3 in *R. ovata*, 2.2 in *C. odorifera*, 0.4 in *S. gilliesii*, and 1.8 in *L. caustica*. Thus, environmental constraints that lead to different foliage area indices may lead to different leaf longevities and different rates of nutrient cycling.

Deciduous shrubs can support higher leaf area indices than evergreen shrubs, first because of the lower cost of producing leaves, about 1.3 g glucose g^{-1} dry weight for deciduous leaves versus 1.7 g for evergreen leaves, and second because of the higher photosynthetic rates in deciduous shrubs. However, the deciduous form requires at least an 80-day period with temperatures and moisture favorable for photosynthesis to regain the carbon costs of leaf growth. Such a favorable period is not usually available in the midelevations of the mediterranean shrub region in southern California.

Leaf longevity is related to nitrogen availability. Small (1972) suggested that the ratio of photosynthesis accumulated through the life of the leaf to leaf nitrogen content should be optimized in nitrogen-limited sites. As nitrogen becomes deficient, the production system can be expected to shift from forms with high leaf turnover to forms with low turnover. Even with evergreen leaves, this can mean increased leaf longevities. As nitrogen becomes deficient, the competing plants can be expected to converge to similar ratios of accumulated photosynthesis to leaf nitrogen content throughout the life of the leaf. Nitrogen availability is also influenced by the vegetation type. When evergreen leaves are shed, they have carbon:nitrogen ratios of about 100, which should inhibit rates of decomposition and nitrogen mineralization. Evergreen leaves also often have secondary compounds that inhibit decomposition and herbivory. A site with nitrogen relations that favor deciduous species but with climatic conditions, such as drought or the timing of favorable and unfavorable conditions for growth, that favor evergreenness may develop low rates of nitrogen turnover because of low rates of mineralization or leaching losses. Such sites can appear nitrogen limited as the vegetation develops. Competition, measured by the ratio of resource incorporated to resource made available, should be more intense for nitrogen than for water, with competition for light being least intense.

The constraints, which relate to the utilization of water and light energy, also affect the apparent availability and utilization of nitrogen. Evergreen leaves require strengthening by ligneous compounds and herbivore defense compounds, both of which reduce the quality of the substrate for decomposition, thus decreasing mineralization and nutrient availability. The selection of evergreen leaves by climatic factors, such as the length of drought and the seasonal timing of favorable water and temperature conditions, should favor leaves of about 1 year's duration but not necessarily leaves of more than 1 year's duration. The length of the period needed to recover the carbon costs of leaf production depends on climatic and plant factors and may require leaves with longer than 1 year's duration. These drought-selected evergreen leaves should have structural modifications to maintain rigidity as water stress develops and to reduce the intensity of water stress. Such leaves should be rigid, narrow, steeply inclined, and light colored and should have small cell sizes. These characteristics reduce cellular distortion, heat loads, and transpiration rates (Gates, 1962, 1965; Taylor, 1975; Cutler et al., 1977). Nitrogen limitation may favor leaf longevities of more than 1 year (Monk, 1966; Small, 1972), and leaf longevities may increase with the severity of the nitrogen limitation. Although evergreenness caused by nitrogen limitation does not require the adaptations of drought-stressed leaves (Beadle, 1966), the general biochemical composition of nitrogen-limited leaves may be similar to that of water-limited leaves, resulting in similar sclerophyll indices because the sclerophyll index is based on lignin, cellulose, and nitrogen content. Water limitation creates a direct need for sclerophyllous tissue, while the nitrogen limitation may reduce protein formation because of low nitrogen availability, which may increase cellulose and lignin formation because of excess carbohydrate.

The vegetation of the five mediterranean regions of the world may differ because of differences in their nutrient and climatic regimes, as well as differences in their phylogenetic histories (Figure 12.14). The mediterranean region of central Chile is relatively nutrient rich and the vegetation is climatically selected; leaf turnover rates are relatively high. In southern California, soils are less nutrient rich, but climate is still a strong agent; leaf turnover rates are lower than in central Chile. In the mediterranean regions of Australia and South Africa where old, nutrient-poor soils occur and leaf turnover is lower than in California, the climatic effect is strongly reinforced by nutrient limitations (Specht, 1979).

12.4.4 Leaf Size and Orientation

Leaves in mediterranean regions are characteristically small and steeply inclined (Walter, 1973; Cody and Mooney, 1978). In simulations with CAPS for Echo Valley, leaf inclination affected production and water-use efficiency, calculated as photosynthesis/transpiration, only during the summer. Production and water-use efficiency increased in summer-active plants with moderately inclined leaves. In simulations for *C. greggii* and *L. caustica*, photosynthesis increased, transpiration increased, and the photosynthesis:transpiration ratio increased as leaves became more steeply inclined.

Small leaf size increases the potential for convectional exchange of energy. The possible advantages of increasing the convectional heat exchange potential are to avoid high lethal plant temperatures and to reduce transpiration and prolong periods of

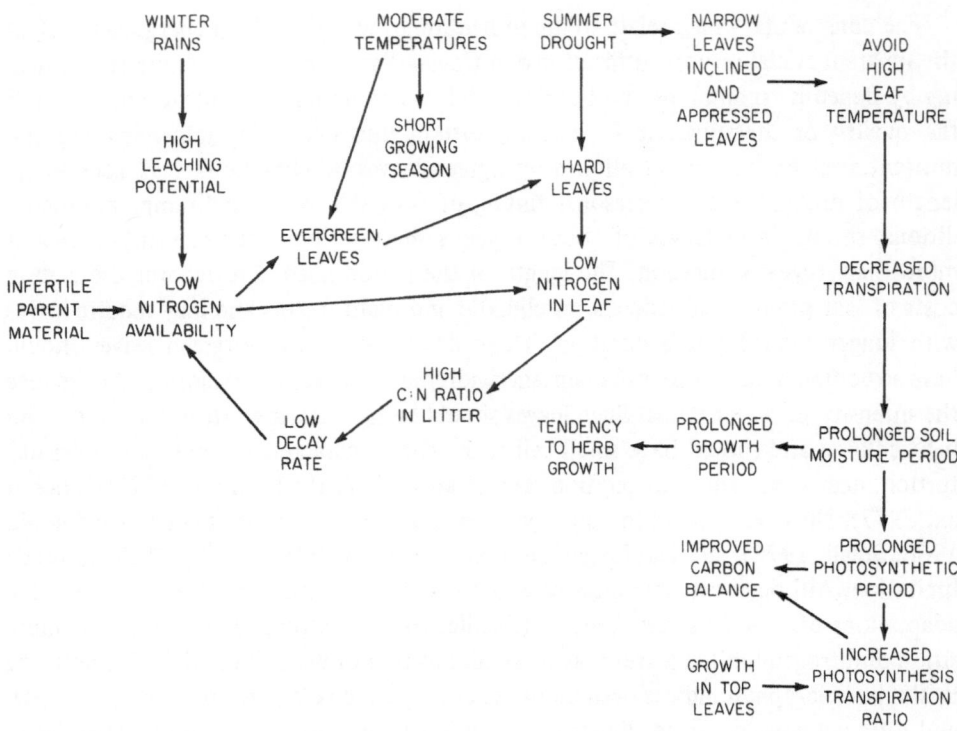

Figure 12.14. Diagram of interactions between climate and soil variables affecting the structure of mediterranean scrub vegetation.

photosynthesis both diurnally and seasonally. The potential effects of leaf width on transpiration, photosynthesis, and water-use efficiency were ascertained with simulations using CAPS. Results of the simulations for *A. fasciculatum* indicated that as leaf width increased, transpiration increased, photosynthesis decreased, and water-use efficiency decreased. The advantage appeared to be with narrow leaves. Simulations for *C. greggii* indicated that as leaf width increased, transpiration decreased, and photosynthesis decreased in February and May but increased in September. The water-use efficiency was constant in February but increased in May and September. The advantage of narrow leaves was not as marked for *C. greggii* as for *A. fasciculatum*. In *C. odorifera*, as leaf width increased, transpiration increased and photosynthesis was constant in March but decreased in July and November. The photosynthesis:transpiration ratio decreased in all months. Narrow leaves seem to be advantageous for *C. odorifera*. Increasing leaf width in *L. caustica* caused transpiration to increase in March but decrease in July and November. Photosynthesis was constant in March and July but decreased in November. The water-use efficiency decreased in March and November and increased in July. The simulations indicated that *L. caustica* was more advantaged with wider leaves than were the other species. The results of the simulations are consistent with the actual leaf sizes of the species.

Both leaf width and leaf inclination had less effect on production and water-use efficiency than did height within the canopy, location with respect to the hot soil surface, and leaf duration-carbon balance considerations. Leaf temperatures were usually not critical. Modifications of width and inclination only constituted fine tuning of an already-surviving plant system. The presence of the broad-leaved shrubs in mediterranean type ecosystems may be due to the unavailability of needle-leaved species, although in several species the leaf size is reduced to needle-like proportions. The effect of increased leaf clustering was similar to that of increased stem area and resulted in decreased photosynthesis and decreased transpiration. Increased leaf clustering had a variable effect on the water-use efficiency.

12.5. Relative Importance of Plant Adaptations Versus Environment in Affecting Resource Use and Resource-Use Efficiency

The factors recognized in this study as important in affecting resource use and resource-use efficiency indicate that transpiration efficiency is largely controlled by the environment, although several plant properties are also important, while photosynthetic efficiency and nitrogen-uptake efficiency are largely controlled by the plant. The plant properties most influential in affecting resource use and resource-use efficiency are expected to be most strongly selected for, if the use of resources is important.

The transpiration efficiency is mainly determined by the annual precipitation, the seasonal distribution of precipitation, the solar or net total irradiance in winter, and the soil depth. This efficiency will be affected to a lesser degree by the air temperature, vapor pressure, and wind, which affect potential and actual evapotranspiration and the water-use efficiency. Transpiration efficiency is also influenced by the successional stage of the vegetation, which is affected by the rapidity of shrub regrowth. Plant variables, including leaf conductance, also affect transpiration efficiency. The maximum leaf conductance should increase as soil evaporation becomes an increasingly important contribution to water loss. However, the high leaf conductance of *A. glauca* reduced soil moisture earlier, which shortened the growing period and increased the length of drought. The lower leaf conductances of *R. ovata* reduced the rate of water loss, which lengthened the growing period and decreased the length of drought. Decreasing the minimum water potential that can be endured and increasing survival during drought increases the period when stomates can be open and photosynthesis is possible. Survival during drought can be increased by smaller cell size and cell wall strengthening, but apparently at the cost of increased resistance of the cell wall to carbon dioxide diffusion. Concentrating roots near the soil surface should increase the transpiration efficiency, especially if soil evaporation is important. However, the rapid drying of the surface soil may cause high turnover rates in shallow roots, resulting in high carbon and nitrogen costs to the plant. Decreasing the turnover rate of leaves, stems, and roots increases the possible biomass and increases shading and, when biomasses are low, can increase transpiration efficiency. Increasing the shade tolerance by decreasing the light compensation level, leaf clustering, or stem area can increase the possible biomass. The shade tolerance of the species affected the optimal and maximal leaf area indices attainable. As the leaf area index increased, the length of drought increased.

The photosynthetic efficiency is mainly determined by height in the canopy, length of the photosynthetic season, leaf and stem area indices, water-use efficiency, and maximum photosynthetic rate. The water-use efficiency and maximum photosynthetic rates are affected by the cellular and wall resistances to carbon dioxide diffusion.

The nitrogen-uptake efficiency is determined by both nitrogen availability and nitrogen loss. Nitrogen availability is affected by soil temperature, soil water, and by substrate quality, which decreases with evergreen leaves. Both soil temperature and soil water are affected by the vegetation cover. Nitrogen uptake is affected by the total root length or surface area and by the distribution of roots through the soil profile. Root length is affected by the root extension per unit biomass and by the root mass. The root mass is influenced by the carbon available for root growth.

The timing of the initiation of shoot and root growth and the temperature response of growth were also critical to plant carbon balance and survival. Neither the Californian nor the Chilean species survived in simulations in which the species were grown with the climatic conditions of the other country. Actual transplants do survive. The problem with the simulated transplants was largely in the temperature differences between the two countries and the influences of temperature on the initiation and growth of shoots and roots. The initiation of root growth must allow the plant to maintain adequate water and nitrogen reserves while it also maintains sufficient carbohydrate reserves for shoot growth. If leaf growth occurs ahead of root growth, the plant may develop water and nitrogen stresses. The problems of timing root and shoot growth should not be as critical for evergreen shrubs as for seedlings and deciduous plants, which do not have large reserves.

12.6. Conclusions

Ecological generalities are difficult to test. A set of simulation experiments conducted in this study indicate that the range of natural variation in an ecosystem may make it impossible to completely disprove reasonable generalities. An implicit assumption often made in community theory is that competition and niche specialization among the species in a community results in optimal or maximal resource use by the community (Odum and Pinkerton, 1955; Lewontin, 1969; Cohen, 1970; MacArthur, 1970, 1972; Rosenzweig, 1974; Orians, 1975). The assumption of optimal resource use asserts that the natural mix of species in a community produces a higher resource use than any other mix of the same species. In simulation experiments for chaparral and matorral using root distributions measured at Echo Valley and Fundo Santa Laura, the resource use by mixed species communities was always less than resource use by single species communities (Miller et al., 1978). However, in earlier simulations, which used the distribution of roots in the soil given by Hellmers et al. (1955b) for San Dimas, resource use in the mixed species community was greater. Factors other than root distributions also allow alternate conclusions. Simulations using older measurements of root extension per unit dry weight indicated that *A. glauca* would use resources more fully than *A. fasciculatum* or than the two together. Later simulations, which used newer data on root extension, indicated that *A. fasciculatum* could use resources more fully than *A. glauca* or than a mixture of the two. The range of variation encountered in the natural system indicates caution in accepting or rejecting ecological theories.

As discussed in this and other chapters of this book, the plant properties described above are important factors in resource use and resource-use efficiency in chaparral and matorral. The existence of many plant properties that increase resource use and resource-use efficiency indicates that the species are not merely "staying in the game" (Chapter 1). They also show adaptations to the climate that increase resource use, but it cannot be said that the mature mediterranean shrub vegetation uses all resources fully. This ecological generality may not be true in mature mediterranean shrub vegetation because the generality is false or because resource use is more important in earlier successional stages or over the total successional sequence after fire.

The resource use and resource-use efficiency in the two predominant chaparral types in California, chamise chaparral and mixed chaparral, differ more than do the resource use and resource-use efficiency in the mixed chaparral type in California and the matorral in Chile. Resource use and resource-use efficiency in mediterranean climates are environmentally constrained by interactions and feedbacks between several environmental and vegetational properties. Low irradiance in winter, rather than plant or soil properties, limits the use of water and light in central Chile. Even with climatic limitations, the vegetation appears nutrient limited because the turnover rates of the vegetation are adjusted to the rates at which nutrients become available. Leaf strengthening is necessary if the leaves are evergreen and if they must survive the summer drought. The relatively high lignin and cellulose contents of evergreen leaves is related to the low nitrogen content in the leaves and may not be due to low nitrogen availability but to dilution of the cytoplasm by supporting materials. The low nitrogen content of leaves increases the carbon:nitrogen ratio of the litter and decreases the rates of nitrogen release by litter decomposition. This low rate of release in turn favors evergreen leaves and low nitrogen contents. Infertile parent materials and leaching losses may also contribute to low nitrogen availability in the soil. Thus, the sclerophyllous evergreen form is reinforced by low available nutrients. The summer drought also may lead to narrow, steeply inclined leaves that are closely appressed to the stems or are clustered in order to reduce leaf temperatures in spring and summer and to decrease transpiration. The decreased water loss rate prolongs the period of available soil moisture, the growth period, and the photosynthetic period. Growth at the top of the canopy leads to increased ratios of photosynthesis:transpiration and photosynthesis:solar irradiance, which creates a more favorable carbon balance for a plant with leaves at the top of the canopy. Physiological and morphological compensations result in the general similarity of vegetation form in spite of dissimilar phyllogenetic histories.

Literature Cited

Acock, B., Thornley, J.H.M., Warren Wilson, J. (1970) Spatial variation of light in the canopy. In: Prediction and Measurement of Photosynthetic Productivity. Wageningen: Centre for Agricultural Publishing and Documentation, pp. 91-102

Adams, M.S., Strain, B.R., Ting, I.P. (1967) Photosynthesis in chlorophyllous stem tissue and leaves of *Cercidium floridum*: Accumulation and distribution of ^{14}C from ^{14}CO$_2$. Plant Physiol. *42*, 1797-1799

Ahlgren, I.F., Ahlgren, C.E. (1960) Ecological effects of forest fires. Bot. Rev. *26*, 483-533

Alexander, M. (1961) Introduction to Soil Microbiology. New York: John Wiley and Sons

Alexander, M. (1976) Introduction to Soil Microbiology, 2d ed. New York: John Wiley and Sons

Allen, L.H., Jr. (1974) Model of light penetration into a wide-row crop. Agron. J. *66*, 41-47

Anderson, D.J., Perry, R.A., Leigh, J.H. (1972) Some perspectives on shrub/environment interactions. pp. 172-181. In: Wildland Shrubs—Their Biology and Utilization. McKell, C.M., Blaisdell, J.P., Goodin, J.R. (eds.). U.S. Forestry Service Gen. Tech. Rep. INT-1

Anderson, H.W., Coleman, G.B., Zinke, P.J. (1959) Summer Slides and Winter Scour, Dry-wet Erosion in Southern California Mountains. U.S. Forestry Service Tech. Pap. PSW-36

Anderson, M.C. (1964) Light relations of terrestrial plant communities and their measurement. Biol. Rev. *39*, 425-486

Anderson, M.C. (1966) Stand structure and light penetration. II. A theoretical analysis. J. Appl. Ecol. *3*, 41-54

Anderson, M.C., Denmead, O.T. (1969) Shortwave radiation on inclined surfaces in model plant communities. Agron. J. *61*, 867-872

Armesto, J., Martinez, J.A. (1978) Relations between vegetation structure and slope aspect in the mediterranean region of Chile. J. Ecol. *66*, 881-889

Art, H.W., Bormann, F.H., Voigt, G.K., Woodwell, G.M. (1974) Barrier Island forest ecosystem: Role of meteorologic nutrient inputs. Science *184*, 60-62

Aschmann, H. (1973a) Man's impact on the several regions with mediterranean climates. In: Mediterranean Type Ecosystems: Origin and Structure. Castri, F. di, Mooney, H.A. (eds.). New York: Springer-Verlag, pp. 363-371

Aschmann, H. (1973b) Distribution and peculiarity of mediterranean ecosystems. In: Mediterranean Type Ecosystems: Origin and Structure. Castri, F. di, Mooney, H.A. (eds.). New York: Springer-Verlag, pp. 11-19

Aschmann, H. (1977) Aboriginal use of fire. In: Proceedings of the Symposium on the Environmental Consequences of Fire and Fuel Management in Mediterranean Ecosystems, 1-5 August 1977, Palo Alto, California. Mooney, H.A., Conrad, C.E. (eds.). Washington, D.C.: U.S. Department of Agriculture, Forest Service and U.S. Forestry Service Gen. Tech. Rep. WO-3, pp. 132-141

Aschmann, H., Bahre, C. (1977) Man's impact on the wild landscape. In: Convergent Evolution in Chile and California Mediterranean Climate Ecosystems. Mooney, (ed.). Stroudsburg, Pa.: Dowden, Hutchinson and Ross, pp. 73-84

Ashton, D.H. (1976) Phosphorus in forest ecosystems at Beenak, Victoria. J. Ecol. 64, 171-186

Avila, G., Aljaro, M., Araya, S., Montenegro, G. (1975) The seasonal cambium activity of Chilean and Californian shrubs. Am. J. Bot. 62, 473-478

Avila, G., Araya, S., Riveros, F., Kummerow, J. (1978) Secondary root and stem growth in a Chilean matorral shrub (Colliguaya odorifera Mol.). Oecol. Plant. 13, 367-373

Axelrod, D.I. (1973) History of the mediterranean ecosystem in California. In: Mediterranean Type Ecosystems: Origin and Structure. Castri, F. di, Mooney, H.A. (eds.). New York: Springer-Verlag, pp. 225-277

Azevedo, J., Morgan, D.L. (1974) Fog precipitation in coastal California forests. Ecology 55:1135-1141

Bailey, R.W. (1973) Structural carbohydrates. In: Chemistry and Biochemistry of Herbage. Butler, G.W., Bailey, R.W. (eds.). New York: Academic Press, pp. 157-211

Baker, D.N., Hesketh, J.D., Duncan, W.G. (1972) Simulation of growth and yield in cotton. I. Gross photosynthesis, respiration, and growth. Crop Sci. 12, 431-435

Ball, J.T. (1977) Effects of fire on soil mineral nutrients in a southern California soft chaparral community. M.A. Thesis, Santa Barbara: University of California

Bancroft, T.A. (1968) Topics in Intermediate Statistical Methods. Vol. I. Ames: Iowa State University Press

Barbour, M.G. (1973) Desert dogma reexamined: Root/shoot productivity and plant spacing. Am. Midl. Nat. 89, 41-57

Barkley, S.A., Barel, D., Stoner, W.A., Miller, P.C. (1978) Controls on decomposition and mineral release in wet meadow tundra—a simulation approach. In: Environmental Chemistry and Cycling Processes. Adriano, D.C., Brisbin, I.L., Jr. (eds.). Washington, D.C.: U.S. Department of Energy, Technical Information Center, CONF-760429, pp. 754-778

Barrow, N.J. (1977) Phosphorus uptake and utilization by tree seedlings. Aust. J. Bot. 25, 571-584

Barrs, H.D. (1968) Determination of water deficits in plant tissues. In: Water Deficits and Plant Growth. Vol. 1. Kozlowski, T.T. (ed.). New York: Academic Press, pp. 235-347

Bate, G.C., Canvin, D.T. (1972) Simulation of the daily growth of an aspen population from the measured CO_2-exchange rates of the components. Can. J. Bot. 50, 205-214

Beadle, N.C.W. (1954) Soil phosphate and the delimitation of plant communities in eastern Australia. Ecology 35, 370-375

Beadle, N.C.W. (1962) Soil phosphate and the delimitation of plant communities in eastern Australia. Part II. Ecology 43, 281-288

Beadle, N.C.W. (1966) Soil phosphate and its role in molding segments of the Australian flora and vegetation, with special reference to xeromorphy and sclerophylly. Ecology 47, 992-1007

Beadle, N.C.W. (1968) Some aspects of the ecology and physiology of Australian xeromorphic plants. Aust. J. Sci. 30, 348-355

Beek, J., Frissel, M.J. (1973) Simulation of Nitrogen Behavior in Soils. Wageningen: Centre for Agricultural Publishing and Documentation

Benacchio, S.S., Blair, R.O. (1972) A new approach to phenological relationship between environmental factors and day to the appearance of the first leaf in four perennial species. Agron. J. *74*, 297-301

Billés, G., Cortez, J., Lossaint, P. (1971a) L'activité biologique des sols dans les écosystèmes méditerranéens. I. Minéralisation du carbone. Rev. Ecol. Biol. Sol *8*, 375-395

Billés, G., Cortez, J., Lossaint, P. (1975) Étude comparative de l'activité biologique des sols sous peuplements arbustifs et herbacés de la garrigue méditerranéene. I. Minéralisation du carbon et de l'azote. Rev. Ecol. Biol. Sol *12*, 115-139

Billés, G., Lossaint, P., Cortez, J. (1971b) L'activité biologique des sols dans les écosystèmes méditerranéens. II. Minéralisation de l'azote. Rev. Ecol. Biol. Sol *8*, 533-552

Billings, W.D. (1973) Arctic and alpine vegetations: Similarities, differences, and susceptibility to disturbance. BioScience *23*, 697-704

Billings, W.D., Godfrey, P.J., Chabot, B.F., Bourque, D.P. (1971) Metabolic acclimation to temperature in arctic and alpine ecotypes of *Oxyria digyna*. Arct. Alp. Res. *3*, 277-289

Billings, W.D., Morris, R.J. (1951) Reflection of visible and infrared radiation from leaves of different ecological groups. Am. J. Bot. *38*, 327-331

Billings, W.D., Shaver, G.R., Trent, A.W. (1973) Temperature effects of growth and respiration of roots and rhizomes in tundra graminoids. In: Primary Production and Production Processes, Tundra Biome. Bliss, L.C., Wielgolaski, F.E. (eds.). Edmonton: University of Alberta Printing Services, pp. 57-63

Birch, H.F. (1958) The effect of soil drying on humus decomposition and nitrogen availability. Plant Soil *10*, 9-31

Birkebak, R.C., Birkebak, R. (1964) Solar radiation characteristics of tree leaves. Ecology *45*, 646-649

Biswell, H.H. (1974) Effects of fire on chaparral. In: Fire and Ecosystems. Kozlowski, T.T., Ahlgren, C.E. (eds.). New York: Academic Press, pp. 321-364

Biswell, H.H., Schultz, A.M. (1957) Surface runoff and erosion as related to prescribed burning. J. For. *55*, 372-374

Blackman, G.E. (1968) The application of the concepts of growth analysis to the assessment of productivity. In: Functioning of Terrestrial Ecosystems at the Primary Production Level. Eckardt, F.E. (ed.). Paris: UNESCO, pp. 243-259

Bormann, F.H., Likens, G.E. (1967) Nutrient cycling. Science *155*, 424-429

Bormann, F.H., Likens, G.E., Siccama, T.G., Pierce, R.S., Eaton, J.S. (1974) The export of nutrients and recovery of stable conditions following deforestation at Hubbarb Brook. Ecol. Monogr. *44*, 255-277

Botkin, D.B. (1975) A functional approach to the niche concept in forest communities. In: New Directions in the Analysis of Ecological Systems. Innis, G.S. (ed.). La Jolla, Ca.: Society for Computer Simulation, pp. 149-158

Bottner, P., Peyronel, A. (1977) Dynamique de la matière organique dans deux sols méditerranéens étudiée à partir de techniques de datation par le radiocarbone. Rev. Ecol. Biol. Sol *14*, 385-393

Bray, J.R., Curtis, J.T. (1957) An ordination of the upland forest communities of southern Wisconsin. Ecol. Monogr. *27*, 325-349

Bray, J.R., Gorham, E. (1964) Litter production in forests of the world. Adv. Ecol. Res. *2*, 101-157

Bresler, E., Kemper, W.D., Hanks, R.J. (1969) Infiltration, redistribution and subsequent evaporation of water from soil as affected by wetting rate and hysteresis. Soil Sci. Soc. Am. Proc. *33*, 832-839

Brooks, F.A. (1959) An Introduction to Physical Microclimatology. Davis: University of California Press

Brouwer, R., Wit, C.T. de (1969) A simulation model of plant growth with special attention to root growth and its consequences. In: Root Growth. Whittington, W.J. (ed.). New York: Plenum Press, pp. 224-244

Brown, R.W. (1970) Measurement of Water Potential with Thermocouple Psychrometers: Construction and Application. U.S. Forestry Service Res. Pap. INT-80

Brunt, D. (1932) Notes on radiation in the atmosphere. Int. Qtly. J. Roy. Meteorol. Soc. *58*, 389-420

Buckman, H.O., Brady, N.C. (1969) The Nature and Properties of Soil. New York: Macmillan Co.

Budyko, M.T. (1956) The heat balance of the earth's surface. [Translation PB131692, Washington, D.C.: U.S. Department of Commerce, Office of Technical Services] Gidrometeorologicheskoe izdatel'stro (Leningrad). Izv. Akad. Nauk U.S.S.R. Series B3, 17(1954)

Bunnell, F.L., Tait, D.E.N. (1974) Mathematical simulation models of decomposition processing. In: Soil Organisms and Decomposition in Tundra. Holding, A.J., Heal, O.W., McLean, S.F., Jr., Flanagan, P.W. (eds.). Stockholm: Tundra Biome Steering Committee, pp. 207-225

Byers, H.R. (1953) Coast redwoods and fog drip. Ecology *34*, 192-193

Byrne, R., Michaelsen, J., Soutar, A. (1977) Fossil charcoal as a measure of wildfire frequency in southern California: A preliminary analysis. In: Proceedings of the Symposium on the Environmental Consequences of Fire and Fuel Management in Mediterranean Ecosystems, 1-5 August 1977, Palo Alto, California. Mooney, H.A., Conrad, C.E. (eds.). Washington, D.C.: U.S. Department of Agriculture, Forest Service, and U.S. Forestry Service Gen. Tech. Rep. WO-3, pp. 361-367

Caldwell, M.M., Camp, L.B. (1974) Belowground productivity of two cool desert communities. Oecologia *17*, 123-130

Caldwell, M.M., Fernandez, O.A. (1975) Dynamics of Great Basin shrub root systems. In: Environmental Physiology of Desert Organisms. Hadley, N.F. (ed.). New York: Halsted Press, pp. 38-51

Caldwell, M.M., White, R.S., Moore, R.T., Camp, L.B. (1977) Carbon balance, productivity, and water use of cold-winter desert shrub communities dominated by C_3 and C_4 species. Oecologia *29*, 275-300

Carter, S.I. (1973a) A comparison of the patterns of herb and shrub growth in comparable sites in Chile and California. M.S. Thesis, San Diego, Ca.: San Diego State University

Carter, S.I. (1973b) Observations on the Role of the Herbaceous Vegetation in the Chilean Matorral. U.S. IBP Tech. Rep. 73-6

Carter, S.I. (1973c) Data on the Herbaceous and Shrubby Vegetation at Study Sites in California and Chile, 1971-73. U.S. IBP Tech. Rep. 73-3

Castri, F. di. (1973) Climatological comparisons between Chile and the western coast of North America. In: Mediterranean Type Ecosystems: Origin and Structure. Castri, F. di, Mooney, H.A. (eds.). New York: Springer-Verlag, pp. 21-36

Castri, F. di, Mooney, H.A. (eds.) (1973) Mediterranean Type Ecosystems: Origin and Structure. New York: Springer-Verlag, 405 pp.

Challa, H. (1976) An Analysis of the Diurnal Course of Growth, Carbon Dioxide Exchange and Carbohydrate Reserve Content of Cucumbers. Wageningen: Agric. Res. Rep. 861; Publ. Cent. Agrobiol. Res. 020; and Centre for Agricultural Publishing and Documentation

Chang, C. (1968) Climate and Agriculture, an Ecological Survey. Chicago: Aldine Publishing Co.

Christensen, N.L. (1973) Fire and the nitrogen cycle in California chaparral. Science *181*, 66-68

Christensen, N.L. (1977) Fire and soil-plant nutrient relations in a pine-wiregrass savanna on the coastal plain of North Carolina. Oecologia *31*, 27-44

Christensen, N.L., Muller, C.H. (1975) Effects of fire on factors controlling plant growth in *Adenostoma* chaparral. Ecol. Monogr. *45*, 29-55

Clayton, J.L. (1972) Salt spray and mineral cycling in two California coastal ecosystems. Ecology *53*, 74-81

Clayton, J.L. (1976) Nutrient gains to adjacent ecosystems during a forest fire: An evaluation. For. Sci. *22*, 162-166

Clements, F.E., Shelford, V.E. (1939) Bio-ecology. New York: John Wiley and Sons

Cocheme, J. (1968) FAO/UNESCO/WHO agroclimatology survey of a semiarid area of West Africa south of the Sahara. In: Agroclimatological Methods. Paris: UNESCO, pp. 235-248

Cocheme, J., Franquini, P. (1967) An Agroclimatology Survey of a Semiarid Area in Africa South of the Sahara. World Meteorol. Organ. Notes 86 and 210, and World Meteorol. Bur. Tech. Pap. 110

Cody, M.L., Mooney, H.A. (1978) Convergence versus nonconvergence in mediterranean-climate ecosystems. Annu. Rev. Ecol. Syst. 9, 265-321

Cohen, D. (1966) Optimizing reproduction in a randomly varying environment. J. Theor. Biol. 12, 119-129

Cohen, D. (1967) Optimizing reproduction in a randomly varying environment when a correlation may exist between the conditions at the time a choice has to be made and the subsequent outcome. J. Theor. Biol. 16, 1-14

Cohen, J.E. (1970) A Markov contingency table mode for replicated Lotka-Volterra systems near equilibrium. Am. Nat. 104, 547-559

Cole, G.V., Innis, G.S., Stewart, J.W.B. (1977) Simulation of phosphorus cycling in semiarid grasslands. Ecology 58, 1-15

Coleman, D.C. (1976) A review of root production processes and their influence on soil biota in terrestrial ecosystems. In: The Role of Terrestrial and Aquatic Organisms in Decomposition Processes. Anderson, J.M., Macfadyen, A. (eds.). Oxford: Blackwell Scientific Publications, pp. 417-434

Cooke, H.B.S. (1965) The Pleistocene environment in southern Africa. In: Ecological Studies in Southern Africa. Davis, D.H.S. (ed.). The Hague: Dr. W. Junk Publishers, pp. 1-23

Cooper, C. (1976) Vegetation map of San Diego County for the Sun Desert Nuclear Project. San Diego, Ca.: Center for Regional Environmental Studies, San Diego State University

Cooper, J.P. (ed.) (1975) Photosynthesis and Productivity in Different Environments. Cambridge: Cambridge University Press

Corbett, E.S., Crouse, R.P. (1968) Rainfall Interception by Annual Grass and Chaparral . . . Losses Compared. U.S. Forestry Service Res. Pap. PSW-48

Cortez, J., Lossaint, P., Billès, G. (1972) L'activite biologique des sols dans les écosystèmes méditerranéens. III. Activités enzymatiques. Rev. Ecol. Biol. Sol 9, 1-19

Cowan, I.R. (1965) Transport of water in the soil-plant-atmosphere system. J. Appl. Ecol. 2, 221-239

Crocker, R.L., Major, J. (1955) Soil development in relation to vegetation and surface age at Glacier Bay, Alaska. J. Ecol. 43, 427-448

Cromack, K., Monk, C.D. (1975) Litter production, decomposition, and nutrient cycling in a mixed hardwood watershed and a white pine watershed. In: Mineral Cycling in Southeastern Ecosystems. Howell, F.G., Gentry, J.B., Smith, M.H. (eds.). Springfield, Va.: National Technical Information Service, CONF-740513, pp. 609-624

Cunningham, G.L., Reynolds, J.F. (1978) A simulation model of primary production and carbon allocation in the creosotebush (Larrea tridentata [DC] COV.). Ecology 59, 37-52

Cunningham, G.L., Syvertsen, J.P. (1977) The effect of nonstructural carbohydrate levels on dark CO_2 release in creosote bush. Photosynthetica 11, 291-295

Cutler, J.M., Rains, D.W., Loomis, R.S. (1977) The importance of cell size in the water relations of plants. Physiol. Plant 40, 255-260

Dagnelie, P. (1960) Contribution a l'étude des communautes vegetales per l'analyse factorielle. Bull. Serv. Carte. Phytogeogr. CNRS 5, 7-71, 93-195

Dainty, J. (1965) Osmotic flow in the State and Movement of Water in Living Organisms. New York: Academic Press

Davidson, R.L. (1969) Effects of edaphic factors on the soluble carbohydrate content of roots of Lolium perenne L. and Trifolium repens L. Ann. Bot. 33, 579-589

Davidson, J.L., Philip, J.R. (1958) Light and pasture growth. In: Climatology and Microclimatology. Paris: UNESCO, pp. 181-187

DeBano, L.F., Conrad, C.E. (1976) Nutrients lost in debris and runoff water from a burned chaparral watershed. In: Proceedings of the Third Federal Inter-agency Sedimentation Conference, March 1976, Denver, Co., pp. 3-13 to 3-27

DeBano, L.F., Conrad, C.E. (1978) The effect of fire on nutrients in a chaparral ecosystem. Ecology *59*, 489-497

DeBano, L.F., Dunn, P.H., Conrad, C.E. (1977) Fire's effect on physical and chemical properties of chaparral soils. In: Proceedings of the Symposium on the Environmental Consequences of Fire and Fuel Management in Mediterranean Ecosystems, 1-5 August 1977, Palo Alto, California. Mooney, H.A., Conrad, C.E. (eds.). Washington, D.C.: U.S. Department of Agriculture, Forest Service, and U.S. Forestry Service Gen. Tech. Rep. WO-3, pp. 65-74

DeBano, L.F., Eberlein, G.E., Dunn, P.H. (1979) Effects of burning on chaparral soils. I. Soil nitrogen. Soil Sci. Soc. Am. J. *43*, 504-509

DeBano, L.F., Osborn, J.F., Krammes, J.S., Letey, J. (1967) Soil Wettability and Wetting Agents . . . Our Current Knowledge of the Problem. U.S. Forestry Service Res. Pap. PSW-43

DeBell, D.S., Ralston, C.W. (1970) Release of nitrogen by burning light forest fuels. Soil Sci. Soc. Am. Proc. *34*, 936-938

de Jong, E., Schappert, H.J.V., MacDonald, K.B. (1974) Carbon dioxide evolution from virgin and cultivated soil as affected by management practices and climate. Can. J. Soil Sci. *54*, 299-307

del Moral, R., Muller, C.H. (1969) Fog drip: A mechanism of toxin transport from *Eucalyptus globulus*. Bull. Torrey Bot. Club *96*, 467-475

Delwiche, C.C., Zinke, P.J., Johnson, C.M. (1965) Nitrogen fixation by *Ceanothus*. Plant Physiol. *40*, 1045-1047

Dement, W.A., Mooney, H.A. (1974) Seasonal variation in the production of tannins and cyanogenic glucosides in the chaparral shrub, *Heteromeles arbutifolia*. Oecologia *15*, 65-76

Denmead, O.T. (1964) Evaporation sources and apparent diffusivities in a forest canopy. J. Appl. Meteorol. *3*, 383-389

Denmead, O.T. (1973) Relative significance of soil and plant evaporation in estimating evapotranspiration. In: Plant Response to Climatic Factors. Slatyer, R.O. (ed.). Paris: UNESCO, pp. 505-511

Dibblee, T.W. (1966) Geology of the Central Santa Ynez Mountains, Santa Barbara County, California. Calif. Div. Mines Geol. Bull. 186.

Dodge, J.M. (1975) Vegetational changes associated with land use and fire history in San Diego County. Ph.D. Dissertation, Riverside: University of California

Duncan, W.C., Loomis, R.S., Williams, W.A., Hanau, R. (1967) A model for simulating photosynthesis in plant communities. Hilgardia *38*, 181-205

Dunn, E.L. (1975) Environmental stresses and inherent limitation affecting CO_2 exchange in evergreen sclerophylls in mediterranean climates. In: Perspectives of Biophysical Ecology. Gates, D.M., Schmerl, R.B. (eds.). New York: Springer-Verlag, pp. 159-181

Dunn, P.H., DeBano, L.F. (1977) Fire's effect on biological and chemical properties of chaparral soils. In: Proceedings of the Symposium on the Environmental Consequences of Fire and Fuel Management in Mediterranean Ecosystems, 1-5 August 1977, Palo Alto, California. Mooney, H.A., Conrad, C.E. (eds.). Washington, D.C.: U.S. Department of Agriculture, Forest Service, and U.S. Forestry Service Gen. Tech. Rep. WO-3

Dunn, P.H., DeBano, L.F., Eberlein, G.E. (1979) Effects of burning on chaparral soils. II. Soil microbes and nitrogen mineralization. Soil Sci. Soc. Am. J. *43*, 509-514

Dutt, G.R., Shaffer, M.J., Moore, W.J. (1972) Computer Simulation Model of Dynamic Bio-physico-chemical Processes in Soils. Ariz. Agric. Exp. Stn. Tech. Bull. 196

Duvigneaud, P., Denaeyer-DeSmet, S. (1970) Biological cycling of minerals in temperate deciduous forests. In: Analysis of Temperate Forest Ecosystems. Reichle, D.E. (ed.). New York: Springer-Verlag, pp. 199-225

Edwards, N.T., Harris, W.F. (1977) Carbon cycling in a mixed deciduous forest floor. Ecology *58*, 431-437

Ehleringer, J.R., Miller, P.C. (1975a) A simulation model of plant water relations and production in the alpine tundra, Colorado. Oecologia *19*, 177-193

Ehleringer, J.R., Miller, P.C. (1975b) Water relations of selected plant species in the alpine tundra, Colorado. Ecology *56*, 370-380

Elton, C. (1958) The Ecology of Invasions by Animals and Plants. London: Methuen and Co.

Epstein, E. (1972) Mineral Nutrition of Plants: Principles and Perspectives. New York: John Wiley and Sons

Evans, G.C. (1972) The Quantitative Analysis of Plant Growth. Berkeley and Los Angeles: University of California Press

Evans, L.T. (ed.) (1975a) Crop Physiology. Cambridge: Cambridge University Press

Evans, L.T. (1975b) The physiological basis of crop yield. In: Crop Physiology. Evans, L.T. (ed.). Cambridge: Cambridge University Press

Fick, G.W., Loomis, R.S., Williams, W.A. (1975) Sugar beet. In: Crop Physiology. Evans, L.T. (ed.). Cambridge: Cambridge University Press

Fischer, R.A., Turner, N.C. (1978) Plant productivity in the arid and semiarid zones. Annu Rev. Plant Physiol. *29*, 277-317

Fishbeck, K., Kummerow, J. (1976) Flower bud differentiation in *Ceanothus greggii.* (Abstr.) Bull. Ecol. Soc. Am. *57*, 14

Fleming, G.A. (1973) Mineral composition of herbage. In: Chemistry and Biochemistry of Herbage. Vol, 1. Butler, G.W., Bailey, R.W. (eds.). New York: Academic Press, pp. 529-566

Fogel, R., Cromack, K. (1977) Effect of habitat and substrate quality on douglas fir litter decomposition in western Oregon. Can. J. Bot. *55*, 1632-1640

Foster, N.W., Morrison, I.K. (1976) Distribution and cycling of nutrients in a natural *Pinus banksiana* ecosystem. Ecology *57*, 110-120

Frank, E.C., Lee, R. (1966) Potential Solar Beam Irradiation on Slopes, Tables for 30° to 50° Latitude. U.S. Foestry Service Res. Pap. RM-18.

French, N., Sauer, R.H. (1974) Phenological studies and modeling in grasslands. In: Phenology and Seasonality Modeling. Lieth, H. (ed.). New York: Springer-Verlag, pp. 227-236

Frissel, M.J., Reinger, P. (1974) Simulation of Accumulation and Leaching in Soils. Wageningen: Centre for Agricultural Publishing and Documentation

Fritschen, L.J., Doraiswamy, P. (1973) Dew: An addition to the hydrologic balance of douglas fir. Water Resour. Res. *9*, 891-894

Fukai, S., Loomis, R.S. (1976) Leaf display and light environment in row-planted cotton communities. Agric. Meteorol. *17*, 353-379

Furman, T.E. (1959) The structure of the root nodules of *Ceanothus saguineus* and *Ceanothus velutinus* with special reference to the endophyte. Am J. Bot. *46*, 698-703

Galloway, J.N., Likens, G.E. (1976) Calibration of collection procedures for the determination of precipitation chemistry. Water Air Soil Pollut. *6*, 241-258

Garcia-Moya, E., McKell, C.M. (1970) Contribution of shrubs to the nitrogen economy of a desert-wash plant community. Ecology *51*, 81-88

Gates, D.M. (1962) Energy Exchange in the Biosphere. New York: Harper and Row

Gates, D.M. (1965) Energy, plants and ecology. Ecology *46*, 1-13

Gates, D.M. (1968) Transpiration and leaf temperature. Annu. Rev. Plant Physiol. *19*, 211-238

Gates, D.M., Keegan, H.J., Schleter, J.C., Weidner, V.R. (1965b) Spectral properties of plants. Appl. Opt. *4*, 11-20

Gates, D.M., Tibbals, E.C., Kreith, F. (1965a) Radiation and convection for ponderosa pine. Am. J. Bot. *52*, 66-71

Gauch, H.G. (1972) Inorganic Plant Nutrition. Stroudsburg, Pa.: Dowden, Hutchinson and Ross

Gigon, A. (1979) CO_2-gas exchange, water relations and convergence of mediterranean shrub-types from California and Chile. Oecol. Plant. *14*, 129-150

Giliberto, J., Estay, H. (1978) Seasonal water stress in selected shrub species in the mediterranean region of central Chile. Bot. Gaz. *139*, 236-240

Giliberto, J., Mooney, H.A., Kummerow, J. (1977) Shrub structure analysis. In: Chile-California Mediterranean Scrub Atlas. Thrower, N.J.W., Bradbury, D.E. (eds.). Stroudsburg, Pa.: Dowden, Hutchinson and Ross, pp. 144-147

Goering, H.K., Van Soest, P.J. (1970) Forage Fiber Analyses (Apparatus, Reagents, Procedures, and Some Applications). U.S. Dep. Agric., Agric. Handbook 379

Golley, F.B., McGinnis, J.T., Clements, F.G., Child, C.I., Duever, M.J. (1975) Mineral Cycling in a Tropical Moist Forest Ecosystem. Athens, Ga.: University of Georgia Press

Gorham, E. (1961) Factors influencing supply of major ions to inland waters, with special reference to atmosphere. Bull. Geol. Soc. Am. *72*, 795-840

Gosz, J.R., Likens, G.E., Bormann, F.H. (1976) Organic matter and nutrient dynamics of the forest and forest floor in the Hubbard Brook Forest. Oecologia *22*, 305-320

Goudriaan, J., Waggoner, P.E. (1972) Simulating both aerial microclimate and soil temperature from observations above the foliage canopy. Neth. J. Agric. Sci. *20*, 104-124

Gray, J.T., Schlesinger, W.H. (1981) Biomass, production, and litterfall in the coastal sage scrub of southern California. Am. J. Bot. *68*, 24-33

Greig-Smith, P. (1964) Quantitative Plant Ecology, 2d ed. London: Butterworths

Griffin, J. R. (1973) Xylem sap tensions in three woodland oaks of central California. Ecology *54*, 152-159

Griesebach, A.H.R. (1872) Die vegetation der Erde nach ihrer klimatischen Anordnung. Leipzig: W. Engelmann

Groves, R.H. (1977) Fire and nutrients in the management of Australian vegetation. In: Proceedings of the Symposium on the Environmental Consequences of Fire and Fuel Management in Mediterranean Ecosystems, 1-5 August 1977, Palo Alto, California. Mooney, H.A., Conrad, C.E. (eds.). Washington, D.C.: U.S. Department of Agriculture, Forest Service, and U.S. Forestry Service Gen. Tech. Rep. WO-3, pp. 220-229

Hamilton, E.L., Rowe, P.B. (1949) Rainfall Interception by Chaparral in California. Washington, D.C. and Berkeley: U.S. Forest Service and California Forest and Range Experiment Station

Hanes, T.L. (1965) Ecological studies on two closely related chaparral shrubs in southern California. Ecol. Monogr. *35*, 213-235

Hanes, T.L. (1971) Succession after fire in the chaparral of southern California. Ecol. Monogr. *41*, 27-52

Hanes, T.L. (1977) Chaparral. In: Terrestrial Vegetation of California. Barbour, M.G., Major, J. (eds.). New York: John Wiley and Sons, pp. 417-469

Hanks, R.J., Gardner, H.R., Florian, R.L. (1969a) Plant-growth evapotranspiration relations for several crops in the central Great Plains. Agron. J. *61*, 30-34

Hanks, R.J., Klute, A., Bresler, E. (1969b) A numerical method for estimating infiltration, redistribution, drainage, and evaporation of water from soil. Water Resour. Res. *5*, 1065-1069

Hansen, G.K., Jensen, C.R. (1977) Growth and maintenance respiration in whole plants, tops, and roots of *Lolium multiflorum*. Physiol. Plant. *39*, 155-164

Hardwick, K., Wood, M., Woolhouse, H.W. (1968) Photosynthesis and respiration in relation to leaf age in *Perilla futescens* (L.) Britt. New Phytol. *67*, 79-86

Harpaz, Y. (1975) Simulation of the nitrogen balance in semi-arid regions. Ph.D. Dissertation, Jerusalem: Hebrew University

Harrison, A.T. (1971) Temperature related effects on photosynthesis in *Heteromeles arbutifolia* M. Roem. Ph.D. Dissertation, Palo Alto, Ca.: Stanford University

Harrison, A.T., Small, E., Mooney, H.A. (1971) Drought relationships and distribution of two mediterranean climate California plant communities. Ecology *52*, 869-875

Head, G.C. (1973) Shedding of roots. In: Shedding of Plant Parts. Kozlowski, T.T. (ed.). New York: Academic Press, pp. 237-293

Hellmers, H., Bonner, J.F., Kelleher, J.M. (1955a) Soil fertility: A watershed management problem in the San Gabriel Mountains of southern California. Soil Sci. *80*, 189-197

Hellmers, H., Horton, J.S., Juhren, G., O'Keefe, J. (1955b) Root systems of some chaparral plants in southern California. Ecology *36*, 667-678

Hellmers, H., Kelleher, J.M. (1959) *Ceanothus leucodermis* and soil nitrogen in southern California mountains. For. Sci. *5*, 275-278

Henderson, G.S., Swank, W.T., Waide, J.B., Grier, C.C. (1978) Nutrient budgets of Appalachian and Cascade region watersheds: A comparison. For. Sci. *24*, 385-397

Heusser, L. (1978) Pollen in Santa Barbara Basin, California: A 12,000-yr record. Geol. Soc. Am. Bull. *89*, 673-678

Hicklenton, P.R., Oechel, W.C. (1976) Physiological aspects of the ecology of *Dicranum fuscescens* in the subarctic. I. Acclimation and acclimation potential of CO_2 exchange in relation to habitat, light, and temperature. Can. J. Bot. *54*, 1104-1119

Hill, L.W. (1963) The San Dimas Experimental Forest. (Mimeo). Berkeley, Ca.: U.S. Department of Agriculture, Forest Service, Pacific Southwest Forest and Range Experiment Station

Hoffmann, A. (1972) Morphology and histology of *Trevoa trinervis* (Rhamnaceae), a drought deciduous shrub from the Chilean matorral. Flora *161*, 527-538

Hoffmann, A., Hoffmann, A.E. (1976) Growth pattern and seasonal behavior of buds of *Colliguaya odorifera* a shrub from the Chilean mediterranean vegetation. Can. J. Bot. *54*, 1767-1774

Hoffmann, A., Kummerow, J. (1978) Root studies in the Chilean matorral. Oecologia *32*, 57-69

Hoffmann, A., Mooney, H.A., Kummerow, J. (1977) Qualitative phenology. In: Chile-California Mediterranean Scrub Atlas. Thrower, N.J.W., Bradbury, D.E. (eds.). Stroudsburg, Pa.: Dowden, Hutchinson and Ross, pp. 102-120

Holt, D.A., Bula, R.J., Miles, G.A., Schreiber, M.M., Peart, R.M. (1975) Environmental Physiology, Modeling and Simulation of Alfalfa Growth. II. Conceptual Development of SIMED. Purdue Univ. Agric. Exp. Stn. Res. Bull. 907

Honert, T.H. van den (1948) Water transport in plants as a catenary process. Disc. Faraday Soc. *3*, 146-153

Horowitz, J.L. (1969) An easily constructed shadow-band for separating direct and diffuse solar radiation. Solar Energy *12*, 543-545

Horowitz, L. (1958) Some simplified mathematical treatments of translocation in plants. Plant Physiol. *33*, 81-93

Horton, J.S. (1941) The sample plot as a method of quantitative analysis of chaparral vegetation in southern California. Ecology *22*, 457-468

Horton, J.S., Kraebel, C.J. (1955) Development of vegetation after fire in the chamise chaparral of southern California. Ecology *36*, 244-262

Horwitz, W. (1960) Official methods of the Association of Official Agricultural Chemists, 9th ed. Washington, D.C.: Association of Official Agricultural Chemists.

Huck, M.G., Hagemann, R.H., Hansen, J.B. (1962) Diurnal variation in root respiration. Plant Physiol. *37*, 371-375

Huck, M.G., Klepper, B., Taylor, H.M. (1970) Diurnal variations in root diameter. Plant Physiol. *45*, 529-530

Hunt, H.W. (1977) A simulation model for decomposition in grasslands. Ecology *58*, 469-484

Hunt, W.F., Loomis, R.S. (1979) Respiration modelling and hypothesis testing with a dynamic model of sugar beet growth. Ann. Bot. *44*, 5-17

Hynum, B.G. (1974) Spatial and temporal variation in temperature, precipitation, and potential biological response in two regions in mediterranean scrub. M.S. Thesis, San Diego, Ca.: San Diego State University

Idso, S.B., Wit, C.T. de (1970) Light relations in plant canopies. Appl. Opt. *9*, 177-184

Innis, G. (1974) Numbers of species and optimization in biology. In: Structure, Functioning and Management of Ecosystems. Proceedings of the First International Congress of Ecology, The Hague, 8-14 September 1974. Wageningen: Centre for Agricultural Publishing and Documentation, pp. 384-387

Jaksic, F.M., Montenegro, G. (1979) Resource allocation of Chilean herbs in response to climatic and microclimatic factors. Oecologia *40*, 81-89

Jarvis, P.G., Jarvis, M.S. (1963) The water relations of tree seedlings. IV. Some aspects of the tissue water relations and drought resistance. Physiol. Plant *16*, 501-516

Jeffery, D.W. (1964) The formation of polyphosphate in *Banksia ornata*, an Australian heath plant. Aust. J. Biol. Sci. *17*, 845-854

Jeffery, D.W. (1967) Phosphate nutrition of Australian heath plants. I. The importance of proteoid roots in *Banksia* (Proteaceae). Aust. J. Bot. *15*, 403-412

Jeffery, D.W. (1968) Phosphate nutrition of Australian heath plants. II. The formation of polyphosphate by five heath species. Aust. J. Bot. *16*, 603-613

Jenny, H., Vlamis, J., Martin, W.E. (1950) Greenhouse assay of fertility of California soils. Hilgardia *20*, 1-8

Jensen, M.E. (1972) Programming inrrigation for greater efficiency. In: Optimizing the Soil Physical Environment Towards Greater Crop Yields. Hillel, D. (ed.). New York: Academic Press, pp. 133-161

Johnson, F.L., Risser, P.G. (1974) Biomass, annual net primary production, and dynamics of six mineral elements in a post oak-blackjack oak forest. Ecology *55*, 1246-1258

Jones, M.B., Laude, H.M. (1960) Relationships between sprouting in chamise and the physiological condition of the plant. J. Range Manage. *13*, 210-214

Jow, W., Bullock, S.H., Kummerow, J. (1980) Leaf turnover rates of *Adenostoma fasciculatum* (Rosaceae). Am. J. Bot. *67*, 256-261

Junge, C.E. (1963) Air Chemistry and Radioactivity. New York: Academic Press

Jungk, A., Barber, S.A. (1974) Phosphate uptake rate of corn roots as related to the proportion of the roots exposed to phosphate. Agron. J. *66*, 554-557

Kaskurewicz, A., Fogg, P.J. (1967) Growing seasons of cottonwood and sycamore as related to geographic and environmental factors. Ecology *48*, 785-793

Katz, P.L., Bartnick, M.W. (1974) Instantaneous (static) vs. long-term (dynamic) optimization in ecosystems. In: Structure, Functioning and Management of Ecosystems. Proceedings of the First International Congress of Ecology, The Hague, 8-14 September 1974. Wageningen: Centre for Agricultural Publishing and Documentation, pp. 395-400

Keeley, J.E. (1975) Longevity of nonsprouting *Ceanothus*. Am. Midl. Nat. *93*, 504-507

Keeley, J.E., Zedler, P.H. (1978) Reproduction of chaparral shrubs after fire: A comparison of sprouting and seeding strategies. Am. Midl. Nat. *99*, 142-161

Keeley, S.C. (1977) The relationship of precipitation to post-fire succession in the southern California chaparral. In: Proceedings of the Symposium on the Environmental Consequences of Fire and Fuel Management in Mediterranean Ecosystems, 1-5 August 1977, Palo Alto, California. Mooney, H.A., Conrad, C.E. (eds.). Washington, D.C.: U.S. Department of Agriculture, Forest Service, and U.S. Forestry Service Gen. Tech. Rep. WO-3, pp. 387-390

Kira, T. (1975) Primary production of forests. In: Photosynthesis and Productivity in Different Environments. Cooper, J.P. (ed.). Cambridge: Cambridge University Press, pp. 5-40

Kittredge, J. (1955) Litter and forest floor of the chaparral in parts of the San Dimas Experimental Forest, California. Hilgardia *23*, 563-596

Kluge, M., Ting, I.P. (1979) Crassulacean Acid Metabolism. Analysis of an Ecological Adaptation. New York: Springer-Verlag

Klute, A. (1965) Laboratory measurement of hydraulic conductivity of saturated soil. In: Methods of Soil Analysis. Black, C.A. (ed.). Madison, Wis.: American Society of Agronomy, pp. 210-211

Knievel, D.P. (1973) Procedure for estimating ratio of live to dead root dry matter in root core samples. Crop Sci. *13*, 124-126

Köppen, W., Geiger, R. (eds.) (1930) Handbuch der Klimatologie. 5 vols. Berlin: Gebrüder Borntraeger

Kramer, P.J. (1969) Plant and Soil Water Relationships: A Modern Synthesis. New York: McGraw-Hill

Krammes, J.S. (1965) Seasonal debris movement from steep mountainside slopes in southern California. In: Proceedings of the Federal Inter-agency Sedimentation Conference 1963. U.S. Dep. Agric. Misc. Publ. 970, pp. 85-88

Krause, D., Kummerow, J. (1977a) Root and shoot growth rates of chaparral shrubs. Bot. Soc. Am. Misc. Ser. *154*, 25

Kruase, D., Kummerow, J. (1977b) Xeromorphic structure and soil moisture in the chaparral. Oecol. Plant. *12*, 133-148

Kuiper, P.J.C. (1964) Water uptake of higher plants as affected by root temperature. Meded Landbouwhogesch. Wageningen *63*, 1-11

Kummerow, J. (1962) Quantitative Messungen des Nebelniederschlages im Walde von Fray-Jorge an der nordchilenischen Küste. Naturwissenschaften *49*, 203-204

Kummerow, J., Alexander, J.V., Neel, J.W., Fishbeck, K. (1978b) Symbiotic nitrogen fixation in *Ceanothus* roots. Am. J. Bot. *65*, 63-69

Kummerow, J., Fishbeck, K. (1977) Part 2. In: Chile-California Mediterranean Scrub Atlas: A Comparative Analysis. Thrower, N.J.W., Bradbury, D.E. (eds.). Stroudsburg, Pa.: Dowden, Hutchinson and Ross, pp. 78-237

Kummerow, J., Krause, D. (1976) Root systems of chaparral shrubs. Bull. Ecol. Soc. Am. *57*, 28

Kummerow, J., Krause, D., Jow, W. (1977) Root systems of chaparral shrubs. Oecologia *29*, 163-177

Kummerow, J., Krause, D., Jow, W. (1978a) Seasonal changes of fine root density in the southern California chaparral. Oecologia *37*, 201-212

Kvĕt, J., Ondok, J.P., Nečas, J., Jarvis, P.G. (1971) Methods of growth analysis. In: Plant Photosynthetic Production. Sestak, Z., Catsky, J., Jarvis, P.G. (eds.). The Hague: Dr. W. Junk Publishers, pp. 343-391

Lambers, H., Steingröver, E. (1978) Efficiency of root respiration of a flood-tolerant and a flood-intolerant *Senecio* species as affected by low oxygen tension. Physiol. Plant. *42*, 179-184

La Mont, B. (1972) The effect of soil nutrients on the production of proteoid roots by *Hakea* species. Aust. J. Bot. *20*, 27-40

La Mont, B. (1973) Factors affecting the distribution of proteoid roots within the root systems of two *Hakea* species. Aust. J. Bot. *21*, 165-187

Lang, G.E., Forman, R.T.T. (1978) Detrital dynamics in a mature oak forest: Hutcheson Memorial Forest, New Jersey. Ecology *59*, 580-595

Larcher, W. (1975) Physiological Plant Ecology. New York: Springer-Verlag

Laude, H.M., Jones, M.B. (1961) Annual variability in indicators of sprouting potentials in chamise. J. Range Manage. *14*, 323-326

Lawrence, B.A., Lewis, M.C., Miller, P.C. (1978) A simulation model of population processes of arctic tundra graminoids. In: Vegetation and production Ecology of an Alaskan Arctic Tundra. Tieszen, L.L. (ed.). New York: Springer-Verlag, pp. 599-619

Lawrence, W.T. (1975) A radiation model for chaparral canopies. M.S. Thesis, San Diego, Ca.: San Diego State University

Leach, G.J. (1969) Shoot numbers, shoot size, and yield of regrowth in three lucerne cultivars. Aust. J. Agric. Res. *20*, 425-434

Ledig, F.T., Drew, A.P., Clark, J.G. (1976) Maintenance and constructive respiration, photosynthesis, and net assimilation rate in seedlings of pitch pine. Ann. Bot. *40*, 289-300

Lemon, E., Stewart, D.W., Shawcroft, R.W. (1971) The sun's work in a cornfield. Science *174*, 371-378

Lemon, E.R. (1960) Photosynthesis under field conditions. II. An aerodynamic method for determining the turbulent carbon dioxide exchange between the atmosphere and a cornfield. Agron. J. *52*, 697-703

Lemon, E.R. (1967) The impact of the atmospheric environment of the integument of plants. Int. J. Biometeor. *3*, 57-69

Lepper, M.G., Fleschner, M. (1977) Nitrogen fixation by *Cercocarpus ledifolius* (Rosaceae) in pioneer habitats. Oecologia *27*, 333-338

Lewis, W.M. (1974) Effects of fire on nutrient movement in a south California pine forest. Ecology *55*, 1120-1127

Lewis, W.M. (1975) Effects of forest fires on atmospheric loads of soluble nutrients. In: Mineral Cycling in Southeastern Ecosystems. Howell, F.G., Gentry, J.B., Smith, M.H. (eds.). Springfield, Va.: National Technical Information Service, CONF-740513, pp. 833-846

Lewontin, R.C. (1969) The meaning of stability. In: Diversity and Stability in Ecological Systems. Brookhaven Symposia in Biology no. 22. Upton, N.Y.: Brookhaven National Laboratory, pp. 13-24

Leyton, L., Reynolds, E.R.C., Thompson, F.B. (1967) Rainfall interception in forest and moorland, In: Forest Hydrology. Sopper, W.E., Lull, H.W. (eds.). New York: Pergamon Press, pp. 163-178

Likens, G.E., Bormann, F.H. (1972) Nutrient cycling in ecosystems. In: Ecosystem Structure and Function. Weins, J.A. (ed.). Corvallis: Oregon State University Press, pp. 25-67

Likens, G.E., Bormann, F.H. (1974) Linkages between terrestrial and aquatic ecosystems. BioScience *24*, 447-456

Likens, G.E., Bormann, F.H., Pierce, R.S., Eaton, J.S., Johnson, N.M. (1977) Biogeochemistry of a Forest Ecosystem. New York: Springer-Verlag

Likens, G.E., Bormann, F.H., Pierce, R.S., Reiners, W.A. (1978) Recovery of a deforested ecosystem. Science *199*, 492-496

List, R.J. (1968) Smithsonian Meteorological Tables, 6th rev. ed. Smithson. Misc. Collect. 114. Washington, D.C.: Smithsonian Institution Press

Lommen, P.W., Schwintzer, C.R., Yocum, C.S., Gates, D.M. (1971) A model describing photosynthesis in terms of gas diffusion and enzyme kinetics. Planta *98*, 195-220

Loomis, R.S., Rabbinge, R., Ng, E. (1979) Explanatory models in crop physiology. Annu. Rev. Plant Physiol. *30*, 339-367

Lossaint, P. (1973) Soil-vegetation relationships in mediterranean ecosystems of southern France. In: Mediterranean Type Ecosystems: Origin and Structure. Castri, F. di, Mooney, H.A. (eds.). New York: Springer-Verlag, pp. 199-210

Lossaint, P., Rapp, M. (1971) Repartition de la matière organique, productivité et cycles des elements mineraux dans des écosystèmes de climat méditerranéen. In: Productivity of Forest Ecosystems. Duvigneaud, P. (ed.). Paris: UNESCO, pp. 597-617

Loveless, A.R. (1962) Further evidence to support a nutritional interpretation of sclerophylly. Ann. Bot. *26*, 551-561

Ludlow, M.M., Wilson, G.L. (1971) Photosynthesis of tropical pasture plants. III. Leaf age. Aust. J. Biol. Sci. *24*, 1077-1087

Lull, H.W., Sopper, W.E. (1969) Hydrologic Effects from Urbanization of Forested Watersheds in the Northeast. U.S. Forestry Service Res. Pap. NE-146

Lyr, H., Hoffmann, G. (1967) Growth rates and growth periodicity of tree roots. Intern. Rev. For. Res. *1*, 181-236

Lyttleton, J.W. (1973) Proteins and nucleic acids. In: Chemistry and Biochemistry of Herbage. Butler, G.W., Bailey, R.W. (eds.). New York: Academic Press, pp. 63-103

McAlpine, J.R. (1970) Estimating pasture growth periods and drought from simple water balance models. In: Proceedings XI International Grassland Congress. Norman, M.J. (ed.). St. Lucia: University of Queensland Press, pp. 484-498

MacArthur, R.H. (1957) On the relative abundance of bird species. Proc. Nat. Acad. Sci. *43*, 293-295

MacArthur, R.H. (1970) Species packing and competitive equilibrium for many species. Theor. Pop. Bio. *1*, 1-11

MacArthur, R.H. (1972) Geographical Ecology. New York: Harper and Row

McColl, J.G., Bush, D.S. (1978) Precipitation and throughfall chemistry in the San Francisco Bay area. J. Environ. Qual. *7*, 352-357

McColl, J.G., Grigal, D.F. (1977) Nutrient changes following a forest wildfire in Minnesota: Effects in watersheds with differing soils. Oikos *28*, 105-112

McCree, K.J. (1974) Equations for the rate of dark respiration of white cloves and grain sorghum as a function of dry weight, photosynthetic rate, and température. Crop Sci. *14*, 509-514

McKinion, J.M., Baker, D.N., Hesketh, J.D., Jones, J.W. (1975) Part 4–SIMCOT II: A simulation of cotton growth and yield. In: Computer Simulation of a Cotton Production System. Users Manual. Washington, D.C.: U.S. Department of Agriculture, Agricultural Research Service, pp. 27-82

McPherson, J.K., Chou, C., Muller, C.H. (1971) Allelopathic constituents of the chaparral shrub *Adenostoma fasciculatum*. Phytochemistry *10*, 2925-2933

McPherson, J.K., Muller, C.H. (1967) Light competition between *Ceanothus* and *Salvia* shrubs. Bull. Torrey Bot. Club *94*, 41-55

Maggs, J., Pearson, C.J. (1977a) Minerals and dry matter in coastal scrub and grassland at Sydney, Australia. Oecologia *31*, 227-237

Maggs, J., Pearson, C.J. (1977b) Litterfall and litter layer decay in coastal scrub at Sydney, Australia. Oecologia *31*, 239-250

Major, J. (1973) Preface. In: Vegetation of the Earth. Walter, H. (Transl. from 2d German ed. by J. Wiesera.) New York: Springer-Verlag, pp. v-viii

Margaris, N.S. (1977) Decomposers and the fire cycle in mediterranean ecosystems. In: Proceedings of the Symposium on the Environmental Consequences of Fire and Fuel Management in Mediterranean Ecosystems, 1-5 August 1977, Palo Alto, California. Mooney, H.A., Conrad, C.E. (eds.). Washington, D.C.: U.S. Department of Agriculture, Forest Service, and U.S. Forestry Service Gen. Tech. Rep. WO-3, pp. 37-45

Marks, P.L., Bormann, F.H. (1972) Regeneration following forest cutting: Mechanisms for return to steady state nutrient cycling. Science *176*, 914-915

Matuszkiewicz, W., Traczyk, T. (1958) Zur Systematik der Bruchwald gesellschaften (*Alnetalia glutinosa*) in Polen. Acta Soc. Bot. Pol. *27*, 21-44

Meentemeyer, V. (1978) Macroclimate and lignin control of litter decomposition rates. Ecology *59*, 465-472

Meidner, H., Mansfield, T.A. (1968) Physiology of Stomata. New York: McGraw-Hill

Meigs, P. (1953) World distribution of arid and semi-arid homo-climates. In: Reviews of research on arid zone hydrology. Paris: UNESCO, pp. 203-210

Mengel, D.B., Barber, S.A. (1974) Rate of nutrient uptake per unit of corn root under field conditions. Agron. J. *66*, 399-402

Miller, P.C. (1969a) Comparison of models of direct solar radiation in vegetation canopies with observations in three forest stands. Ecology *50*, 878-885

Miller, P.C. (1969b) Solar radiation profiles in openings in canopies of aspen and oak. Science *164*, 308-309

Miller, P.C. (1972a) Bioclimate, leaf temperature, and primary production in red mangrove canopies in south Florida. Ecology *53*, 22-45

Miller, P.C. (1972b) A model to incorporate minerals into tundra plant production. In: Proceedings of 1972 Tundra Biome Symposium, July 1972, Lake Wilderness Center, University of Washington, Seattle. Hanover, N.H.: U.S. International Biological Program, Tundra Biome, pp. 51-54

Miller, P.C. (1973) A model of temperatures, transpiration rates and photosynthesis of sunlit and shaded leaves in vegetation canopies. In: Plant Response to Climatic Factors. Proceedings of Uppsala Symposium, 1970. Paris: UNESCO, pp. 427-434

Miller, P.C. (1975) A comparison of short-term effects of thermal addition on photosynthesis and plant-water stress in three ecosystems. In: Environmental Effects of Cooling Systems at Nuclear Power Plants. Proceedings of Symposium on the Physical and Biological Effects on the Environment of Cooling Systems and Thermal Discharges at Nuclear Power Stations, Oslo, 26-30 August 1974. Vienna: International Atomic Energy Agency, pp. 623-636

Miller, P.C. (1978) Problems of synthesis of mineral cycling studies: The tundra as an example. In: Environmental Chemistry and Cycling Processes. Adriano, D.C., Brisbin I.L., Jr. (eds.). Washington, D.C.: U.S. Department of Energy, Technical Information Center, CONF-760429, pp. 59-71

Miller, P.C. (1979) Quantitative plant ecology. In: Analysis of Ecosystems. Horn, D., Stairs, G.R., Mitchell, R.D. (eds.). Columbus: Ohio State University Press, pp. 179-232

Miller, P.C., Bradbury, D.E., Hajek, E., LaMarche, V., Thrower, N.J.W. (1977) Past and present environment. In: Convergent Evolution in Chile and California, Mediterranean Climate Ecosystems. Mooney, H.A. (ed.). Stroudsburg, Pa.: Dowden, Hutchinson and Ross, pp. 27-72

Miller, P.C., Mooney, H.A. (1976) The origin and structure of American arid-zone ecosystems. The producers: Interactions between environment form, and function. In: Critical Evaluation of Systems Analysis in Ecosystems Research and Management. Wit, C.T. de, Arnold, G.W. (eds.). Wageningen: Centre for Agricultural Publishing and Documentation, pp. 38-56

Miller, P.C., Ng, E. (1977) Root:shoot biomass ratios in shrubs in southern California and central Chile. Madroño 24, 215-223

Miller, P.C., Poole, D.K. (1979) Patterns of water use by shrubs in southern California. For. Sci. 25, 84-98

Miller, P.C., Poole, D.K. (1981) Partitioning of solar and net irradiance in mixed and chamise chaparral in southern California. Oecologia

Miller, P.C., Stoner, W.A. (1979) Canopy structure and environmental interactions. In: Plant Population Biology. Solbrig, O., Jain, S., Johnson, G.B., Raven, P.H. (eds.). New York: Columbia University Press, pp. 163-173

Miller, P.C., Stoner, W.A., Hom, J., Poole, D.K. (1976b) Potential influence of thermal effluents on the production and water use efficiency of mangrove species in south Florida. In: Thermal Ecology II. Proceedings of the Thermal Ecology Conference. Esch, G.W., McFarland, R.W. (eds.). Washington, D.C.: U.S. Energy Research and Development Agency, pp. 39-45

Miller, P.C., Stoner, W.A., Richards, S.P. (1978) MEDECS, a simulator for mediterranean ecosystems. Simulation 30, 173-190

Miller, P.C., Stoner, W.A., Tieszen, L.L. (1976a) A model of stand photosynthesis for the wet meadow tundra at Barrow, Alaska. Ecology 57, 411-430

Miller, P.C., Tieszen, L. (1972) A preliminary model of processes affecting primary production in the arctic tundra. Arct. Alp. Res. 4, 1-18

Millington, R.J., Quirk, J.P. (1959) Permeability of porous media. Nature 183, 387-388

Millington, R.J., Quirk, J.P. (1960) Transport in porous media. In: Transactions of the Seventh International Congress on Soil Science. Baren, F.A. van (ed.). Madison, Wis.: Soil Science Society of America, pp. 97-106

Millington, R.J., Quirk, J.P. (1961) Permeability of porous solids. Trans. Faraday Soc. 57, 97-106

Monk, C.D. (1966) An ecological significance of evergreenness. Ecology 47, 504-505

Monsi, M., Murata, Y. (1970) Development of photosynthetic systems as influenced by distribution of matter. In: Prediction and Measurement of Photosynthetic Productivity. Wageningen: Centre for Agricultural Publishing and Documentation, pp. 115-129

Monsi, M., Saeki, T. (1953) Uber den Lichtfaktor in den Pflanzengesellschaften und seine Bedeutung fur die Stoffproduktion. Jap. J. Bot. 14, 22-52

Monteith, J.L. (1965) Light distribution and photosynthesis in field crops. Ann. Bot. N.S. *29*, 17-37

Monteith, J.L. (1973) Principles of Environmental Physics. London: Edward Arnold

Montenegro, G., Aljaro, M.E., Kummerow, J. (1979) Growth dynamics of Chilean matorral shrubs. Bot. Gaz. *140*, 114-119

Montenegro, G., Jordan, M., Aljaro, M.E. (In press) Interactions between Chilean matorral shrubs and phytophagous insects. Oecologia

Montenegro, G., Rivera, O., Bas, F. (1978) Herbaceous vegetation in the Chilean matorral. Dynamics of growth and evaluation of allelopathic effects of some dominant shrubs. Oecologia *36*, 237-244

Mooney, H.A. (1972) The carbon balance of plants. Annu. Rev. Ecol. Syst. *3*, 315-346

Mooney, H.A. (1973) Plant Forms in Relation to Environment. U.S. IBP Tech. Rep. 73-4

Mooney, H.A. (ed.) (1977a) Convergent Evolution in Chile and California Mediterranean Climate Ecosystems. Stroudsburg, Pa.: Dowden, Hutchinson and Ross

Mooney, H.A. (1977b) Southern coastal scrub. In: Terrestrial Vegetation of California. Barbour, M.G., Major, J. (eds.). New York: John Wiley and Sons, pp. 471-489

Mooney, H.A., Bartholomew, B. (1974) Comparative carbon balance and reproductive modes of two Californian *Aexsculus* species. Bot. Gaz. *135*, 306-313

Mooney, H.A., Chu, C. (1974) Seasonal carbon allocation in *Heteromeles arbutifolia*, a California evergreen shrub. Oecologia *14*, 295-306

Mooney, H.A., Conrad, C.E. (eds.) (1977) Proceedings of the Symposium on the Environmental Consequences of Fire and Fuel Management in Mediterranean Ecosystems, 1-5 August 1977, Palo Alto, California. Washington, D.C.: U.S. Department of Agriculture, Forest Service, and U.S. Forestry Service Gen. Tech. Rep. WO-3

Mooney, H.A., Dunn, E.L. (1970a) Convergent evolution of mediterranean climate evergreen sclerophyll shrubs. Evolution *24*, 292-303

Mooney H.A., Dunn, E.L. (1970b) Photosynthetic systems of mediterranean climate shrubs and trees of California and Chile. Am Nat. *104*, 447-453

Mooney, H.A., Dunn, E.L., Shropshire, F., Song, L. (1970) Vegetation comparisons between the mediterranean climate areas of California and Chile. Flora *159*, 480-496

Mooney, H.A., Ehleringer, J., Berry, J.A. (1976) High photosynthetic capacity of a winter annual in Death Valley. Science *194*, 322-325

Mooney, H.A., Gulmon, S.L., Parsons, D.J., Harrison, A.T. (1974) Morphological changes within the chaparral vegetation type as related to elevational gradients. Madrono *22*, 281-316

Mooney, H.A., Harrison, A.T., Morrow, P.A. (1975) Environmental limitations of photosynthesis on a California evergreen shrub. Oecologia *19*, 293-301

Mooney, H.A., Hays, R.I. (1973) Carbohydrate storage cycles in two California mediterranean climate trees. Flora *162*, 295-304

Mooney, H.A., Kummerow, J., Johnson, A.W., Parsons, D.J., Keeley, S., Hoffmann, A., Hays, R.I., Giliberto, J., Chu, C. (1977) The producers—their resources and adaptive responses. In: Convergent Evolution in Chile and California Mediterranean Climate Ecosystems. Mooney, H.A. (ed.). Stroudsburg, Pa.: Dowden, Hutchinson and Ross, pp. 85-143

Mooney, H.A., Parsons, D.J. (1973) Structure and function of the California chaparral—an example from San Dimas. In: Mediterranean Type Ecosystems: Origin and Structure. Castri, F. di, Mooney, H.A. (eds.). New York: Springer-Verlag, pp. 83-112

Mooney, H.A., Parsons, D.J., Kummerow, J. (1973) Plant Development in Mediterranean Climates. U.S. IBP Tech. Rep. 73-6

Mooney, H.A., Rundel, P.W. (1979) Nutrient relations of the evergreen shrub, *Adenostoma fasciculatum*, in the California chaparral. Bot. Gaz. *140*, 109-113

Moore, J.J., Fitzsimons, P., White, J. (1970) A comparison and evaluation of some phytosociological techniques. Vegetatio *20*, 1-20

Morgan, J.M. (1976) A simulation model of the growth of the wheat plant. Ph.D. Dissertation. Sydney, Australia: Macquerie University

Morrow, P.A. (1971) The eco-physiology of drought adaptation of two mediterranean climate evergreens. Ph.D. Dissertation. Palo Alto, Ca.: Stanford University

Morrow, P., Mooney, H.A. (1974) Drought adaptations in two Californian evergreen sclerophylls. Oecologia *15*, 205-222

Muller, C.H. (1966) The role of chemical inhibition (allelopathy) in vegetational composition. Bull. Torrey Bot. Club *93*, 332-351

Muller, W.H., Muller, C.H. (1964) Volatile growth inhibitors produced by *Salvia* species. Bull. Torrey Bot. Club *91*, 327-330

Muñoz Pizarro, C. (1966) Sinopsis de la Flora Chilena, 2d ed. Santiago: Ediciones de la Universidad de Chile

Munz, P.A. (1974) A Flora of Southern California. Berkeley: University of California Press

Murphy, C.E., Knoerr, K.R. (1970) Modeling the Energy Balance Processes of Natural Ecosystems. U.S. IBP Analysis of Ecosystems, Eastern Deciduous Forest Biome Subproject Final Research Report 1969-1970, Oak Ridge, Tenn.: Oak Ridge National Laboratory

Murphy, C.E., Knoerr, K.R. (1972) Modeling the Energy Balance Processes of Natural Ecosystems. U.S. IBP Analysis of Ecosystems, Eastern Deciduous Forest Biome Sub Subproject Research Report EDFB-IBP 72-10, Oak Ridge, Tenn.: Oak Ridge National Laboratory

Mustafa, J. (1978) The effect of growth and species specific variability on photosynthesis along an elevational gradient in the chaparral. M.S. Thesis, Montreal: McGill University

Nakano, N., Brown, J. (1972) Mathematical modeling and validations of the thermal regimes in tundra soils, Barrow, Alaska. Arct. Alp. Res. *4*, 19-38

Naveh, Z. (1967) Mediterranean ecosystems and vegetation types in California and Israel. Ecology *48*, 445-459

Naveh, Z. (1974) Effects of fire in the mediterranean region. In: Fire and Ecosystems. Kozlowski, T.T., Ahlgren, C.E. (eds.). New York: Academic Press, pp. 401-434

Naveh, Z. (1977) The role of fire in the mediterranean landscape of Israel. In: Proceedings of the Symposium on the Environmental Consequences of Fire and Fuel Management in Mediterranean Ecosystems, 1-5 August 1977, Palo Alto, California. Mooney, H.A., Conrad, C.E. (eds.). Washington, D.C.: U.S. Department of Agriculture, Forest Service, and U.S. Forestry Service Gen. Tech. Rep. WO-3, pp. 299-306

Nazar, J., Hajek, E.R., Castri, F. di (1966) Determinación para Chile de algunas analogías bioclimáticas mundiales. Bol. Prod. Anim. (Chile) *4*, 103-173

Neales, T.F., Incoll, L.D. (1968) The control of leaf photosynthesis rate by the level of assimilate concentration in the leaf: A review of the hypothesis. Bot. Rev. *34*, 107-125

Ng, E. (1974) Soil moisture relations in chaparral. M.S. Thesis, San Diego, Ca.: San Diego State University

Ng, E., Miller, P.C. (1975) A model of the effect of tundra vegetation on soil temperatures. In: Climate of the Arctic. Weller, G., Bowling, S.A. (eds.). Fairbanks: Geophysical Institute, University of Alaska, pp. 222-226

Ng, E., Miller, P.C. (1980) Soil moisture relations in the southern California chaparral. Ecology *61*, 98-107

Nye, P.H., Tinker, P.B. (1969) The concept of a root demand coefficient. J. Appl. Ecol. *6*, 293-300

Oberlander, G.T. (1956) Summer fog precipitation on the San Francisco peninsula. Ecology *37*, 851-852

Odum, E. (1960) Organic production and turnover in old field succession. Ecology *41*, 34-48

Odum, E.P. (1969) The strategy of ecosystem development. Science *164*, 262-270

Odum, H.T., Pinkerton, R.C. (1955) Time's speed regulator. Am. Sci. *43*, 331-343

Oechel, W.C. (1976) Seasonal patterns of temperature response of CO_2 flux and acclimation in arctic mosses growing *in situ*. Photosynthetica *10*, 447-456

Oechel, W.C., Hicklenton, P., Sveinbjörnsson, Miller, P.C., Stoner, W.A. (1975) Physiological adaptations of arctic and subarctic mosses. In: Proceedings of the Circumpolar Conference on Northern Ecology, 15-18 September 1975, Ottawa. Ottawa: National Research Council of Canada, pp. I-131 to I-144

Oechel, W.C., Lawrence, W. (1979) Energy utilization and carbon metabolism in mediterranean scrub vegetation of Chile and California. I. Methods: A transportable for *Ceanothus greggii.* Oecologia *39*, 321-335

Oechel, W.C., Mustafa, J. (1979) Energy utilization and carbon metabolism in mediterranean scrub vegetation of Chile and California. II. The relationship between photosynthesis and cover in chaparral evergreen shrubs. Oecologia *41*, 305-315

Oechel, W.C., Strain, B.R., Odening, W.R. (1972) Tissue water potential, photosynthesis, ^{14}C-labelled photosynthate utilization, and growth in the desert shrub *Larrea divaricata* Cav. Ecol. Monogr. *42*, 127-141

Oficina Meteorológica de Chile. (1930-1970) Anuarios meteorológicos de Chile. Santiago, Chile

Olsen, S.R., Kemper, W.D. (1968) Movement of nutrients to plant roots. In: Advances in Agronomy, vol. 20. Norman, A.G. (ed.). New York: Academic Press, pp. 91-151

Olson, J.S. (1963) Energy storage and the balance of producers and decomposers in ecological systems. Ecology *44*, 322-331

Omar, M.H. (1968) Potential evapotranspiration in a warm arid climate. In: Agroclimatological Methods. Paris: UNESCO, pp. 347-353

Oppenheimer, H.R. (1936) Études sur développement des racines de quelques plantes méditerranéenes (une simple galerie a racines). Bull. Silva Medit. *10*, 142-162

Orians, G.H. (1975) Diversity, stability and maturity in natural ecosystems. In: Unifying Concepts in Ecology. van Dobben, W.H., Lowe-McConnell, R.H. (eds.). The Hague: Dr. W. Junk Publishers, and Wageningen: Centre for Agricultural Publishing and Documentation, pp. 139-150

Orians, G.H., Solbrig, O.T. (1977) A cost-income model of leaves and roots with special reference to arid and semiarid areas. Am. Nat. *111*, 677-690

Orshan, G. (1938) Seasonal leaf dimorphism in *Ononis natrix* L. Palest. J. Bot. Jerus. Ser. *I*, 233-234

Parsons, D.J. (1973) A comparative study of vegetation structure in the mediterranean scrub communities of California and Chile. Ph.D. Dissertation, Palo Alto, Ca.: Stanford University

Parsons, D.J. (1976a) The role of fire in natural communities: An example from the southern Sierra Nevada, California. Environ. Cons. *3*, 91-99

Parsons, D.J. (1976b) Vegetation structure in the mediterranean scrub communities of California and Chile. J. Ecol. *64*, 435-447

Parsons, D.J., Moldenke, A.R. (1975) Convergence in vegetation structure along analogous climatic gradients in the mediterranean climate ecosystems of California and Chile. Ecology *56*, 950-957

Paskoff, R.P. (1973) Geomorphological processes and characteristic landforms in the mediterranean regions of the world. In: Mediterranean Type Ecosystems: Origin and Structure. Castri, F. di, Mooney, H.A. (eds.). New York: Springer-Verlag, pp. 53-60

Patric, J.H. (1974) Water relations of some lysimeter-grown wildland plants in southern California. (Mimeo). Upper Darby, Pa.: U.S. Forest Service, Northeastern Forest Experiment Station

Peevy, W.J., Norman, A.G. (1948) Influence of composition of plant materials on properties of decomposed residues. Soil Sci. *65*, 209-226

Penman, H.L. (1948) Natural evaporation from open water, bare soil and grass. Proc. Roy. Soc. (A) *193*, 120-145

Penning de Vries, F.W.T. (1972a) A model for simulating transpiration of leaves with special attention to stomatal functioning. J. Appl. Ecol. *9*, 57-77

Penning de Vries, F.W.T. (1972b) Respiration and growth. In: Crop Processes in Controlled Environments. Rees, A.R., Cockshull, K.E., Hand, D.W., Hurd, R.G. (eds.). New York: Academic Press

Penning de Vries, F.W.T. (1973) Substrate utilization and respiration in relation to growth and maintenance in higher plants. Ph.D. Dissertation, Wageningen: Agricultural University

Penning de Vries, F.W.T. (1974) Substrate utilization and respiration in relation to growth and maintenance in higher plants. Neth. J. Agric. Sci. *22*, 40-44

Penning de Vries, F.W.T. (1975) Use of assimilates in higher plants. In: Photosynthesis and Productivity in Different Environments. Cooper, J.P. (ed.). Cambridge: Cambridge University Press, pp. 459-480

Penning de Vries, F.W.T., Brunsting, A.H.M., van Laar, H.H. (1974) Products, requirements, and efficiency of biosynthesis: A quantitative approach. J. Theor. Biol. *45*, 339-377

Penning de Vries, F.W.T., Murphy, C.E., Jr., Wells, C.G., Jorgensen, J.R. (1975) Simulation of nitrogen distribution in time and space in even-agral loblolly pine plantations and its effect on productivity. In: Mineral Cycling in Southeastern Ecosystems. Howell, F.G., Gentry, J.B., Smith, M.H. (eds.). Washington, D.C.: Energy Research and Development Administration, CONF-740513

Persson, H. (1978) Root dynamics in a young Scots pine stand in central Sweden. Oikos *30*, 508-519

Phillip, J.R. (1966) Plant water relations: Some physical aspects. Annu. Rev. Plant Physiol. *17*, 245-268

Phillips, M., Weihe, H.D., Smith, N.R. (1930) The decomposition of lignified materials by soil microorganisms. Soil Sci. *30*, 383-390

Phillips, R.E., NaNagara, T., Zartman, R.E., Leggett, J.E. (1976) Diffusion and moss flow of nitrate-nitrogen to plant roots. Agron. J. *68*, 63-66

Pisek, A., Winkler, E. (1959) Licht- und Temperaturabhängigkeit der CO_2-Assimilation von Fichte (*Picea excelsa* Link), Zirbe (*Pinus cembra* L.) und Sonnenblume *Helianthus annuus* L.). Planta *53*, 532-550

Poole, D.K., Miller, P.C. (1975) Water relations of selected species of chaparral and coastal sage communities. Ecology *56*, 1118-1128

Poole, D.K., Miller, P.C. (1978) Water related characteristics of some evergreen sclerophyll shrubs in central Chile. Oecol. Plant. *13*, 289-299

Poole, D.K., Miller, P.C. (1981) The distribution of plant water stress and vegetation characteristics in southern California chaparral. Am. Midl. Nat. *105*, 32-43

Purnell, H.M. (1960) Studies of the family Proteaceae. I. Anatomy and morphology of the roots of some Victorian species. Aust. J. Bot. *8*, 38-50

Radosevich, S.R., Conrad, S.G., Adams, D.R. (1977) Regrowth responses of chamise following fire. In: Proceedings of the Symposium on the Environmental Consequences of Fire and Fuel Management in Mediterranean Ecosystems, 1-5 August 1977, Palo Alto, Calif. Mooney, H.A., Conrad, C.E. (eds.). Washington, D.C.: U.S. Department of Agriculture, Forest Service, and U.S. Forestry Service Gen. Tech. Rep. WO-3, pp. 378-382

Randall, E.L. (1974) Improved method for fat and oil analysis by a new process of extraction. J. HOAC *57*, 1165-1168

Raunkiaer, C. (1934) The Life Forms of Plant and Statistical Plant Geography. Oxford: Clarendon Press

Raven, P.H. (1973) The evolution of mediterranean floras. In: Mediterranean Type Ecosystems: Origin and Structure. Castri, F. di, Mooney, H.A. (eds.). New York: Springer-Verlag, pp. 213-224

Rawlins, S.L. (1963) Resistance to water flow in the transpiration stream. In: Stomata and Water Relations in Plants. Zelitch, I. (ed.). New Haven: Conn. Agric. Exp. Stn. Bull. 664, pp. 69-84

Reid, J.S.G., Wilkie, U.K.C.B. (1969) Total hemicelluloses from oat plants at different stages of growth. Phytochemistry *8*, 2059-2065

Reiners, W.A. (1968) Carbon dioxide evolution from the floor of three Minnesota forests. Ecology *49*, 471-483

Reuss, J.O., Innis, G.S. (1977) A grassland nitrogen flow simulation model. Ecology 58, 379-388

Reynolds, E.R.C. (1975) Tree rootlets and their distribution. In: The Development and Function of Roots. Torrey, J.G., Clarkson, D.T. (eds.). New York: Academic Press, pp. 163-177

Rice, R.M., Foggin, G.T. (1971) Effect of high intensity storms on soil slippage on mountainous watersheds in southern California. Water Resour. Res. 7, 1485-1496

Rice, R.M., Corbett, E.S., Bailey, R.G. (1969) Soil slips related to vegetation, topography, and soil in southern California. Water Resour. Res. 5, 647-659

Richards, L.A. (1954) Diagnosis and improvement of saline and alkali soils. U.S. Dept. Agric., Agric. Handbook 60

Riveros de la Puente, G.F. (1973) Ritmo annual del crecimiento en raices de Quillaja saponaria y Cryptocarya alba. Thesis, Santiago: Universidad Católica de Chile

Roberts, S.W. (1974) The morphology and solar radiation environment of two mediterranean shrubs. M.S. Thesis, San Diego, Ca.: San Diego State University

Roberts, S.W., Miller, P.C. (1977) Interception of solar radiation as affected by canopy organization of two mediterranean shrubs. Oecol. Plant. 12, 273-290

Rochow, J.J. (1975) Mineral nutrient pool and cycling in a Missouri forest. J. Ecol. 63, 985-994

Rodin, L.E., Bazilevich, N.I. (1967) Production and Mineral Cycling in Terrestrial Vegetation. (transl. by G.E. Fogg). Edinburgh: Oliver and Boyd

Rosenzweig, M.L. (1974) On the evolution of habitat selection. In: Structure, Functioning and Management of Ecosystems. Proceedings of the First International Congress of Ecology, The Hague, 8-14 September 1974. Wageningen: Centre for Agricultural Publishing and Documentation, pp. 405-409

Rowe, P.B., Colman, E.A. (1951) Disposition of Rainfall in Two Mountain Areas of California. U.S. Dept. Agric. Tech. Bull. 1018

Rowe, P.B., Reiman, L.F. (1961) Water use by brush, grass, and grass-forb vegetation. J. For. 59, 175-181

Rowe, P.B., Storey, H., Hamilton, E.L. (1951) Some results of hydrologic research. In: Some Aspects of Watershed Management in Southern California. U.S. Forestry Service, Calif. For. Range Exp. Stn. Misc. Pap. 1, pp. 19-29

Rubin, J., Steinhardt, R. (1964) Soil water relations during rain infiltration. III. Water uptake at incipient ponding. Soil Sci. Soc. Am. Proc. 28, 614-619

Rubin, J., Steinhardt, R., Reiniger, P. (1964) Soil water relations during rain infiltration. II. Moisture content profiles during rains of low intensities. Soil Sci. Soc. Am. Proc. 28, 1-5

Rundel, P.W., Mahu, M. (1976) Community structure and diversity in a coastal fog desert in northern Chile. Flora 165, 493-505

Rundel, P.W., Neel, J.W. (1978) Nitrogen fixation by Trevoa trinervis (Rhamnaceae) in the Chilean matorral. Flora 167, 127-132

Rundel, P.W., Parsons, D.J. (1979) Structural changes in chamise (Adenostoma fasciculatum) along a fire induced age gradient. J. Range Manage. 32, 462-466

Rundel, P.W., Parsons, D.J. (1980) Nutrient changes in two chaparral shrubs along a fire-induced age gradient. Am. J. Bot. 67, 51-58

Running, S.W., Waring, R.H., Rydell, R.A. (1975) Physiological control of water flux in conifers. A computer simulation model. Oecologia 18, 1-16

Russell, S.A., Evans, H.J. (1966) The nitrogen fixing capacity of Ceanothus velutinus. For. Sci. 12, 164-169

Sampson, A.W. (1944) Plant Succession on Burned Chaparral Lands in Northern California. Berkeley: Calif. Agric. Exp. Stn. Bull. 685

Sanders, J.L., Brown, D.A. (1978) A new fiber optic technique for measuring root growth of soybeans under field conditions. Agron. J. 70, 1073-1076

San Diego County Department of Sanitation and Flood Control (1966-1973) Hydrology Reports. Loose-leaf pub. San Diego, California

Sauer, R.H. (1978) A simulation model for grassland primary producer phenology and biomass dynamics. In: Grassland Simulation Model. Innis, G.S. (ed.). New York: Springer-Verlag, pp. 55-87

Schaefer, R. (1973) Microbial activity under seasonal conditions of drought in mediterranean climates. In: Mediterranean Type Ecosystems: Origin and Structure. Castri, F. di, Mooney, H.A. (eds.). New York: Springer-Verlag, pp. 191-198

Schilling, G. (1975) Mapa fitogeográfico de Chile. In: Crónica de Una Expedición a Chile 2(11). Loose-leaf pub. Santiago: Empresa Editora Gabriela Mistral

Schimper, A.F.W. (1898) Pflanzen-geographie auf physiologischer Grundlage. Jena: G. Fisher [Engl. ed. transl. by W.R. Fisher, rev. and ed. by P. Groom and B. Balfour, 3d ed. by F.C.V. Faber, Vol. I, 1-588, Vol. II, 589-1612 (1935)].

Schlesinger, W.H. (1977) Carbon balance in terrestrial detritus. Annu. Rev. Ecol. Syst. 8, 51-81

Schlesinger, W.H., Gill, D.S. (1978) Demographic studies of the chaparral shrub, *Ceanothus megacarpus*, in the Santa Ynez Mountains, California. Ecology *58*, 1256-1263

Schlesinger, W.H., Gill, D.S. (1980) Biomass, production, and changes in the availability of light, water, and nutrients during the development of pure stands of the chaparral shrub, *Ceanothus megacarpus*, after fire. Ecology *61*, 781-789

Schlesinger, W.H., Hasey, M.M. (1980) The nutrient content of precipitation, dry fallout, and intercepted aerosols in the chaparral of southern California. Am. Midl. Nat. *103*, 114-122

Schoener, T.W. (1971) Theory of feeding strategies. Annu. Rev. Ecol. Syst. *2*, 369-404

Schultz, A.M., Biswell, H.H., Vlamis, J. (1958) Responses of brush seedlings to fertilizers. Calif. Fish Game *44*, 335-348

Schulze, E.-D., Lange, O.L., Buschbom, U., Kappen, L., Evenari, M. (1972a) Stomatal responses to changes in humidity in plants growing in the desert. Planta *108*, 259-270

Schulze, E.-D., Lange, O.L., Kappen, L., Evenari, M., Buschbom, U. (1975) The role of air humidity and leaf temperature in regulating stomatal resistance of *Prunus armeniaca* L. under desert conditions. II. The significance of leaf water status and internal carbon dioxide concentration. Oecologia *18*, 219-233

Schulze, E.-D., Lange, O.L., Koch, W. (1972b) Okophysiologische Untersuchungen an Wild- und Kulturphlanzen der Negev-Wuste. II. Die Wirkung der Aubenfaktoren auf CO_2-Gaswechsel und Transpiration an Ende der Trockenzeit. Oecologia *8*, 334-355

Scott, K.M., Williams, R.P. (1978) Erosion and Sediment Yields in the Transverse Ranges, Southern California. U.S. Geol. Surv. Prof. Pap. 1030

Seligman, N.G., van Keulen, H., Goudriaan, J. (1975) An elementary model of nitrogen uptake and redistribution by annual plant species. Oecologia *21*, 243-261

Sellers, W.D. (1965) Physical Climatology. Chicago: University of Chicago Press

Shachori, A., Rosenzweig, D., Poljakoff-Mayber, A. (1967) Effect of mediterranean vegetation on the moisture regime. In: Forest Hydrology. Sopper, W.E., Lull, H.W. (eds.). Oxford: Pergamon Press, pp. 291-311

Shapiro, A.A., deForest, H. (1932) A comparison of transpiration rates in chaparral. Ecology *13*, 290-295

Shaver, G.R. (1978) Leaf angle and light absorptance of *Arctostaphylos* species (Ericaceae) along environmental gradients. Madroño *25*, 133-138

Shaver, G.R., Billings, W.D. (1975) Root production and root turnover in a wet tundra ecosystem, Barrow, Alaska. Ecology *56*, 401-409

Shaver, G.R., Chapin, F.S. (1980) Response to fertilization by various plant growth forms in an Alaskan tundra: Nutrient accumulation and growth. Ecology *61*, 662-675

Sierra Ràfols, E. (1977) Analog tree and shrub species. In: Chile-California Mediterranean Scrub Atlas. Thrower, N.J.W., Bradbury, D.E. (eds.). Stroudsburg, Pa.: Dowden, Hutchinson and Ross, pp. 129-143

Simpson, G.G. (1969) The first three billion years of community evolution. In: Diversity and Stability in Ecological Systems. Brookhaven Symposia in Biology no. 22. Upton, N.Y.: Brookhaven National Laboratory, pp. 162-176

Sims, R.L., Singh, J.S. (1978) The structure and function of ten western North American grasslands. III. Net primary production, turnover and efficiencies of energy capture and water use. J. Ecol. *66*, 573-597

Sinclair, J.D. (1954) Erosion in the San Gabriel Mountains of California. Am. Geophy. Union Trans. *35*, 264-268

Skre, O. (1972) High temperature demands for growth and development in Norway spruce (*Picea abies* (L.) Karst.) in Scandinavia. Meld. Nor. Landbrukshogsk *51*, 1-29

Skre, O. (1975) CO_2 exchange in Norwegian tundra plants studied by infrared gas analyzer technique. In: Fennoscandian Tundra Ecosystems. Part 1. Plants and Microorganisms. Wielgolaski, F.E. (ed.). New York: Springer-Verlag, pp. 168-183

Small, E. (1972) Photosynthetic rates in relation to nitrogen recycling as an adaptation to nutrient deficiency in peat bog plants. Can. J. Bot. *50*, 2227-2233

Small, E. (1973) Xeromorphy in plants as a possible basis for migration between arid and nutritionally-deficient environments. Bot. Not. *126*, 534-539

Smith, D. (1969) Removing and Analyzing Total Nonstructural Carbohydrates From Plant Tissue. Madison: University of Wisconsin, Coll. Agric. Life Sci., Res. Div. Res. Rep. 41

Smith, D.W. (1970) Concentrations of soil nutrients before and after fire. Can J. Soil Sci. *50*, 17-29

Smith, E.M., Hadley, E.B. (1974) Photosynthetic and respiratory acclimation to temperature in *Ledum goenlandicum* populations. Arct. Alp. Res. *6*, 13-27

Smith, O.L., Shugart, H.H., O'Neill, R.V., Booth, R.S., McNaught, D.C. (1975) Resource competition and an analytical model of zooplankton feeding on phytoplankton. Am. Nat. *109*, 571-591

Sokal, R.R., Rohlf, F.J. (1969) Biometry. San Francisco: W.H. Freeman

Sokal, R.R., Sneath, P.H. (1963) Principals of Numerical Taxonomy. San Francisco: W.H. Freeman

Solbrig, O.T., Orians, G.H. (1977) The adaptive characteristics of desert plants. Am. Sci. *65*, 412-421

Sørensen, L.H. (1974) Rate of decomposition of organic matter in soil as influenced by repeated air drying-rewetting and repeated additions of organic material. Soil Biol. Biochem. *6*, 287-292

Specht, R.L. (1957a) Dark Island heath (Ninety-Mile Plain, South Australia). IV. Soil moisture patterns produced by rainfall interception and stem-flow. Aust. J. Bot. *5*, 137-150

Specht, R.L. (1957b) Dark Island heath (Ninety-Mile Plain, South Australia). V. The water relationships in heath vegetation and pastures on the Makin Sand. Aust. J. Bot. *5*, 151-172

Specht, R.L. (1963) Dark Island Heath (Ninety-Mile Plain, South Australia). VII. The effect of fertilizers on composition and growth, 1950-60. Aust. J. Bot. *11*, 67-94

Specht, R.L. (1966) The growth and distribution of mallee-broombush (*Eucalyptus incrassata-Melaleuca uncinata* association) and heath vegetation near Dark Island Soak, Ninety-Mile Plain, South Australia. Aust. J. Bot. *14*, 361-371

Specht, R.L. (1969a) A comparison of the sclerophyllous vegetation characteristic of mediterranean type climates in France, California, and southern Australia. I. Structure, morphology, and succession. Aust. J. Bot. *17*, 277-292

Specht, R.L. (1969b) A comparison of the sclerophyllous vegetation characteristic of mediterranean type climates in France, California, and southern Australia. II. Dry matter, energy, and nutrient accumulation. Aust. J. Bot. *17*, 293-308

Specht, R.L. (1972a) The Vegetation of South Australia. Adelaide: A.B. Jones

Specht, R.L. (1972b) Water use by perennial evergreen plant communities in Australia and Papua New Guinea. Aust. J. Bot. *20*, 273-299

Specht, R.L. (1973) Structure and functional response of ecosystems in the mediterranean climate of Australia. In: Mediterranean Type Ecosystems: Origin and Structure. Castri, F. di, Mooney, H.A. (eds.). New York: Springer-Verlag, pp. 113-120

Specht, R.L. (ed.) (1979) Heathlands and Related Shrublands. Part A: Dexcriptive Studies. Amsterdam: Elsevier Scientific Publishing Co.

Specht, R.L. (ed.) (In press) Mediterranean-Type Shrublands. Amsterdam: Elsevier Scientific Publishing Co.

Specht, R.L., Groves, G.H. (1966) A comparison of the phosphorus nutrition of Australian heath plants and introduced economic plants. Aust. J. Bot. *14*, 201-221

Specht, R.L., Rayson, P. (1957) Dark Island Heath (Ninety-Mile Plain, South Australia). III. The root systems. Aust. J. Bot. *5*, 103-114

Specht, R.L., Rayson, P. (1958) Dark Island Heath (Ninety-Mile Plain, South Australia). VI. Pyric succession: Changes in composition, coverage, dry weight, and mineral nutrient status. Aust. J. Bot. *6*, 59-88

Stanhill, G. (1961) The accuracy of meteorological estimates of evapotranspiration in arid climates. J. Inst. Water Engrs. *15*, 477-482

Stevenson, I.L. (1956) Some observations on the microbial activity of remoistened air-dried soils. Plant Soil *8*, 170-182

Stewart, D.W., Lemon, E.R. (1969) The Energy Budget at the Earth's Surface: A Simulation of Net Photosynthesis of Field Corn. Fort Huachuca: ECOM Atmospheric Sciences Lab., Tech. Rep. 2-68

Stewart, D.W., Lemon, E.R. (1972) The Energy Budget at the Earth's Surface: A Simulation of Net Photosynthesis of Field Corn. Fort Huachuca: ECOM Atmospheric Sciences Lab., Int. Rep. 69-3

Stoner, W.A., Miller, P.C. (1975) Water relations of plant species in the wet coastal tundra at Barrow, Alaska. Arct. Alp. Res. *7*, 109-124

Stoner, W.A., Miller, P.C., Miller, P.M. (1978a) A test of a model of irradiance within vegetation canopies at northern latitudes. Arct. Alp. Res. *10*, 761-767

Stoner, W.A., Miller, P.C., Oechel, W.C. (1978b) Simulation of the effect of the tundra vascular plant canopy on the productivity of four moss species. In: Vegetation and Production Ecology of an Alaskan Arctic Tundra. Tieszen, L.L. (ed.). New York: Springer-Verlag, pp. 371-387

Stoner, W.A., Miller, P.C., Richards, S.P., Barkley, S.A. (1978d) Internal nutrient cycling as related to plant life-form: A simulation approach. In: Environmental Chemistry and Cycling Processes. Adriano, D.C., Brisbin, I.L., Jr. (eds.). Washington, D.C.: U.S. Dept. of Energy, Technical Information Center, CONF–760429, pp. 165-181

Stoner, W.A., Miller, P.C., Tieszen, L.L. (1978c) A model of plant growth and phosphorus allocation for *Dupontia fischeri* in coastal wet meadow tundra. In: Vegetation and Production Ecology of an Alaskan Arctic Tundra. Tieszen, L.L. (ed.). New York: Springer-Verlag, pp. 559-576

Strain, B.R. (1969) Seasonal adaptations in photosynthesis and respiration in four desert shrubs growing *in situ*. Ecology *50*, 511-513

Sutton, O.G. (1953) Micrometeorology. New York: McGraw-Hill

Sveinbjörnsson, B. (1979) Controls on the carbon metabolism of arctic and subarctic populations of two moss species. Ph.D. Dissertation, Montreal: McGill University

Switzer, G.L., Nelson, L.E. (1972) Nutrient accumulation and cycling in loblolly pine (*Pinus taeda* L.) plantation ecosystems: The first twenty years. Soil Sci. Soc. Am. Proc. *36*, 143-147

Syvertsen, J.P. (1974) Relative stem water potentials of three woody perennials in a southern oak woodland community. Bull. So. Calif. Acad. Sci. *73*, 108-113

Syvertsen, J.P., Cunningham, G.L. (1977) Rate of leaf production and senescence and effect of leaf age on net gas exchange in Creosote bush. Photosynthetica *11*, 161-166

Talbot, A.J.B., Tyree, M.T., Dainty, J. (1975) Some notes concerning the measurement of water poentials of leaf tissue with specific reference to *Tsuga canadensis* and *Pices albies*. Can. J. Bot. *53*, 784-788

Tarrant, R.F., Lu, K.C., Chen, C.S., Bollen, W.B. (1968) Nitrogen content of precipitation in a coastal Oregon forest opening. Tellus *20*, 554-556

Taylor, F.G., Jr. (1974) Phenodynamics of production in a mesic deciduous forest. In: Phenology and Seasonality Modeling. Lieth, H. (ed.). New York: Springer-Verlag, pp. 237-254

Taylor, S.E. (1975) Optimal leaf form. In: Perspectives of Biophysical Ecology. Gates, D.M., Schmerl, R.B. (eds.).New York: Springer-Verlag, pp. 73-86

Tenhunen, J.D., Yocum, C.S., Gates, D.M. (1976) Development of a photosynthesis model with an emphasis on ecological applications. I. Theory. Oecologia 26, 89-100

Terent'ev, V.M., Petrovich, Z.I. (1968) Formation and metabolism of hemicellulose A and B in barley stalks. (Abstr.) Chem. Abst. 68, 9011

Terjung, W.H., Kickert, R.N., Potter, G., Swarts, S. (1969) Energy and moisture balances of an alpine tundra in mid-July. Arct. Alp. Res. 1, 247-266

Thornley, J.H.M. (1972b) A model to describe the partitioning of photosynthate during vegetative plant growth. Ann. Bot. 36, 419-430

Thrower, N.J.W., Bradbury, D.E. (es.) (1977) Chile-California Mediterranean Scrub Atlas: A Comparative Analysis. Stroudsburg, Pa.: Dowden, Hutchinson and Ross

Tieszen, L.L. (1973) Photosynthesis and respiration in arctic tundra grasses: Field light intensities and temperature responses. Arct. Alp. Res. 5, 239-251

Tieszen, L.L. (1979) Vegetation and Production Ecology of the Alaskan Arctic Tundra. New York: Springer-Verlag

Tieszen, L.L., Johnson, D.A., Caldwell, M.M. (1974) A portable system for the measurement of photosynthesis using 14-carbon dioxide. Photosynthetica 3, 151-160

Tieszen, L.L., Lewis, M.C., Miller, P.C., Mayo, J., Chapin, F.S., III, Oechel, W.C. (In press) An analysis of processes of primary production in tundra growth forms. In: The Ecology of Tundra and Related Habitats. Heal, O.W. (ed.). Cambridge: Cambridge University Press

Turner, J., Singer, M.J. (1976) Nutrient distribution and cycling in a sub-alpine coniferous forest ecosystem. J. Appl. Ecol. 13, 295-301

Turner, N.C. (1974) Stomatal behavior and water status of maize, sorghum, and tobacco under field conditions. II. At low soil water potential. Plant Physiol. 53, 360-365

Tyree, M.T., Dainty, J., Benis, M. (1973) The water relations of hemlock (Tsuga canadensis). I. Some equilibrium water relations as measured by the pressure bomb. Can. J. Bot. 51, 1471-1480

Tyree, M.T., Dainty, J., Hunger, D.M. (1974) The water relations of hemlock (Tsuga canadensis). IV. The dependence of the balance pressure and the water relations of plants by the pressure-bomb technique. J. Exp. Bot. 23, 267-282

Tyree, M.T., Hammel, H.T. (1972) The measurement of the turgor pressure and the water relations of plants by the pressure-bomb technique. J. Exp. Bot. 23, 267-282

U.S. Department of Agriculture, Soil Conservation Service and Forest Service (1973) Soil Survey. San Diego Area California. Loose-leaf pub. Washington, D.C.: U.S. Government Printing Office

U.S. National Oceanic and Atmospheric Administration (1950-1970) Local Climatological Data: San Diego, California. Loose-leaf pub. Washington, D.C.

van Keulen, H. (1975) Simulation of Water Use and Herbage Growth in Arid Regions. Wageningen: Centre for Agricultural Publishing and Documentation

van Keulen, H., Seligman, N.G., Goudriaan, J. (1975) Availability of anions in the growth medium of roots of an actively growing plant. Neth. J. Agric. Sci. 23, 131-138

Viro, P.J. (1974) The effects of forest fire on soil. In: Fire and Ecosystems. Kozlowski, T.T., Ahlgren, C.E. (eds.). New York: Academic Press, pp. 7-45

Visvalingam, M., Tandy, J.B. (1972) The neutron method for measuring soil moisture content—a review. J. Soil Sci. 23, 499-511

Vitousek, P.M., Reiners, W.A. (1975) Ecosystem succession and nutrient retention: A hypothesis. BioScience 25, 376-381

Vlamis, J., Gowans, K.D. (1961) Availability of nitrogen, phosphorus, and sulfur after brush burning. J. Range Manage. 14, 38-40

Vlamis, J., Schultz, A.M., Biswell, H.H. (1958) Nitrogen fixation by deerbrush. Calif. Agri. *12*, 11,15

Vlamis, J., Schultz, A.M., Biswell, H.H. (1964) Nitrogen fixation by root nodules of western mountain mahogany. J. Range Manage. *17*, 73-74

Vlamis, J., Stone, E.C., Young, C.L. (1954) Nutrient status of brushland soils in southern California. Soil Sci. *78*, 51-55

Vogl, R.J. (1977) Fire frequency and site degradation. In: Proceedings of the Symposium on the Environmental Consequences of Fire and Fuel Management in Mediterranean Ecosystems, 1-5 August 1977, Palo Alto, California. Mooney, H.A., Conrad, C.E. (eds.). Washington, D.C.: U.S. Department of Agriculture, Forest Service, and U.S. Forestry Service Gen. Tech. Rep. WO-3, pp. 193-201

von Humboldt, A. (1806) Ideen zur einer Physiognomik der Gervachse. Tübingen

Vowinckle, T., Oechel, W.C., Boll, W.G. (1975) The effect of climate on the photosynthesis of *Picea mariana* at the subarctic tree line. 1. Field measurements. Can J. Bot. *53*, 604-620

Waggoner, P.E., Furnival, G.M., Reifsnyder, W.E. (1969) Simulation of microclimate in a forest canopy. For. Sci. *15*, 37-45

Waggoner, P.E., Reifsnyder, W.E. (1968) Simulation of temperature, humidity, and evaporation profiles in a leaf canopy. J. Appl. Meteorol. *7*, 400-409

Waksman, S.A. (1929) Chemical and microbiological principles underlying the decomposition of green manures in the soil. J. Am. Soc. Agron. *21*, 1-18

Wallén, C.C. (1968) Agroclimatological studies in the Levant. In: Agroclimatological Methods. Proceedings of the Reading Symposium. Paris: UNESCO, pp. 225-233

Wallén, C.C., Brichambaut, G.P. de (1962) A Study of Agroclimatology in Semiarid and Arid Zones of the Near East. Rome: FAO

Walter, H. (1973) Vegetation of the Earth in Relation to Climate and the Eco-physiological Conditions. (Trans. from the 2nd German ed. by Joy Wieser). New York: Springer-Verlag

Wareing, P.F., Patrick, J. (1975) Source-sink relations and the partition of assimilates in the plant. In: Photosynthesis and productivity in Different Environments. Cooper, J.P. (ed.). Cambridge: Cambridge University Press, pp. 481-499

Warming, E. (1909) Oecology of Plants. (Trans. by P. Groom and T.T. Balfour.) Oxford: Clarendon Press

Warren Wilson, J. (1964) Annual growth of *Salix arctica* in the high-arctic. Ann. Bot. N.S. *28*, 71-76

Warren Wilson, J. (1966) Effect of temperature on net assimilation rate. Ann. Bot. *30*, 753-761

Watkins, V.M., deForest, H. (1941) Growth in some chaparral shrubs of California. Ecology *22*, 79-83

West, N.E., Skujins, J. (1977) The nitrogen cycle in North American cold-winter semi-desert ecosystems. Oecol. Plant. *12*, 45-53

Westhoff, V., Maarel, E. van der (1973) The Braun-Blanquet approach. In: Handbook of Vegetation Science. Tüxen, R. (ed.). Part V: Ordination and Classification of Communities. Whittaker, R.H. (ed.). The Hague: Dr. W. Junk Publishers, pp. 617-726

White, E.M., Thompson, W.W., Gartner, F.R. (1973) Heat effects on nutrient release from soils under ponderosa pine. J. Range Manage. *26*, 22-24

White, J. (1979) The plant as a metapopulation. Annu. Rev. Ecol. Syst. *10*, 109-145

Whitehead, H.C., Feth, J.H. (1964) Chemical composition of rain, dry fallout, and bulk precipitation at Menlo Park, California, 1957-1959. J. Geophys. Res. *69*, 3319-3333

Whittaker, R.H. (1965) Dominance and diversity in land plant communities. Science *147*, 250-260

Whittaker, R.H. (1970) The population structure of vegetation. In: Gesellshafts Morphologie (Strukturforschung). Tüxen, R. (ed.). The Hauge: Dr. W. Junk Publishers, pp 39-62

Whittaker, R.H. (ed.) (1973) Ordination and classification of vegetation. The Hague: Dr. W. Junk Publishers, pp. 737-

Whittaker, R.H. (1975) Communities and Ecosystems, 2d ed. New York: Macmillan Co.

Whittaker, R.H., Likens, G.E. (1975) The biosphere and man. In: Primary Productivity of the Biosphere. Lieth, H., Whittaker, R.H. (eds.). New York: Springer-Verlag, pp. 305-328

Wiebe, H.H., Campbell, G.S., Gardner, W.H., Rawlins, S.L., Cary, J.W., Brown, R.W. (1971) Measurement of Plant and Soil Water Status. Utah Agric. Exp. Stn. Bull. 484

Wielgolaski, F.E. (1974) Phenology in agriculture. In: Phenology and Seasonality Modeling. Lieth, H. (ed.). New York: Springer-Verlag, pp. 369-381

Williams, S.T., Gray, T.R.G. (1974) Decomposition of litter on the soil surface. In: Biology of Plant Litter Decomposition. Dickinson, C.H., Pugh, G.J.F. (eds.). New York: Academic Press, pp. 611-632

Wilson, B.F. (1975) Distribution of secondary thickening in tree root systems. In: The Development and Function of Roots. Torrey, J.G., Clarkson, D.T. (eds.). New York: Academic Press, pp. 197-219

Wit, C.T. de (1965) Photosynthesis of leaf canopies. Wageningen: Agric. Res. Rep. 663

Wit, C.T. de, et al. (1978) Simulation of Assimilation, Respiration and Transpiration of Crops. New York: John Wiley and Sons

Wit, C.T. de, Brouwer, R., Penning de Vries, F.W.T. (1970) The simulation of photosynthetic systems. In: Prediction and Measurement of Photosynthetic Productivity. Wageningen: Centre for Agricultural Publishing and Documentation, pp. 47-70

Wit, C.T. de, van Keulen, H. (1972) Simulation of Transport Processes in Soils. Wageningen: Centre for Agricultural Publishing and Documentation

Wolman, M.G., Schick, A.P. (1967) Effects of construction of fluvial sediment, urban and suburban areas of Maryland. Water Resour. Res. *3*, 451-464

Woodmansee, R.G. (1978) Additions and losses of nitrogen in grassland ecosystems. BioScience *28*, 448-453

Yeilding, L. (1977) Decomposition in chaparral. In: Proceedings of the Symposium on the Environmental Consequences of Fire and Fuel Management in Mediterranean Ecosystems, 1-5 August 1977, Palo Alto, California. Mooney, H.A., Conrad, C.E. (eds.). Washington, D.C.: U.S. Department of Agriculture, Forest Service, and U.S. Forestry Service Gen. Tech. Rep. WO-3, pp. 419-425

Yim, Y., Ogawa, H., Kira, T. (1969) Light interception by stems in plant communities. Jap. J. Ecol. *19*, 233-238

Youngberg, C.T., Wollum, A.G. (1976) Nitrogen accretion in developing *Ceanothus velutinus* stands. Soil Sci. Soc. Am. J. *40*, 109-112

Zartman, R.E., Phillips, R.E., Leggett, J.E. (1976) Comparison of simulated and measured nitrogen accumulation in burley tobacco. Agron. J. *68*, 406-410

Zinke, P.J. (1973) Analogies between the soil and vegetation types of Italy, Greece, and California. In: Mediterranean Type Ecosystems: Origin and Structure. Castri, F. di, Mooney, H.A. (eds.). New York: Springer-Verlag, pp. 61-80

Zinke, P.J. (1977) Mineral cycling in fire-type ecosystems. In: Proceedings of the Symposium on the Environmental Consequences of Fire and Fuel Management in Mediterranean Ecosystems, 1-5 August 1977, Palo Alto, California. Mooney, H.A., Conrad, C.E. (eds.). Washington, D.C.: U.S. Department of Agriculture, Forest Service, and U.S. Forestry Service Gen. Tech. Rep. WO-3, pp. 85-94

Resumen

El objetivo general de la investigación descrita en este volumen es analizar la hipótesis siguiente: la vegetación de regiones que tienen un ambiente similar, a pesar de tener distinta historia filogenética, presenta modelos similares en la utilización de los recursos y modelos similares de distribución de carbono y de nutrientes hacia las estructuras vegetativas y reproductivas. La metodología consistió en usar modelos de procesos de producción primaria para sintetizar datos relacionados con la utilización de energía solar, agua y nitrógeno en el sistema suelo-vegetación. Estos modelos se usaron para generalizar los datos a patrones diurnos, estacionales y anuales de utilización de recursos. La hipótesis fue examinada en la vegetación de ecosistemas mediterráneos de California del Sur y de Chile Central, durante un lapso de cuatro años, desde Septiembre de 1975 hasta Enero de 1980. La mayor parte de los experimentos y mediciones en terreno se realizaron en Echo Valley (116°40'W, 32°55'N), ubicado a 45 km al Este de San Diego, California, y en el Fundo Santa Laura (71°00'W, 33°04'S) 30 km al Noroeste de Santiago, Chile. También se realizaron mediciones en Camp Pendleton y Mount Laguna en el distrito de San Diego, California.

En base a diversas características climáticas – tales como temperatura, precipitaciones, radiación solar, y evapotranspiración potencial – el Fundo Santa Laura ubicado a 1000 metros de altura equivale a lugares con 1400 a 1500 metros de altura ubicados en California del Sur, entre Echo Valley y Mount Laguna. El Fundo Santa Laura es más fresco y más lluvioso que Echo Valley, pero recibe la misma radiación solar. Por otra parte, la evapotranspiración potencial del Fundo Santa Laura es más alta que la de Echo Valley, Mount Laguna, y Camp Pendleton. La influencia de la profundidad del suelo y de la progresión estacional de las precipitaciones sobre la disponibilidad de agua indica que los arbustos semideciduos de sequía se darían con más frequencia en las alturas intermedias de Chile Central que en California del Sur.

Se estudiaron con más detalle las especies comunes en el chaparral de California del Sur y en el matorral de Chile Central, especies que definen tales comunidades vegetales. La baja afinidad florística indica historia filogenética distinta para las dos vegetaciones. La cobertura arbustiva resultó ser más alta en California del Sur que en Chile Central, aún cuando la diversidad de arbustos es mayor en esta última región. Sin embargo la cobertura herbácea es más baja en el chaparral de California que en el matorral de Chile Central. Las tasas de crecimiento de hojas, tallos y raíces son similares en ambas regiones. La producción aérea es más alta en Echo Valley que en el Fundo Santa Laura, pero la producción herbácea es más alta en Chile. Se encontró que la biomasa de raíces finas era más baja en Echo Valley que en el Fundo Santa Laura. Se apoyó la hipótesis de que los patrones de distribución de carbono a los sistemas raíz-tallo son similares cuando las condiciones de ambiente son similares.

Aunque en invierno las temperaturas atmosféricas en diversas laderas de Echo Valley son similares a las del Fundo Santa Laura, en verano son más altas. En ambos lugares, las temperaturas promedio del suelo, especialmente en zonas de baja cobertura vegetal, son mayores que las temperaturas atmosféricas promedio.

La transpiración y la eficiencia de captación de aguas son más altas en el chaparral mixto que en el matorral, pero el chaparral de *Adenostoma* arrojó valores similares a los del matorral. El control del uso del agua en el chaparral difiere del que se observa en el matorral, debido a que las máximas conductancias foliares son mayores en el chaparral. El contenido relativo de agua en que se encontró la mínima conductancia foliar era similar en arbustos esclerófilos y malacófilos, tanto en California como en Chile, pero este contenido era diferente entre un tipo de arbusto y otro, dentro del mismo país. Los modelos estacionales de uso del agua resultaron similares dentro de un mismo tipo de arbusto en cada país, pero eran distintos para los diferentes tipos de arbustos. La menor eficiencia en la captación de agua en Santa Laura se debe a los menores valores de máxima conductancia foliar, a la menor radiación neta durante el invierno y a las precipitaciones más abundantes.

Las especies de arbustos siempreverdes mediterráneos de Echo Valley y del Fundo Santa Laura exhiben modelos semejantes en la respuesta fotosintética y en la eficiencia de captación y uso de energía solar. La fotosíntesis es relativamente insensible a la temperatura; las temperaturas óptimas para la fotosíntesis se aproximan a los valores teóricos que producen el máximo de absorción de carbono. La acción limitante de la temperatura sobre la fotosíntesis es mínima. Las tasas de captación de carbono no parecen corroborar la sugerencia de que la transición desde chaparral a matorral costero en el sur de California se debería al efecto de la sequía sobre la producción. Las especies siempreverdes presentan tasas considerables de asimilación de carbono a fines del verano, tanto en la costa como en otras áreas.

La convergencia que se había encontrado en los modelos de asimilación de carbono entre grupos de especies del chaparral y del matorral no se observó en los modelos de distribución del carbono. Las tasas de respiración y las sensibilidades de la respiración a la temperatura de las especies del chaparral difieren de las del matorral. También es diferente la distribución de carbono a las fracciones químicas en tallos y hojas. Las tasas de respiración en hojas y tallos, y los valores de Q_{10} tienden a ser más bajos en las especies del chaparral. Las proporciones entre tasas de respiración foliar y tasas máximas de fotosíntesis eran inferiores en el chaparral comparadas con las del matorral. Las tasas de respiración foliar eran generalmente más variables en las especies del chaparral.

Los nutrientes minerales son más limitantes del crecimiento de arbustos en Echo Valley, comparados con los de Santa Laura. El contenido total de hidratos de carbono no estructurales y de lípidos es mayor en Echo Valley. En esta región, el crecimiento anual se ve favorecido por el reciclaje de fósforo, y posiblemente de magnesio, que ocurre dentro de la planta. En Santa Laura, la mayor parte de los nutrientes minerales necesarios al crecimiento anual son captados del suelo durante la estación de crecimiento. Los ciclos estacionales de nutrientes minerales y el almacenamiento de hidratos de carbono no indican convergencia de estas funciones en la región mediterránea de California del Sur y de Chile Central.

Los resultados experimentales se sintetizaron en dos modelos de simulación, el Simulador de Procesos de la Capa Vegetal (CAPS) y el Simulador de Ecosistemas Mediterráneos (MEDECS). Ambos son modelos complejos de los procesos de la vegetación y del clima que afectan a la producción primaria. El CAPS simula los procesos con incremento de una hora durante un período diurno, para una sola especie. El MEDECS simula los procesos en intervalos diurnos durante un período de un año, para una comunidad con un máximo de cuatro especies. El CAPS predijo en forma precisa el ambiente físico dentro de la capa vegetal, los potenciales hídricos de las plantas, y las tasas de fotosíntesis. Se usó para generar relaciones simplificadas para el MEDECS. Las simulaciones de MEDECS comparaban favorablemente con los datos de terreno sobre humedad del suelo y biomasa, y se les consideró como válidos para comparar uso de recursos entre los ecosistemas de tipo mediterráneo de California y de Chile.

Las mediciones y simulaciones indicaron que la absorción y eficiencia de absorción de agua, energía solar y nitrógeno variaban según las especies, laderas, y país, de tal modo que había mayor diferencia entre absorción y eficiencia de absorción del chaparral de las laderas de orientación polar y ecuatorial de Echo Valley, que entre el chaparral ubicado en la ladera de orientación polar de Echo Valley y la del matorral del Fundo Santa Laura. Se determinaron tres patrones de eficiencia en el uso de recursos. La limitación por agua, radiación solar y nitrógeno se destaca por las eficiencias relativamente altas en el uso de estos tres recursos, en las laderas de orientación polar, tanto en Echo Valley como en el Fundo Santa Laura. En la ladera de orientación equatorial del Fundo Santa Laura sólo uno de estos recursos, el nitrógeno, aparece como factor limitante.

En el chaparral de *Adenostoma* (chamise) de la ladera de orientación ecuatorial de Echo Valley, la limitación causada por estos recursos es bastante modesta, como lo indica la eficiencia relativamente baja en el uso de los tres recursos. El chaparral de *Adenostoma* es el tipo más abundante en el distrito de San Diego y puede reemplazar a otras especies del chaparral después de un incendio. Aún cuando la eficiencia para el uso de recursos del chaparral de *Adenostoma* es baja, su eficiencia de crecimiento es alta, de modo que su crecimiento es comparable al chaparral mixto.

El uso de recursos y la eficiencia en el uso de recursos de los climas mediterráneos estan ambientalmente constreñidos por interacciones y retroalimentaciones entre varias propiedades ambientales y vegetales. Comparando matorral con chaparral, es la radiación relativamente baja de invierno, más que las propiedades de planta y suelo, la que limita el uso de agua y luz en el matorral. Aún con limitaciones climáticas, la vegetación puede presentarse como limitada de nutrientes. Las tasas de recambio de la vegetación se ajustan a las tasas de disponibilidad de los nutrientes. La inversión en materiales estructurales es requisito para las hojas siempreverdes que tienen que sobrevivir a la sequía del verano. El contenido relativamente alto de lignina y celulosa de las hojas siempreverdes guarda relación con su bajo contenido de nitrógeno y es posible que no se deba a la baja disponibilidad de nitrógeno, sino más bien a la dilución del citoplasma por materiales estructurales. El bajo contenido de nitrógeno de las hojas aumenta la proporción carbono-nitrógeno de la hojarasca, y puede disminuir las tasas de liberación de nitrógeno por descomposición de la hojarasca. A su vez, esta baja tasa de liberación favorece a las hojas siempreverdes y al bajo contenido de nitrógeno. El origen de los suelos de material de baja fertilidad y las pérdidas por filtración son factores que también pueden contribuir a la baja disponibilidad de nitrógeno del suelo. De este modo la forma siempreverde-esclerófila se ve reforzada por baja disponibilidad de nutrientes. La sequía de verano también puede redundar en hojas angostas, muy incli- nadas que se disponen estrechamente adosadas a los tallos o en forma de racimos, para así poder reducir la temperatura foliar en primavera y verano, y contribuir a una dis- minución de la transpiración. La tasa reducida de pérdida de agua prolonga el período de disponibilidad de humedad del suelo, el período de crecimiento y el de fotosíntesis. El crecimiento en la parte superior de la capa vegetal conduce a valores más altos en las proporciones fotosíntesis-transpiración y fotosíntesis-radiación solar, lo cual crea un equilibrio de carbono más favorable para las plantas con hojas en la parte superior de la capa. Las compensaciones fisiológicas y morfológicas se traducen en una similitud de vegetación a pesar de orígenes filogenéticos disímiles.

Las características de la vegetación mediterránea, incluyendo hojas esclerofíticas y siempreverdes, se seleccionan a través de las condiciones de clima y de nutrientes. Un sitio en que la condición de siempreverde ha sido seleccionada por el clima puede aparecer limitado por nutrientes, especialmente en cuanto al nitrógeno. Un sitio en el cual la condición de siempreverde ha sido seleccionada a causa de la baja fertilidad del material de origen aparecerá limitado en cuanto a nutrientes, pero fundamentalmente en lo que respecta a fósforo o cationes, y sólo en forma secundaria por nitrógeno. Los sitios con vegetación siempreverde de las regiones mediterráneas de Australia y Sud Africa se ven limitados por la baja fertilidad del material de origen, en tanto que sitios similares de las regiones mediterráneas de California, Chile, y la cuenca Medi- terránea estan, en primer lugar, limitados por el clima y sólo en forma secundaria por los nutrientes.

Gloria Montenegro
Santiago, Chile

Index

Ecological Studies

Springer-Verlag
New York Heidelberg Berlin